CHEMISTRY OF WASTEWATER TECHNOLOGY

CHEMISTRY OF WASTEWATER TECHNOLOGY

edited by

ALAN J. RUBIN
Professor of Civil Engineering
The Ohio State University
Columbus, Ohio

230 Collingwood, P. O. Box 1425, Ann Arbor, Michigan 48106

Library of Congress Catalog Card No. 76-50991
ISBN 0-250-40185-1

Manufactured in the United States of America

PREFACE

This book deals with the physical chemistry of particle separations and other chemical aspects of wastewater processes. *Chemistry of Wastewater Technology* presents the latest research of interest to engineers and scientists as well as others concerned with environmental protection. Some of the topics covered in its 26 chapters include coagulation, precipitation, filtration, nutrient removal, adsorption and disinfection. The original papers were presented at a symposium sponsored by the Environmental Chemistry Division of the American Chemical Society. The highlight of the symposium was the keynote address of Dr. Werner Stumm on the occasion of his being presented the coveted ACS Pollution Control Award. His award paper, in this book as at the parent symposium, is introduced by Dr. James J. Morgan, a close associate of Dr. Stumm for many years. We, the co-authors of this book who have been so profoundly influenced by him, are pleased to acknowledge the many contributions of Werner Stumm.

The cooperation and efforts of the co-authors and the publisher's staff, particularly Jan Carter, in producing *Chemistry of Wastewater Technology* are greatly appreciated. The support of the Civil Engineering Department and the College of Engineering of The Ohio State University is also acknowledged.

Alan J. Rubin
Columbus, Ohio

Alan J. Rubin holds the bachelor's degree in civil engineering from the University of Miami and a master's degree in sanitary engineering as well as a PhD (environmental chemistry) from the University of North Carolina. The author of many publications and papers, Dr. Rubin is Professor of Civil Engineering and is associated with the Water Resources Center at The Ohio State University. He is also the editor of two other books published by Ann Arbor Science.

CONTENTS

Introduction
to the
Award Address by the Recipient of the 1977
American Chemical Society Pollution Control Award
Werner Stumm

It is a pleasure both as a member of the American Chemical Society and as a colleague and friend of the author to provide a few introductory remarks for Professor Werner Stumm's Award Address, the keynote paper for this symposium on the *Chemistry of Wastewater Technology.* It is fitting that Professor Stumm be honored on the occasion of this symposium, in which chemistry figures so largely, for his career in aquatic chemistry has been dedicated to directing the attention of the profession to the effects of chemical parameters on water processes both in the natural environment and in technology. The Pollution Control Award, sponsored by Monsanto, recognizes Werner Stumm for his research contributions in the field of water pollution control. He has authored over 150 papers in his field. Examples of his innovative ideas and experimental results are to be found in such important areas as water and wastewater treatment, control of eutrophication, acid mine drainage, trace metals, and hydrocarbon pollution. In the wastewater treatment area Professor Stumm was one of the first to advocate and research the wider use of physical-chemical techniques to enhance the traditional modes of biological waste treatment. In this regard one can cite his work on flocculation of bacterial sludges and on the precipitation removal of phosphate nutrients. What has distinguished his work in all areas is that it has been new, it has been as quantitative as possible, and has been directed to areas of great practical concern in water quality management.

Werner Stumm is known as an inspirational teacher and colleague. He excites those with whom he works through the clarity of his ideas and the enormous energy and enthusiasm with which he approaches research problems. Without exception his students and collaborators have admired his didactic abilities and his research leadership. He communicates his enthusiasm to others. To quote from a recent editorial in *Environmental Science & Technology*, "It is also important to recognize Dr. Stumm's contribution to the education of environmental scientists. These include his role in establishing aquatic chemistry as a subject with academic rigor and environmental utility, and also his inspirational teaching at Harvard and the Swiss Federal Institute of Technology."

Professor at the Swiss Federal Institute of Technology in Zürich, Switzerland, and Director of the Institute of Water Resources and Water Pollution Control, EAWAG, since 1970, Werner Stumm was, for the fifteen years prior, Professor of Applied Chemistry at Harvard University. He received his PhD in chemistry from the University of Zürich, where his major areas of interest were analytical and physical chemistry. He quite early saw the possibilities for applying ideas and methodologies for analytical and physical chemistry to problems in water chemistry, first as a research associate at EAWAG, then as a postdoctoral fellow in the Sanitary Engineering Department at Harvard in 1954-55. He joined the Harvard faculty in 1956. His research activities over his career have embraced studies in such areas as corrosion rates, kinetics of metal-ion oxidations, electrochemistry, coagulation and flocculation mechanisms, surface chemistry of metal oxides, nucleation processes, chemical oceanography, global chemical cycles, and chemical ecology of pollution. The impact of his work has been great.

Werner Stumm has been the PhD professor of more than a dozen students at Harvard and at EAWAG, and in addition he has worked with many more research fellows and visiting scientists and engineers in his laboratories. Since returning to Switzerland in 1970 he has established the chemistry department of EAWAG as a focal point for aquatic chemistry research and has attracted many, many visitors in the fields of chemistry, engineering, biology, geochemistry and oceanography.

This symposium comprises an impressive group of papers dealing with research and technical innovations in wastewater technology. The importance of chemical concepts in improving wastewater technology is evident from the subject matter represented in the symposium. It is an additional tribute to the keynote author that a dozen former students and research colleagues, or their students, are participants in the symposium.

J. J. Morgan
Professor of Environmental Engineering Science
California Institute of Technology
Pasadena, California

CHAPTER 1

CHEMICAL INTERACTIONS IN PARTICLE SEPARATION

Werner Stumm

Swiss Federal Institute of Technology and
Federal Institute for Water Resources and
Water Pollution Control
Zurich, Switzerland

INTRODUCTION

The solid-solution interface is of particular importance in natural waters and in water treatment systems. Suspended particles in natural and wastewaters vary in diameter from 0.005 to about 100 μm (5×10^{-9} to 5×10^{-4} m). For particles smaller than 10 μm, terminal gravitational settling will be less than *ca.* 10^{-2} cm/sec. As suggested in Figure 1, filter pores of sand filters are typically larger than 500 μm. The smaller particles (colloids) can become separated either by settling, if they aggregate, or by filtration, if they attach to filter grains. Particle separation is of importance in the following processes: (1) aggregation of suspended particles (clays, hydrous oxides, phytoplankton, biological debris) in natural waters; (2) coagulation (and flocculation) in water supply and wastewater treatment; (3) bioflocculation (aggregation of bacteria and other suspended solids) in biological treatment processes; (4) sludge conditioning (dewatering, filtration); (5) filtration, groundwater infiltration; and (6) removal of precipitates (*e.g.,* phosphate elimination).

In the past few decades water supply and wastewater treatment have had to depend essentially on the same unit processes and unit operations; innovation and process improvement, especially improvement in process dynamics, had to come from a better engineering design rather than from new tools. Progress in engineering design of coagulation (followed by sedimentation, flotation or filtration) and filtration processes has been

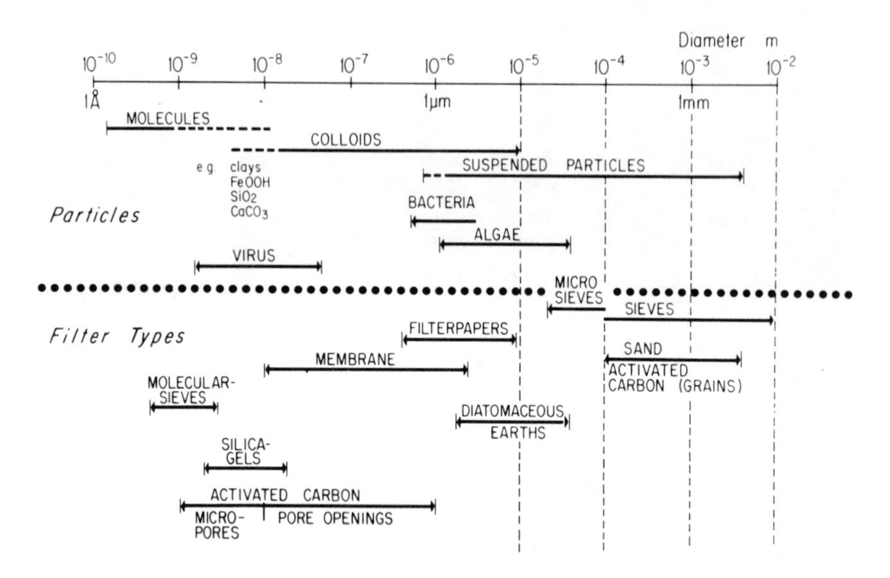

Figure 1. Size spectrum of waterborne particles and of filter pores.

achieved from a more thorough understanding of the process concepts. Considering separately the chemical (especially colloid and surface chemical) and the physical (including mass transport and fluid mechanic) aspects is of great help in assessing the variables and parameters that influence these processes. Coagulation and filtration may be rendered more dynamic or more efficient by modifying the chemical variables in such a way as to improve the efficiency of the collision between particles and particles and filter grains. Higher filtration rates may often be achieved without impairing the quality of the filtrate if a reduction in contact opportunities between suspended particles and filter grains (*e.g.*, by a decrease of filter depth and medium diameter) is compensated for by a chemical improvement of the collision efficiency.

This chapter is organized as follows:

1. It shows how solution variables may influence the charge of colloidal surfaces. A simple case history will serve for illustration. The specific adsorption of cations (*e.g.*, Mg^{2+}, Ca^{2+}, heavy metal cations and H^+) and of anions [*e.g.*, SO_4^{2-}, HPO_4^{2-}, $SiO(OH)_3^-$ and OH^-] on the surface of hydrous oxides of Si, Al^{3+} or Fe^{3+} affects the resultant net charge of these oxide particles.

2. It compares the effectiveness of particle aggregation in coagulation and particle removal in filtration, both in natural and in treatment processes,

and illustrates how they depend on contact opportunities, on the one hand, and on the efficiency of the contacts, on the other hand. The designer of a treatment process can choose among various chemical or physical variables.

3. It gives an application in practice. The removal of phosphate in complementary biological and chemical sewage treatment exemplifies some of the concepts discussed.

SPECIFIC CHEMICAL INTERACTION AND COLLOID STABILITY

The presence of an electrical charge on the surface of particles is often essential for their existence as colloids; the electrical double layer on their surface hinders the attachment of colloidal particles to each other and to filter grains. The resultant net charge is sensitive to the composition of the aqueous phase because adsorption or binding of solutes to the surface of the colloids may increase, decrease, or reverse the effective charge on the solid. Adsorption occurs as a result of a variety of binding mechanisms: electrostatic attraction or repulsion, covalent bonding, hydrogen-bond formation, van der Waal's interaction or hydrophobic interaction. One speaks of specific chemical interaction if binding mechanisms other than electrostatic interaction are significantly involved in the adsorption process.

Chemical destabilization of colloids is achieved in coagulation (or flocculation) and contact filtration by adding substances which enhance the aggregation or attachment tendency of these colloids. Natural and synthetic macromolecules (often polyelectrolytes) have a strong tendency to accumulate at interfaces and have been used successfully as aggregating agents. The Fe^{3+} and Al^{3+} salts used as coagulants and destabilizers belong also to this category because they form polynuclear hydrolysis products, $Me_q(OH)_n{}^{z+}$, which are adsorbed readily at particle-water interfaces. That specific chemical interactions contribute significantly to the adsorption and colloid destabilization is evident from the observation that these coagulants, at proper dosage, can reverse the charge of the colloid. Colloids also become less stable with increasing concentrations of indifferent (not specifically interacting) electrolytes because the diffuse part of the electrical double layer becomes compressed by counter-ions[1-4]

Interactions of Cations and Anions with Hydrous Oxides

Oxides, especially those of Si, Al and Fe, are abundant components of the earth's crust; they participate in many chemical processes in natural waters and often occur as colloids in water and waste treatment systems. The colloid stability of hydrous oxides may be affected by electrolytes in a different way than that of hydrophobic colloids. Specific adsorption of cations and anions on hydrous oxide surfaces may be interpreted as surface coordination reactions.[3,5-8] As indicated in Figure 2, hydrous metal oxides exhibit amphoteric behavior and can, at least operationally, be compared with amphoteric polyelectrolytes.

$$\equiv MeOH_2^+ \rightleftharpoons \equiv MeOH + H^+; \quad K^s_{a_1} \tag{1}$$

$$\equiv MeOH \rightleftharpoons \equiv MeO^- + H^+; \quad K^s_{a_2} \tag{2}$$

Since H^+ and OH^- ions are primarily the potential determining ions for hydrous oxides, alkalimetric or acidimetric titration curves provide a quantitative explanation for the manner in which the charge depends on the pH of the medium,[6-8] as indicated in Figure 3.

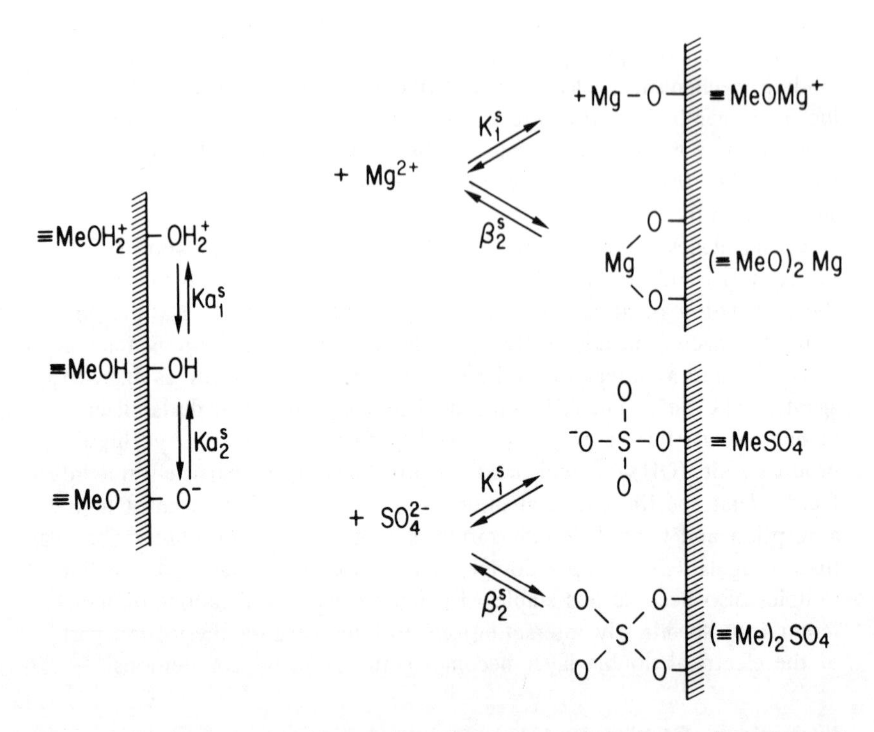

Figure 2. Interactions of hydrous oxides with cations and anions can be interpreted in terms of surface complex formation and ligand exchange equilibria.

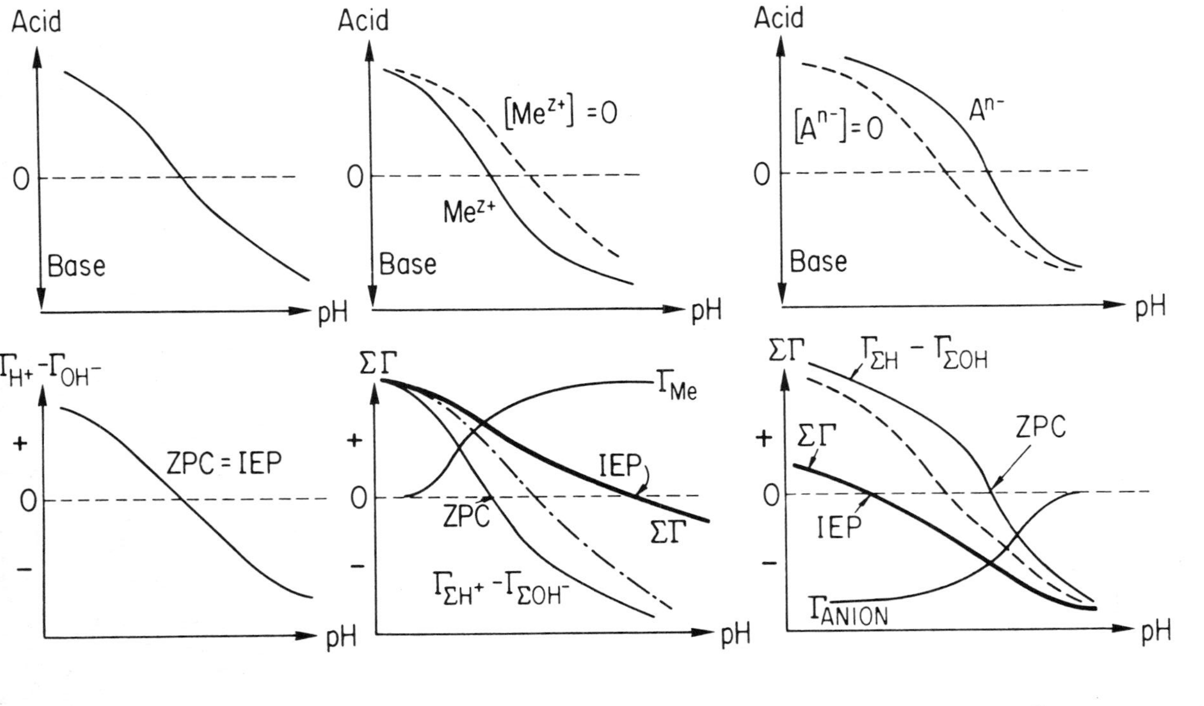

Figure 3. The net charge at the hydrous oxide surface is established by the proton balance (adsorption of H or OH^- and their complexes) at the interface and specifically bound cations or anions. This charge can be determined from an alkalimetric-acidimetric titration curve and from a measurement of the extent of adsorption of specifically adsorbed ions. Specifically adsorbed cations (anions) increase (decrease) the pH of the isoelectric point (IEP) but lower (raise) the pH of the zero point of charge (ZPC).[6,8]

Operationally, there is a similarity between H^+, metal ions and other Lewis acids. The OH groups on a hydrous oxide surface have a complex-forming O-donor group like OH^- or OH groups attached to other elements (phosphate, silicate, polysilicate). Proton and metal ions compete with each other for the available coordinating sites on the surface:

$$\equiv MeOH + M^{z+} \rightleftharpoons \equiv MeOM^{(z-1)} + H^+ \; ; \quad {}^*K_1^s \tag{3}$$

$$2\equiv MeOH + M^{z+} \rightleftharpoons (\equiv MeO)_2M^{(z-2)} + 2H^+ ; \quad {}^*\beta_2^s \tag{4}$$

The extent of coordination is related to the exchange of H^+ by M^{z+} ions; *i.e.*, it can be measured by the displacement of an alkalimetric titration curve. Similarly, ligand exchange with coordinating anions leads to a release of OH^- from the surface and a displacement of the titration curve in the other direction

$$\equiv Me\text{-}OH + A^{z-} \rightleftharpoons \equiv Me\text{-}A^{z-1} + OH^- \; ; \quad K_1^s \tag{5}$$

$$2\equiv Me\text{-}OH + A^{z-} \rightleftharpoons (\equiv MeO)_2A^{z-2} + 2OH^- ; \quad \beta_2^s \tag{6}$$

For protonated anions the ligand exchange may be accompanied by a deprotonation of the ligand at the surface. For example, in the case of HPO_4^{2-}

$$\begin{aligned} \equiv MeOH + HPO_4^{2-} &\rightleftharpoons \equiv MeHPO_4^- + OH^- \\ &\rightleftharpoons \equiv MePO_4^{2-} + H_2O \end{aligned} \tag{7}$$

it is also conceivable that at high pH, adsorption of a metal ion may be accompanied by hydrolysis. The processes described in Figure 2 and in Equations 1 to 7 can be characterized by equilibrium constants. These constants can be used to estimate the extent of adsorption and the resultant net charge of the particle surface as a function of pH and solute activity. Some equilibrium constants with cations and anions have been determined for a few representative oxides.[6-8]

As Figure 3 illustrates, the net resultant charge at the surface of an oxide is experimentally accessible from the proton balance at the solid-solution interface (obtainable from an alkalimetric or acidimetric titration curve in presence of specifically bound cations and/or anions) and from a measurement of the bound unhydrolyzed cation and deprotonated anion. Specifically adsorbed cations will shift the proton condition in such a way as to lower the pH of zero point of charge (ZPC); *i.e.*, the pH at which that portion of the charge which is due to H^+ or OH^- ions or their complexes becomes zero. Because of the binding of M^{z+}, the resultant net charge increases, or becomes less negative, and the pH at which the resultant net charge becomes zero, the isoelectric point (IEP), is shifted to higher pH values. Correspondingly, specifically adsorbing anions increase the pH of the ZPC but lower the pH of the IEP.

Figure 4 illustrates that cations and anions typically present in a natural water become specifically adsorbed at an aluminum oxide surface and modify its charge. The adsorption of SO_4^{2-} causes a reduction,

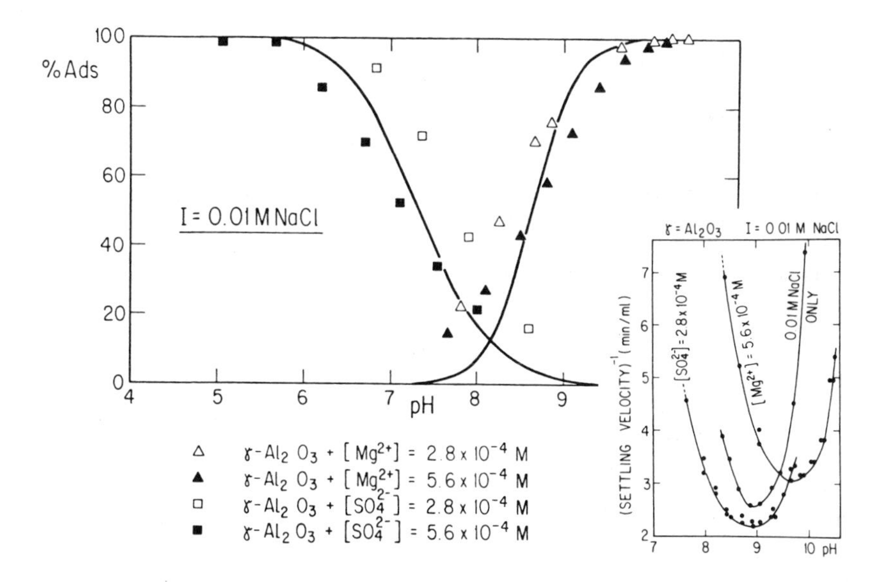

Figure 4. Extent of specific adsorption of SO_4^{2-} and Mg^{2+} on Al_2O_3. The drawn out lines have been calculated from experimentally determined equilibrium constants for the adsorption from a 5.6×10^{-4} *M* $MgSO_4$ solution. As shown by the insert, Mg^{2+} causes an increase and SO_4^{2-} a decrease in the isoelectric point (maximum settling rate).

while the adsorption of Mg^{2+} causes an increase in charge (compare Figure 2). The concomitant shift in the pH of the isoelectric point becomes apparent in sedimentation experiments because maximum settling velocities of suspended particles are observed at zero charge.

PARTICLE ATTACHMENT IN COAGULATION AND FILTRATION

The kinetics of particle transport and particle attachment determine to a large extent the various particle removal processes in natural waters and in water and waste treatment. O'Melia and Stumm and others[9-11] pointed out that filtration is analogous to coagulation in many respects. This is illustrated by juxtaposing the basic kinetic equations on particle removal:

$$\text{coagulation:} \quad -dn/dt = (4/\Pi)\, \alpha G \phi n \tag{8)*}$$

$$\text{filtration:} \quad -dn/dL = (3/2)\, \frac{(1-f)}{d}\, \alpha \eta n \tag{9}$$

where

n	=	the concentration of particles (number per cm^3)
α	=	the collison efficiency factor that reflects the chemistry of the system
ϕ	=	the volume of solid material (suspended particles) per unit volume of solution
G	=	the mean velocity gradient ($time^{-1}$), which reflects the rate at which contacts occur between particles by mass transport
t	=	the time
$(1-f)$	=	the volume of filter media per unit volume of filter bed where f is porosity
η	=	a "single collector efficiency" that reflects the rate at which particle contacts occur between suspended particles and the filter bed by mass transport
L	=	the bed depth
d	=	the diameter of the filter grain

The single collector efficiency is an inverse function of filter medium size, filtration rate and water temperature, and also depends on the size and density of the particles to be filtered.[9-11] The effectiveness of particle aggregation in coagulation and particle removal in filtration (attainable on integration of Equations 8 and 9) depends on a dimensionless product of the variables shown in Figure 5, that, as shown by O'Melia, have comparable meanings for both processes. Either equation can be used to evaluate the performance of upflow sludge-blanket clarifiers. Particle transport is required in both processes either to move the particles toward each other or to transport them to the surface of the filter grain or to the surface of a previously deposited particle. The contact opportunities of the particles to collide with one another or with filter grains depend on Gt or $\eta(L/d)$. The detention time, t, is somewhat related to (L/d), the ratio of bed depth to medium diameter.

Figure 6 illustrates how coagulation in natural systems and in water and waste treatment systems depends on the variables of Figure 5.

---

*The rate of coagulation for particles of small diameter ($d < 1\mu$) occurring mainly by Brownian interparticle contact is given in Equation 10.

COAGULATION	α	ϕ	G	t
	COLLISION EFFICIENCY	VOLUMETRIC CONCENTRATION OF SUSPENDED PARTICLES	VELOCITY GRADIENT	TIME
FILTRATION	α	$1-f$	η	$\frac{L}{d}$
		VOLUMETRIC CONCENTRATION OF FILTER MEDIUM	SINGLE COLLECTOR EFFICIENCY (v, d)	NUMBER OF COLLECTORS
			CONTACT OPPORTUNITIES	
DESIGN AND OPERATIONAL VARIABLES	CHEMICALS	COAGULATION AIDS MEDIA SIZE SLUDGE RECIRCULATION	ENERGY INPUT MASS TRANSPORT	RESIDENCE TIME FILTER LENGTHS AND MEDIA DIAMETER

Figure 5. Comparison between coagulation and filtration. The effectiveness of both processes depends on a related dimensionless product.

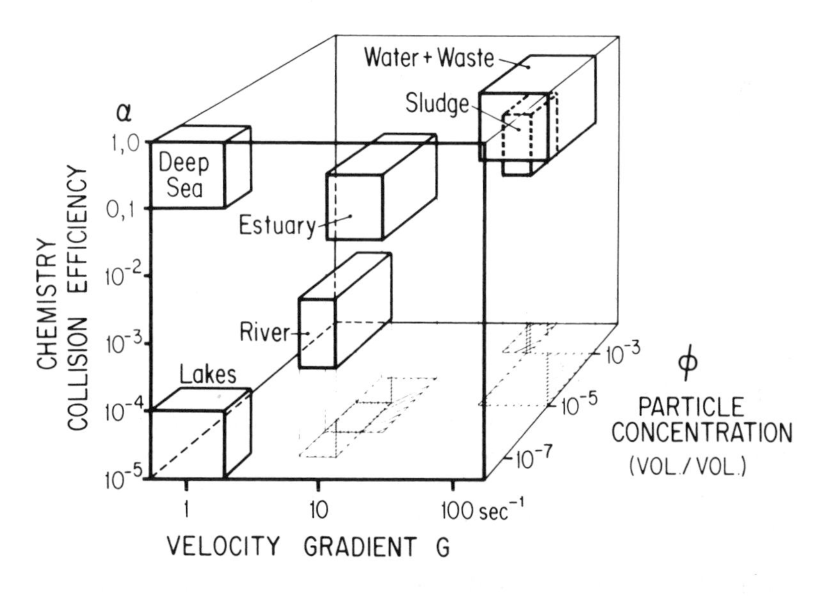

Figure 6. Variables that typically determine efficiency of coagulation in natural waters and in water and waste treatment systems (see Equation 8 and Figure 5).

In natural waters long detention times may provide sufficient contact opportunities despite very small collision frequencies (small G and small ϕ). In freshwater the collision efficiency is usually also low ($\alpha \sim 10^{-4}$ to 10^{-6}; *i.e.*, only 1 out of 10^4 to 10^6 collisions leads to a successful agglomeration). In seawater, colloids are less stable ($\alpha \cong 0.1$-1) because the salinity compresses their double layer. Estuaries with their salinity gradients and tidal movements represent gigantic natural coagulation tanks where much of the dispersed colloidal matter of the rivers becomes settled. In water and waste treatment systems we can reduce detention time (volume of tank) by adding coagulants at proper dosage ($\alpha \rightarrow 1$) and by adjusting power input (G). If the concentration, ϕ, is too small, it can be increased by adding additional colloids as so-called coagulation aids.

O'Melia has shown that similar trade offs in design and operation exist for optimizing particle removal in filtration.[10] We will restrict this discussion to the efficiency of particle removal, although other considerations, especially avoidance of excessive buildup of head loss in the filter bed, are, of course, also very important.

Figure 7 shows how, in the development from very slow filtration in groundwater percolation to ultrahigh-rate contact filtration, a relatively constant efficiency in particle removal [constant product $\alpha\eta(L/d)$] is maintained despite a dramatic increase in filtration rate. This is achieved by counterbalancing decreased contact opportunities (decreasing η and L and increasing d) by improving (through addition of suitable chemical destabilizing agents, aids that increase α up to 1) the effectiveness of particle attachment.[10]

Kinetic Considerations Determine the Type of Process to be Used for Particle Removal

The ideas discussed above and some other considerations permit us to select the particle removal process best suited for a practical problem. Figure 8[12] shows that the range of possible applications of various particle removal processes depends on the concentration of suspended particles and their diameters. In this figure, the various boundaries can be readily explained. If particles are sufficiently large ($d > 30$-50 μm), in accordance with Stokes' law, they can readily be removed by sedimentation (large concentrations) or by straining, *e.g.*, by a microstrainer (small concentrations). Coagulation, followed by sedimentation, can be accomplished within reasonable detention times only if the particle concentration is high enough to provide sufficient contact opportunities. This is especially important in perikinetic coagulation in which colloids

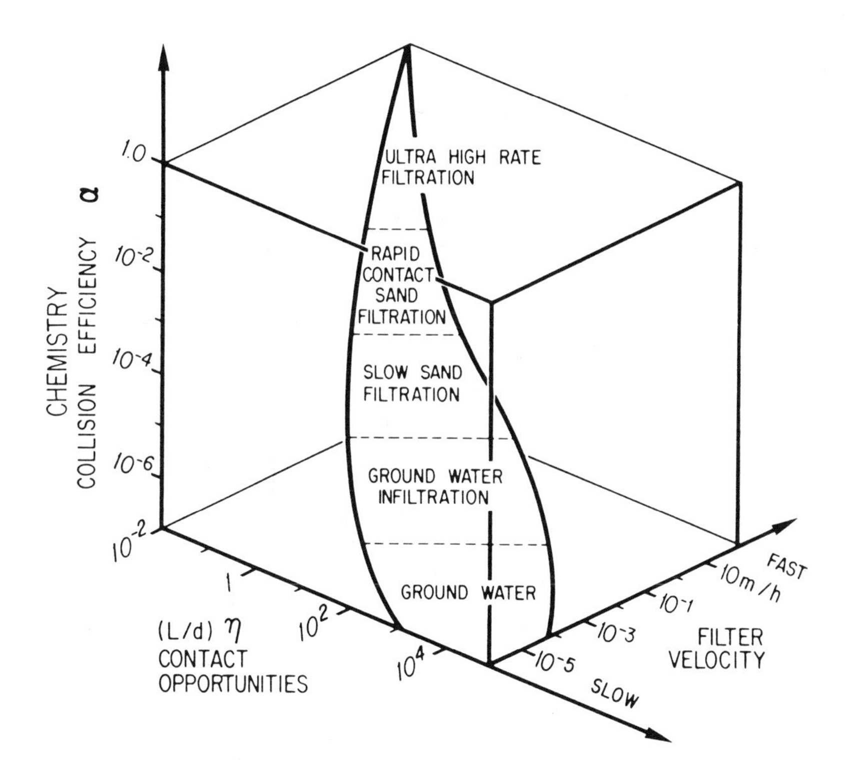

Figure 7. Filtration processes in natural waters and water and wastewater systems. Similar efficiency in particle removal, but marked increase in filtration rate, can be achieved by counterbalancing a reduction in contact opportunities by chemically improving the contact efficiency.

are sufficiently small ($d < 1$ μm) that interparticle contact occurs predominantly by Brownian motion (diffusion). In this case, the rate of particle removal may be given by

$$-dn/dt = (4\alpha\bar{k}T/3\mu)n^2 \tag{10}$$

where

$\bar{k}$ = Bolzmann's constant
T = the absolute temperature
μ = the fluid viscosity

With Equation 10 one can calculate that the detention time becomes very large ($t > 1$ hr) even for $\alpha = 1$ when the concentration is smaller than 10^9 particles cm^{-3}. At these low concentrations there are not enough interparticle collisions. However, contact filtration can, over the

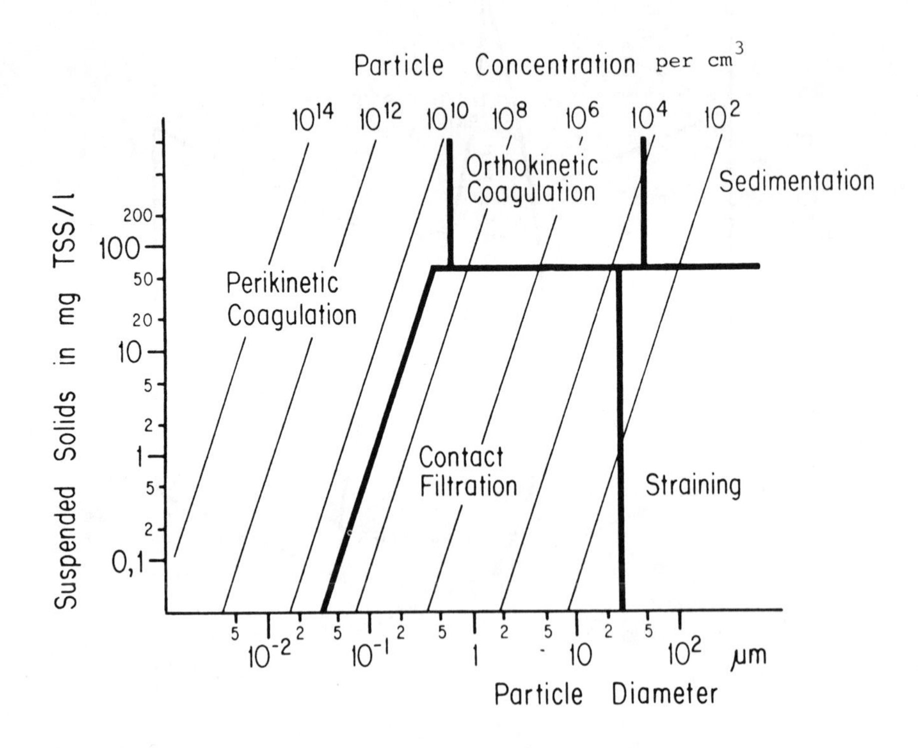

Figure 8. Most suitable ranges for application of various processes of particle removal in water and waste treatment.[12]

depth of the filter bed, provide sufficient contact opportunities in this case ($n < 10^9$ cm^{-3}). If particles are larger ($d > 0.1$ to 1 μm), shear forces enhance the interparticle contact (orthokinetic coagulation; Equation 8). These larger particles may also be removed conveniently by filtration, but because of limited retention capacity of a depth filter, orthokinetic coagulation, followed by sedimentation or upflow sludge-blanket clarification, are more suitable processes at suspended solids concentrations larger than *ca.* 50 mg/l.

PHOSPHATE REMOVAL IN SEWAGE TREATMENT: A CASE HISTORY

Phosphate can be removed from sewage by precipitation with Fe^{3+} or Al^{3+} salts; precipitation may be carried out in the aeration tank of the biological treatment. The following processes occur upon addition of Fe^{3+} to the wastewater: (1) precipitation of soluble phosphate to a nonstoichiometric solid phosphate of the general composition $Fe(PO_4)_x(OH)_{3-3x}$; (2) formation of particular $Fe(OH)_3$ or FeOOH to the surface of which phosphate becomes strongly bound; and (3) coagulation of suspended solid and of suspended P by polynuclear Fe^{3+} hydrolysis products. These processes are very efficient for the elimination of P from solution, but because of the strong adsorbability of phosphate to precipitated iron phosphate and to Fe^{3+} oxide hydroxide, some highly dispersed P-bearing colloids are formed. These colloids do not settle and may not even be retained by a membrane filter.

Pilot experiments carried out by Boller and Kavanaugh[12] in the experimental station of our Institute have shown that contact filtration for removal of P precipitated with Fe^{3+} following mechanical-biological treatment (Figure 9) is a process of practical feasibility leading to very low residual P and suspended solids concentrations (Figure 10).

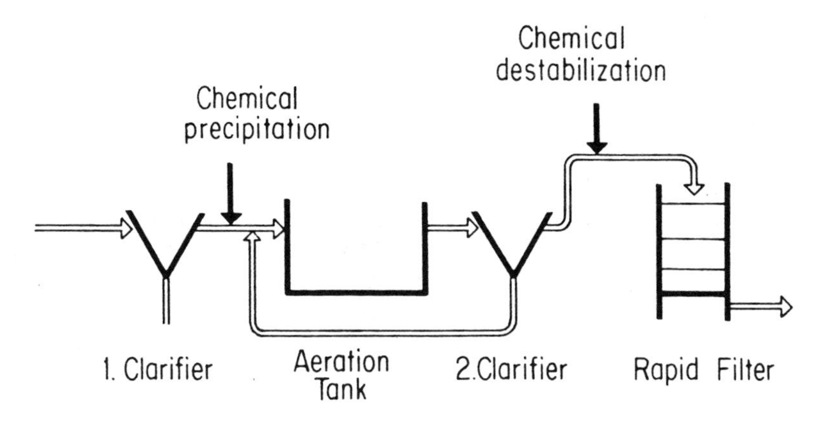

Figure 9. Removing phosphate by precipitation with Fe^{3+} simultaneous with biological treatment and by removing residual P by contact filtration.

Contact filtration with appropriate chemical doses and media design appears to be a potential alternative to coagulation/sedimentation or flotation systems for phosphorus removal following conventional treatment. The permissible solids loading must be determined for each case. For the chemical conditions tested, and filter design constraints, the solids limit was about 50 mg TSS/1 at flow rates up to 4 $gal/min/ft^2$ (10 m/hr), equivalent to about 3-4 mg P/l dissolved total phosphorus in the filter influent.[12]

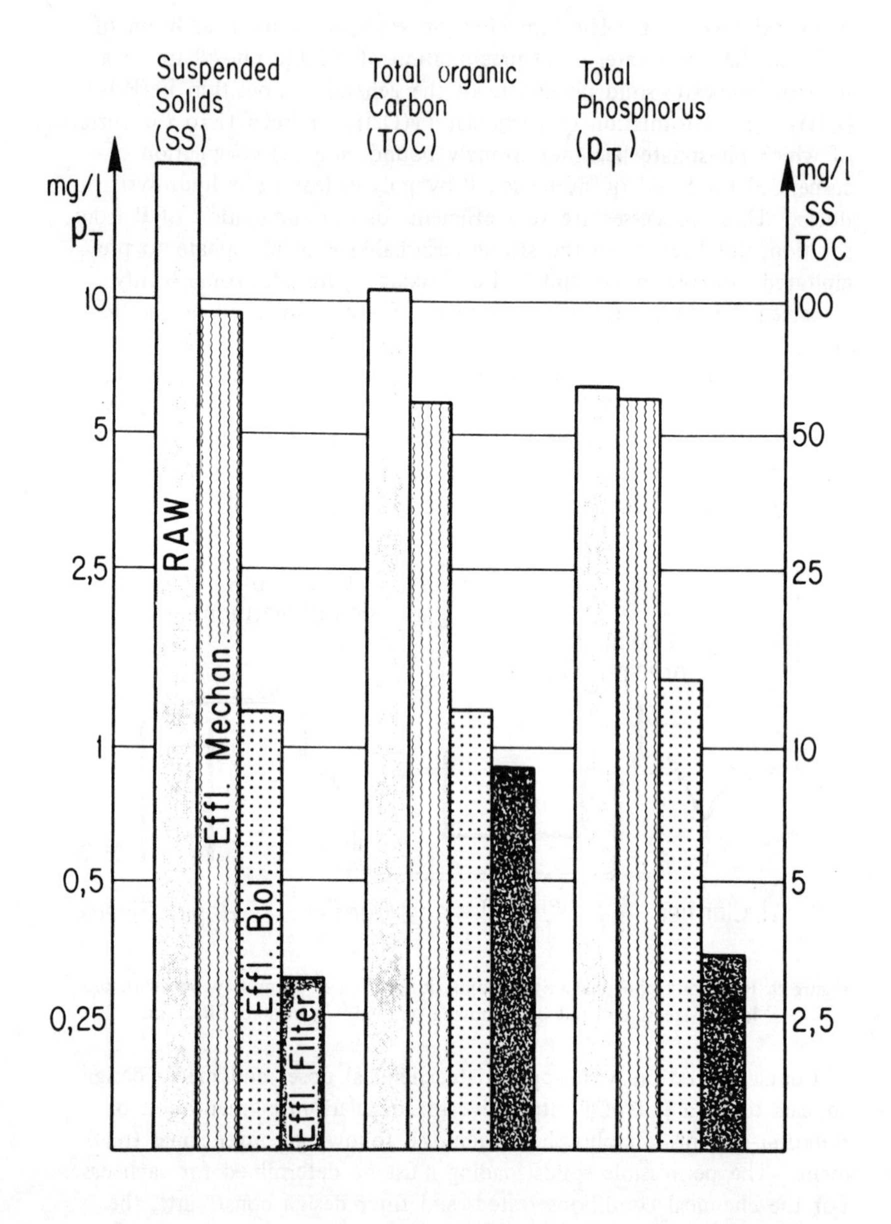

Figure 10. Treatment efficiency achieved in complementary chemical (precipitation of phosphate with Fe^{3+}) and biological treatment followed by contact filtration. Dosage for precipitation: 13.5 mg/l Fe^{3+}; for destabilization: 2.5 mg/l Fe^{3+} and 0.1 mg/l nonionic polyelectrolyte.[12]

ACKNOWLEDGMENTS

This chapter is taken, in part, from an address by the author to the American Chemical Society, Division of Environmental Chemistry meeting in New Orleans, Louisiana, on March 21, 1977, in response to his receiving the 1977 American Chemical Society Award for Pollution Control, sponsored by Monsanto Co. In reviewing some of the work that led to the recognition by this award, the author would like to express his gratitude to all his doctoral students and collaborators. He is especially indebted to James J. Morgan, Charles R. O'Melia and Elisabeth Stumm-Zollinger.

REFERENCES

1. Stumm, W., and J. J. Morgan. *J. Am. Water Works Assoc.* 54:971 (1962).
2. Stumm, W., and C. R. O'Melia. *J. Am. Water Works Assoc.* 60:514 (1968).
3. Stumm, W., C. P. Huang and S. R. Jenkins. *Croat. Chim. Acta*, 42: 223 (1970).
4. Matijević, E., M. B. Abramson, K. F. Schulz and M. Kerker. *J. Phys. Chem.* 64:1157 (1960).
5. Morgan, J. J., and W. Stumm. In: *Proc. Internat. Conf. Water Pollution Research* (New York: Pergamon Press, 1965), p. 103.
6. Stumm, W., H. Hohl and F. Dalang. *Croat. Chim. Acta* 48:491 (1976).
7. Schindler, P. W., B. Fuerst, R. Dick and P. Wolf. *J. Colloid Interface Sci.* 55:469 (1976).
8. Sigg, L., H. Hohl and W. Stumm. "Ligand Exchange on Hydrous Oxides," (to be published).
9. O'Melia, C. R., and W. Stumm. *J. Am. Water Works Assoc.* 59:1393 (1967).
10. O'Melia, C. R. "The Role of Polyelectrolytes in Filtration Processes," EPA 670/2-74-032, U.S. Environmental Protection Agency, Washington D.C. (1974).
11. Yao, K. M., M. T. Habibian and C. R. O'Melia. *Environ. Sci. Technol.* 5:1105 (1971).
12. Boller, M., and M. C. Kavanaugh. "Contact Filtration from the Elimination of P in Waste Water Treatment," Internat. Assoc. Water Pollution Research, Vienna Workshop, 1975.

CHAPTER 2

PRECIPITATION OF CALCIUM PHOSPHATES: THE INFLUENCE OF TRICARBOXYLIC ACIDS, MAGNESIUM AND PHOSPHONATE

G. H. Nancollas, M. B. Tomson, G. Battaglia, H. Wawrousek and M. Zuckerman

Department of Chemistry
State University of New York at Buffalo
Buffalo, New York

INTRODUCTION

The precipitation and dissolution of calcium phosphates constitute important problems in biology, oceanography, wastewater treatment, cooling tower water chemistry and the production of fertilizer from phosphate rock.[1-3] However, there is little agreement concerning the mechanisms of calcium phosphate precipitation. Considerable attention has been paid to the spontaneous precipitation process,[4] but such experiments are seldom quantitatively reproducible by different workers. To induce spontaneous precipitation in reasonably short times, unrealistically high concentrations of ions and pH are often used in the laboratory. Under such conditions, however, it is difficult to separate the processes of nucleation and ripening from the growth of already existing nuclei. The development of industrial precipitation processes, therefore, is seldom based upon laboratory models. Instead, expensive field testing is normally required. The availability of laboratory methods of testing which would yield reproducible results under conditions close to those in the field have obvious attractions for the optimization of the precipitation processes.

The precipitation of calcium phosphates can be divided into four quite well-defined regions of concentration shown in Figure 1. With a total calcium, T_{Ca}, to total phosphate, T_P, ratio of $1.6\overline{6}$ and a pH of 7.40, at calcium concentrations of about 10 m*M*, spontaneous precipitation sets in in less than a few seconds. In the concentration range $3\ mM < T_{Ca} < 10\ mM$, precipitation will take place through homogeneous nucleation if the solutions are sufficiently free from foreign particles.

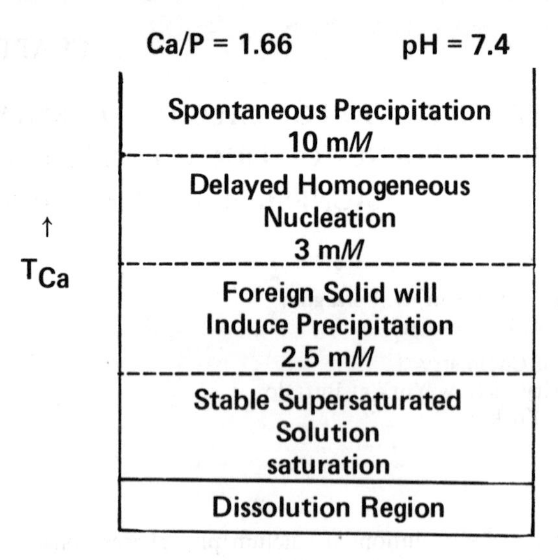

Figure 1. Various precipitation processes at pH 7.4 as the supersaturation (T_{Ca} at a constant Ca/P) is increased. The concentrations listed are approximate.

This can be delayed for periods of from minutes to hours by the presence of various trace additives in the supersaturated solutions. In the range 2.5 to 3.0 m*M* in T_{Ca}, there is a poorly defined region in which dust particles or irregularities on the walls of the vessel or electrode surfaces may act as nucleation sites. At concentrations of calcium ion below a about 2.5 m*M*, stable supersaturated solutions can be prepared by slowly adding calcium chloride solution to dilute phosphates and adjusting the final pH with dilute base or acid. These supersaturated solutions will remain stable for hours or days unless specific precipitation nucleators are added.

All experiments to be reported in this chapter have been performed in this region of supersaturation. At a pH of 7.40, the thermodynamically stable calcium phosphate phase is hydroxyapatite [HAP, $Ca_5(PO_4)_3OH$] with a solubility value at the conditions specified above corresponding to T_{Ca} = 0.03 m*M*. Stoichiometric HAP probably never precipitates at the pH, temperature, calcium and phosphate levels encountered in most natural systems. The eventual hydrolysis of the more acidic solid phases first precipitated into HAP would probably take considerable periods of time. In these slightly basic solutions, three other distinct calcium phosphate phases have been suggested as important precursors to HAP precipitation: (1) dicalcium phosphate dihydrate (DCPD, $CaHPO_4 \cdot 2H_2O$), (2) octacalcium phosphate [OCP, $Ca_4H(PO_4)_3 \cdot 1.5\ H_2O$], and (3) tricalcium phosphate [TCP, $Ca_3(PO_4)_2$]. DCPD is the thermodynamically stable phase at pH values below about 6. Under more alkaline conditions, HAP becomes thermodynamically stable up to pH values of 13-14 where calcium hydroxide precipitates. Nevertheless, in the pH range 4.5-6.0, calcium phosphate precipitation is predominantly stoichiometric DCPD even if HAP seed material is used to induce precipitation.[5] It has been suggested[6] that at pH values above 6.0, OCP is the kinetically stable material. The supporting experimental data for this hypothesis are not as convincing as those for the formation of DCPD at lower pH values. If OCP is the material which separates initially from solution, it hydrolizes, probably within a few minutes under most conditions, into a material of a higher molar Ca/P ratio. This phase slowly ages into a more basic and apatite-like material. It has been suggested that TCP, which is normally formed at temperatures of about 1,150°C, is never the stable phase in aqueous solutions unless stabilized by some foreign ion such as magnesium or other trace components.[7]

Unfortunately, the crystals that precipitate in slightly basic solutions (pH 7.4) are too small to be detected by X-ray analysis or normal scanning electron microscopy (SEM). Since the Ca/P ratios range only from 1.33 for OCP through 1.66 for HAP, an error of only 1% in the analysis of T_{Ca} and T_P can often preclude the differentiation between HAP, TCP and OCP. One method to identify the material that precipitates from solution is to dissolve these surface phases selectively. The results of such dissolution kinetic studies[8] support the suggestion that OCP is the material favored kinetically at these slightly basic conditions. It has also been shown that magnesium ion stabilizes the initially precipitated material.[8]

A number of components normally found in natural waters and physiological fluids are known to alter or inhibit calcium phosphate precipitation. In natural waters supporting biological acitvity, components of the citric acid cycle are present. The calcium citrate complexes in solution have been characterized, but the mechanism of the action of citric acid on the rate of crystallization of calcium phosphate has not been elucidated.[9] Aside from its complexing properties, which will effectively reduce the concentration of free calcium ions participating in the crystal growth reaction, the citrate ions could be adsorbed onto the crystal surface, thus blocking growth sites. In the present study, the kinetics of crystallization of HAP seed material have been studied in the presence of a number of tricarboxylic acids of the Kreb's cycle: citric acid(I), isocitric acid(II), *cis*-aconitic acid(III), *trans*-aconitic acid(IV), and tricarballylic acid(V) (Figure 2).

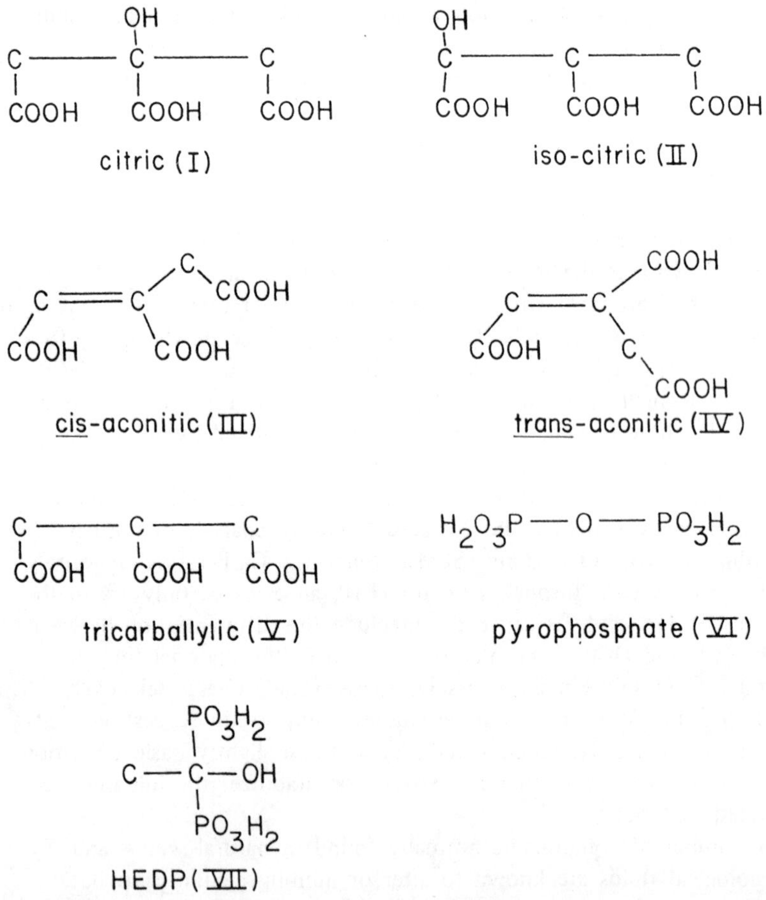

citric (I)

iso-citric (II)

cis-aconitic (III)

trans-aconitic (IV)

tricarballylic (V)

$H_2O_3P—O—PO_3H_2$

pyrophosphate (VI)

HEDP (VII)

Figure 2. Some tricarboxylic acids of the Kreb's cycle.

The influence of these additives is compared with that of other ions such as magnesium, pyrophosphate(VI), and 1-hydroxy-1,1-ethanediphosphonic acid, HEDP(VII), which have been studied previously.[8-10]

EXPERIMENTAL

Materials and Analyses

Chemicals, reagent grade or better, included calcium chloride tetrahydrate (Alfa Ventron, Ultrahigh purity) potassium monobasic phosphate (Baker Chemicals Ultrex quality), citric acid (Fisher Scientific), tricarballylic acid (Aldrich Chemical), dl-isocitric acid (Calbiochemical, allo-free) and *cis*- and *trans*-aconitic acids (Sigma Chemical). Deionized water, triple distilled with a final distillation under continuous stream of purified nitrogen, was used to prepare the solutions in grade A volumetric glassware. Beckman glass electrodes (#39301) with silver-silver chloride internal elements and Beckman calomel reference electrodes (#39071) were standardized at 25 or 37°C using standard buffers.[11]

Hydroxyapatite seed material, prepared by slowly adding phosphoric acid to calcium hydroxide solution at 80°C, was aged for at least four months before use.[12] The molar Ca/P ratio of the seed material was 1.66 ± 0.01. The specific surface area was 35 m^2g^{-1} (Monosorb Surface Analyzer, Quantachrome Corp., Greenville, NY).

Calcium and phosphate concentrations were determined by a precise combined method developed in our laboratory. Following the addition of vanadophosphomolybdate for the development of the yellow phosphomolybdate, using the normal phosphate analytical procedures,[13] the calcium content was determined by atomic absorption (± 0.3%) using a Model 503, Perkin Elmer atomic absorption spectrophotometer. Since the calcium hollow cathode lamp emits at 422.37 nm and the phosphate yellow difference maximum is at 420 nm, a cell holder was mounted in the burner head position and the atomic absorption instrument was used as a spectrophotometer to measure the absorbance of the phosphomolybdate complex. The absorbance/concentration data for a typical set of phosphate standards could be represented by a straight line with a correlation coefficient of $0.\overline{9}(5)$ to $0.\overline{9}(6)$ and a standard error of estimate of about 0.1%. The relative error in T_{Ca}/T_P is reduced using this method because determinations are made on the same sample.

Crystal Growth Experiments

Precipitation experiments were made in a double-walled, water-jacketed Pyrex® vessel at 25.0 ± 0.02°C. Stable supersaturated solutions were prepared by slowly adding calcium chloride solution to a potassium dihydrogen phosphate solution and adjusting the pH with dilute potassium hydroxide. Crystallization was induced by adding HAP seed slurry, and during the reaction the pH was maintained constant (± 0.01) by the pH-stat- (Model 3D, Brinkmann Instruments, Westbury, NY) controlled addition of dilute potassium hydroxide. At various times during the reactions, samples were withdrawn, filtered through 0.22 μm Millipore filter pads and analyzed for calcium and phosphate. The solid phases were also examined by scanning electron microscope (Etek Scanning Electron Microscopy, Hayward, CA).

Radio-labeled citric acid (1-5-^{14}C, ICN, Irvine, CA) was also used to determine the amount adsorbed by the HAP seed crystals. ^{14}C-citrate analysis was made with a Packard Tri-Carb scintillation counter with a resultant expected standard deviation of 0.1%.

RESULTS AND DISCUSSION

The tricarboxylic acids form stable solution complexes with the calcium ion; the citrate complex, CaCit, is the most stable with an association constant of 7.94×10^4 at zero ionic strength. To be able to separate the effects due to complexing and those reflecting the adsorption of tricarboxylate ions at the crystal surface, it is necessary to calculate the concentrations of each ionic species in solution. These computations were made as described previously[14] from mass balance, electroneutrality, proton dissociation and calcium and magnesium phosphate and tricarboxylate ion-pair association constants[15] by successive approximations for the ionic strength, I. The activity coefficients of z-valent ionic species, f_z, were calculated from the extended form of the Debye-Huckel equation proposed by Davies:[16]

$$-\log f_z = Az^2[I^{1/2}/(1 + I^{1/2}) - 0.3I$$

Activity products, IP, for each of the calcium phosphate phases could be calculated from Equations 1-4

$$IP_{HAP} = [Ca^{2+}]^5[PO_4^{3-}]^3[OH^-]\ f_1^{48} \quad (1)$$

$$IP_{TCP} = [Ca^{2+}]^3[PO_4^{3-}]^2\ f_1^{30} \quad (2)$$

$$IP_{OCP} = [Ca^{2+}]^4[H][PO_4^{3-}]^3\ f_1^{44} \quad (3)$$

$$IP_{DCPD} = [Ca^{2+}][HPO_4^{2-}]\ f_1^{8} \quad (4)$$

in which the square brackets enclose molar concentrations. The negative Briggsian logarithms, $pIP_{(solid\ phase)}$, are reported in Table I and describe the initial conditions for each of the experiments. At an ionic strength of 0.16 *M*, the introduction of the corresponding f_1 value, 0.76, changes the pIP_{HAP} by 5.8 units. This illustrates the importance of making activity coefficient corrections in assessing degrees of supersaturation.

Table I. Crystal Growth in the Presence of Citrate Ions (initial conditions, 25°C)

Experiment No.	T_{Ca} (m*M*)	T_P (m*M*)	T_{cit} (m*M*)	$[Ca^{2+}]$ (m*M*)	$[PO_4^{3-}]$ (10^5 m*M*)	$[Cit^{3-}]$ (m*M*)	pIP OCP	pIP TCP	pIP HAP	Growth
1	1.666	1.023	–	1.403	1.296	–	43.48	25.06	46.24	Yes
2	1.562	0.973	–	1.320	1.230	–	43.53	25.16	46.39	Yes
3	1.829	1.050	3.00	0.082	2.267	1.205	48.04	28.64	52.54	No
4	1.609	1.010	1.00	0.551	1.524	0.059	44.80	26.14	48.07	No
5	1.723	1.020	0.50	1.032	1.382	0.016	43.83	25.40	46.81	Yes
6	1.660	1.000	1.660	0.216	1.716	0.248	46.37	27.34	50.12	No
7	1.000	0.600	1.00	0.154	0.931	0.165	47.55	28.15	53.29	No
8	0.885	0.547	0.500	0.367	0.733	0.033	46.21	27.10	49.41	After 2 hr.
9	0.886	0.554	0.200	0.612	0.708	0.008	45.39	26.49	48.39	Yes
10	0.925	0.560	–	0.820	0.692	–	44.93	26.14	47.83	Yes
11	1.826	0.589	1.000	0.781	0.870	0.044	44.98	26.21	48.12	After 2.5 hr.
12	3.00	0.600	1.000	1.862	0.880	0.025	43.70	25.27	46.71	Yes
13	1.695	1.017	0.500	^{14}C-citrate (see Experiment # 5)						Yes

At equilibrium, the ion activity products for DCPD, OCP, TCP and HAP are 6.73, 46.90, 28.94 and 59.04, respectively.[17-20]

Crystal growth experiments made in the presence of citric acid are summarized in Table I. Experiments 6 and 7 show that when calcium and citrate are present at equal concentrations, 80 to 90% of the calcium is complexed. The solutions are still supersaturated with respect to HAP and TCP, however, and Experiment 6 is also supersaturated with respect to OCP. Experiment 4 is clearly supersaturated with respect to all three phases, and yet the plot of T_{Ca} against time in Figure 3 indicates that no precipitation takes place for several hours. It can be concluded that citric acid is an effective inhibitor of calcium phosphate crystal growth.

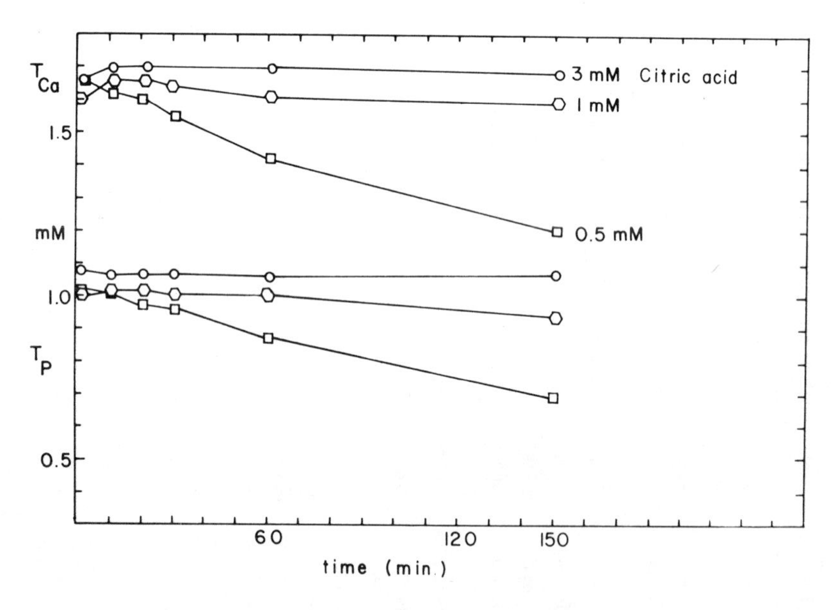

Figure 3. Plots of total calcium and total phosphate against time for various levels of citric acid; ○ Experiment 3, ⬡ Experiment 4, and □ Experiment 5.

To determine the importance of calcium citrate complexing, in Experiment 11 the supersaturated solution of calcium phosphate was prepared with a higher initial T_{Ca}/T_P ratio such that the concentration of free calcium ion was approximately the same as that for the control Experiment 10 in the absence of citrate. In Experiment 10, 50% of the calcium and phosphate have precipitated in 16 min, whereas in Experiment 11 less than 1% had precipitated in this period of time. In Experiment 12, the free calcium ionic concentration was approximately twice that of the control, Experiment 10, and the crystal growth was still substantially slower.

To determine the extent of adsorption of citrate ion on the surface of the HAP seed material, Experiment 13 was performed using ^{14}C citrate. In other respects, the experiment was closely similar to Experiment 5, which had identical calcium and phosphate concentration profiles as a function of time. In 20 min the citrate concentration had decreased by about 2%, and yet it can be seen in Figure 3 that the rate is decreased by about 50-75% at this stage of the experiment. This adsorption corresponds to about 1 Cit^{3-} ion per 100 A^2 of seed surface.

Crystallization experiments made in the presence of other tricarboxylic acids II, III, IV and V were performed at 3.0, 1.0 and 0.5 m*M* under conditions similar to those in Table I. Plots of T_{Ca} against time in the presence of 1.0 m*M* of each of the additives are shown in Figure 4.

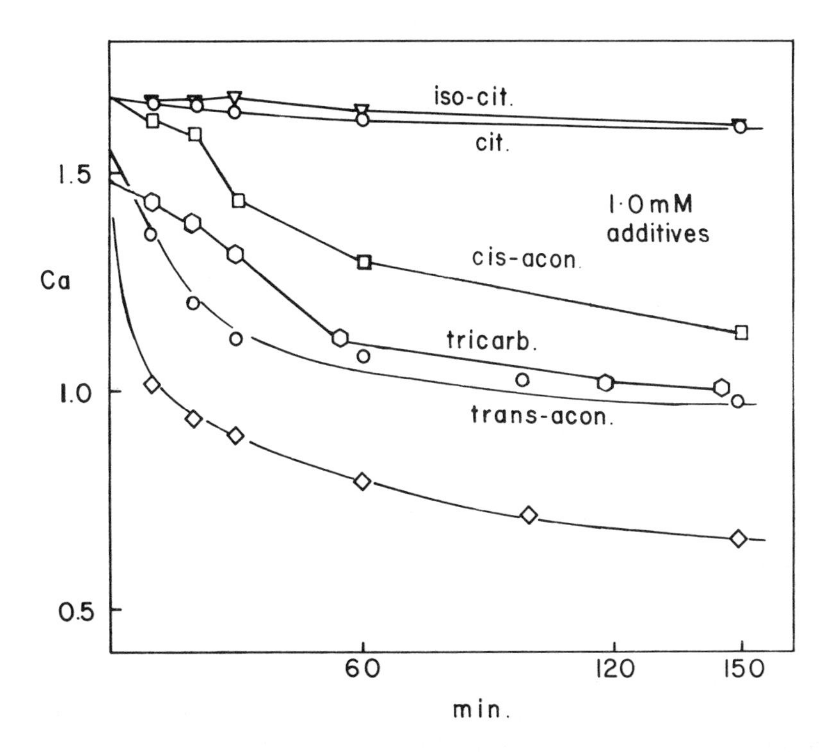

Figure 4. Plots of total calcium against time for the five tricarboxylic acids–I, II, III, IV and V–each at 1 m*M* concentration.

As can be seen, those acids having a hydroxyl group, citric and isocitric, are effective precipitation inhibitors. The other tricarboxylic acids, however, have much less influence on the rate of precipitation. At a higher concentration of 3.0 m*M*, citrate, isocitrate and *cis*-aconitic acids completely prevent precipitation for 5 hr. In the case of *trans*-aconitic acid and tricarballylic acid, however, precipitation takes place under these conditions at a rate that is considerably less than that of the control. It is clear from these results that the hydroxyl groups present in citrate and isocitrate ions are important in determining their adsorption at the crystal surface. The inhibiting effect cannot be explained simply in terms of the formation of solution calcium complexes of these ligands; a relatively small degree of surface adsorption is sufficient to retard the crystal growth reaction markedly.

Magnesium

The magnesium ion is a cofactor for numerous enzymatic processes and is involved in biological calcification. This ion also occurs in millimolar levels in some natural waters. Previous work has shown that the magnesium ion retards the precipitation of calcium phosphate.[8] Plots of T_{Ca} against time for various magnesium concentrations are shown in Figure 5. It can be seen that the magnesium ion is less effective than the citrate ion in retarding the crystallization rate. Since magnesium forms relatively weak complexes with the phosphate ions, it does not appreciably lower the thermodynamic driving force for precipitation. The retarding influence illustrated in Figure 5 must therefore be attributed to specific seed surface interaction with the magnesium ion. It has been shown that the magnesium ion is not incorporated into the calcium phosphate solid when precipitation reactions are carried out in its presence.[8] Analysis of the solid phases withdrawn at various times during the crystallization reaction also indicated that the magnesium ion was not incorporated during precipitation. The initial concentration of the magnesium ion was maintained to within less than 1% during the whole of the reactions.

Phosphonate

In Figure 6, the fraction of calcium remaining in solution is plotted against time for various levels of HEDP additive. It can be seen that the precipitation is completely inhibited at a concentration of additive of 10^{-5} *M*. The results of experiments in the presence of 10^{-6} *M* ^{14}C-HEDP indicate that 95% or more of the HEDP in solution is adsorbed onto the seed material. If the concentration of this additive is increased to 2.5×10^{-5} *M*, spontaneous precipitation sets in. This precipitate probably

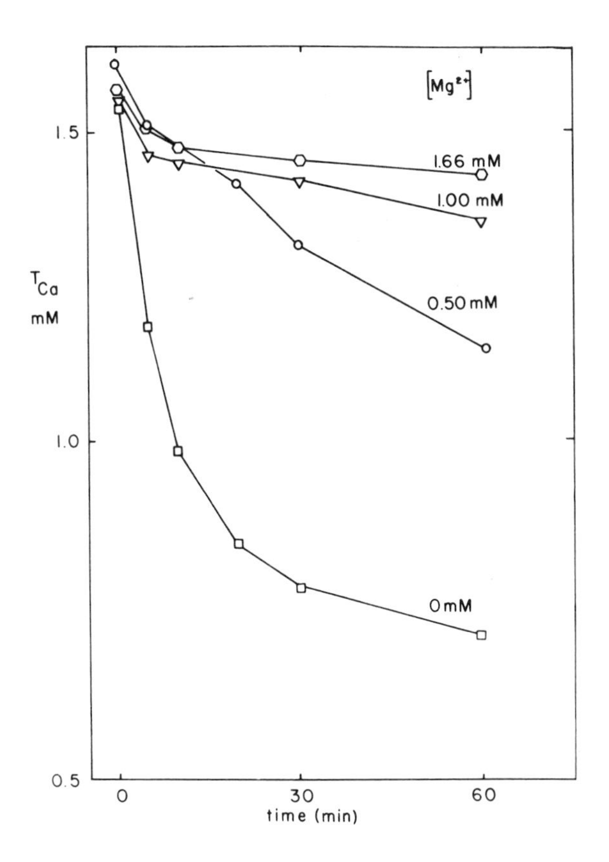

Figure 5. Plots of total calcium against time for various levels of added magnesium. The control, Experiment 1 (□), is plotted for comparison.

consists of a calcium HEDP salt, although this has not yet been confirmed.[21] Assuming that one HEDP molecule will cover approximately 10 A^2, 10^{-5} M will cover about 0.6 $m^2 l^{-1}$ for the experiments reported here. Since the HAP seed surface was about 6 $m^2 l^{-1}$ (30 $m^2 g^{-1}$), this would correspond to about 10% coverage of the surface for complete inhibition of crystal growth. This represents a lower estimate since the effective HEDP surface area coverage could be as high as 30 A^2 (molecule)$^{-1}$. The specific surface area of the solid phases was measured by nitrogen adsorption in the dry state; the effective surface area for the solid-liquid interface could therefore be in error by at least a factor of 2. Thus it is anticipated that a minimum of about 10% and a maximum of

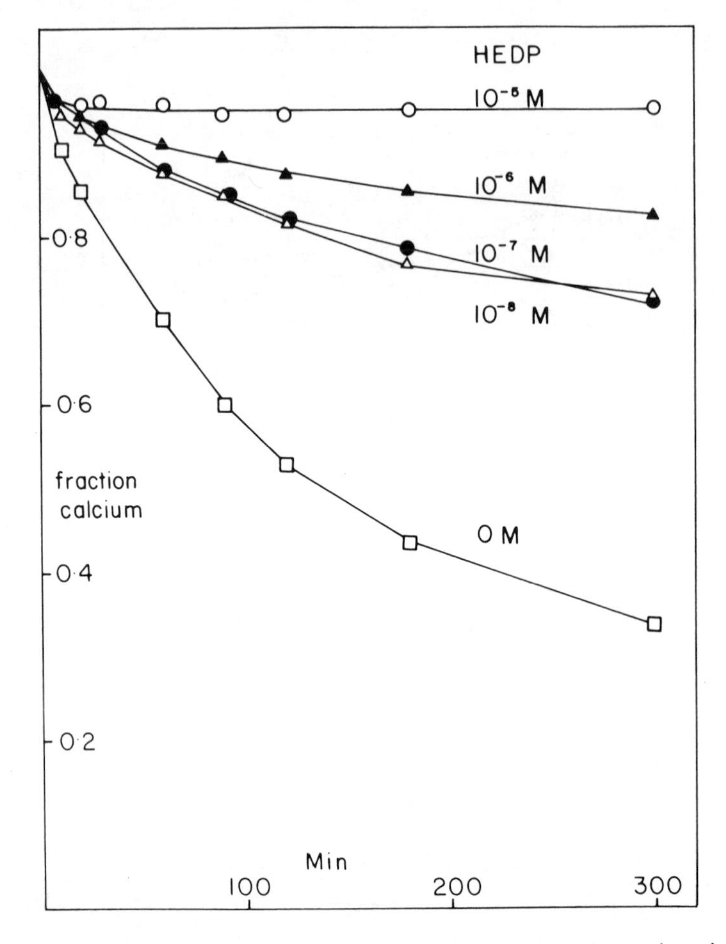

Figure 6. Plots of the fraction of total calcium remaining in solution against time for various levels of HEDP. The control for this set, Experiment 10 (□), is plotted.

about 50% of the seed surface must be covered for complete inhibition of precipitation. Assuming that about half of the surface sights are available for phosphonate adsorption, it follows that monolayer coverage will correspond to about 1 ± 3 x 10^{-5} *M* HEDP, which is the observed effective level for complete inhibition. The results of this work show that with only 1 to 10% or less of the surface covered by HEDP (Figure 6) the rate of precipitation is markedly reduced. It is significant that for magnesium and citrate ions the amount adsorbed onto the surface for *complete* inhibition appears to be about the same as for HEDP. For these ions, when the concentrations are reduced by less than 1/10 of these levels the effect on the rate of precipitation is almost eliminated. In the light of these findings, it is possible to speculate that these results are consistent with a Langmuir-type adsorption isotherm, along with the assumption that there are relatively few active growth sites on the HAP seed material. Similar suggestions have been made for the growth of seed crystals of other sparingly soluble salts.[22] Movement up the isotherm in each case will be limited by spontaneous precipitation of calcium citrate, magnesium phosphate or calcium HEDP.

CONCLUSIONS

Wherever biological activity occurs in nature it is expected that tricarboxylic acids and magnesium ions will be present along with various calcium phosphate phases. Calcium phosphate precipitation is also important in the recycling of boiler and heat exchanger waters and in the treatment of municipal wastewater by lime. The results of the present work indicate that although the citrate ion forms stable complexes with calcium ions in solution, its effect on the precipitation of calcium phosphates is a surface adsorption phenomenon. Comparisons of the effect of citrate, isocitrate and tricarballylic acids suggest that the hydroxyl group in the molecular backbone is a key factor in the effectiveness of these tricarboxylate ions as inhibitors. Although the tricarboxylates, magnesium ion and HEDP exert their retarding influence on the rate of crystallization at very different concentrations, a comparison of their adsorption properties indicates that approximately the same fraction of the surface is covered when complete inhibition sets in.

ACKNOWLEDGMENTS

This research was supported in part by the National Institute of Dental Research, NIH, Research Grant # DE03223, and the National Institute of Arthritic Metabolic and Digestive Diseases, NIH, Research Grant R1AM-19048.

REFERENCES

1. Stumm, W., and J. J. Morgan. *Aquatic Chemistry,* (New York: John Wiley and Sons, Inc., 1971).
2. Cowan, J. C., and D. J. Weintritt. *Water-Formed Scale Deposits,* (Houston: Gulf Publishing Co., 1976).
3. Slack, A. V., Ed. *Phosphoric Acid,* Vol. 1, Part 1 (New York: Marcel Dekker, 1968).

4. Boskey, A. L., and A. S. Posner. *J. Phys. Chem.* 77:2313 (1973).
5. Barone, J. P., G. H. Nancollas and M. B. Tomson. *Calcif. Tiss. Res.* 21:171 (1976).
6. Brown, W. E. *Nature* 196:1048 (1962).
7. Newesley, H. In: *Advances in Oral Biology,* Vol. 4, H. P. Staple, (Ed.), (New York: Academic Press, 1970), pp.11-42.
8. Tomazic, B., M. B. Tomson and G. H. Nancollas. *Arch. Oral Biol.* 20:803 (1975).
9. Dallemagne, M. J., and L. J. Richelle. In: *Biological Mineralization,* I. Zipkin, (Ed.), (New York: John Wiley and Sons, Inc., 1972), Chap. 2.
10. Nancollas, G. H., and M. B. Tomson. *Faraday Disc. Chem. Soc.* 2976 (61):24 (1976).
11. Bates, R. G. *Determination of pH* (New York: John Wiley and Sons, Inc., 1963).
12. Nancollas, G. H., and M. S. Mohan. *Arch. Oral Biol.* 15:731 (1970).
13. Deitz, V. R., and A. Gee. *Anal. Chem.* 25:1320 (1953).
14. Nancollas, G. H. *Interactions in Electrolyte Solutions* (Amsterdam: Elsevier Publishing Co., 1966).
15. Schubert, J., and A. Lindenbaum. *J. Am. Chem. Soc.* 74:3529 (1952).
16. Davies, C. W. *Ion Association* (London: Butterworths, 1962).
17. Moreno, E. C., T. M. Gregory and W. E. Brown. *J. Res. Nat. Bur. Stand. Sect. A* 70:545 (1966).
18. Moreno, E. C., W. E. Brown and G. Osborn. *Soil Sci. Soc. Amer.* 24:99 (1960).
19. Moreno, E. C. In: *Structural Properties of Hydroxyapatite and Related Compounds,* W. E. Brown and R. Y. Young, Eds., (New York: Gordon and Breach, in press).
20. Clark, J. S. *Can. J. Chem.* 33:1969 (1955).
21. Tomson, M. B., M. Zuckerman, H. Wawrousek and G. H. Nancollas. Unpublished results.
22. Liu, S. T., and G. H. Nancollas. *J. Colloid Interface Sci.* 52:593 (1975).

CHAPTER 3

KINETIC INHIBITION OF CALCIUM CARBONATE FORMATION BY WASTEWATER CONSTITUENTS

Michael M. Reddy

Environmental Health Center
New York State Department of Health
Albany, New York

INTRODUCTION

Lime addition, with the subsequent formation of calcium carbonate and calcium phosphate, is currently employed for wastewater coagulation and phosphorus removal at a number of sites.[1-11] In actual operations the degree of kinetic inhibition of calcium carbonate formation by wastewater constituents may determine the suitability of lime treatment. Successful application of the technology over a range of operating conditions requires a detailed knowledge of the basic chemical principles which govern calcium carbonate formation and interaction in wastewater systems.[12-19] For example, lime treatment of wastewater is frequently complicated by problems of excessive dosage, incomplete precipitation and poor settleability. In practice, these difficulties are circumvented by using a pH as high as 10 or 11 and by using large dosages of lime for satisfactory phosphorus removal and good clarification.

This chapter examines the effect of several typical components of wastewater on precipitation reactions which normally occur during lime treatment, and shows that the effect of several ions on calcium carbonate formation can be expressed as a function of their concentration using a simple Langmuir adsorption isotherm model. This relationship can be used to optimize lime addition when ions inhibiting crystallization are known to be present.

Chemical Treatment of Wastewater

Merrill and Jorden[10] have predicted that lime will be used increasingly in wastewater treatment because its addition removes trace metals as well as phosphorus. Maruyama *et al.*,[9] in examining the removal of trace metals by a number of treatment processes including lime addition, observed that kinetic factors often determine the final composition of a treated wastewater and that solubility products do not generally give an accurate estimate of the final trace metal concentrations. Thus, in each lime treatment process the precipitation reaction kinetics must be evaluated to determine realistically the concentrations of effluent metals.

Ferguson *et al.*[8] have shown that it is possible to reduce phosphate concentration to less than 2 mg/l-P in finished wastewater with lime and subsequent calcium phosphate precipitation at a pH less than 9. Optimum conditions are pH 7.5-8.5 with 80 mg/l of CaO. The process can be used with wastewater in which the alkalinity is less than 350 mg/l as $CaCO_3$ and magnesium is less than 24 mg/l. Under these conditions the addition of lime raises the calcium concentration from 0.5 to 2 m*M*, increases the pH from 7.5 to 8.0, and removes typically 85 to 95% of the phosphorus. It is noteworthy that as the pH after dosing is increased from 8 to 9 in this process calcium carbonate is often formed, precipitating at lower pH values in higher alkalinity wastewaters. In most situations this interferes with calcium phosphate precipitation, and the treated water has higher phosphate content than is desirable.

In their discussion of phosphate removal by lime treatment, Merrill and Jorden[10] emphasize that more phosphate is removed if calcium carbonate formation is inhibited. These authors report anomalous calcium carbonate solubility values for the solids formed during addition of lime to wastewaters. They conclude, nevertheless, that equilibrium or steady-state models can be used to characterize the precipitation reactions caused by lime additions. As will be shown, the use of such models to calculate calcium carbonate formation in wastewaters may lead to large errors. Indeed, normal levels of certain constituents, *e.g.*, phosphate, can reduce the rate constant for calcium carbonate formation by several orders of magnitude and may make a critical difference in the success or failure of lime for wastewater treatment.

Calcium Carbonate Formation

Recent studies have shown that the growth of calcium carbonate (calcite) follows a parabolic kinetic equation.[12-19] The reaction rate is controlled by an interfacial process rather than by bulk diffusion of lattice ions through the boundary layer adjacent to the crystal surface. Calcium carbonate formation during water treatment processes such as softening, where the concentration of growth-inhibiting substances is generally low, can often be adequately described by reaction rate parameters obtained by analyzing the process in pure solutions. Bench-scale or pilot-plant evaluation of calcium carbonate formation during potable water treatment requires knowledge of the solid phase in contact with the solution during the treatment, the reactor flow conditions, the rate of approach to equilibrium, and the identity and amounts of ionic complexes in solution. The application of this information to water treatment has been discussed.[17]

In each specific wastewater application the kinetic factors affecting calcium carbonate formation must be carefully evaluated to determine the probability of success. Since several recent publications have considered them in detail,[20,21] the kinetic factors associated with crystal-phase transformation (*e.g.*, the transformation of an initially formed thermodynamically less stable phase to a thermodynamically more stable form) or with carbon dioxide transfer to or from the precipitating solution will not be discussed here. In aqueous heterogeneous reactions the particulate surface area available for growth is often important, and this kinetic factor has been examined recently as it applies to calcium carbonate growth in a supersaturated solution.[17]

Three different types of experimental procedures can be employed to investigate heterogeneous chemical reactions in wastewater systems: (1) spontaneous precipitation from a highly supersaturated solution, (2) nucleation and growth of a thermodynamically more stable phase from a solution saturated with a thermodynamically less stable phase, and (3) seeded growth of a thermodynamically stable crystalline phase from a metastable supersaturated solution. While each of these techniques may offer certain advantages in particular experiments, the seeded growth procedure is the most useful method for kinetic analysis of growth inhibition. By using preformed crystals, seeded growth eliminates uncertainties about the varying roles of nucleation and growth in the overall formation of calcium carbonate.

The most important kinetic factor affecting calcium carbonate formation is the reaction rate constant, which can be dramatically reduced by traces of certain other components.[14] Simkiss[22] employed spontaneous precipitation measurements to relate the chemical properties of a number of ions to their effectiveness as calcium carbonate growth inhibitors. These experiments were conducted at high ionic strength (seawater) and showed a wide range of inhibition by several ions containing phosphorus. Kitano[23] has demonstrated the importance of added ions in the regulation of the polymorphic composition of spontaneously precipitated calcium carbonate. Ferguson and co-workers,[8] seeking to optimize lime treatment conditions for wastewater, noted the importance of employing seed material via a solids recycle step for efficient phosphorus removal.

The study reported in this chapter examined the inhibition of calcium carbonate crystallization by several wastewater components. To avoid the problems associated with spontaneous precipitation, the investigation employed seeded growth in metastable supersaturated solutions. These solutions can be prepared under carefully controlled experimental conditions, after which they are stable for several days. Crystal growth was initiated in the stable supersaturated solution by inoculation with a standardized seed crystal suspension. This procedure is analogous to backmixing precipitated crystals or to a solids recycle step in a waste treatment process. The rate of calcium carbonate formation was determined by measuring the concentration of lattice ions remaining in solution as a function of time. Thus the crystal growth rate could be reliably measured in the presence and absence of added wastewater constituents. Moreover, dilute electrolyte solutions were used, allowing a chemical thermodynamic approach for description of the solution phase composition. Published equilibrium constants for solution complexes were used, when available, to estimate the extent of ion pair formation in the experimental solutions.

EXPERIMENTAL

Materials

Analytical reagent-grade chemicals were used except where noted. Doubly distilled, deionized and filtered (0.22 μm Millipore) water and grade A glassware were also used throughout.

Phosphatidylinositol (phosphatidyl inositide, beef brain, Pfaltz and Bauer), stearic acid (Fisher Scientific), humic acid (Lot 8985A, K and K Laboratories), albumin (bovine albumin powder, fraction V, from bovine plasma, lot no. C25208, Armour Pharmaceutical Co.), and nucleic acid (desoxyribonucleic acid, thymonucleinsauer Na-Salz hochmolekular nach Hammaisten, Carl Roth, Karlsruhe) were used without further purification.

A Corning Model 12 pH meter with a Sargent Combination Electrode and a Sargent model SP recorder allowed continuous measurement of solution pH during each crystallization experiment. Seed crystal surface areas were determined with a Quantachrome Monosorb surface area analyzer. X-Ray powder diffraction patterns of seed material were obtained with a Phillips powder diffraction apparatus with copper K_α

radiation and a nickel filter. An ETEC scanning electron microscope was used for detailed crystal examination.

Methods

Preparation of seed crystals

Calcite seed crystals were prepared by rapidly adding 1 liter of 0.5 *M* $CaCl_2$ solution to 1 liter of 0.5 *M* Na_2CO_3 solution at 5°C. The viscous suspension formed was gradually warmed to 25°C and stirred overnight. The solids were allowed to settle and the supernatant was decanted and replaced with distilled water. The suspension was stirred thoroughly and the process of settling and decanting repeated. In this way the seed was washed three times. It was then aged in distilled deionized water for six months before use. In previous work (unpublished results) improper aging of seed crystals resulted in erratic growth-rate constants, largely because changes in carbon dioxide partial pressure in the seed solution led to inconsistent seed cementation and secondary nucleation of the seed material.

Shortly before the start of a series of growth experiments, the aged seed suspension was diluted with 4×10^{-3} *M* sodium bicarbonate solution to yield a convenient seed concentration.

The aged seed crystals, examined by scanning electron microscopy, were found to consist of 2-μm aggregates of small flat crystal plates. A sample of the suspension was dried at room temperature and atmospheric pressure. X-Ray powder diffraction analysis showed the material to be exclusively calcite. The specific surface area for the seed crystals was 1.71 m^2/g.

Preparation of supersaturated solutions

Stable supersaturated calcium carbonate solutions were prepared by drop-by-drop addition of 200 ml of calcium chloride (5.0×10^{-4} *M*) to 200 ml of a sodium bicarbonate (8.0×10^{-3} *M*)–sodium carbonate (4.0×10^{-4} *M*) solution at 25°C which was stirred continuously at 200 rpm in the double-walled Pyrex® glass vessel (Figure 1) used for the growth experiments.

The concentration of calcium in aliquots filtered through a Millipore filter was determined by EDTA titration with a micrometer burette in the presence of calcein indicator. Total carbonate concentrations were calculated from the solution pH and from titration of the filtrate with 0.01 *N* sulfuric acid in the presence of a methyl purple indicator (pH 4.8-5.4); pH was measured at 25.00 ± 0.02°C. The electrodes were standardized before and after each experiment with National Bureau of Standards standard buffer solutions at pH 6.865 (0.025 *M* KH_2PO_4 + 0.025 *M* Na_2HPO_4) and pH 9.180 (0.01 *M*

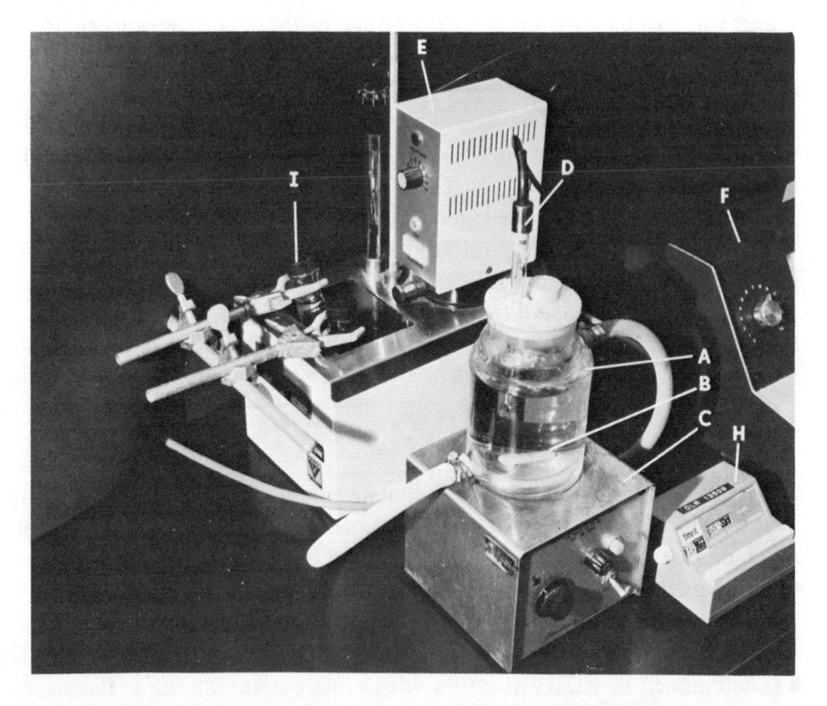

Figure 1. Experimental apparatus used in seeded crystal growth experiments in the presence and absence of added ions: (A) thermostatted, double-walled, Pyrex reaction cell; (B) magnetic stirring bar; (C) magnetic stirrer; (D) combination glass-reference electrode; (E) constant temperature bath; (F) pH meter; (G) recorder; (H) timer; (I) buffer solution vessels.

sodium borate) at 25°C. The calcium concentration obtained by analysis of the stable supersaturated solution shortly before the addition of seed crystals always agreed closley with its calculated concentration.

A typical seeded growth experiment was started by inoculating 380 ml of stirred supersaturated solution with 5 ml of stock calcite seed suspension, yielding a seed concentration in the supersaturated solution of approximately 100 mg/l. Prior to inoculation the stock seed suspension was shaken thoroughly to obtain a representative aliquot. During the growth experiments calcium carbonate seed crystals were separated from 10 ml aliquots of the supernatant by filtration through a Millipore filter (BDWP-01200).

RESULTS AND DISCUSSION

Growth in the Absence of Wastewater

Plots of the concentration of calcium in solution and of pH against time in experiments with only calcium and carbonate ions are shown in Figure 2. Addition of inoculating seed crystals to the stable supersaturated solution at zero time induced immediate crystal growth.

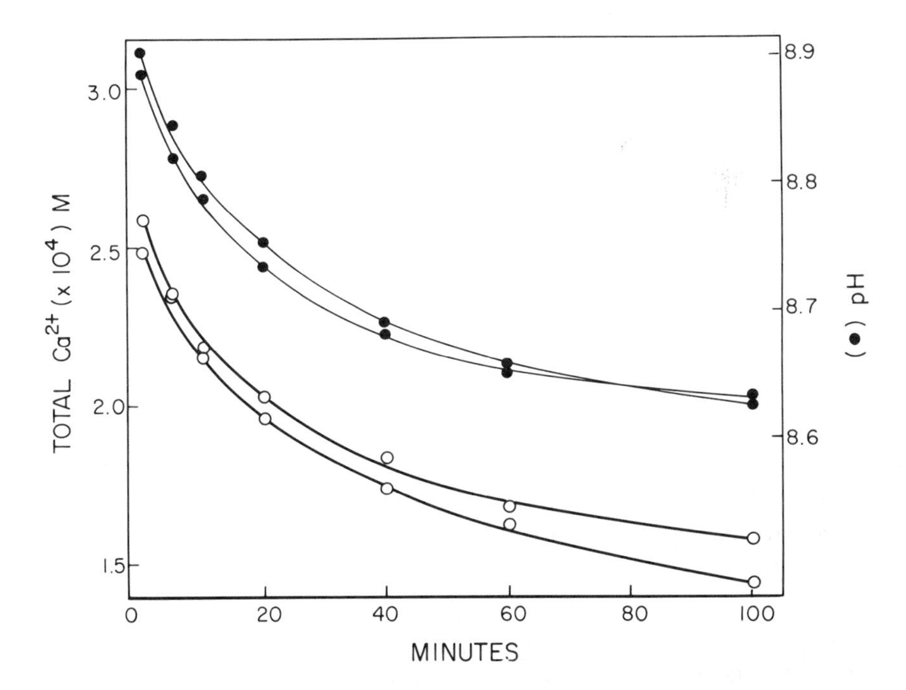

Figure 2. pH and concentration of calcium ion in solution as a function of time in the absence of added ions.

Solution-phase ionic concentrations were calculated from the measured solution pH and from total calcium and carbonate concentrations using the mass action and mass balance equations in Table I. The calculation consisted of successive approximations for ionic strength, I, and activity coefficients by the modified Debye-Hückel equation proposed by Davies[24]:

$$\log_{10} f_z = -Az^2 \left(\frac{I^{0.5}}{I + I^{0.5}} \right) -0.3I \quad (1)$$

where f_z is the activity coefficient for ion of charge z and A is a constant. In the experimental solutions bicarbonate ion comprised more than 95% of the total carbonate concentration. Ion pair concentrations were significant (Table II) and were considered in the solubility calculations.

Table I. Equations used for Calculation of Ionic Species at 25°C and 1 atm[a]

Equation	Reference
Mass action	
1. $[H^+]\ [HCO_3^-]/[H_2CO_3] = 10^{-6.35}$	25
2. $[H^+]\ [CO_3^{2-}]/[HCO_3^-] = 10^{-10.33}$	26
3. $[H^+]\ [OH^-] = 10^{-14.00}$	27
4. $[CaCO_3]/[Ca^{2+}]\ [CO_3^{2-}] = 10^{3.2}$	28
5. $[CaHCO_3^+]/[Ca^{2+}]\ [HCO_3^-] = 10^{1.4}$	29
6. $[CaOH^+]/[Ca^{2+}]\ [OH^-] = 10^{1.4}$	30
7. $([Ca^{2+}]\ [CO_3^{2-}])$ calcite $= 10^{-8.4}$	31
Mass balance[b]	
8. $T_{Ca} = [Ca^{2+}]/f_2 + [CaOH^+]/f_1 + [CaHCO_3^+]/f_1 + [CaCO_3]/f_0$	
9. $T_{CO_3} = [CO_3^{2-}]/f_2 + [HCO_3^-]/f_1 + [H_2CO_3]/f_0 + [CaHCO_3^+]/f_1$	

[a]Bracketed terms denote thermodynamic activities in solution.
[b]The term f_2 denotes the activity coefficient for ion of charge 2.

Crystallization rate data (Figure 2) followed a rate law found to be applicable in a previous study[18]

$$dN/dt = -ksN^2 \quad (2)$$

where N = calcium carbonate concentration (mol/l) at time t (min) to be precipitated from solution before equilibrium is reached,
k = the crystal growth rate constant (liter/mol-min per mg/l), and
s (mg/l) = the seed crystal concentration which is proportional to the surface area available for growth.

Table II. Experimental Conditions and Calculated Rate Constants for Calcite Growth Experiments in the Absence of Additives

			Initial Concentrations (mol/l x 10^4)					
Initial pH	Total Ca^{2+}	Total CO_3^{2-}	Ca^{2+}	CO_3^{2-}	$CaCO_3$	$CaHCO_3^+$	s (mg/l)	k
8.850	2.603	41.77	2.172	1.651	0.3081	0.1223	94.5	1.21
8.886	2.593	41.57	2.143	1.781	0.3286	0.1200	95.2	0.82
8.905	2.496	41.02	2.057	1.830	0.3250	0.1135	94.3	1.20
8.805	2.594	40.83	2.194	1.461	0.2773	0.1219	94.3	0.790
8.798	2.255	42.01	1.902	1.483	0.2435	0.1089	56.6	1.98
8.895	2.274	41.97	1.871	1.837	0.2960	0.1059	74.3	2.14
8.910	2.489	42.21	2.039	1.910	0.3339	0.1154	98.5	2.40
8.870	2.580	41.78	2.142	1.725	0.3173	0.1202	94.4	1.36
8.808	2.618	41.78	2.207	1.505	0.2856	0.2480	94.0	1.56
8.862	2.625	41.24	2.188	1.676	0.3150	0.1216	185.6	1.55
8.837	2.615	42.11	2.187	1.618	0.3034	0.1241	37.9	1.01

As defined above, N is equivalent to the "theoretical crystal yield," described by Mullin[32] at any time during the crystallization reaction. It is calculated from the measured total calcium concentration at any given time by

$$N = T_{ca}(t) - T_{ca}(\infty) \tag{3}$$

where $T_{ca}(t)$ equals the total molar concentration of all calcium-containing species in solution at time t. The calcite solubility product is used to calculate the $T_{ca}(\infty)$ value using the solution pH and total carbonate concentration at time t. An integrated form of Equation 2 facilitates analyses of the growth data

$$N^{-1} - N_0^{-1} = kst \tag{4}$$

where N_0 is calcium carbonate (mol/l) to be precipitated from a supersaturated solution at zero time. The linear plot of $N^{-1} - N_0^{-1}$ as a function of time in Figure 3 confirms the validity of Equation 2 for interpretation of the experimental results.

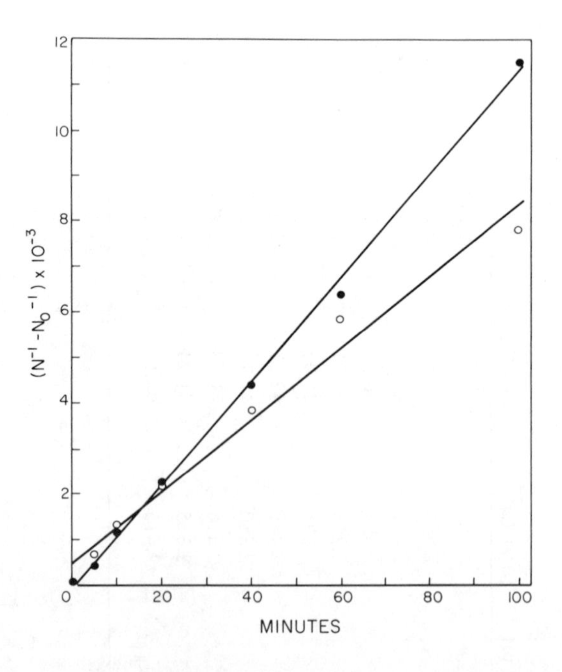

Figure 3. Rate function plotted against time in the absence of added ions.

Rate constants, k, for calcite crystallization in the absence of added ions are shown in Table II. These rate constants are independent of seed crystal concentration, s, over the range employed in this study (40 to 185 mg/l). They were obtained over a period of two years with the same stock calcite seed suspension diluted to a working solution shortly before the start of each series of experiments. The relatively small variations of these rate constants over the life of the experimental program is strong evidence for the reliability and reproducibility of the seeded growth technique. Minor changes in rate constants for pure solutions from one series of growth experiments to the next reflect changes in the stock seed suspension during the investigation.

Preliminary Experiments with Certain Wastewater Components

Studies with chemically defined solutions containing a single wastewater component showed a striking difference in the ability of each additive to inhibit calcium carbonate crystallization. Figure 4 shows the percentage of the theoretical crystal yield from the seeded supersaturated solutions one day after inoculation. The control experiments, in which no test substances were added, reached nearly 90% theoretical crystal yield after 24-hr reaction time. Phosphatidylenositol, a naturally occurring organophosphate,[33] had almost no effect on calcium carbonate growth, whereas orthophosphate and metaphosphate inhibited growth almost completely.

Naturally occurring organic acids, humic and stearic, which are significant constituents of many wastewaters, also dramatically retard calcite formation. This inhibition probably involves adsorption of the acid anion, the predominant ionic form of these compounds under the experimental conditions. The finding by Suess[33] and Myers and Quinn[34] that stearic acid and natural organic materials can strongly adsorb at a calcium carbonate-seawater interface supports this hypothesis. Such adsorption has been proposed to explain the inhibition of calcium carbonate formation in surface oceanwaters. In the presence of 10^{-4} M stearic acid there is a small but measurable crystallization reaction (Figure 4) which indicates that this anion does not inhibit the crystallization reaction as completely as does metaphosphate.

Except in the presence of metaphosphate ion, there appeared to be no induction period for calcite growth since crystallization was initiated immediately after the supersaturated solution was inoculated with seed suspension. Thus this system is not analogous to crystallization of barium sulfate and calcium sulfate dihydrate in the presence of additives.[35,36]

The pronounced inhibition by metaphosphate ion at 10^{-4} M is similar to that produced by other phosphorus-containing ions.[14] At lower

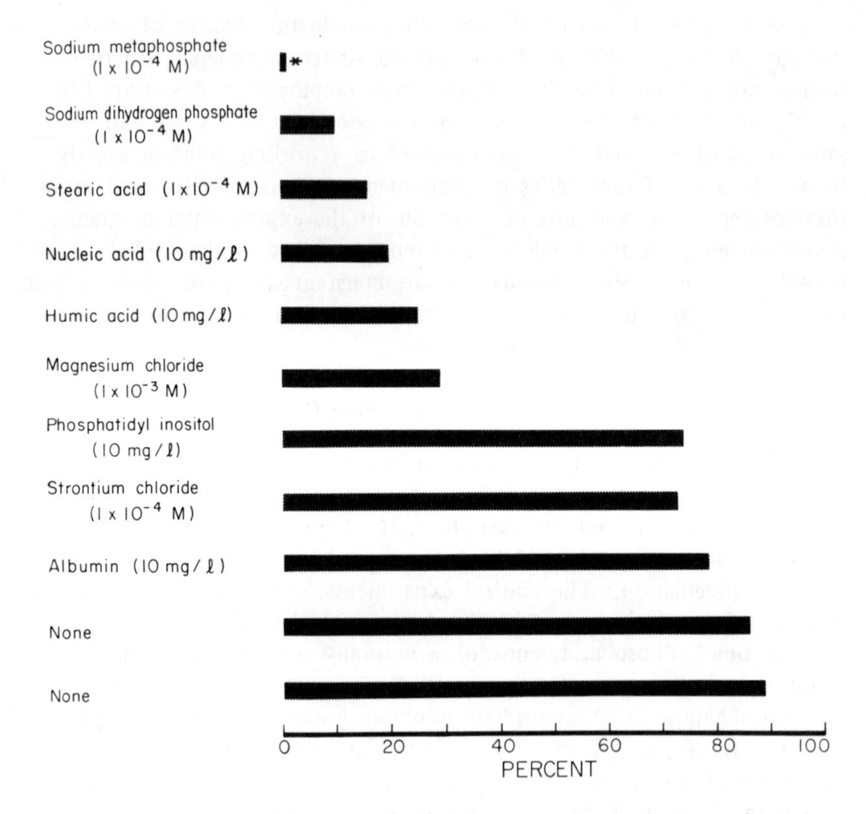

Figure 4. Percentage of calcium carbonate precipitated one day after seed suspension was added to a stable supersaturated solution in the presence and absence of added ions. This experiment was completed 11 days after seed crystals were added.

concentrations (10^{-6} to 10^{-8} *M*) metaphosphate may reduce the rate of calcite formation without an induction period. For orthophosphate ion over a wide range of concentrations, a systematic reduction in the rate constant with no induction period was found.

The substances tested at the concentrations shown can be ranked in order of their ability to inhibit calcite formation as follows: sodium metaphosphate > sodium dihydrogen phosphate ≅ stearic acid ≅ magnesium chloride ≅ nucleic acid ≅ humic acid >> strontium chloride ≅ phosphatidylinositol ≅ albumin. Albumin had the smallest effect on calcite formation; in the presence of 10 mg/l the amount formed was only slightly lower than that in pure solution. In marked contrast, phosphate and metaphosphate ions at a concentration of 10^{-4} *M* reduced by 90% the amount of calcite formed in one day. Differences in the degree to which these additives inhibit may be due to additive-calcium precipitate formation, additive-calcium solution-phase complex formation, differences in specific adsorption, or differences in orientation of adsorbed molecules on the crystal surface. It appears that additive ions which are negatively charged in the experimental solutions (*e.g.,* metaphosphate, phosphate, stearic acid, humic acid and nucleic acid) are able to interact with the calcite surface, which is near the zero point of charge under the conditions employed in this study.

Specific interaction between calcite growth sites and certain phosphorus-containing anions, especially the straight-chain polymers such as metaphosphate, results in very effective growth inhibition.[35] Rastrick[37] attributed the inhibition by straight-chain polyphosphate to precision of fit between the calcium ions in a calcite crystal face and oxygen atoms in the polyphosphate chain. A general description of growth inhibition by this type of specific adsorption requires evaluation of the crystal growth rates at several crystallographic faces and of the differences in surface adsorption exhibited by each face for the ion being added.

Ion association can significantly alter solution supersaturation and thus the calcite precipitation reaction. The extent of ion association in the experimental solutions can be estimated when the total solution composition, pH and thermodynamic stability constants are known. For supersaturated solutions containing phosphate, glycerophosphate or magnesium ions, additive ion pair formation in solution did not measurably affect the reaction kinetics. Furthermore, the added ion did not form an insoluble product, since analysis of added ion in solution, total calcium and carbonate concentrations before and after addition of seed crystal showed no concentration changes other than those associated with formation of calcium carbonate.

Because of its relatively highly soluble carbonate salts, magnesium ion was added at a significantly higher concentration (10^{-3} *M*) than the other additives tested. Only at this concentration did it measurably affect the calcite formation process. The other cation tested, strontium ion, was not a growth inhibitor. Thus, at the experimental pH, cationic species appear to interact weakly with growth sites on the calcite surface. Indeed, the inhibition by magnesium ion probably occurs because the similar crystal structure requirements of calcite and mixed magnesium-calcium carbonates enhance specific adsorption of magnesium ions at the calcite growth sites and override the charge effect. There is not, however, the same structural similarity between calcite and mixed calcium-strontium carbonates.

It is particularly significant that two natural organic components of wastewater, phosphatidylinositol and albumin, do not inhibit calcite growth. At 10 mg/l, each of these polar compounds adsorbs at the solution-calcite interface. Their charge in solution and/or their failure to interact with the growth sites on the calcite surface may prevent their mediation in the crystallization process. Nor was calcite seeded growth inhibited by concentrations of gelatin up to 25 mg/l at a similar solution pH and higher solution alkalinity [Reddy, unpublished results]. On the other hand, gelatin is a strong growth inhibitor in other systems such as calcium sulfate dihydrate seeded growth. It seems that, like phosphatidylinositol and albumin, gelatin does not specifically interact with the calcite growth sites. Therefore, it is not a growth inhibitor under the experimental conditions.

Growth in the Presence of Phosphate, Glycerophosphate and Magnesium

In view of the varying capacities of certain wastewater components to inhibit calcite formation, additional studies were performed with a combination of three such ions in chemically defined solutions: (1) phosphate ion, a major pollutant of domestic wastewater and an ion often removed from wastewater by lime addition; (2) glycerophosphate ion, a typical organophosphate compound found in wastewater and generally not removed by lime addition; and (3) magnesium ion, a divalent cation with a simple solution chemistry, which is a model for the interactions of other divalent toxic metals often removed by lime addition, such as Pb^{2+} and Cd^{2+}.

Experimental data and rate constants for calcium carbonate growth in the presence of the additive ions are presented in Table III. The marked inhibition of calcium carbonate crystallization by all three ions examined is clearly seen in the plot of total calcium concentration as a function of time during several of these experiments (Figure 5). There is a significant difference in the concentration at which each of the three additive ions is effective: phosphate ion inhibits at 0.1 the concentrations of glycerophosphate ion and 0.01 that of magnesium ion.

When the integrated rate funciton is plotted as shown in Figure 6, the slope, ks, decreases as the concentration of each additive ion increases. As found for calcite crystallization from pure solutions (Table II and Figure 3), the best-fit linear relationship between the integrated rate function ($N^{-1} - N_0^{-1}$) and time as shown in Figure 6 confirms the validity of Equation 2 and its integrated form, Equation 4 for the interpretation of calcite crystallization in the presence of additives. When the integrated rate function is plotted in this way, the slopes of the best-fit straight lines decrease systematically as the concentration of each

Table III. Experimental Conditions and Calculated Rate Constants for Calcite Growth Experiments in the Presence of Additives

Reaction Mixture (mol/l x 10^4)				Initial Concentration of Ions in Solution (mol/l x 10^4)		s (mg/l)	k
Added Ion	Total Ca^{2+}	Total CO_3^{2-}	pH	Ca^{2+}	CO_3^{2-}		
None							
	2.593	41.57	8.886	2.143	1.781	95.2	0.82
	2.496	41.02	8.905	2.057	1.830	94.3	1.20
Orthophosphate							
0.01	2.634	40.81	8.971	2.131	2.104	95.0	0.628
0.02	2.620	40.62	8.959	2.129	2.039	94.5	0.423
0.04	2.680	40.22	8.982	2.166	2.121	93.5	0.374
0.10	2.262	39.38	8.972	2.136	2.029	90.7	0.150
Glycerophosphate							
0.10	2.592	41.67	8.847	2.165	1.637	93.8	0.826
0.20	2.611	41.57	8.853	2.179	1.565	93.5	0.562
0.30	2.606	41.47	8.832	2.187	1.578	93.3	0.413
0.40	2.618	41.37	8.852	2.187	1.646	93.0	0.152
1.0	2.567	39.91	8.855	2.155	1.600	91.6	0.155
Magnesium							
1.0	2.752	40.82	8.948	2.243	2.003	99.5	0.30
10.0	2.578	40.82	8.793	2.196	1.449	99.5	0.028

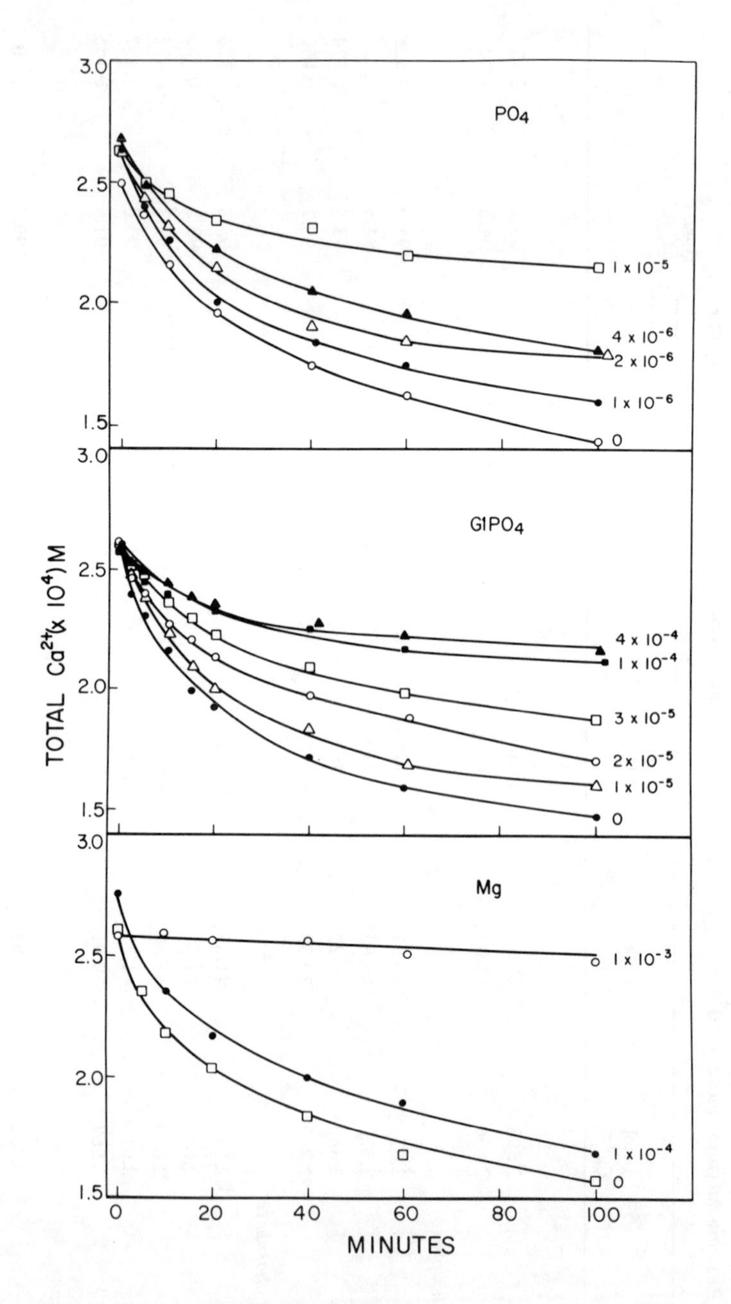

Figure 5. Concentration of calcium ion remaining in solution during growth of seed crystals in the presence and absence of added ions. Molar concentrations of added ions are indicated besides the curves.

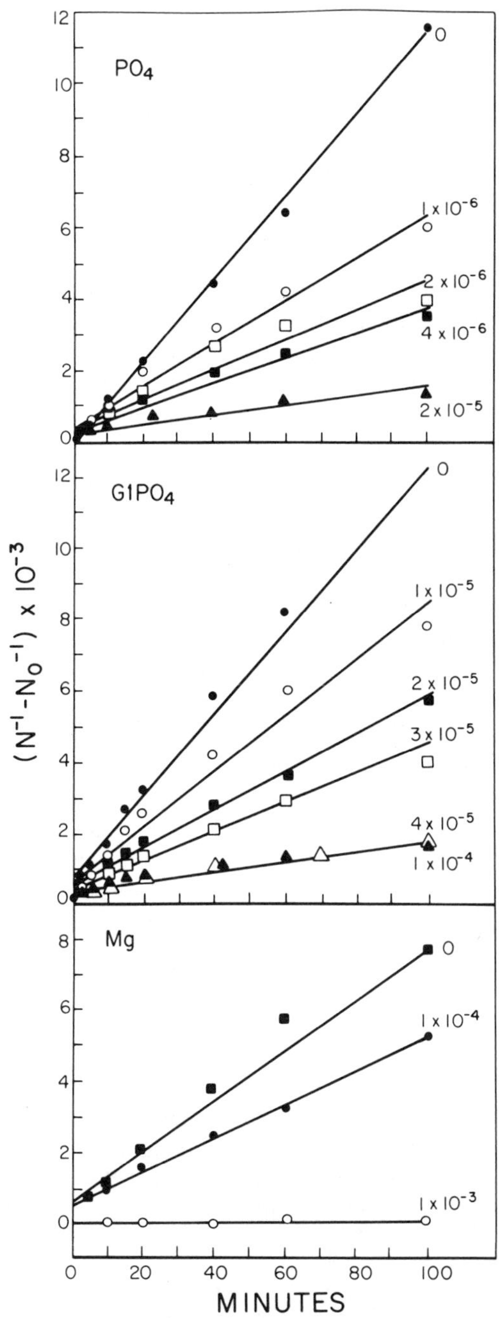

Figure 6. Calcite crystal growth in the presence and absence of added ions, as expressed by the rate function plotted against time. Molar concentrations of added ions are indicated beside the curves.

additive ion increases. This relationship further supports the usefulness of the seeded growth procedure for the characterization of crystallization inhibition.

Mechanism of Growth Inhibition

There are several possible mechanisms by which an additive can inhibit a crystallization reaction. Ion-pair formation is often responsible for a decrease in solution supersaturation when the concentration of the added ion is sufficiently high to form a stable complex with one of the precipitating ions. Phosphate and glycerophosphate ions form moderately stable ion pairs with calcium ion, while magnesium ion forms somewhat weaker complexes with carbonate and bicarbonate ions in solution. To calculate the extent of such ion-pair formation, stability constants from the literature can be used. Table IV summarizes the constants relevant to the systems tested here. The results of these calculations, given in Table V, show that in supersaturated solutions with 10^{-4} *M* glycerophosphate or 10^{-5} *M* phosphate, less than 1% of the total dissolved calcium is bound as a phosphorus-containing ion pair. Therefore this mechanism cannot account for the effect of phosphorus-containing compounds on the rate of calcite growth.

When calcium carbonate crystallization occurs in the presence of magnesium ion, there is a substantial amount of ion-pair formation. Magnesium carbonate and bicarbonate concentrations in the experimental supersaturated solution containing 10^{-3} *M* total magnesium ion are significant: 20% of the total dissolved magnesium ion exists as carbonate-containing ion pairs (Table V). This ion-pair formation reduces the free carbonate concentration by 4% and slightly alters the solution supersaturation. However, such a small reduction in supersaturation does not explain the dramatic decrease in the calcite growth rate. Thus the effect of magnesium ion-pair formation on reaction kinetics is virtually negligible.

If the inhibition of crystallization by added ions is due to surface adsorption of the ions at growth sites, some form of adsorption isotherm should be applicable. A simple Langmuir-type adsorption has been found to account for the inhibition of calcium carbonate growth in the presence of phosphonate ions,[14] and the rate constant for the inhibited reaction can be calculated from the relationship

$$\frac{k_o}{k_o - k} = 1 + \frac{k_1}{(k_2 C)} \qquad (5)$$

where k_o = the crystal growth rate constant in the absence of additive ion,
k_1 and k_2 = the adsorption and desorption rate constants, respectively, for the inhibiting ions
C = the concentration of the additive.

Table IV. Equations Used for Calculation of Ionic Species at 25°C and One atm for Calcium Carbonate Crystallization Reactions in the Presence of Phosphate, Glycerophosphate or Magnesium Ion

Equation	Reference
Mass action	
1. $[H^+]\ [H_2PO_4^-]/[H_3PO_4] = 10^{-2.148}$	38,39
2. $[H^+]\ [HPO_4^{2-}]/[H_2PO_4^-] = 10^{-7.198}$	38,39
3. $[H^+]\ [PO_4^{3-}]/[HPO_4^{2-}] = 10^{-12.37}$	40
4. $[CaH_2PO_4^+]/[Ca^{2+}]\ [H_2PO_4^-] = 10^{1.408}$	41
5. $[CaHPO_4]/[Ca^{2+}]\ [HPO_4^{2+}] = 10^{2.739}$	41
6. $[CaPO_4^-]/[Ca^{2+}]\ [PO_4^{3-}] = 10^{6.46}$	41
7. $[H^+]\ [G1P^-]/[HG1P] = 10^{-6.28}$	42
8. $[CaG1P^+]/[Ca^{2+}]\ [G1P^-] = 10^{1.92}$	42
9. $[MgCO_3]/[Mg^{2+}]\ [CO_3^{2-}] = 10^{2.98}$	43
10. $[MgHCO_3^+]/[Mg^{2+}]\ [HCO_3^-] = 10^{1.06}$	43

Table V. Concentrations of Some Major and Minor Ions, and Ion Pairs, in Supersaturated Solutions Exhibiting Reduced Calcite Growth Rate Constants

	Added Ion					
	Phosphate		Glycerophosphate		Magnesium	
Initial pH	8.971		8.853		8.948	
	Total Ca^{2+}	2.6340	Total Ca^{2+}	2.6110	Total Mg^{2+}	1.000
	Total PO_4^{3-}	0.0100	Total G1P	0.2000	Total CO_3^{2-}	40.82
	$H_2PO_4^-$	0.0001	HG1P	0.0005	HCO_3^-	38.22
Initial	HPO_4^{2-}	0.0074	$G1P^-$	0.1933	CO_3^{2-}	2.003
Concentrations	PO_4^{3-}	0.0000	$CaG1P^+$	0.0062	$MgCO_3^0$	0.1196
(mol/l x 10^4)	$CaHPO_4^0$	0.0008			$MgHCO_3^+$	0.0296
	$CaPO_4^-$	0.0018			Mg^{2+}	0.8500

In Figure 7, $k_O/(k_O - k)$ is plotted against the reciprocal of the concentration of the added ions. Since two sets of experiments with glycerophosphate as inhibitor having two different k_O values lie on the best-fit line, 0.944 (±0.041) + 0.162 (±0.011) x $10^{-4}/[G1PO_4]$, and three sets of experiments with phosphate as inhibitor having three different k_O values lie on a second best-fit line, 0.992 (±0.012) + 0.0196 (±0.001) x $10^{-4}/[PO_4]$, the use of the Langmuir adsorption model seems justified. This agreement also demonstrates the reproducibility and reliability of the seeded growth technique for testing the inhibition of crystallization by wastewater constituents. The best-fit linear relation and the intercept (within experimental uncertainty) of unity strongly suggest that the mechanism of inhibition is the same as that proposed for the Langmuir adsorption isotherm; namely, the formation of a monomolecular layer of additive ions at growth sites on the crystal face which blocks further growth.

A plot of $k_O/(k_O - k)$ against the reciprocal of the total magnesium ion concentration for the data presented in Table IV is a straight line which has the equation $k_O/(k_O - k) = 0.85 + 1.87 \times 10^{-4}/[Mg]$. An intercept of 0.85 for this limited data set is probably not different from unity and suggests that the Langmuir isotherm also describes the magnesium ion inhibition of calcite growth. The suggestion has been confirmed by more recent data.

Comparison of slopes for phosphate (0.020 x 10^{-4}), glycerophosphate (0.162 x 10^{-4}) and magnesium ions (1.87 x 10^{-4}), as shown in Figure 7, indicates that a 50% reduction in the rate constant for calcite formation occurs at a phosphate concentration 0.1 that for glycerophosphate and 0.01 that for magnesium ion. These differences may reflect stronger equilibrium adsorption of phosphate than of glycerophosphate or magnesium ions at the interface, probably due to a reduction in the desorption rate constant, k_2.

Analysis of the solutions before and after growth experiments showed that there was virtually no uptake or loss of the added ions during crystallization. After an experiment in which crystallization was inhibited by the presence of glycerophosphate, the seed crystal was found to contain between 0.05 and 0.10 mg of P per gram of $CaCO_3$ (N = 4), which is consistent with the adsorption of natural organic material by calcite surfaces.[34] However, there was a negligible concomitant change in the glycerophosphate in solution.

Calcite growth inhibition in solutions containing either phosphate, glycerophosphate or magnesium ion is similar to that observed in supersaturated solutions containing polyphosphates[17] and phosphonic acid derivatives.[14] In these experiments an abrupt change in crystal growth

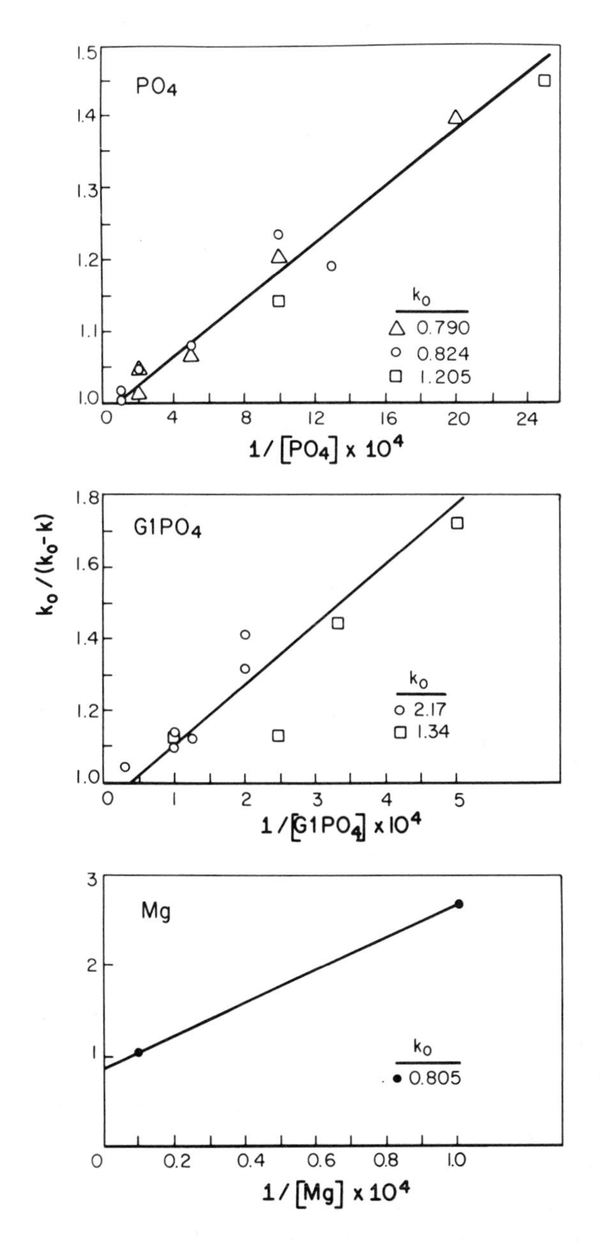

Figure 7. Langmuir isotherm plots. k_0 is the calcite growth rate constant in the absence of added ions and k is the rate constant in the presence of added substances.

rate which occurred over a narrow range of additive concentrations could not be attributed to an induction period, as could inhibition of calcium sulfate dihydrate crystal growth.[44] This type of crystal growth inhibition is similar to that ascribed by Sears[45] to trace inhibitors, which he defines as substances that reduce the crystallization rate at low concentrations but are not incorporated into the crystallizing material. Under the experimental conditions employed in this study, separate phases may form at higher additive concentrations, or ion-pair formation may become a significant kinetic factor.

An adsorbed inhibitor hypothesis to explain calcite growth inhibition by phosphate and glycerophosphate ions is compatible with published data for calcium carbonate inhibition. In one set of experiments,[46] growth of calcium carbonate spherulitic seed crystals from highly supersaturated solutions was said to be completely inhibited in the presence of 25 mg/l sodium phosphate (added as a mixture of polymetaphosphate and polyphosphate). A significant percentage of the phosphorus originally in the solution had been adsorbed onto the seed crystals. The high solution supersaturation and polyphosphate concentration employed may have led to the formation of separate calcium phosphate phases, which effectively removed phosphorus from the solution.

Adsorption of additive ions has been reported to account for inhibition of spontaneous precipitation and crystal growth in supersaturated calcium carbonate solutions containing 0 to 1.5 mg/l metaphosphate.[47,48] Several phosphorus-containing anions at trace concentrations (0.001 of calcium) prevented the spontaneous precipitation of calcium carbonate. Muira[49] found that pyrophosphate was adsorbed in significant amounts on calcium carbonate crystals and was an effective growth inhibitor. He concluded that surface adsorption is a major factor in the mechanism of growth inhibition. Adsorption of phosphorus by spontaneously precipitating calcium carbonate in a modified lakewater has been reported by Otsuki and Wetzel.[50]

Simkiss[22] has studied three phosphorus-containing ionic metabolites which inhibit the spontaneous precipitation of calcium carbonate in artificial and natural seawater. Of the three ions glycerophosphate had the smallest measurable effect. This observation is in agreement with experimental results shown in Table III. Simkiss has shown that a 10% reduction in the amount of calcium carbonate precipitated in a given time interval occurred at a glycerophosphate concentration of 10^{-4} to 10^{-5} *M*, a figure in agreement with the present data. The spontaneous precipitation results suggest that calcium carbonate formation in highly supersaturated solutions, whether by nucleation or by crystal growth, is markedly inhibited by phosphorus-containing ions. Moreover, it appears that during the nucleation significant amounts of phosphorus are incorporated into insoluble calcium carbonate. This conclusion is in accord with the composition of the complex precipitates which crystallize from solutions containing calcium, carbonate and phosphate ions[51] and with the absence of calcium carbonate precipitation from solutions where the amount of calcium is exceeded by the sum of carbonate and phosphate concentrations.[7] A careful study of calcium phosphate-seeded growth in the absence of carbonate has shown that the reaction occurs in several steps.[52] In one of these, two or more different calcium phosphate phases are formed, while a thermodynamically unstable phase, octacalcium phosphate, which has been found in the initial stages of reaction, simultaneously dissolves.

In the seeded growth experiments reported here there was no evidence for the formation of any phase other than calcium carbonate on the surface of the calcite crystals during their growth in the presence of added ions. The experimental solutions which contained the highest added phosphate concentrations were supersaturated with respect to hydroxyapatite, the thermodynamically stable calcium phosphate phase under these conditions. In these solutions, as well as in all others examined, the phosphate concentration remained essentially unchanged during the calcite crystallization reaction. This absence of a separate calcium phosphate phase limits the conditions under which the phosphate-carbonate substitution reaction may occur (*e.g.*, calcite to hydroxyapatite).

The experimental conditions adopted in this study simulate those encountered in a wastewater treatment process, where a phosphate substitution reaction may take place. Leckie and Stumm[53] have shown that a calcium phosphate phase can nucleate and grow from a supersaturated solution on a calcium carbonate substrate. Formation of such a phase may occur, as discussed previously, at much higher calcium phosphate supersaturations than those employed in this study. Indeed, Ferguson and McCarthy[7] demonstrated that the marked inhibition of calcium carbonate growth may be accompanied by the formation of a calcium phosphate phase in a solution highly supersaturated with both calcium carbonate and phosphate. During spontaneous precipitation from complex, highly supersaturated solutions,[54,55] inhibition of calcite formation may occur after stable nuclei are formed in solution. For example, the observation that magnesium ion has no effect on the growth rate of aragonite crystals[56] suggests that the varied polymorphic composition of some calcium carbonate spontaneous precipitates may be regulated by crystal growth rates established after nucleation. In our seeded growth experiments, however, no magnesium-containing phase appears to form on the surface of calcite seed crystals even though magnesium ion can substitute for calcium ion in the calcite lattice with minor crystallographic distortion. Solutions containing the highest magnesium ion concentration were supersaturated with respect to the thermodynamically stable mixed carbonate, dolomite, but even here the magnesium ion concentration was essentially unchanged during the crystallization reaction. This observation supports the interfacial inhibition mechanism indicated by the Langmuir function plot. It is also in agreement with the data of Berner,[56] which show that direct evidence of magnesium ion incorporation into growing calcite seed crystals was obtained only during extended growth experiments (10-50 hr) in highly supersaturated solutions (*e.g.*, 0.50 *M* Na_2CO_3 plus 0.50 *M* $CaCl_2$).

The trace inhibition mechanism proposed here for the reduction in the rate of calcite formation by added ions is applicable to calcium carbonate polymorph transformation kinetics. Surface nucleation of calcite on the unstable polymorph has been proposed as a rate-limiting step. Exchange of magnesium ion for calcium ion on the initial polymorph surface and on the nucleated calcite surface leads to changes in the concentration of magnesium in solution.[57] In addition, such an exchange reaction can lead to alteration in the chemical potential of both the surface phase[56] and the bulk carbonate phase.[58] The exchange process alters the rate of reaction because changes in the solubility of the solid change the solution supersaturation, whereas the trace inhibition mechanism found for magnesium ion in this study alters the rate of reaction by reducing the reaction rate constant.

The proposed adsorption mechanism for the inhibition of calcite growth by magnesium ion (Figure 5) is compatible with much of the published data dealing with the interaction of aqueous magnesium ion and heterogeneous calcium carbonate solutions. Möller and Rajagopalan,[59] using a spontaneous precipitation technique, attributed magnesium-ion inhibition of calcium carbonate nucleation to destabilization of the critical nuclei and thought that destabilization resulted from incorporation of the magnesium ion into the calcium carbonate nuclei. Bischoff[57] also reported that magnesium ion reduced the nucleation rate of calcite and, in doing so, was incorporated into the crystal nuclei, but he did not examine the role of magnesium ion in the subsequent calcite crystal growth. It seems likely that at least part of the crystallization inhibition observed in these experiments was due to the inhibition by magnesium ion of the growth of stable crystal nuclei. Direct comparison of the results presented in Table I with the data of Möller and Rajagopalan or Bischoff is not possible because different experimental conditions were employed in these investigations.

CONCLUSIONS

The results reported here can be applied to a number of lime-addition wastewater treatment processes. It has been shown that several ions inhibit calcium carbonate formation at concentrations occurring in conventional wastewaters. Calcium carbonate formation in the presence and absence of these growth inhibitors follows a second-order rate expression, consistent with an interfacially controlled rate-determining step. The kinetics of the inhibition can be derived from a simple Langmuir adsorption isotherm model. This model yields a mathematical function which can be used to calculate the reduction in the overall formation rate in the presence of added ions.

During lime addition to a wastewater which contains 5 mg/l of polyphosphates (derived from phosphorus-based detergents), calcium carbonate formation would be completely inhibited, and phosphorus would be efficiently removed as calcium phosphate. If the polyphosphate initially present were degraded to orthophosphate, the initial concentration of orthophosphate (10 mg/l) would be high enough to reduce significantly the rate of calcium carbonate formation. If the organic phosphorus component in the wastewater reacts like the organophosphates employed in this study, its concentration would not be sufficiently high to influence the kinetics of calcium carbonate formation. In areas where phosphorus-containing detergents are prohibited, the concentration of polyphosphates will be very low and will probably have little influence on the kinetics. Orthophosphate concentrations will also be lower, and calcium carbonate will be more likely to precipitate during lime addition for phosphorus removal, thus lowering the efficiency of the process. As a corrective it may be appropriate to add a calcium carbonate growth inhibitor to the input wastewater.

Lime treatment of industrial wastewaters to remove toxic metals requires a different procedure. Unlike phosphorus removal, optimum metal removal requires the formation of calcium and other metal carbonates. The results with magnesium indicate the probability that there will be little kinetic interference from divalent metal ions, but additional investigations are needed, especially for specific metals of interest such as cadmium, copper, lead and zinc. Since the chemical reactions associated with lime addition to wastewaters containing metals never approach equilibrium, it seems likely that in some instances anionic crystallization inhibitors such as fatty acids or humic acids (shown here to be effective kinetic inhibitors) may prevent optimal removal of metals.

As pointed out previously, a very attractive feature of lime addition as a treatment process is that it can be used to simultaneously remove two of the most significant wastewater pollutants: phosphorus and heavy metals. However, as shown here, the kinetic considerations for optimization of each type of removal process are antagonistic, and considerable research may be required to establish a lime-treatment scheme which will remove both of these pollutants efficiently and economically.

ACKNOWLEDGMENT

G. W. Fuhs, L. N. Plummer and J. M. Cohen provided constructive criticism of an earlier version of this manuscript. This work was supported, in part, by Grant No. 35043.75 from Health Research Incorporated, Albany, New York.

REFERENCES

1. Peirano, L. E. *J. Water Poll. Control Fed.* 49:568 (1977).
2. LeBoice, J. N., and J. F. Thomas. *J. Water Poll. Control Fed.* 47: 2246 (1975).
3. Minton, G. R., and D. A. Carlson. *J. Water Poll. Control Fed.* 48: 1697 (1976).
4. Schmid, L. A., and R. E. McKinney. *J. Water Poll. Control Fed.* 41:1259 (1969).
5. Hautala, E., J. Randall, A. Goodban and A. Waiss, Jr. *Water Res.* 11:243 (1977).
6. Weber, W. J., and H. S. Posselt. In: *Aqueous-Environmental Chemistry of Metals*, A. J. Rubin, Ed. (Ann Arbor, MI: Ann Arbor Science Publishers, Inc., 1974).
7. Ferguson, J. F., and P. L. McCarty. *Environ. Sci. Technol.* 5:534 (1971).
8. Ferguson, J. F., D. Jenkins and J. Eastman. *J. Water Poll. Control Fed.* 45:620 (1973).
9. Maruyama, T. S., A. Hannah and J. M. Cohen. *J. Water Poll. Control Fed.* 47:926 (1975).
10. Merrill, D. T., and R. M. Jorden. *J. Water Poll. Control Fed.* 47:2783 (1975).
11. Wuhrmann, K. In: *Advances in Water Quality Improvement*, E. F. Gloyna and W. W. Eckenfelder, Eds. (Austin, TX: University of Texas Press, 1968).
12. Reddy, M. M., and G. H. Nancollas. *J. Colloid Interface Sci.* 37: 166 (1971).
13. Nancollas, G. H., and M. M. Reddy. *J. Colloid Interface Sci.* 37: 824 (1971).
14. Reddy, M. M., and G. H. Nancollas. *Desalination* 12:61 (1973).
15. Reddy, M. M., and G. H. Nancollas. *J. Crystal Growth* 35:33 (1976).
16. Nancollas, G. H., and M. M. Reddy. *Soc. Petrol. Eng. J.* (April 1974), p. 117.

17. Nancollas, G. H., and M. M. Reddy. In: *Aqueous-Environmental Chemistry of Metals*, A. J. Rubin, Ed. (Ann Arbor, MI: Ann Arbor Science Publishers, Inc., 1974).
18. Reddy, M. M. *Proc. Internat. Assoc. Theoret. Appl. Limnol.* 19: 429 (1975).
19. Reddy, M. M. *J. Crystal Growth* 41:287 (1977).
20. Brocker, W. S. *Chemical Oceanography* (New York: Harcourt Brace Jovanovich, Inc., 1974).
21. Bischoff, J. L., and W. S. Fyfe. *Am. J. Sci.* 226:65 (1968).
22. Simkiss, K. J. *International Council Exploration Sea* 29:6 (1964).
23. Kitano, Y. *Bull. Chem. Soc. Japan* 35:1973 (1967).
24. Davis, C. W. *Ion Association* (Washington, DC: Butterworths, 1963).
25. Harned, H. S., and R. Davis. *J. Am. Chem. Soc.* 65:2030 (1943).
26. Harned, H. S., and S. R. Scholes. *J. Am. Chem. Soc.* 63:1706 (1941).
27. Harned, H. S., and W. J. Hamer. *J. Am. Chem. Soc.* 55:2194 (1933).
28. Garrels, R. M., and M. E. Thompson. *Am. J. Sci.* 260:57 (1962).
29. Greenwald, I. *J. Biol. Chem.* 141:789 (1941).
30. Nancollas, G. H. *Interactions in Electrolyte Solutions* (Amsterdam: Elsevier Publishing Co., 1966).
31. Langmuir, D. *Geochim. Cosmochim. Acta* 32:835 (1968).
32. Mullin, J. W. *Crystallization* (Cleveland, OH: CRC Press, Inc., 1972).
33. Suess, E. *Geochim. Cosmochim. Acta* 37:2435 (1973).
34. Myers, P. A., and J. G. Quinn. *Limnol. Oceanog.* 16:992 (1971).
35. Liu, T. S., and G. H. Nancollas. *J. Colloid Interface Sci.* 52:593 (1975).
36. Liu, T. S., and G. H. Nancollas. *J. Colloid Interface Sci.* 52:582 (1975).
37. Rastrick, B. *Discuss. Faraday Soc.* 5:243 (1949).
38. Bates, R. G., and S. F. Acree. *J. Res. Natl. Bur. Stds.* 30:129 (1943).
39. Bates, R. G. *J. Res. Natl. Bur. Stds.* 47:127 (1951).
40. Vanderzee, C. E., and A. S. Quist. *J. Phys. Chem.* 65:118 (1961).
41. Chughtai, A., R. W. Marshall and G. H. Nancollas. *J. Phys. Chem.* 72:208 (1968).
42. Schwarzenbach, G., and G. Anderegg. *Helvet. Chim. Acta* 40:1229 (1957).
43. Plummer, L. N., B. F. Jones and A. H. Truesdell. *U.S. Geological Survey, Water-Resources Investigations*, 76-13 (September 1976).
44. Liu, T. S., and G. H. Nancollas. *J. Colloid Interface Sci.* 44:422 (1973).
45. Sears, G. W. In: *Growth and Perfection of Crystals*, R. H. Doremus, B. W. Roberts and D. Turnbull, Eds. (New York: John Wiley and Sons, Inc., 1958).
46. Brooks, R., L. M. Clark and E. F. Thurston. *Phil. Trans. Roy. Soc. (London)* A243:145 (1950-51).
47. Buehrer, T. F., and R. F. Reitemeier. *J. Phys. Chem.* 44:535 (1940).
48. Buehrer, T. F., and R. F. Reitemeier. *J. Phys. Chem.* 44:552 (1940).

49. Miura, M., H. Naono and S. Otani. *Kogyo Koguku Zasshi* 66:597 (1963).
50. Otsuki, A., and R. G. Wetzel. *Limnol. Oceanog.* 17:763 (1972).
51. Malone, Ph. G., and K. M. Towe. *Marine Geol.* 9:301 (1970).
52. Nancollas, G. H., and B. Tomažič. *J. Phys. Chem.* 78:2218 (1974).
53. Stumm, W., and J. O. Leckie. In: *Water Quality Improvement by Physical and Chemical Processes*, E. F. Gloyna and W. W. Eckenfelder, Jr., Eds. (Austin, TX: University of Texas Press, 1970).
54. Barber, D. M., Ph. G. Malone and R. J. Larson. *Chem. Geol.* 16: 239 (1975).
55. Towe, K. M., and Ph. G. Malone. *Nature* 226:348 (1970).
56. Berner, R. A. *Geochim. Cosmochim. Acta* 39:489 (1975).
57. Bischoff, J. L. *J. Geol. Res.* 73:3315 (1968).
58. Windland, H. D. *J. Sediment. Petrol.* 39:1529 (1969).
59. Möller, P., and G. Rajagopalan. *Z. Phys. Chem.* 94:297 (1975).

CHAPTER 4

AQUEOUS CHEMISTRY AND PRECIPITATION OF ALUMINUM PHOSPHATE

Tanhum Goldshmid and Alan J. Rubin
Water Resources Center
College of Engineering
Ohio State University
Columbus, Ohio

INTRODUCTION

Aluminum salts are used extensively for the clarification of raw water for drinking purposes and recently have been used in physical-chemical wastewater treatment processes for suspended solids removal and phosphate precipitation. Because the chemical interactions between aluminum and phosphate have not been well defined, particularly with respect to the solubility of the condensed phase in metastable solutions, it has been difficult to develop a rational model of phosphate precipitation using aluminum salts. Furthermore, since the ability of aluminum to coagulate suspended solids and to precipitate phosphate are interrelated, it is evident that the aluminum phosphate system must be carefully delineated to achieve higher efficiencies and a better control of wastewater treatment processes.

The aqueous chemistry of phosphorus is well defined. The element occurs in nature primarily in the fully oxidized state as a phosphate with formal oxidation number five. Phosphate may be of the simple ortho form as phosphoric acid and its salts or in a condensed form. The latter are referred to as polyphosphates if they have a linear structure and as metaphosphates if they possess a ring structure. The relative concentrations of the various inorganic species of orthophosphate are pH-dependent.

Each ion is in equilibrium with protons and other acidic or basic species of the system

$$H_3PO_4 \rightleftharpoons H_2PO_4^- \rightleftharpoons HPO_4^{2-} \rightleftharpoons PO_4^{3-} \quad \xrightarrow{\text{Increasing pH}}$$

In contrast to the relatively simple and clearly defined aqueous reactions of the orthophosphate ion, those of aluminum(III) are more complicated since they involve both mononuclear and polynuclear ionic species as well as insoluble phases. The free, unhydrolyzed metal ion, Al^{3+}, which exists only at very low pH, is highly hydrated, octahedrally coordinating six water molecules.[1] As the pH increases, the hydrated aluminum ion hydrolyzes, yielding soluble and insoluble hydroxy-aluminum products. Some of these are well established while the stoichiometry of others is more speculative, being based on somewhat circumstantial evidence. Chemical species of the latter type include polynuclear complexes which have been proposed by several investigators.[2,3] Recent studies based on solubility data as well as coagulation and titration results have suggested the existence of an octamer, $Al_8(OH)_{20}^{4+}$.[4,5] The hydrolytic reactions of aluminum(III) can be summarized by:

$$Al^{3+} \rightleftharpoons AlOH^{2+} \rightleftharpoons Al_8(OH)_{20}^{4+} \rightleftharpoons Al(OH)_3(s) \rightleftharpoons Al(OH)_4^- \quad \xrightarrow{\text{Increasing pH}}$$

The interaction between aluminum(III) and phosphate in acid solutions was studied by Bjerrum and Dahm[6] using conductivity and pH measurements. They reported the formation of complexes with the general stoichiometry of $Al(H_2PO_4)_x^{(3-x)+}$, in which x ranges between one and three depending on the ratio of the applied concentrations of aluminum and phosphate. In addition, these complexes dissociate to yield phosphato-aluminum complexes of the general formula $AlH_y(PO_4)_x^{(3+y-3x)+}$, in which y ranges from 1 to 2x depending on pH. The formation of phosphato-aluminum complexes in acid solutions has also been reported by other investigators.[7-9]

A large number of aluminum phosphate solid phases have been identified, the majority by X-ray diffraction of natural samples. However, only a few of those solid phases have been reproduced experimentally. Among these have been variscite, $Al(OH)_2H_2PO_4(s)$, and sterrettite, $(Al(OH)_2)_3HPO_4H_2PO_4(s)$, identified by Cole and Jackson.[10] Variscite is essentially the hydrated form of the tertiary salt, $AlPO_4(s)$. The solubilities of variscite and the tertiary salt are related by

$$K_{so}\ (AlPO_4) = \frac{K_{so(var)}K_2K_3}{K_w^2} \quad (1)$$

where K_2 and K_3 are the second and third dissociation constants of phosphoric acid and K_w is the ion product of water.

The existence of a family of minerals in nature, known as the taranakites, $M_3Al_5H_6(PO_4)_8 \cdot 18H_2O(s)$ where M is any monovalent ion including ammonium ion, has been reported.[11,12] However, only the potassium and ammonium analogs have been reproduced in the laboratory.[13]

This chapter examines the interaction between aluminum(III) and phosphate over a wide range of concentrations and pH to develop a rational model for the system in metastable solution. The solubility of aluminum-phosphate precipitates was studied as a function of aluminum, phosphate and hydrogen ion concentrations to define the equilibrium reactions between soluble species and insoluble phases. Also identified were the soluble and insoluble products of the reactions between aluminum(III) and phosphate, and their concentration distribution was determined over a wide range of pH and P and Al concentrations. The study was limited to pure systems of well-defined solutions. No dispersed solid phases were present other than those precipitated through the interaction between aluminum and phosphate or as a result of changing the pH.

EXPERIMENTAL

Preparation of Solutions

Stock nitric acid solutions were prepared by diluting the concentrated acid; they were standardized titrametrically against *tris*(hydroxymethyl)-aminomethane. Endpoints were detected potentiometrically as with all the other determinations, unless otherwise stated. Stock base solutions were prepared from sodium or potassium hydroxide pellets and standardized by titration with the standard acid. Aluminum nitrate stock solutions were prepared in concentrations greater than 0.1 *M* to prevent hydrolysis and subsequent aging.[4] Diluted working solutions of the metal salt were prepared as needed just prior to each experiment. The concentrations of the aluminum(III) solutions were determined by alkalimetric titrations, or colorimetrically using calgamite as the chromogenic reagent in an extraction procedure.[14] The method was found to be virtually unaffected by the presence of phosphate. Reproducible results were obtained even when the phosphate concentration was a thousand times greater than that of the metal. The extraction process was carried out in 250-ml separatory funnels instead of in vials as proposed in the original method. This modification allowed the determination of soluble aluminum in very dilute solutions by taking larger volumes for the analysis. The absorbance of the uncomplexed calgamite reagent increases with its concentration and with decreasing wavelength. Readings were taken at 602.5 nm to minimize the absorbance of the uncomplexed dye instead of at 550 nm, as originally reported. The molar absorbtivity of the complex is 42,000 at that wavelength.

Phosphate solutions were standardized by alkalimetric titrations or colorimetrically using the modified ascorbic acid-molybdate method.[15] The molar absorbtivity of the colored complex was calculated to be 29,800 at 890 nm. Sodium nitrate solutions were prepared by dissolving accurately weighed quantities of the dried reagent-grade salt. Determination of the sodium ion content of the solutions was by atomic absorption spectrophotometry.[16]

Solubility Studies

Stock solutions containing known concentrations of aluminum nitrate and sodium dihydrogen phosphate were split into 100-ml aliquots and stored in 4-oz polyethylene bottles. The pH of the solutions was systematically varied with sodium hydroxide solution, and the samples were shaken for 24 hr in a mechanical shaker at room temperature. The solutions were vacuum filtered through 0.2 μm microporous membrane filters, and the filtrates were examined for the absence of precipitate by comparing their light scattering intensities to that of distilled-deionized and membrane-filtered water. The samples were refiltered if precipitate was detected. The filtrates were then analyzed for aluminum or phosphate, as required, and their pH measured. Figure 1 shows a typical plot of residual soluble phosphate concentration against pH. The results displayed were obtained from solutions containing applied concentrations of aluminum nitrate and sodium dihydrogen phosphate of 1.0×10^{-3} *M*.

The precipitates were washed five times with 5 ml of distilled water and dried in a dessicator over calcium sulfate (Drierite) to constant weight. Samples of approximately 50 mg were accurately weighed and dissolved with 2 ml of 1 *M* nitric acid in 50-ml volumetric flasks on a hot plate. The solutions were cooled and diluted to the mark with distilled water, and aliquots were taken for aluminum and phosphate analysis.

Aluminum phosphate precipitate was also prepared by the procedure for taranakite described by Taylor and Gurney.[13] Dried and ground 500-mg samples were equilibrated with 100-ml solutions of varying aluminum(III), phosphate, sodium and hydrogen ion concentrations. These components were analyzed in the supernatants weekly until two successive identical measurements were obtained.

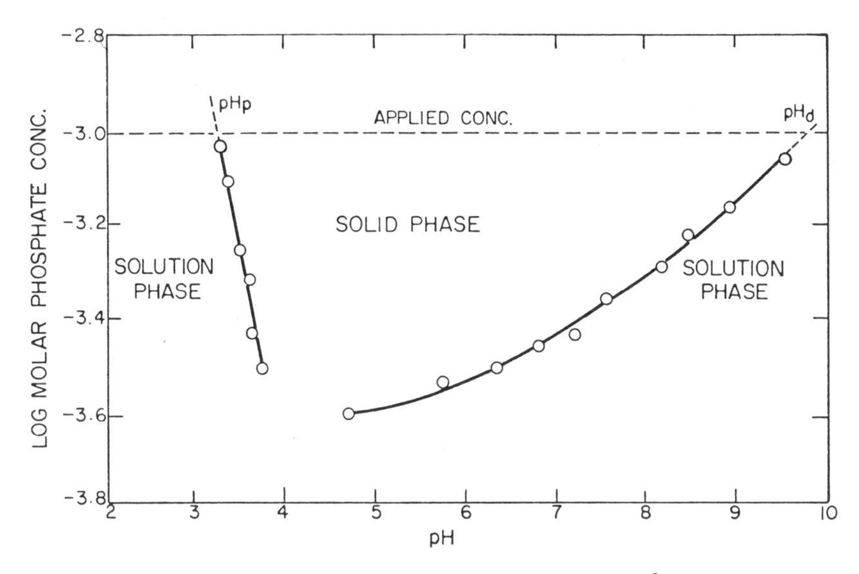

Figure 1. Soluble phosphate concentration–pH curve for 1 x 10^{-3} *M* aluminum nitrate and 1 x 10^{-3} *M* sodium dihydrogen phosphate solution.

Light Scattering Studies

Light scattering was used to detect the formation of solid phase after adding aluminum(III) to phosphate. Scattering intensities were measured with unfiltered light at an angle of 90° to the incident light with a nephelometer. The measurements are expressed as relative scattering units on a 0 to 100 scale. The solutions were placed in round 19x105-nm cuvettes, which were previously matched to within 0.5 relative scattering units with distilled water. The nephelometer was calibrated with turbidity standards of relative scattering values ranging between 0 and 81. A series of samples, each containing the same concentrations of aluminum nitrate, sodium dihydrogen phosphate, and sodium nitrate, was prepared for each experiment. The pH was systematically varied with sodium hydroxide or nitric acid as needed and the solutions were brought to a constant volume of 10 ml with distilled water. The cuvettes were stoppered, vigorously mixed for 30 sec and left undisturbed for 60 min before being measured for pH and scattering intensity. The same measurements were repeated after incubation for 24 hr but without further agitation. A typical relative scattering-pH curve for equal applied concentration of 1 x 10^{-3} *M* aluminum nitrate and sodium dihydrogen phosphate is shown in Figure 2. The results were used to obtain the limiting pH of precipitation, pH_p, and pH of dissolution, pH_d, which define the boundary of precipitation or the conditions under which the applied concentrations of the reactants are in equilibrium with precipitate. This was accomplished by extrapolating the steepest segments of the curve back to the initial scattering value.

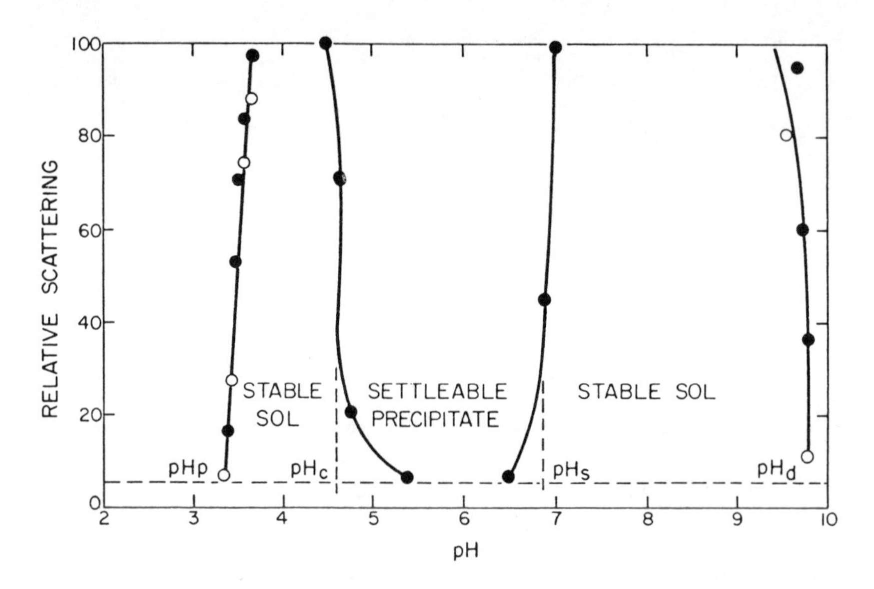

Figure 2. Relative scattering–pH curve for 1 x 10^{-3} *M* aluminum nitrate and 1 x 10^{-3} *M* dihydrogen phosphate solution: 1-hr data (open circles); 24-hr data (blackened circles).

EXPERIMENTAL RESULTS

Excess Aluminum in Acid Solutions

The solubility limits of aluminum phosphate precipitates in acid solutions, pH_p, were determined by light scattering. The results were obtained at several applied phosphate concentrations over the range of 1 x 10^{-5} *M* to 1 x 10^{-2} *M*, each in the presence of three applied metal concentrations of 1 x 10^{-2} *M*, 1 x 10^{-3} *M* and 5 x 10^{-4} *M*. The results are summarized in Table I and plotted in Figure 3 in terms of the logarithm of applied phosphate concentration against pH. Each of the data points marked by a square above pH 3 represents a common point being the average of three critical pH values. As indicated by Table I, these pH values were almost identical, although each was determined in the presence of a different applied aluminum(III) concentration.

Table I. Summary of Critical pH of Precipitation Values as a Function of the Applied Concentrations of Aluminum Nitrate and Sodium Dihydrogen Phosphate

Concentration	pH_p			
	$Log[Al(NO_3)_3]$			
$Log[NaH_2PO_4]$	- 2.00	- 3.00	3.30	Average
- 5.00	4.37	4.39	4.40	4.39
- 4.70	4.23	4.24	4.24	4.24
- 4.52	4.17	4.17	4.15	4.16
4.30	4.06	4.06	4.03	4.06
- 4.00	3.90	3.87	3.85	3.87
- 3.70	3.66	3.71	3.69	3.69
- 3.53	3.53	3.60	3.57	3.57
- 3.30	3.47	3.45	3.45	3.46
- 2.70	3.12	3.16	3.36	3.30
- 2.70	3.12	3.16	3.36	3.14[a]
- 2.52	2.95	3.14	3.37	
- 2.30	2.74	3.14	3.37	
- 2.00	2.52	3.21	3.38	

[a]An average of two critical pH values.

The precipitation boundary so determined consisted of four linear segments. Three segments were found within the lower end of the examined pH range, indicating that the values of the critical pH of precipitation depend on the applied concentrations of both aluminum(III) and phosphate. As the pH increased, the segments converged into a straight line where the pH_p were independent of the applied concentration of the metal. The slopes and intercepts of the various segments were calculated by least squares. The slope and intercept of the common boundary of precipitation were - 1.86 and 3.37, respectively, whereas those calculated for the segment that corresponds to applied aluminum(III) concentration of 1 x 10^{-2} M and within the concentration ranges of phosphate of 1 x 10^{-2} M and 1 x 10^{-3} M were - 1.36 and 0.40, respectively. The remaining pair of segments was nearly vertical to the pH axis, indicating that the critical pH points were independent of the applied phosphate concentration.

The absolute value of the slope of the boundary of precipitation decreased with pH and with increasing applied phosphate concentration. If the slope is indicative of the charge of the ionic species in equilibrium with precipitate, then there must be at least two phosphato-aluminum species in equilibrium with the precipitate because of the two calculated slopes of the precipitation boundaries. The first is dominant in the very low pH region, whereas the second complex is formed within the upper end of the examined pH range of this boundary.

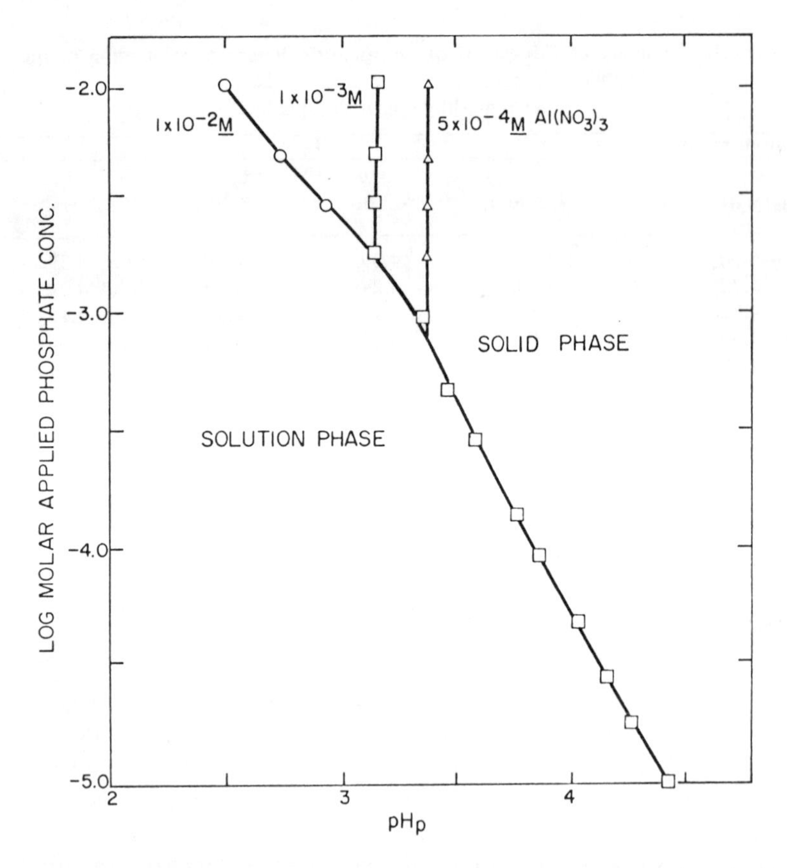

Figure 3. Solubility limits of aluminum phosphate precipitate as a function of pH and applied phosphate concentration at three applied aluminum nitrate concentrations.

The boundary of precipitation of aluminum phosphate solid phases was also studied by measuring the concentration of soluble phosphate remaining after 24 hr of mixing and after separating the precipitate from the solution. The results are shown in Table II. The slope, intercept and standard deviation of the regression line were calculated to be - 2.01, 3.64 and 0.08, respectively. In general, there is a good agreement between the parameters of the regression line and those of the common segment of the boundary of precipitation shown in Figure 3, although the data points were obtained by different analytical approaches. The smaller values calculated for the slope of the common boundary of precipitation are apparently due to the shift of the critical pH points of precipitation, corresponding to the lower end of the phosphate concentration range,

Table II. Solubility of Phosphate and Acid Solutions as a Function of pH and Applied Concentration of Aluminum Nitrate

Applied Aluminum Concentration							
2×10^{-3} *M*		3×10^{-3} *M*		5×10^{-3} *M*		1×10^{-2} *M*	
$-\log[P]_s$	pH	$-\log[P]_s$	pH	$-\log[P]_s$	pH	$-\log[P]_s$	pH
3.02	3.31	3.02	3.34	3.14	3.45	3.80	3.81
3.06	3.35	3.11	3.40	3.38	3.57	4.53	4.20
3.14	3.39	3.23	3.48	3.72	3.74	4.89	4.31
3.23	3.44	3.41	3.56	4.03	3.95	4.85	4.39
3.34	3.49	3.63	3.64	4.31	4.25		
3.49	3.58	3.90	3.80	4.34	4.35		
3.59	3.62	4.16	3.93	4.42	4.38		
3.85	3.74	4.38	4.04				
		4.62	4.14				
		4.95	4.22				
		5.29	4.35				

toward a higher pH range. The amount of precipitate formed at the true boundary of precipitation was apparently insufficient to be detected by the nephelometer. Consequently, the lower end of the precipitation boundary was shifted toward a higher pH range where more precipitate was formed.

Excess Aluminum in Alkaline Range

The concentrations of soluble phosphate in high pH solutions containing excess aluminum(III) remaining after precipitation and subsequent filtration through a membrane were determined colorimetrically. Typical results obtained after 24 hr of mixing are shown in Figure 1. The solubility of phosphate in the alkaline range increased with pH and decreased with increased applied aluminum(III) concentration.

As the soluble concentration of phosphate depended on the amount of precipitate present, its adsorption onto the solid phase surface is suggested. The mathematical treatment developed by Kurbatov *et al.*[17] in studies of the sorption of cobalt on ferric hydroxide was used to analyze the solubility data. The mathematical model assumes that adsorption can be described quantitatively by the mass action law. Thus, the sorption of phosphate on aluminum hydroxide can be represented by:

$$v[Al(OH)_3{:}Al(OH)_2]\text{-}OH_{(surf)} + PO_4^{3-} \rightleftharpoons$$

$$[Al(OH)_3{:}Al(OH)_2]_v\text{-}PO_{4(surf)}^{(3-v)-} + vOH^- \qquad (2)$$

where $[Al(OH)_3:Al(OH)_2]\text{-}OH_{(surf)}$ and $[Al(OH)_3:Al(OH)_2]_V\text{-}PO_{4(surf)}^{(3-v)-}$ are the free and phosphate-sorbed surfaces, respectively. The mass action expression is given by:

$$K_{ad} = \frac{([Al(OH)_3:Al(OH)_2]_V\text{-}PO_{4(surf)}^{(3-v)-})\,[OH^-]^V}{([Al(OH)_3:Al(OH)_2]\text{-}OH_{(surf)})^V\,[PO_4^{3-}]} \qquad (3)$$

According to Kurbatov's approach, Equation 3 can be simplified by assuming that the concentration of the phosphate-sorbed surface is equal to the amount of phosphate sorbed, and that the concentration of the free surface is a linear function of the applied aluminum concentration. Introducing these assumptions into Equation 3, substituting $[P]_S\alpha_3$ for $[PO_4^{3-}]$, and $K_W/[H^+]$ for $[OH^-]$, and taking logarithms and rearranging into a linear form, gives

$$\log \frac{\Delta P}{[P]_S\alpha_3} = -VpH + \log \frac{K_{ad}(Q[Al]_t)^V}{K_W^{\ V}} \qquad (4)$$

where

ΔP = amount of phosphate sorbed given by $[P]_t - [P]_S$, the difference between applied and the soluble phosphate concentrations
α_3 = distribution coefficient of PO_4^{3-}
$[Al]_t$ = applied aluminum concentration
Q = coefficient of linearity.

Plots of Equation 4 are shown in Figure 4. The slopes and intercepts were calculated by the method of least squares and the results are given in Table III. The results indicate that the slope of the lines increased with the applied concentration of the metal. This in turn suggests that the activity of the sorbent depends on the amount of phosphate sorbed. If ΔP is very small compared with $Q[Al]_t$, the magnitude of ΔP will not affect the value of the slope since the activity of the sorbent is apparently unchanged by sorption. If, on the other hand, ΔP is not very small with respect to $Q[Al]_t$, a linear relationship is still obtained, but the slope lacks the theoretical significance of indicating the hydroxide ion displacement per equivalent of phosphate sorbed. This assumption was tested by examining the variation of sorbed phosphate according to Equation 4 for solutions containing applied molar ratios of aluminum(III) to phosphate of 10. The results are shown in Figure 5 and the statistical parameters in Table III.

Effect of Excess Phosphate

The solubility of aluminum in solutions containing excess phosphate was studied by measuring the soluble concentration of the metal as a

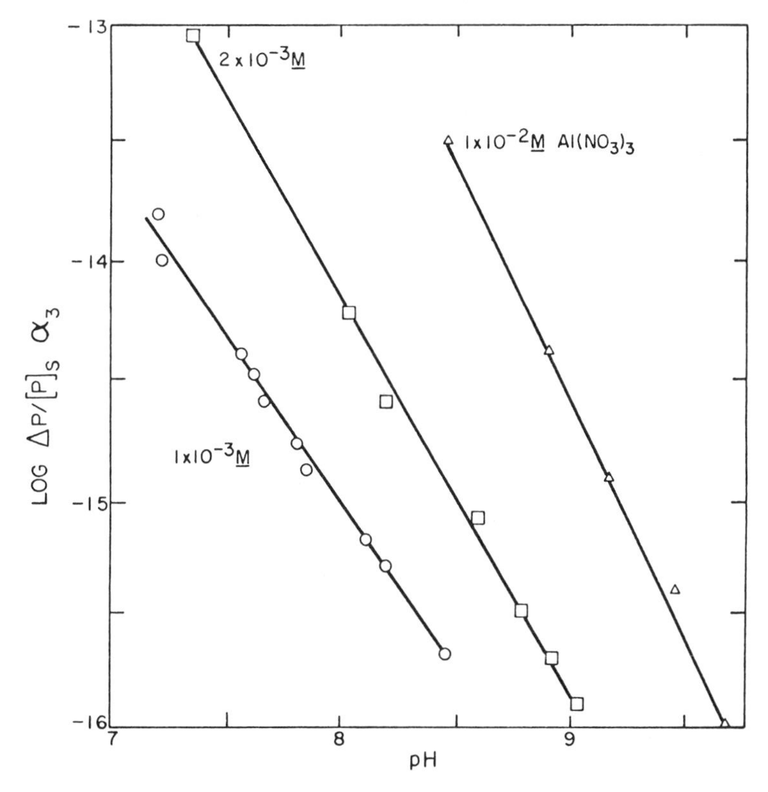

Figure 4. Plots showing the pH-dependent sorption of phosphate on aluminum hydroxide at three applied aluminum nitrate concentrations. Initial phosphate concentration was 0.001 *M*.

Table III. Parameters of Adsorption Isotherms of Aluminum Phosphate in Alkaline Solutions

Initial Concentrations		Parameters	
$[Al]_t$	$[P]_t$	Slope	Intercept
1×10^{-3} *M*	1×10^{-3} *M*	1.41	-3.77
2×10^{-3} *M*	1×10^{-3} *M*	1.69	-0.66
1×10^{-2} *M*	1×10^{-3} *M*	1.85	2.48
3×10^{-3} *M*	3×10^{-4} *M*	1.92	2.20
2×10^{-2} *M*	2×10^{-3} *M*	1.87	2.69

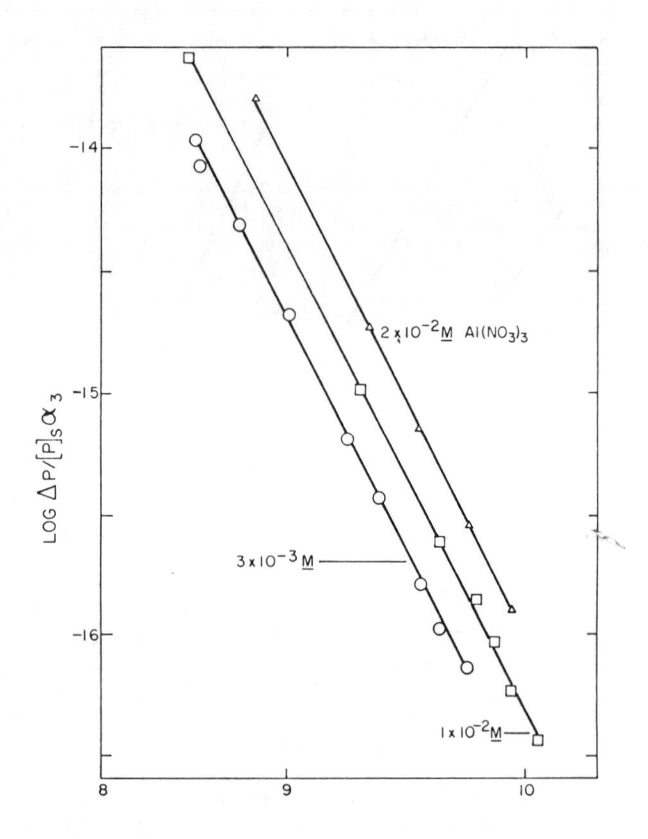

Figure 5. Plots showing the pH-dependent sorption of phosphate on aluminum hydroxide–applied molar ratio of aluminum to phosphate of 10.

function of pH after 24 hr of mixing and subsequent separation of the solid phase through membrane filters. The results are shown in Figure 6. Because of the limited solubility of phosphate in the alkaline range, solutions containing 0.5 *M* and 1 *M* phosphate were oversaturated above pH 9. In general, the solubility of aluminum was affected significantly by both phosphate and hydrogen ion concentrations. Except in the basic end of the pH range of precipitation, the solubility of aluminum increased with the applied phosphate concentration. The solubility curves converged into a single line at the basic end of the pH range of precipitation, suggesting the formation of hydroxy-phosphato-aluminum species.

Each of the solubility curves exhibited three pH points of extreme solubility over the examined range of precipitation. Two pH points of

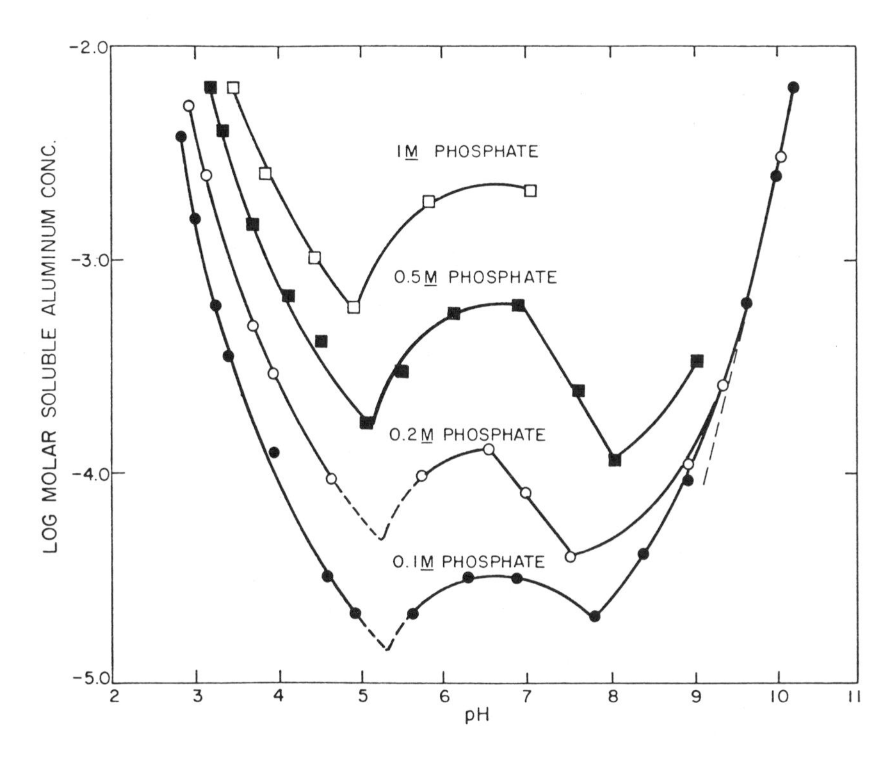

Figure 6. Effect of excess phosphate on the solubility limits of 0.01 *M* aluminum (III).

solubility minimum were found at approximately 5.0-5.5 and 7.5-8.0, whereas a solubility maximum was detected at about pH 6.5. This in turn indicates the formation of two distinct solid phases. The first, being least soluble at around pH 5.3, is dominant in the acid side of the precipitation range. A second solid phase precipitates in the alkaline range and is least soluble at a pH of approximately 7.8. The pH of solubility maximum represents the transition point where a solid phase transformation occurs.

The precipitates were also analyzed chemically for their aluminum and phosphate contents. The results are plotted in Figure 7 in terms of the molar ratio of aluminum to phosphate incorporated in the precipitate as a function of the pH of the solutions measured immediately after the separation of the solid phases. The average molar ratios were approximately 0.62 in acid solutions and 1.0 in the alkaline range. The results tentatively suggest that sodium taranakite, $Na_3Al_5H_6(PO_4)_8(s)$,

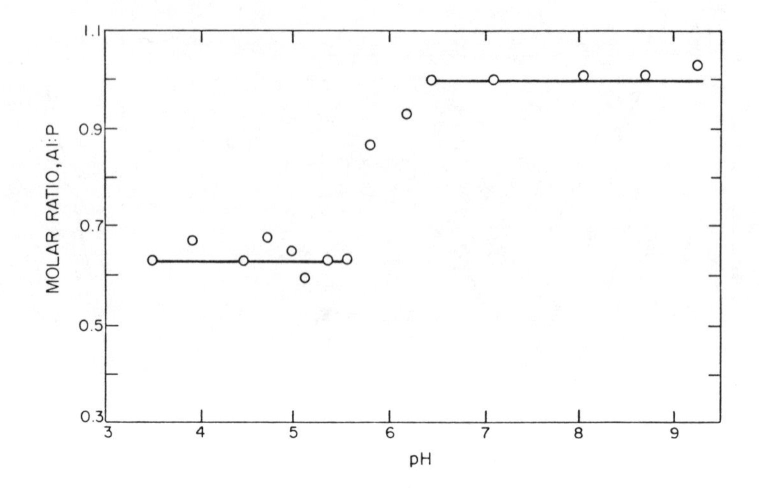

Figure 7. Variation of the molar ratio of aluminum to phosphate incorporated in the precipitate as a function of pH.

precipitates in acid solutions and a tertiary salt of aluminum phosphate, $AlPO_4(s)$, in the alkaline range.

The solubility of the precipitate formed in acid solutions was determined by measuring the soluble concentrations of aluminum(III), phosphate and sodium ions as a function of pH for solutions equilibrated with aluminum phosphate precipitate for approximately 90 days. The precipitate was prepared in accordance with the procedure for the precipitation of taranakites described by Taylor and Gurney.[13] The results are listed in Table IV. Mathematical analysis of the data was based on

$$5AlHPO_4^+ + 3Na^+ + 3H_2PO_4^- \rightleftharpoons Na_3Al_5H_6(PO_4)_8(s) + 5H^+ \quad (5)$$

Assuming that in the presence of excess phosphate the soluble aluminum concentration is equal to $[AlHPO_4^+]$, the logarithmic form of the mass action expression of Equilibrium 5 can be written as

$$\log [Al]_s = -pH - 3/5 [Na^+] + [H_2PO_4^-] + 1/5p^*K_s \quad (6)$$

where

$$[H_2PO_4^-] = \frac{([P]_s - [Al]_s)}{(1 + [H^+]/K_1)} \quad (7)$$

and K_1 is the first dissociation constant of phosphoric acid. A plot of Equation 6 in shown in Figure 8. The slope and intercept were calculated to be -0.58 and 2.33, respectively. The calculated slope is in excellent agreement with the predicted value of -0.60.

Table IV. Equilibrium Concentrations of Soluble Aluminum, Phosphate and Sodium Ion in Acid Solutions Containing Aluminum Phosphate Precipitate

$-\log[Al]_S$	$-\log[Na^+]$	$-\log[P]_S$	$-\log[H_2PO_4^-]$	pH	p^*K_S
2.16	3.23	1.07	1.45	2.69	11.39
2.27	2.90	1.40	1.72	2.71	11.56
2.50	3.04	1.38	1.59	2.82	11.85
2.53	3.01	1.25	1.44	2.87	11.85
2.88	2.65	1.42	1.52	3.04	11.71
2.90	2.72	1.38	1.48	3.05	11.95
3.12	2.58	1.55	1.61	3.29	11.72
3.48	2.52	1.31	1.34	3.45	11.73
3.50	2.41	1.58	1.61	3.56	11.76
4.08	2.24	1.61	1.62	4.03	11.88
				Average	11.74 ± 0.51

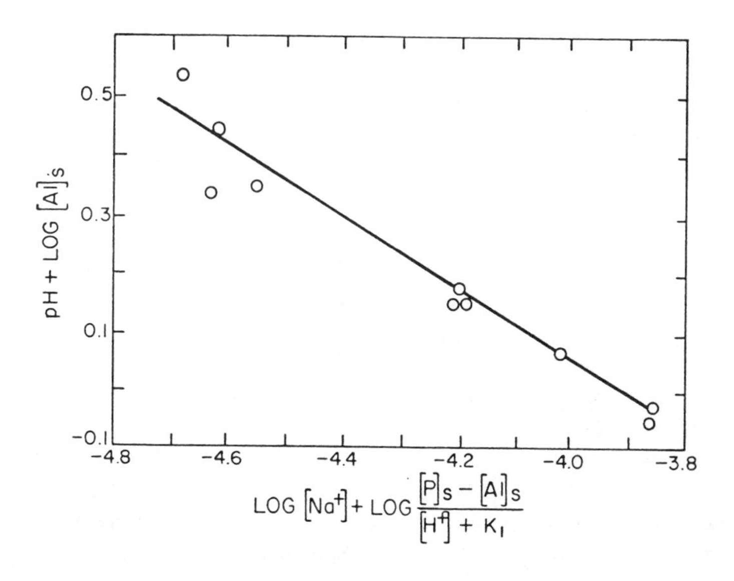

Figure 8. Solubility line for sodium taranakite.

DISCUSSION AND CONCLUSIONS

The solubility of aluminum phosphate precipitated from acid solutions containing excess aluminum(III) concentrations was examined by two analytical approaches. The results of both methods showed that at low phosphate concentrations the solubility of the precipitate is independent of the applied aluminum concentration. This was indicated by the single solubility line obtained with three different applied aluminum concentrations (Figure 3) and by the statistical parameters (slope, intercept and standard deviation) calculated for the regression line of the solubility data shown in Table II. Since the slope of both lines was approximately -2, it can be assumed that a predominant soluble species of aluminum phosphate with a +2 charge exists in equilibrium with the precipitate along these lines. Since the pH limits of precipitation of the common boundary were independent of the applied aluminum concentration, the chemical reaction describing the equilibrium between the soluble species and the precipitate can be viewed as a two-component system in which precipitation occurs when the complex is neutralized by two hydroxide ions. Consequently, the stoichiometry of the precipitate is that of the complex plus two hydroxide ions, or some variable of it.

The relationship between the stoichiometries of the precipitate and the soluble complex can also be interpreted by way of elimination. There are two alternatives other than simple neutralization that must be considered. These include the release of aluminum or phosphate ions from the complex upon precipitation or their incorporation into the complex. However, if that is the case, the mass action expression of the precipitation reaction must include a term for the concentration of the species released or added during the reaction. Since the applied aluminum concentrations were constant, the addition or release of aluminum ions should have affected the intercept of the common boundary of precipitation. This should have been indicated by a series of straight lines, each corresponding to a different applied aluminum concentration. Similarly, since the concentration of the phosphate can be expressed as a function of the aluminum concentration, displacement of the former during the precipitation reaction would have also resulted in a series of boundaries of precipitation, each corresponding to a different aluminum concentration.

The nature of the complex existing in equilibrium with the precipitate was determined from the results of phosphorimetric titrations. It was observed that the pH of aluminum solutions titrated with dihydrogen phosphate decreased with increasing phosphate concentration. This in turn indicates that a phosphato-aluminum complex is formed; otherwise the pH would have increased, since within this pH range the dihydrogen phosphate ion behaves as a base, binding hydrogen ions. Furthermore, this complex must be a basic species because the addition of a base to a chemical system increases the hydroxide ion concentration of the system. Such an increase, when indicated by a decrease in pH as was observed with phosphato-aluminum system, can only be explained by the formation of a basic complex which contains bound hydroxide ions.

The approach used to determine the stoichiometry of the basic aluminum phosphate complex was based on the method of elimination developed by Sillén.[18,19] From the stoichiometries of basic aluminum phosphates documented in the ASTM X-ray powder data file, it was assumed that the stoichiometry of the basic complex should be limited to $Al_2OH(PO_4)^{2+}$, $Al_3OH(PO_4)_2{}^{2+}$, or $Al_4OH(PO_4)_3{}^{2+}$. In general, then, for these complexes

$$(x\text{-}1)\ AlHPO_4^+ + AlOH^{2+} \leftrightharpoons Al_xOH(PO_4)_{x-1}{}^{2+} + (x\text{-}1)H^+ \qquad (8)$$

$$K_{x,1,x-1} = \frac{[H^+]^{x-1}\ [Al_xOH(PO_4)_{x-1}{}^{2+}]}{[AlHPO_4{}^+]^{x-1}\ [AlOH^{2+}]} \qquad (9)$$

Using the phosphorimetric titration data, the concentration of each of these three species was calculated at each point of the titration and a plot of $K_{x,1,x-1}$ against pH was constructed. As shown in Figure 9, the results indicate that the least variation with pH in the plot is for $Al_3OH(PO_4)_2{}^{2+}$. Consequently, the precipitation along the common boundary of precipitation (shown in Figure 3) can be described by

$$Al_3OH(PO_4)_2{}^{2+} + 2H_2O \leftrightharpoons Al_3(OH)_3(PO_4)_2(s) + 2H^+ \qquad (10)$$

the mass action expression being:

$$^*K_{s_{3,1,2}} = \frac{[H^+]^2}{[Al_3OH(PO_4)_2{}^{2+}]} = \frac{2[H^+]^2}{[P]_s} \qquad (11)$$

Taking logarithms and rearranging:

$$\log[P]_s = -2pH + p^*K_{s_{3,1,2}} + \log 2 \qquad (12)$$

Because of the amorphous nature of the precipitate, it was impossible to confirm the exact composition of the basic solid phase by X-ray diffraction. Nevertheless, the results of this study indicate that the molar ratio of aluminum to phosphate incorporated in the precipitate is 3:2.

The formation of a basic trialuminum phosphate precipitate was also suggested by Cole and Jackson.[10] Using electron diffraction, sterretite, $[Al(OH)_2]_3H_2PO_4HPO_4(s)$, was identified as the principal solid phase precipitating at pH 5.5 from solutions containing applied molar ratio of aluminum to phosphate of 1:2. This composition can also be written as $Al_3(OH)_3(PO_4)_2 \cdot 3H_2O(s)$, which is the hydrated form of the basic salt identified in this study.

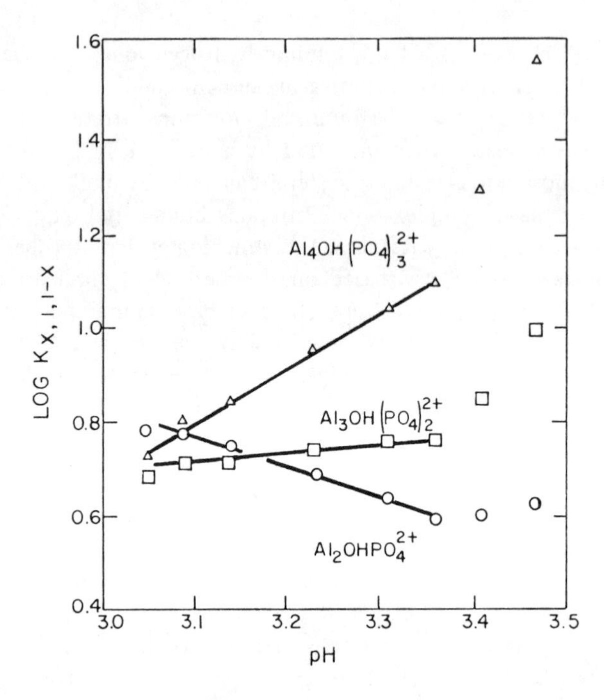

Figure 9. Effect of pH on the formation constants for the various postulated stoichiometries of the basic complex.

The formation of a basic trialuminum phosphate precipitate can also be interpreted from studies on phosphate precipitation in wastewater treatment processes. Such studies have been performed by several investigators.[20,21] They have demonstrated, and it was also found in this study (Figure 1), that only about 70% of the applied phosphate concentration can be precipitated from solutions containing an equimolar concentration of aluminum(III). Since an excess aluminum concentration is required to precipitate the phosphate completely, it is apparent that the number of aluminum ions incorporated into the precipitate is greater than that of phosphate. Furthermore, if it is assumed that a basic trialuminum phosphate precipitate is formed, the percentage removal of phosphate from solutions containing equimolar concentrations of aluminum should be 66.7, a value which is in excellent agreement with the experimentally determined percentage of phosphate removal.

The formation constants of $AlHPO_4^+$ and the basic complex were also calculated from the data using the computer program SCOGS (stability constants of generalized species)[4,22] as follows:

$$\log \beta_{1,1,1} = 19.93 \pm 0.70$$

$$\log \beta_{3,1,2} = 34.94 \pm 0.44$$

where $\beta_{1,1,1}$ and $\beta_{3,1,2}$ are defined by the mass action expressions:

$$\beta_{1,1,1} = \frac{[AlHPO_4^{+}]}{[Al^{3+}][H^+][PO_4^{3-}]} \quad (13)$$

$$\beta_{3,1,2} = \frac{[Al_3OH(PO_4)_2^{2+}][H^+]}{[Al^{3+}]^3[PO_4^{3-}]^2} \quad (14)$$

The solubility product of the basic aluminum phosphate salts can be calculated:

$$\begin{aligned} pK_{so} &= pK_{3,1,2} + 3pK_2 - p{*}K_{s_{3,1,2}} \quad (15) \\ &= 34.9 + 42 - 3.6 \\ &= 73.3 \end{aligned}$$

where $*K_{s_{3,1,2}}$ is the equilibrium constant defined by Equation 11. Using this value the pH of solubility minimum of phosphate can be determined. The experimental results indicated that the pH point of solubility minimum should be within the range of 5-7 (Figure 1). Therefore, the concentrations of phosphoric acid and orthophosphate ion can be ignored. The mass balance equation for phosphate is given by

$$[P]_s = [HPO_4^{2-}] + [H_2PO_4^-] + 2[Al_3OH(PO_4)_2^{2+}] + [AlHPO_4^{+}] \quad (16)$$

Since at the pH of phosphate solubility minimum both the basic aluminum phosphate and the metal hydroxide are equally stable, both solubility product expressions are valid. Expressing each species in terms of the various equilibrium constants and hydrogen ion concentration, differentiating the concentration of phosphate with respect to that of hydrogen ion, equating to zero for the solubility minimum and rearranging gives:

$$10^{23.5}[H^+]^6 + 10^{17.7}[H^+]^3 - 10^{-6.9}[H^+] = 1 \quad (17)$$

This equation was solved by trial and error to give a minimum pH of phosphate solubility of approximately 6. Below this pH the basic phosphato-aluminum salt is the dominant solid phase; at high pH aluminum hydroxide is more stable. The concentration of the various phosphato species and aluminum ion at the pH point of solubility minimum are listed in Table V. Stumm[23] also calculated the pH of the solubility minimum of phosphate to be approximately 6. His calculation, however, was based on a chemical model which involved the tertiary salt.

One of the reasons that has led many investigators to propose the tertiary salt as the dominant solid phase was that its solubility product was found to be constant over a broad pH range.[23-25] However, most of

Table V. Equilibrium Concentrations of Aluminum Ion and the Various Phosphate Species at the pH of Solubility Minimum of Phosphate

Species	Expression	pConc
$[Al^{3+}]$	$[H^+]^3/{}^*K_{so}$	7.6
$[PO_4^{3-}]$	$(K_{so}[H^+]^3/K_w^3[Al^{3+}]_3)^{1/2}$	13.3
$[HPO_4^{2-}]$	$[PO_4^{3-}]\ [H^+]/K_3$	7.0
$[H_2PO_4^-]$	$[HPO_4^{2-}][H^+]/K_2$	5.8
$[Al\ HPO_4^+]$	$[Al^{3+}]\ [H^+]\ [PO_4^{3-}]\ \beta_{1,1,1}$	7.0
$[Al_3OH(PO_4)_3^{2+}]$	$[Al^{3+}]^3[PO_4^{3-}]^2\beta_{3,1,2}/[H^+]$	8.5
$[P]_t$		5.8

their studies were carried out in the presence of aluminum hydroxide precipitate. If it is assumed that the two solid phases are in equilibrium, then the solubility product expression for the basic salt can be written as

$$K_{so[Al_3(OH)_3(PO_4)_2]} = [Al^{3+}]^2[PO_4{}^{3-}]^2[Al^{3+}][OH^-]^3 = [Al^{3+}]^2[PO_4{}^{3-}]^2K_{so[Al(OH)_3]} \quad (18)$$

Equation 18 is valid since the solubility product of the metal hydroxide is constant within the pH range of precipitation of aluminum hydroxide. Consequently, the solubility product of the tertiary salt must also be constant:

$$pK_{so(AlPO_4)} = \frac{pK_{so[Al_3(OH)_3(PO_4)_2]} - pK_{so[Al(OH)_3]}}{2} = \frac{73.3 - 31.6}{2} = 20.9 \quad (19)$$

The calculated value is well within the range reported in the literature.[24-30] Thus, it is evident that the nature and stoichiometry of the precipitate cannot be determined from solubility data alone. Moreover, in several studies in which the solubility data were collected outside the pH range of precipitation of the metal hydroxide, the calculated solubility products of the tertiary salt were considerably higher. For example, Makitie[25] calculated a pK_{so} for the tertiary salt of 18.6 within pH range 4-5 and 23.7 at pH 2.9. In this study, however, the formation of a basic salt was interpreted from solubility data collected outside the pH range of precipitation of aluminum hydroxide and from pH-dependent titration data. Both data sets were obtained within a pH range in which aluminum hydroxide precipitate is unstable and therefore are not subject to the limitations mentioned previously.

SUMMARY

The interaction between aluminum and phosphate in aqueous solutions can lead to the precipitation of several solid phases, depending on the applied concentrations of aluminum and phosphate and the pH. Figure 10 summarizes the distribution of aluminum phosphate solid phases as a function of pH and the ratio of the applied concentrations of aluminum and phosphate. Four different solid phases precipitate within the range of concentration and pH examined in this study.

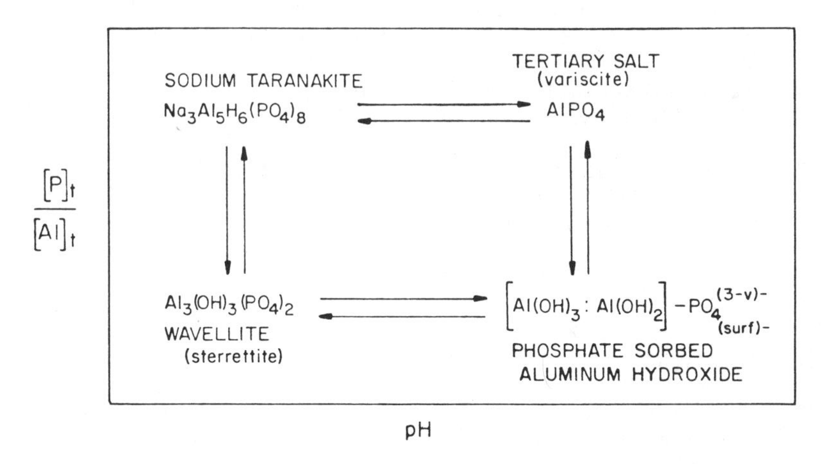

Figure 10. Distribution diagram of solid phases in the phosphato-aluminum system.

In acid solutions containing an excess concentration of aluminum, the most stable solid phase is the basic phosphato-aluminum salt. If the pH is raised beyond the point of solubility minimum of phosphate (pH 6), the basic salt is hydrolyzed to aluminum hydroxide and phosphate is adsorbed on its surface. If, on the other hand, the concentration of phosphate is raised, a taranakite precipitate is formed. The taranakite is transformed into the tertiary salt when the pH of the system is increased in the presence of a high phosphate concentration.

REFERENCES

1. Fiat, S., and R. E. Connick. *J. Am. Chem. Soc.* 90:608 (1968).
2. Biedermann, G. *Svensk Kem. Tidskr.* 76:362 (1964).

3. Brosset, C., G. Biedermann and L. G. Sillen. *Acta Chem. Scand.* 8:1917 (1954).
4. Hayden, P. L., and A. J. Rubin. In *Aqueous-Environmental Chemistry of Metals,* A. J. Rubin, Ed. (Ann Arbor, Michigan: Ann Arbor Science Publishers, Inc., 1954).
5. Matijević, E., G. E. Janauer and M. Kerker. *J. Colloid Sci.* 19:333 (1964).
6. Bjerrum, N., and C. R. Dahm. *Z. Phys. Chem. Bodenstein Festband* 627 (1931).
7. Holroyd, A., and J. E. Salmon. *J. Chem. Soc.* 1956:269 (1956).
8. Jameson, R. F., and J. E. Salmon. *J. Chem. Soc.* 1954:4013 (1954).
9. Salmon, J. E., and J. G. L. Hall. *J. Chem. Soc.* 1958:1128 (1958).
10. Cole, C. V., and M. L. Jackson. *J. Phys. Chem.* 54:128 (1950).
11. Kittrick, J. A., and M. L. Jackson. *J. Soil Sci.* 7:81 (1956).
12. Smith, J. P., and W. E. Brown. *Am. Minerologist* 44:138 (1959).
13. Taylor, A. W., and E. L. Gurney. *J. Phys. Chem.* 66:1613 (1961).
14. Woodward, C., and H. Freiser. *Talanta* 15:321 (1968).
15. Harwood, J. E., R. A. Van Steenderen and A. L. Kuhn. *Water Res.* 3:417 (1969).
16. *Standard Methods for the Examination of Water and Wastewater,* 13th edition, (New York: American Public Health Association, 1971).
17. Kurbatov, M. H., W. B. Gwendolyn and J. D. Kurbatov. *J. Phys. Chem.* 55:258 (1951).
18. Sillén, L. G. *Acta Chem. Scand.* 10:803 (1956).
19. Goldshmid, T. Unpublished Ph.D. Thesis, The Ohio State University, Columbus, 1975.
20. Eberhardt, W. A., and J. B. Nesbitt. *J. Water Poll. Control Fed.* 10:1239 (1968).
21. Tenney, M. W., and W. Stumm. *J. Water Poll. Control Fed.* 35: 1370 (1965).
22. Sayce, I. G. "Computer Calculations of Equilibrium Constants of Species Present in Mixtures of Metal Ions and Complexing Agent," Australian National University, Canberra, Australia (unpublished).
23. Stumm, W. In *Advances in Water Pollution Research,* (London: Pergamon Press, 1969).
24. Chen, Y. S. R., J. N. Butler and W. Stumm. *Environ. Sci. Technol.* 7:4 (1973).
25. Makitie, O. *Ann. Agric. Fenn.* 5:6 (1966).
26. Cole, C. V., and M. L. Jackson. *Soil Sci. Soc. Am. Proc.* 15:84 (1950).
27. Lindsay, W. L., M. Peach and J. S. Clark. *Soil Sci. Soc. Am. Proc.* 23:266 (1959).
28. Bjerrum, J., G. Schwartzenbach and L. G. Sillen. *Stability Constants of Metal Ion Complexes. Part II: Inorganic Ligands,* Special publication. No. 7, The Chemical Society, London (1958).
29. Kittrick, J. A., and M. L. Jackson. *Soil Sci. Soc. Am. Proc.* 19:292 (1955).
30. Backe, B. W. *J. Soil Sci.* 14:113 (1963).

CHAPTER 5

STABILITY OF MIXED COLLOIDAL DISPERSIONS

Alan Bleier* and Egon Matijević
Institute of Colloid and Surface Science
and Department of Chemistry
Clarkson College of Technology
Potsdam, New York

INTRODUCTION

Although many industrial and natural water systems for which methods of treatment are required contain suspended solids differing in size, shape or chemical composition, the stability of mixed dispersions has received little attention as compared to the vast technical literature available on the stability of single colloidal suspensions. However, a number of studies have been conducted on other applied processes involving the interaction of dissimilar interfaces. Some of these have dealt with water treatment in a filter bed,[1] selective flocculation of mineral ores,[2] the destabilization of clay suspensions in which the individual particles exhibit charge heterogeneity,[3,4] flotation processes involving attachment of air bubbles or oil droplets to suspended matter,[5,6] and the dewatering of phosphate slimes by the addition of granular solids to the thickening tanks,[7] to name a few.

Since Derjaguin's[8] papers on heterocoagulation in 1954, a number of theoretical papers[9-12] have examined the problem of overlapping dissimilar electrical double layers, with special emphasis on developing quantitative models to describe the stability of mixtures of unlike particles.[13-16]

*Present address: Union Carbide Corporation, Tarrytown, New York.

However, the application of these theoretical principles to real systems of practical importance, as in wastewater treatment operations, has proved at best to be extremely difficult and, in some instances, futile.

Recently, we undertook a systematic study of the stability of mixed colloidal dispersions in an attempt to define more clearly the roles of various parameters such as the relative concentrations of sol particles, differences in surface potentials and particle sizes, as well as the nature and the composition of chemical additives in determining the overall stability of multicomponent colloidal suspensions. Binary and ternary mixtures containing particles of PVC latex, Ludox HS silica, chromium hydroxide or alumina, respectively, have been investigated.[17-19] These sols were chosen because they either are technologically important or represent model suspensions which have been studied extensively as single-component systems.

This chapter summarizes some of our work on mixed dispersions containing preformed particles, emphasizing the importance of systems differing in chemical composition or surface characteristics. In view of the progress made by colloid scientists in understanding single-component colloidal suspensions, the work presented here demonstrates the feasibility of exploring significantly more complicated mixtures, akin to those which may be encountered in water treatment.

GENERAL CONSIDERATIONS

The development of a quantitative understanding of colloid stability, culminating in what is referred to as the DLVO (Derjaguin-Landau-Verwey-Overbeek) theory of hydrophobic colloids[20,21] has led, since World War II, to a marked increase in the number of studies of well-defined colloidal systems.

The experimental investigations can be put in two categories. The first consists of those works dealing directly with the stability of colloidal suspensions. These studies include measurements of the changes in the physical properties of dispersions, such as optical properties,[22-24] sedimentation rates or volumes,[25] filterability[26] and Coulter counter signals.[27] These measurements allow the establishment of stability domains as well as, under certain conditions, the quantitative evaluation of the coagulation process.[28] The stability domains are particularly useful, as they describe the properties of a colloidal dispersion in terms of the concentrations of coagulating species, pH, or other essential parameters in a manner which is easily adapted to practical uses. The second category consists of works in which either the electrical double layer properties or the dispersion forces have been investigated in detail. For example, the adsorption or desorption of surfactants, counterions or potential determining ions, with special references to surface or Stern-plane potentials, has been the subject of many studies.[29-31] The information so obtained provides a valuable contribution to the understanding of the complex nature of forces relevant to the properties of colloidal dispersions.

In reviewing the problem of colloid stability, one must keep in mind that the electrolyte media in common industrial and natural systems may contain a myriad of solute species capable of substantially affecting the characteristics of suspended matter.[28] Aqueous systems containing hydrolyzable polyvalent electrolytes are of particular interest as these tend to interact strongly with particles of different surface characteristics; also, on changing pH, precipitation of metal hydroxide may take place. These phenomena are utilized, for example, in the purification of water by the addition of aluminum or iron salts.[32,33] While the above-mentioned concepts have been applied mainly to single colloidal systems, they must be considered if one is to interpret properly the stability phenomena exhibited by mixed dispersions.

Hogg *et al.*[14] developed a simple mathematical model describing the interactions of dissimilar electrical double layers. Although their approach does have limitations,[17] it can be used to predict the behavior of binary colloidal mixtures, particularly when dealing with oppositely charged particles. Healy and coworkers[34] also found, when working with binary mixtures of oxides, that to describe their systems completely, it was necessary to take into account the partial dissolution of one of the solids followed by the readsorption of the soluble species onto the surface of the second colloid. Thus, it is evident that one must ascertain the interactions of all species in the solution with each type of interface present when treating the more complicated, multicomponent colloidal mixtures containing two or more types of dissimilar particles.

MATERIALS AND METHODS

PVC Latex

Polyvinyl chloride (PVC) latex, stabilized by sodium dodecyl sulfate (NaDS), was obtained from Monsanto Chemical Company. The total concentration of NaDS was determined to be $\sim 0.5\ \mu mol/g$ of PVC latex.[19] The particle size distribution of this latex sol was established by light scattering techniques described in detail elsewhere[17,19]; these yielded an average modal radius of 255 nm. The particle size distribution of this sample was sufficiently narrow so that the sol particles exhibited brilliant higher order Tyndall spectra (HOTS) when illuminated with a parallel beam of high-intensity white light. This property was used in distinguishing qualitatively between stable and unstable PVC latex suspensions. The presence of DS^- ions at the particle surface renders these latex particles negatively charged for all pH values above ~ 2.

A PVC latex sol having a modal radius of 169 nm was obtained from Uniroyal Chemical Company. The details of the preparation and purification of this latex are given elsewhere.[17] It is similar to the Monsanto latex in that this material is also negatively charged above pH ~ 2.

PVC latex may be considered a classic example of a hydrophobic sol. From measurements of the rates of coagulation in the presence of simple counterions of 1+, 2+ and 3+ charge, it was found that this sol obeys the Schulze-Hardy rule. In contrast, hydrolyzed metal ions, *e.g.*, aluminum complexes,[35] and some metal chelates, such as the tris 2,2′-dipyridyl complex with Co^{3+},[36] have been found to adsorb readily onto similar PVC latex particles. Being cationic species, these adsorbates can neutralize the negative charge of the latex and consequently destabilize the sol. In sufficiently high concentrations the adsorbed counterions may reverse the charge to positive; if this charge is high enough, stable (positive) sols may again result.

Ludox HS Silica

Ludox HS silica, consisting of amorphous, spherical particles, was supplied by E. I. du Pont de Nemours and Company; its surface properties have been described in detail by Bolt.[37] The particle size distribution of this sol is narrow and has an average radius of 6.75 nm[38]; these particles, being small, do not exhibit HOTS. The isoelectric point (i.e.p.) of this silica was found by moving boundary techniques to between pH 1 and 2.[38]

An extensive study of the stability of Ludox HS silica showed that the destabilization by simple salts did not follow the Schulze-Hardy rule but that it was accompanied by counterionic exchange for surface silanolic protons (≡ SiOH).[38-40] It was determined that a minimum, critical equivalent counterionic exchange, which was pH dependent, was required to produce rapid coagulation of these particles. The exchange capacity of Ludox HS for Na^+ was reached at about pH 10.[40] At sufficiently high pH the critical coagulation concentration (c.c.c.) of 1+, 2+ and 3+ counterions became independent of pH, and these limiting values were in reasonable accord with the c.c.c. of the same electrolytes for hydrophobic sols, such as PVC latex.

Chromium Hydroxide

Hydrous chromium(III) oxide sols, consisting of uniform spherical particles, were prepared essentially according to the procedure described earlier[41,42] and were purified by a filtration technique.[17] The particle size distribution of the chromium hydroxide stock suspension was characterized by light scattering, and the modal radius was found to be 186 nm. Owing to the narrow size distribution and to the relatively large particle size as compared to the wavelength of visible light, these sols exhibit strong HOTS. The stability characteristics of such chromium hydroxide sols have been studied extensively.[17,42] This sol has an isoelectric point at approximately pH 7.4 (± 0.2); obviously, H^+ and OH^- must act as potential determining ions. The ratio of the c.c.c. s of NO_3^- and SO_4^{2-} counterions at pH values below the i.e.p. was found to be four times larger than the value predicted by the DLVO theory.[17]

Figure 1 gives the particle size parameter, α, which is proportional to the particle radius, as a function of the time of irradiation of the sol by a laser beam of wavelength, λ_o, of 514 nm. α is defined as $2\pi\ mr/\lambda_o$, where r is the particle radius and m represents the refractive index of the

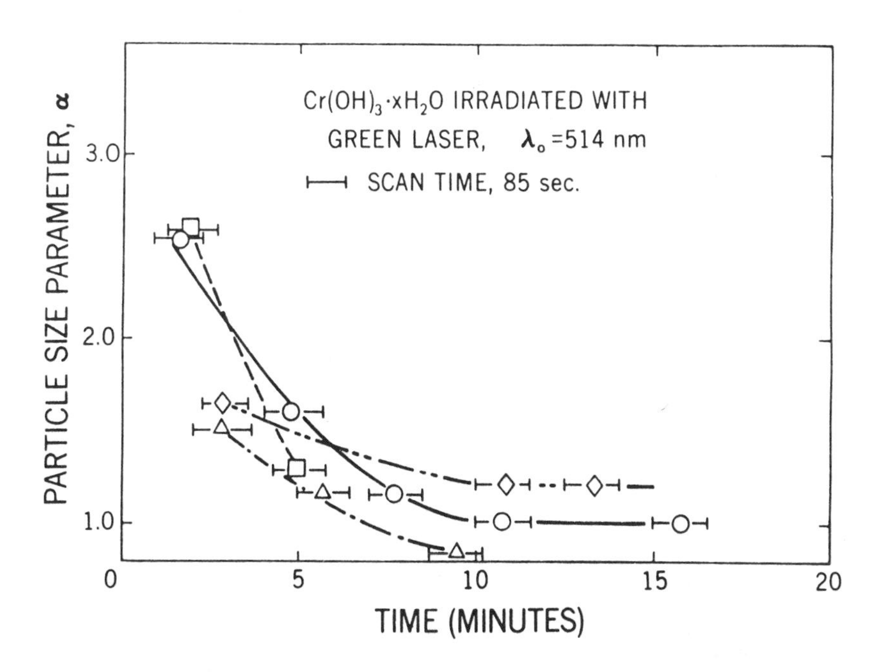

Figure 1. Size parameter, α, for hydrous chromium(III) oxide particles as a function of the time of irradiation using an argon laser.

particles in the suspending medium relative to that in vacuum. Since the particles remain spherical, as inspected in an electron microscope, the data in Figure 1 may be attributed to dehydration of the hydrous oxide.[19]

Alumina

Pseudoboehmite was obtained from Kaiser Chemicals. This material was ground using a mortar and pestle to reduce the particle size so that stable alumina sols could be prepared. The particle size distribution was too broad to give a reliable value for the average particle radius. This pseudoboehmite had an i.e.p. at about pH 9, as determined by electrophoresis.

Experimental Techniques

Light scattering techniques and microelectrophoresis were the principal methods employed to follow the interactions between primary particles in the mixed dispersions. These procedures have been described elsewhere.[19] Electron microscopy and atomic absorption spectroscopy were also used to further characterize the colloidal mixtures discussed here. For PVC latex and hydrous chromium(III) oxide particles the presence or absence of HOTS was used to ascertain whether these sols remained stable and dispersed.

All salts were of the highest purity grade available commercially. Solutions and suspensions were prepared with doubly distilled water. The second distillation was carried out in an all-Pyrex® glass still. All of the experiments described in this chapter were conducted at ambient room temperature (25°C).

RESULTS

PVC Latex–Ludox HS Silica

Figure 2 gives the critical coagulation concentration of two monovalent counterions, Na^+ and K^+, as a function of pH for PVC latex (dash-dot line), Ludox HS silica (dashed line), and a binary mixture of these two dispersions (solid line) obtained 24 hr after mixing the reacting components. Systems are coagulated under conditions on the hatched side of each respective boundary. The error bars for the mixed systems represent the reproducibility associated with the determination of the c.c.c. (vertical bars) or of the corresponding critical pH values (horizontal bars). The sol concentrations were chosen to yield convenient light scattering signals. Similar concentrations have also been employed in

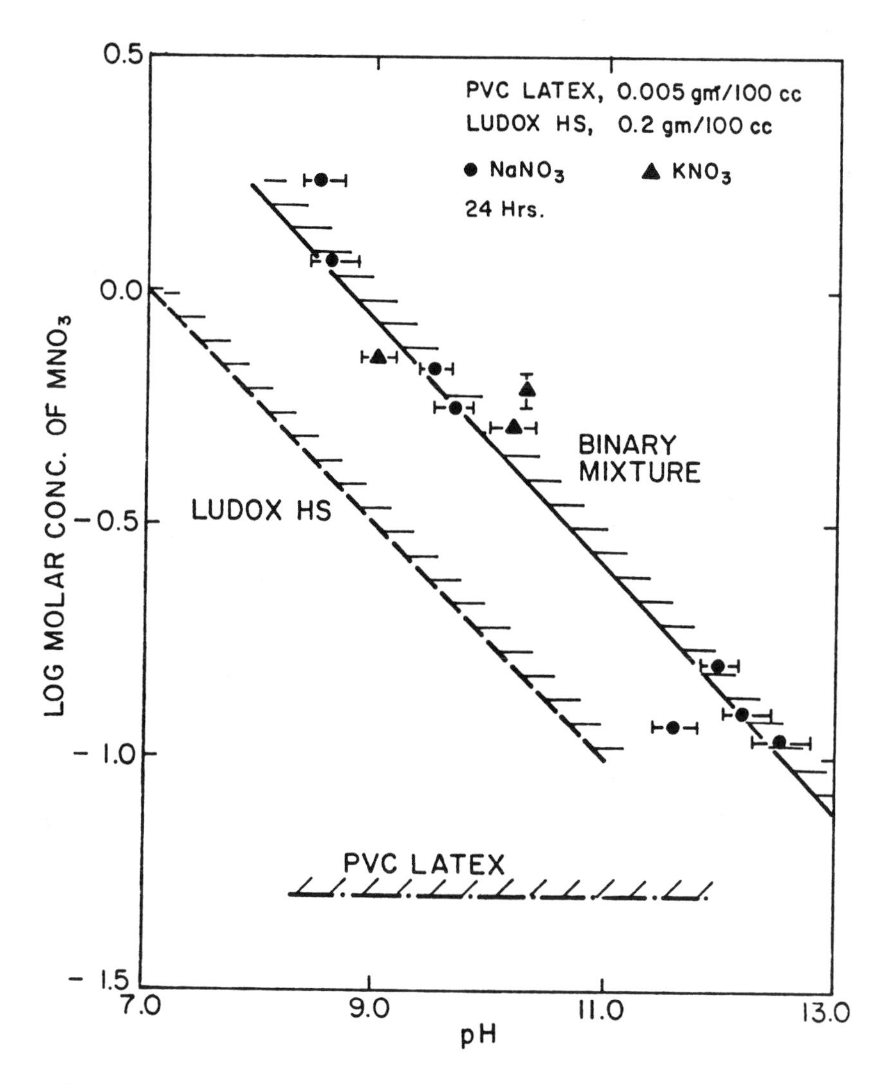

Figure 2. Critical coagulation concentration of $NaNO_3$ and KNO_3 as a function of pH for a mixture of PVC latex and Ludox HS sols 24 hr after sample preparation.

previous studies using the same or similar sols, thus facilitating the interpretation of the complex binary mixture.

The coagulation values for the PVC latex are essentially independent of pH and correspond to a c.c.c. of ~ 0.05 *M* MNO_3, where M is either Na^+ or K^+. On the other hand, the stability boundary for Ludox HS silica is found to have a strong pH dependence. The critical concentrations of Na^+ and K^+ given in Figure 2 for SiO_2 are greater than the corresponding values for the PVC latex, the former approaching 0.05 *M* only under conditions of extremely high pH. It is evident that the transition between stable and unstable mixed sols is also pH-dependent, thus suggesting that the particles of Ludox HS silica have a dominating influence over PVC latex in this mixture.

Moreover, in the presence of either $NaNO_3$ or KNO_3, brilliant, higher ordered Tynall spectra were observed throughout the entire stability region of the mixed dispersions (below the solid line). Since this light scattering effect is, in this mixture, characteristic only of the stable PVC latex particles, the latter remains uncoagulated in the presence of Ludox HS for all combinations of salt concentration and pH below the solid line. This is true even for certain concentrations of monovalent counterions at which rapid destabilization of the PVC latex takes place in the absence of silica, *i.e.,* above the dash-dot line in the figure. Within the coagulation region of the mixture, however, a very low turbidity and the absence of intense HOTS indicated that both sols, the latex and silica, are destabilized. These results suggest that Ludox HS silica can stabilize PVC latex against electrolytic coagulation by $NaNO_3$ and KNO_3. Furthermore, it appears from Figure 2 that in the presence of high concentrations of these counterions the PVC latex is destabilized in the binary mixture under conditions similar to those leading to the rapid destabilization of the Ludox HS silica.

Figure 3 contains analogous coagulation data for PVC latex (dash-dot line), Ludox HS silica (dashed line), and the binary mixture of these sols (solid line) in the presence of $Ca(NO_3)_2$. For conditions above each respective boundary, the corresponding colloidal systems undergo rapid coagulation. In mildly basic media ($pH < 10$), the c.c.c. of Ca^{2+} for the silica sol depends strongly on pH, whereas the coagulation value for latex is again independent of pH. The c.c.c.-pH curve for the binary dispersions, represented by the open circles and the solid line, is similar in both shape and position to that for the pure Ludox silica sol.

Within the coagulation region for the binary colloidal dispersion, two subregions can be distinguished experimentally. The first is one in which both sols are destabilized and is denoted in Figure 3 as showing weak HOTS. The second subarea corresponding to the conditions between the coagulation boundaries for the two pure, single component sols (*i.e.,* between the dash-dot and dashed curves) represents systems which contain destabilized particles, yet exhibit strong HOTS. The latter indicate the presence of stable PVC latex; consequently, within this region selective destabilization of the Ludox HS silica takes place.

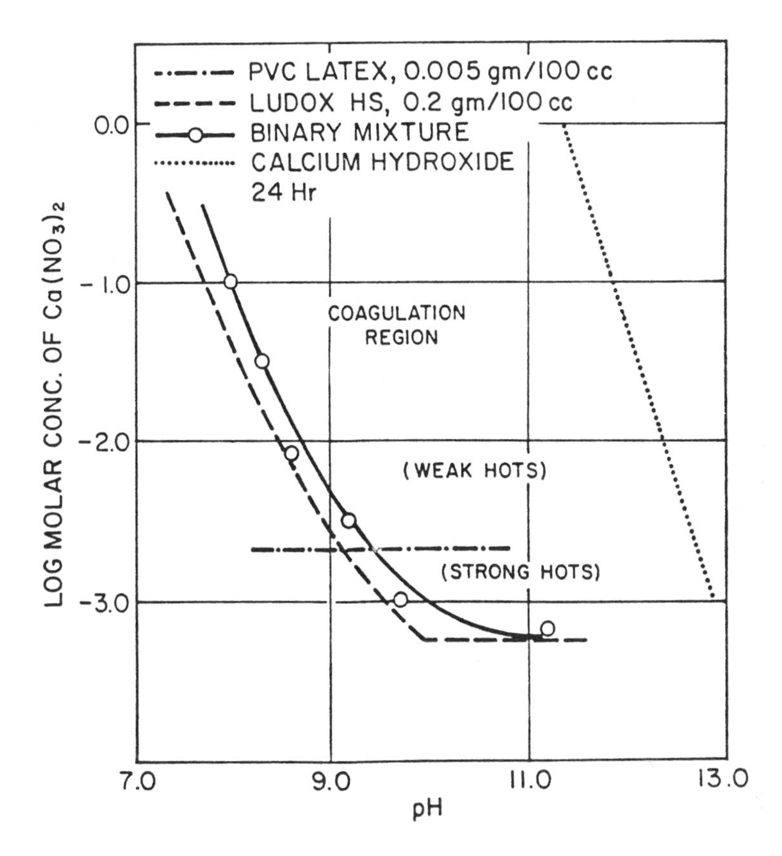

Figure 3. 24-hr stability boundaries for PVC latex, Ludox HS silica and the mixture of these two sols in the presence of $Ca(NO_3)_2$ as a function of pH. Dotted line gives the precipitation boundary for $Ca(OH)_2$.

For the experimental conditions above the c.c.c. of $Ca(NO_3)_2$ for PVC latex but below the stability boundary corresponding to SiO_2, both types of particles remain stable. Thus, it is evident that Ludox HS silica is capable of stabilizing the PVC latex particles under the conditions of pH and concentration of $Ca(NO_3)_2$ for which the latter sol is normally coagulated, which is similar to the observation made in the presence of monovalent counterions (Figure 2).

The dotted line in Figure 3 represents the precipitation boundary of $Ca(OH)_2$.[43] Since the hydrolysis of Ca^{2+} ions is extensive only in the vicinity of this line, *i.e.,* above pH 11, the hydroxylation of Ca^{2+} does not affect the coagulation phenomena described here.

Figures 2 and 3 presented the stability data obtained with simple unhydrolyzed monovalent and divalent counterions, namely, Na^+, K^+ and Ca^{2+}. Similar coagulation experiments were also carried out in the presence of two trivalent cations, Al^{3+} and La^{3+}. The resulting stability curves obtained 24 hr after sample preparation for $Al(NO_3)_3$ and $La(NO_3)_3$ demonstrated[19] that the stability phenomena of these binary component systems are also dominated by the silica particles. That is, for each salt, PVC latex was destabilized only under those conditions for which the Ludox was coagulated. Moreover, in the case of $Al(NO_3)_3$, selective destabilization of silica was observed under the conditions corresponding to the region of overlap of the curves for the two single sols in which the c.c.c. for PVC latex exceeded that for Ludox HS. In the presence of $La(NO_3)_3$, on the other hand, selective coagulation was not observed and the latex was destabilized only under the experimental conditions which lead to the rapid coagulation of Ludox HS silica.

In summary, the binary systems discussed above can be divided into two groups. The first is one in which both sols remain stable until the silica is coagulated. When this occurs, the latex is also destabilized. In addition, the dependence of the stability boundary on pH for the binary mixtures resembles that for the Ludox HS silica alone. This class consists of the systems prepared in the presence of $NaNO_3$, KNO_3 and $La(NO_3)_3$.

The second group contains those colloidal mixtures with added $Ca(NO_3)_2$ and $Al(NO_3)_3$. The stability domains for these systems are more complex than those of the first classification, because Ca^{2+} and Al^{3+} can selectively coagulate Ludox HS silica in the presence of PVC latex. This phenomenon occurred in each case under the conditions corresponding to the region of overlap between the stability curves for the individual sols, such that the instability of the single colloidal Ludox suspension coincides with a region of stability for PVC latex.

Each binary dispersion in the presence of the counterions mentioned above shows an electrolyte concentration-pH region in which both sols remain stable but for which one would anticipate, from the behavior of the single systems, the pure Ludox HS silica to remain stable while the PVC latex component to be coagulated. This observation suggests strongly that PVC latex is stabilized by Ludox HS silica. The details of the protection mechanism of the latex particles by silica in the presence of high concentrations of supporting electrolytes has been treated elsewhere.[19] It was found that this phenomenon could be explained on theoretical grounds by taking into account the current theories of heterocoagulation and by considering the concepts developed earlier[38,39] for the stabilization-destabilization mechanism of small Ludox HS silica particles. In addition, it was established experimentally that this silica indeed adsorbs onto the latex surface yielding a Freundlich-type adsorption isotherm. Apparently, the hydration layer surrounding Ludox HS particles also protects, indirectly, the latex particles from electrolytic coagulation.[19] The electron microscopic picture of PVC latex coated with Ludox HS silica (Figure 4) shows that the surface is rather roughened. In contrast, pure PVC latex particles are spherical and have quite smooth surfaces.

Although the heterocoagulation of negatively charged silica particles with a negative PVC latex could be, in principle, the result of a difference in their surface potentials, Table I shows for the system investigated that the ratio of the zeta potentials is ~ 1 over a wide pH range. Another approach is, therefore, needed to account for the adsorption phenomenon. A kinetic model describing the collisions that can occur in the mixture, consisting of silica and PVC latex particles, can explain the stability behavior exhibited by these mixed dispersions in the presence of mono-, di- and trivalent counterions. The calculations of Pugh and Kitchener[44] illustrate that the large disparity in particle size existing in our binary dispersions, which yields a ratio of particle radii (latex:silica) of ~ 38, is sufficient to lead to heterocoagulation between these two negatively, and approximately equally, charged particles.

Chromium Hydroxide–Ludox HS Silica

A preliminary study was conducted which showed that monodispersed chromium(III) oxide, consisting of spherical particles described earlier, could also be stabilized by Ludox HS silica under conditions which normally lead to the coagulation of the former. The concentrations used in this work were 0.003 g/100 cm^3 chromium hydroxide and 0.2 g/100 cm^3 SiO_2.

Table I. Zeta Potentials and Their Ratios for Ludox HS Silica (Particle 1) and PVC Latex (Particle 2)

pH	ζ_1 (mV)	ζ_2 (mV)	ζ_2/ζ_1	ζ_1/ζ_2
4.0	-28.0	-36.2	1.29	
5.0	-37.0	-42.5	1.15	
6.0	-42.7	-47.5	1.11	
7.0	-48.3	-51.1	1.06	
8.0	-52.1	-53.9	1.03	
9.0	-56.9	-56.7		1.00
10.0	-59.7	-58.9		1.01

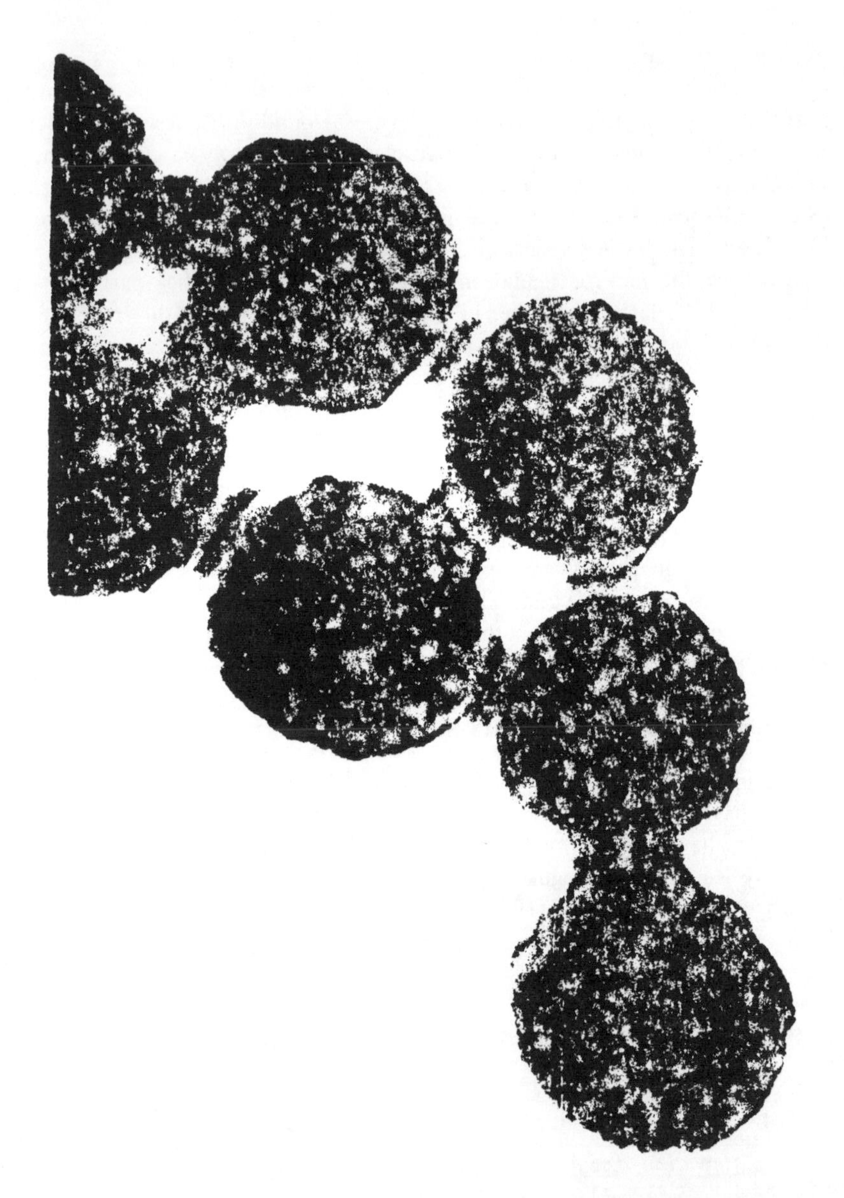

Figure 4. Electron micrograph of PVC latex particles that are coated with Ludox HS silica.

The c.c.c. of NO_3^- (as KNO_3) for positively charged chromium hydroxide sol at pH 3.3 was determined from light scattering data to be 0.06 *M*, whereas the c.c.c. of K^+ (as KNO_3), for the same sol but negatively charged at pH 9-10, is 0.12 *M*. When chromium hydroxide particles were mixed with Ludox HS silica at the concentrations mentioned above, the mixed system remained stable in 0.12 *M* KNO_3 at pH > 9 and exhibited strong HOTS. Furthermore, the chromium hydroxide sol, which is generally unstable over the pH range 7.4-9 (in the vicinity of the isoelectric point) is completely stabilized if Ludox HS silica is added. Microelectrophoretic data, obtained for chromium hydroxide near the i.e.p. of 7.4 or in 0.24 *M* KNO_3 above pH 9, in the presence of 0.2 g/100 cm^3 Ludox HS, gave zeta potentials of -18 to -23 mV. A comparison of this ζ-potential range with the values for pure chromium hydroxide sols, given in Table II, shows that silica can sufficiently increase the negative charge of chromium hydroxide particles to render them stable against coagulation either near the i.e.p. or in the presence of high concentrations of KNO_3 above pH 9. The protection effect must be due to adsorption of silica in a manner similar to what was found for the PVC latex-Ludox HS binary mixture. Furthermore, the data in Table II indicate that unlike in the PVC latex-silica system (Table I), there is a large difference in ζ-potential between chromium hydroxide and silica particles which may contribute to the adsorptivity. This is in addition to the difference in particle size which was considered to be the major reason for the heterocoagulation observed in the latex-silica system.

Table II. Zeta Potentials and Their Ratios for Chromium Hydroxide (Particle 1) and Ludox HS Silica (Particle 2)

pH	ζ_1(mV)	ζ_2(mV)	ζ_2/ζ_1
8.0	-13.0	-53.9	4.15
9.0	-24.8	-56.7	2.29
10.0	-32.9	-58.9	1.79

PVC Latex–Pseudoboehmite

In the mixtures described already, heterocoagulation took place between particles of like charge (negative). The binary sol combination discussed in this section consists of negatively charged PVC latex and positively charged alumina (pseudoboehmite) over the pH range 3 to 9.

Figure 5 gives the electrophoretic mobility of mixed suspensions containing a negatively charged PVC latex (*cf.* Table I) and an expanded pseudoboehmite as a function of pH for three different relative sol concentrations. The data were taken 24 hr after mixing the two dispersions. It is noteworthy that all of the particles in the mixed suspension at a given pH exhibited the same mobility, although the two original sols contained oppositely charged particles. Soluble (cationic) aluminum complexes may be responsible for the mobility behavior exhibited by these mixtures in the range above pH 4. It is unlikely, however, that the positive mobilities shown in Figure 5 for pH values less than 4 are the result of soluble aluminum species. In addition, the small increase in the electrophoretic mobility exhibited by the systems containing 1.4 x 10^{-2} and 2.8 x 10^{-3} g/100 cm^3 Al_2O_3(s) at pH ~ 4 cannot be explained by the adsorption of soluble counterionic species onto negatively charged PVC latex particles.

Finally, since the two types of primary particles discussed in this section are oppositely charged below pH 9, the negative mobility of the particles in the mixed system containing 2.8 x 10^{-4} g/100 cm^3 Al_2O_3(s) is principally due to the negatively charged PVC latex particles. Furthermore, the charge of the latter is somewhat reduced in this case, owing to the formation of latex-alumina aggregates. This conclusion is supported in part by the fact that positively charged particles could not be observed in the systems represented by squares in the figure.

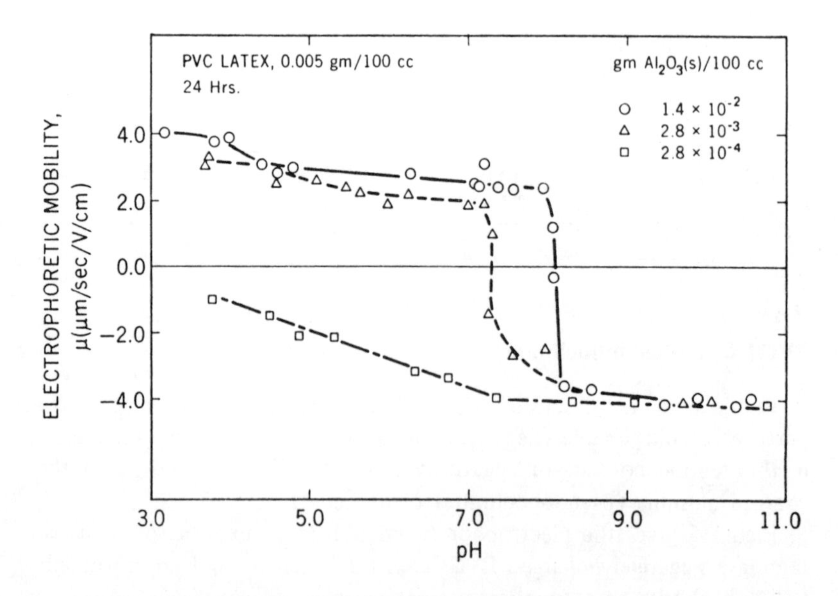

Figure 5. Electrophoretic mobility, μ , of PVC latex-alumina mixtures as a function of pH for various relative concentrations of the two types of particles.

SUMMARY

When reviewing the results and observations reported here, it is evident that it can be difficult to predict whether or not binary dispersions containing unlike particles will be stable. For the mixtures in which both types of particles have similar surface potentials and, therefore, similar electrical double layer properties, as in the case of PVC latex and Ludox HS silica, the resulting stability (or instability) of the mixed dispersion depends to a large extent on the ratio of the particle sizes. If the ratio is large, heterocoagulation of the primary particles can produce an aggregate in which the smaller particles (silica) can effectively coat the larger ones (PVC latex) and protect the latter from electrolytic coagulation, provided, of course, that the relative sol concentrations are favorable. The resultant stability in this case depends upon the interaction between the particles forming the coatings and the surrounding solution medium.

The results obtained with the chromium hydroxide-Ludox HS silica system suggest that, although the zeta potentials differ significantly (Table II) for the two types of primary particles in a binary dispersion, the disparity in particle size still may lead to stabilization of the larger particles (chromium hydroxide) by Ludox HS. Moreover, the silica is capable of stabilizing the metal hydroxide both near the isoelectric point of the latter and in highly concentrated salt solutions.

In contrast to the stability phenomena described above, it was also determined that it is possible under certain conditions to selectively coagulate the small Ludox HS silica particles in the presence of the larger, stable PVC latex particles. Examples of this phenomenon are provided by the latex-silica mixtures containing $Ca(NO_3)_2$ (Figure 3) and $Al(NO_3)_3$.[19] For these systems the salt concentration-pH regions corresponding to selective destabilization of silica are consistent with the stability domains of the single colloidal component PVC latex and Ludox HS silica sols. That is, the conditions which lead to coagulation of a single-component silica sol, yet do not affect the stability of a single-component PVC latex suspension, yield selective destabilization of SiO_2 in the mixed dispersion.

The systems discussed in this chapter consisted only of mixtures in which all colloidal components were preformed. Work in which one component was preformed and the other was formed *'in situ'* is described in a separate publication,[45] which also contains data on the ternary system consisting of alumina formed in the presence of PVC latex and Ludox HS silica. Stability results similar to the ones presented here were found. That is, under specific conditions, the silica particles were capable of stabilizing PVC latex in the presence of either precipitated alumina or the soluble products of the hydrolysis of Al^{3+}. In the ternary system, however, it was found that the alumina ultimately determines the overall stability of the mixed dispersions. Generally, the influence of each sol component was found to follow the sequence: alumina > Ludox HS silica > PVC latex. This order would, in such a complex system, depend upon the relative ratios of the particle concentrations.

CONCLUDING REMARKS

A variety of stability data has been presented in this brief overview of part of our work on mixed dispersions. The particles investigated have many physical and chemical characteristics in common with those of suspended solids commonly encountered in wastewater treatment operations. It is hoped that the principles discussed here will provide a means of studying, understanding, and eventually controlling the properties of the suspensions so prevalent in wastewater systems.

Lastly, in order to understand fully the stability phenomena exhibited by a given binary dispersion, it is necessary to obtain information about the interactions of each solid component with the solute species present. Once this knowledge has been gained, our ability to describe the behavior of mixed dispersions will be greatly enhanced.

ACKNOWLEDGMENT

This research was supported in part by National Science Foundation Grant ENG 75-08403.

REFERENCES

1. Ives, K. J., and J. Gregory. *Proc. Soc. Water Treat. Exam.* 15:93 (1966).
2. Yarar, B., and J. A. Kitchener. *Trans. Inst. Min. Metall., Sec. C.* 79:23 (1970).
3. Van Olphen, H. *An Introduction to Clay and Colloid Chemistry* (New York: Wiley-Interscience, 1963).
4. Schofield, R. K., and H. R. Samson. *Discuss. Faraday Soc.* 18:135 (1954).
5. Ince, C. R. *Trans. Am. Inst. Min. Metall. Pet. Eng.* 87:261 (1930).
6. Bleier, A., E. D. Goddard and R. D. Kulkarni. *J. Colloid Interface Sci.* 59:490 (1977).
7. Somasundaran, P., E. L. Smith, Jr. and C. C. Harris. Paper No. 49, 11th International Mining Procedures Congress, Cagliari, Italy, April 21-26, 1975.
8. Derjaguin, B. V. *Discuss. Faraday Soc.* 18:85 (1954); also see *Colloid J. USSR* 16:403 (1954).
9. Bierman, A. *J. Colloid Interface Sci.,* 10:231 (1955); also see *Proc. Nat. Acad. Sci, U.S.,* 41:245 (1955).

10. Bell, G. M., and G. C. Peterson. *J. Colloid Interface Sci.* 41:542 (1972).
11. Devereux, O. F., and P. L. deBruyn. *Interactions of Plane-Parallel Double Layers* (Cambridge, MA: M. I. T. Press, 1963).
12. Ohshima, H. *Colloid Polym. Sci.* 252:257 (1974).
13. Bell, G. M., S. Levine and L. N. McCartney. *J. Colloid Interface Sci.* 33:335 (1970).
14. Hogg, R., T. W. Healy and D. W. Fuerstenau. *Trans. Faraday Soc.* 62:1738 (1966).
15. Kar, G., S. Chander and T. S. Mika. *J. Colloid Interface Sci.* 44:347 (1973).
16. Wiese, G. R., and T. W. Healy. *Trans. Faraday Soc.* 66:490 (1970).
17. Bleier, A., and E. Matijević *J. Colloid Interface Sci.* 55:510 (1976).
18. Matijević, E., R. Brace and A. Bleier. Paper presented at the Oilfield Chemistry Symposium of the Society of Petroleum Engineers of the Am. Inst. Min. Metall. Pet. Eng., Denver, CO, May 1973.
19. Bleier, A. Ph.D. Thesis, Clarkson College of Technology, Potsdam, NY (1976).
20. Derjaguin, B. V., and L. D. Landau. *Acta Physicochim. URSS* 14:633 (1941).
21. Verwey, E. J., and J. Th. G. Overbeek. *Theory of the Stability of Lyophobic Colloids* (Amsterdam: Elsevier, 1948).
22. Reerink, H., and J. Th. G. Overbeek. *Discuss. Faraday Soc.* 18:74 (1954).
23. Lips, A., C. Smart and E. Willis. *Trans. Faraday Soc.* 67:2979 (1971).
24. Težak, B., E. Matijević and K. Schulz. *J. Phys. Colloid Chem.* 55:1557 (1951).
25. Harding, R. D. *J. Colloid Interface Sci.* 40:164 (1972).
26. Perkins, R., R. Brace and E. Matijević. *J. Colloid Interface Sci.* 48:417 (1974).
27. Matthews, B. A., and C. T. Rhodes. *J. Colloid Interface Sci.* 28:71 (1968).
28. Matijević, E. *J. Colloid Interface Sci.* 43:217 (1973).
29. Bibeau, A. A., and E. Matijević. *J. Colloid Interface Sci.* 43:330 (1973).
30. Eriksson, L., E. Matijević and S. Friberg. *J. Colloid Interface Sci.* 43:591 (1973).
31. Matijević, E. In *Principles and Applications of Water Chemistry,* S. D. Faust and J. V. Hunter, Eds. (New York: John Wiley & Sons, 1967), p. 238.
32. Black, A. P. In *Principles and Applications of Water Chemistry,* S. D. Faust and J. V. Hunter, Eds. (New York: John Wiley & Sons, Inc., 1967), p. 274.
33. O'Melia, C., and W. Stumm. *J. Am. Water Works Assoc.* 59:1393 (1967).
34. Healy, T. W., G. R. Wiese and B. V. Kavanagh. *J. Colloid Interface Sci.* 42:647 (1973).
35. White, D. W. Unpublished data.
36. Lauzon, R. V., and E. Matijević. *J. Colloid Interface Sci.* 38:440 (1972).

37. Bolt, G. H. *J. Phys. Chem.* 61:1166 (1957).
38. Allen, L. H., and E. Matijević. *J. Colloid Interface Sci.* 31:287 (1969).
39. Allen, L. H., and E. Matijević. *J. Colloid Interface Sci.* 33:420 (1970); 35:66 (1971).
40. Allen, L. H., E. Matijević and L. Meites. *J. Inorg. Nucl. Chem.* 33:1293 (1971).
41. Demchak, R., and E. Matijević. *J. Colloid Interface Sci.* 31:457 (1969)
42. Matijević, E., A. D. Lindsay, S. Kratohvil, M. E. Jones, R. I. Larson and N. W. Cayey. *J. Colloid Interface Sci.* 36:273 (1971).
43. Sillén, L. G., and A. E. Martell. *Stability Constants of Metal–Ion Complexes,* The Chemical Society, Special Publication No. 17 (London: Burlington House, 1964); also Supplement No. 1, Special Publication No. 25 (1971).
44. Pugh, R. J., and J. A. Kitchener. *J. Colloid Interface Sci.* 35:656 (1971).
45. Matijević, E., and A. Bleier. *Croat. Chem. Acta* 50:93 (1977).

CHAPTER 6

EFFECTIVENESS OF SAND FILTERS FOR BACTERIOPHAGE *ESCHERICHIA COLI* T_1 REMOVAL USING CALCIUM AS AN AID

Stephen R. Jenkins and James L. Barton*
Department of Civil Engineering
Auburn University
Auburn, Alabama

INTRODUCTION

There is no obvious widespread dissemination of viral diseases through water in this country. However, viruses found in the gastrointestinal tract of man are capable of becoming waterborne. Enteric viruses have been observed in wastewater plants and recovered in rivers receiving effluents from these plants. Water supplies which derive from these virus-laden waters must be entirely safe for public consumption. Consequently, pathogenic viruses must be removed or destroyed.

Some outbreaks of waterborne disease, such as infectious hepatitis and viral gastroenteritis, have been reported. Most of these incidents have resulted from contamination of small water supply systems brought about by improper operation of treatment processes.[1] Although the number has been small, some evidence does exist which indicates that the number of outbreaks of waterborne viral infections is increasing.[1,2]

In an effort to safeguard the health of the public, the effectiveness of water and wastewater unit operations in removing viruses has demanded increasing attention. Anticipated increases in the use of recycled

*Present address: Consulting Engineering Services, Montgomery, Alabama.

wastewater tend to emphasize the need to develop inexpensive but reliable methods of virus removal. Researchers attending a workshop to define and establish priorities and provide direction for a potable reuse research program considered virus removal, or inactivation, as one of the primary public health considerations related to the potable reuse of wastewater.[3]

At present, conflicting evidence exists about the economic effectiveness of water and wastewater operations for virus removal. For example, sand filters have been shown to be ineffective for virus removal unless large doses of alum or polyelectrolytes are used as filter aids.[4] As indicated by Britton,[5] more effort should be given to research dealing with virus removal by filtration.

This investigation dealt with enhancing the effectiveness of sand filtration processes for virus removal. Since filtration processes are common to both wastewater and water treatment, a better understanding of virus removal mechanisms and the effect of filter aids on this removal is of interest in both areas. In particular, this research investigates the effectiveness of model sand filters for removal of *Escherichia coli* T_1 phage when calcium is used as a chemical conditioner. Although the flowrate passing through the filters was more representative of water treatment, the virus titer was more representative of those found in tertiary treatment of wastewater.

SURFACE CHARACTERISTICS OF VIRUSES

Previous researchers[6-8] have shown that metal oxides such as manganese oxides can be effectively removed by sand filtration if a cation such as calcium is used as a filter aid. It has been shown by Stumm *et al.*[6] that manganese oxide surfaces can behave as an acid or base and have positive or negative surface charges depending on solution pH. Above the isoelectric point, the surface hydroxo groups or -Mn-OH groups may disassociate as follows

$$\text{-Mn-OH} + H_2O = \text{-MnO}^- + H_3O^+ \quad (1)$$

The formation of these negative sites results in stable colloidal particles which are difficult to remove by sand filtration in the absence of filter aid. The negatively charged MnO_2 is electrostatically repelled by the negatively charged sand and, thus, particle removal is poor.

Jenkins *et al.*[7,8] showed that these hydrated manganese oxide surface sites may also associate with cations such as Ca^{2+}. The calcium ions act to displace bound H^+ ions

$$\text{-Mn-OH} + Ca^{2+} + H_2O = \text{-Mn-O-Ca}^+ + H_3O^+ \quad (2)$$

Thus, Ca^{2+} chemically associates with the MnO_2 surface, reducing the charge until van der Waals attractive forces overcome any remaining electrostatic repulsion between the oxide and the sand, and attachment occurs.

There are many different viruses, and each may be affected by a given process to a different extent. However, almost all viruses have some type of protein surface coating.[9] The α-amino acids which make up the surface coat of protein have terminal carboxyl groups. It is proposed that the carboxyl surface sites may dissociate in a manner similar to the manganese oxide hydroxide sites to form negatively charged sites

$$-\text{Virus}-\underset{\underset{\text{O}}{\|}}{\text{C}}-\text{OH} + H_2O = -\text{Virus}-\underset{\underset{\text{O}}{\|}}{\text{C}}-\text{O}^- + H_3O^+ \qquad (3)$$

or accept protons

$$-\text{Virus}-\underset{\underset{\text{O}}{\|}}{\text{C}}-\text{OH} + H_3O^+ = -\text{Virus}-\underset{\underset{\text{O}}{\|}}{\text{C}}-\text{OH}_2^+ + H_2O \qquad (4)$$

In the pH range of most natural waters, Equation 3 predominates, and these carboxyl surface sites are negatively charged. Thus, viruses are negatively charged and, if passed through a sand filter, are electrostatically repelled by the negatively charged sand grains. Consequently, in the absence of filter aids, viruses are poorly removed.[10,11]

It is further proposed that certain cations such as calcium ion can specifically react with the hydroxo sites of the protein and form positive sites. Such a reaction can be represented by a complex formation reaction

$$-\text{Virus}-\underset{\underset{\text{O}}{\|}}{\text{C}}-\text{OH} + Ca^{2+} = -\text{Virus}-\underset{\underset{\text{O}}{\|}}{\text{C}}-\text{OCa}^+ + H^+ \qquad (5)$$

The formation of positive sites may reduce the negative surface charge of the virus to a level that the electrostatic repulsive force may be overcome by van der Waals forces of attraction. The virus can then be easily removed by particle-particle attachment or by attachment to sand particles. This hypothesis could be used to explain the results of Kohn and Fuchs,[12] which indicated that Ca^{2+} enhanced the attachment of some virus to host cells, and those of Lefler and Kott,[11] which showed that divalent cations increased considerably the removal efficiency of a sand column for poliovirus 1 and *coli* phage f2. It was the intent of the present study to clarify the effectiveness of addition of calcium in enhancing removal of virus by sand filtration.

EXPERIMENTAL APPARATUS AND METHODS

In preparing the filter experiments, precautions were taken to avoid contamination of virus suspensions, growth media and host cell cultures. All glassware used in the study was soaked in "Micro" cleaner, rinsed several times with tapwater, rinsed in distilled water, and finally rinsed in deionized water. Sterilization was accomplished by autoclaving at 121°C for 30 minutes.

Aseptic techniques were used in pouring all media, in the transfer of host cells and virus suspensions, and in the filtering procedure. The sand used in the filter was not autoclaved prior to the filtering procedure since it was found that deionized water passed through the clean sand did not contain any virus particles. Also, it was feared that the surface characteristics of the sand grains might be changed if the sand were autoclaved.

Test Virus Suspension

No single virus has been shown to serve as an indicator for viruses or to typify all viruses.[13,14] However, *Escherichia coli* bacteriophage T_1 does typify many of the viruses found in domestic wastewater and is relatively easy to assay. Consequently, it was used as the test virus. T_1 is an icosahedral tailed phage of approximatley 500 to 1000 Å. The attachment of this virus to *E. coli* is related to the type and concentration of cations in solution.

Stock suspensions of the virus were prepared by first inoculating 250 ml of nutrient broth with *E. coli* B cells and incubating the inoculated broth at 37°C until the cells were in the log growth phase (about 6 hr). Next, 1 ml of concentrated T_1 phage suspensions was added and the resulting mixture incubated at 37°C until the turbidity caused by the host bacteria cells cleared. The bacteriophage culture was then centrifuged and filtered through a membrane filter (0.45 μ) to remove the lysed bacterial cells. The virus suspension was then frozen, thawed and stored in a refrigerator. Freezing was required to assure a constant high titer (virus number). This suspension yielded a virus titer of about 10^7 to 10^8 plaque forming units per milliliter (PFU/ml).

The virus was assayed using a soft agar overlay method. *E. coli* was grown on an agar surface in the presence of T_1 phage. The assay plates were prepared well in advance of the assay by spreading 1.5% agar in each plate and allowing the agar to harden. At the time of the assay, the virus suspension was first diluted and then mixed with host cell *E. coli* B and a soft agar (0.75%). This suspension was spread evenly over the preprepared hard agar plates, allowed to harden, and then incubated at 37°C for 24 hours. The *E. coli* grew in abundance leaving clear spots only in places infected by virus. The area of phage destruction developed from a single phage was clear and round. After counting the plaques and considering dilutions, the number of PFU/ml for the original suspension could be estimated.

Model Sand Filters

The basic filtration apparatus, as shown in Figure 1, consisted of four 25-cm long, 2.54-cm ID plexiglass tubes filled to a depth of 10 cm with Ottawa sand. The sand was added to the plexiglass tubes and soaked with water overnight before each use. Four interchangeable aspirator bottles containing the virus were located above the filters as shown in the Figure. For a given filter run, only one bottle was used and it provided a fairly constant head. An additional aspirator bottle which contained deionized sterilized water was located as shown and used to estab-

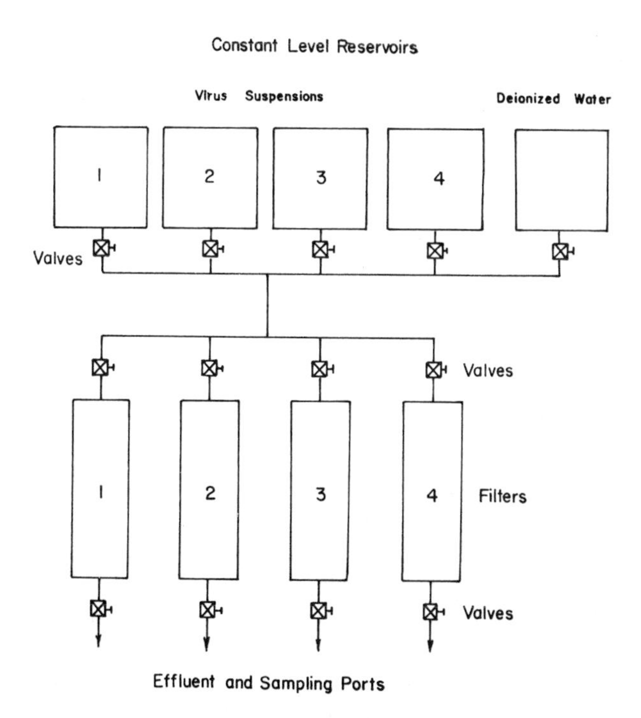

Figure 1. Schematic of sand filter apparatus.

lish the initial flowrate. The flowrate was set at 1.358 liter/sec/m^2 (2 gal/min/ft^2) and maintained constant throughout each filter run by adjustment of the values.

To make a filtration run, stock virus was placed in one of the constant head bottles and diluted to the desired level. Calcium chloride was then added to give the required Ca^{2+} concentration. The ionic strength was adjusted to approximately 5 x 10^{-3} with sodium chloride and the pH was adjusted to 7.0 with sodium hydroxide. After a flash mix, the resulting virus suspension was mixed slowly for 20 min before the filtration run began. This mixing process was used to provide uniformity of experimental procedure and to allow sufficient time to adjust the flowrate through one of the filters with the deionized water. Slow mixing continued during the filter runs which were terminated after 45 min.

Samples (0.1 ml) of filter influent and effluent were taken at various time intervals and the number of virus in each sample determined by the assay technique previously described. After each run, the flow from the bottle containing the test virus was terminated. The bottle and the exhausted sand filter were removed and the process continued using the remaining bottles and filters.

After the virus titers for each run were determined, the influent and effluent values for a given time were compared and the percentage of virus particles passing the filter at various times was calculated. These values were plotted against time to qualify the effect of changing calcium concentration on the removal of virus by a sand filter.

RESULTS

As seen in Figure 2, in the absence of filter aids such as calcium, sand filters have a poor affinity for virus particles. For the three virus titers shown, the viruses passed through the filters in a manner similar to plug flow. The number of viruses present had little effect on the ability of the sand to remove them. After only 10 min, approximately 80% of the virus particles passed through the filter. The poor removal achieved in these tests was not surprising. At pH 7, the virus particles were negatively charged (Equation 3) and were electrostatically repelled by the negatively charged sand grains, and thus were not removed.

Three similar filtration runs utilizing test water containing a virus titer of 138 x 10^4 PFU/ml and a Ca^{2+} concentration of 10^{-3} M at pH 7.0 are shown in Figure 3. This test water was passed through three separate sand filters, and three separate sets of data were collected and plotted. It can be seen from Figure 3 that plots of the data from the three filters varied by only 7 percentage points at 75 min, thus showing good

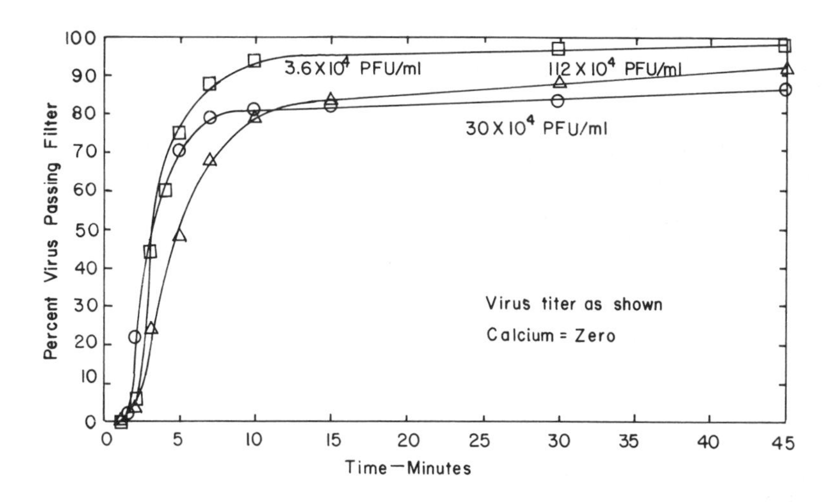

Figure 2. The percentage of virus particles passing the filter vs time in the absence of calcium.

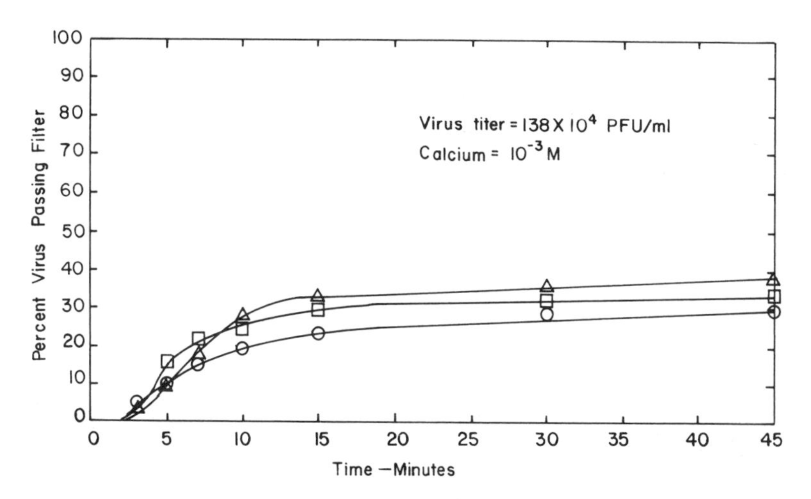

Figure 3. The percentage of virus particles passing the filter vs time for three different filter runs. The filtration conditions (flowrate = 1.3581/sec/m^2, virus titer 138 x 10^4 PFU/ml, [Ca^{2+}] = 10^{-3} *M*, pH = 7.0, I = 5 x 10^{-3}) were the same for all three runs.

consistency in the experimental procedure and the assay technique. Equally reliable data were obtained for most of the other filter runs. The remaining plots are summary (average) plots of two or three runs.

It can also be seen from a comparison of Figures 2 and 3 that when calcium was added to the virus suspension, the removal of the virus by the sand filter was enhanced considerably. For the virus titer of 138 x 10^4 PFU/ml in the presence of only $10^{-3} M$ of Ca^{2+}, approximately 20 to 30% of the virus passed the filter. That is, approximately 70 to 80% of the virus was removed.

In Figure 4, the percentage of virus particles passing the filter vs time is plotted as a function of Ca^{2+} concentration for a constant initial virus titer of 36 x 10^4 PFU/ml. It can be seen, for this virus titer, that the amount of calcium in the system and the efficiency of virus removal by the sand were related. For a concentration of Ca^{2+} of 10^{-5} *M*, approximately 50% of the virus passed through the sand filter. As the concentration of Ca^{2+} is increased, the number of viruses passing through the filter decreased. For concentrations of calcium of 10^{-4} *M*, 10^{-3} *M*, and 10^{-2} *M*, the percentage of viruses passing through the filter was reduced to approximately 25, 15, and 3%, respectively. Increasing the calcium concentration above 10^{-2} *M* did not affect the number of viruses removed.

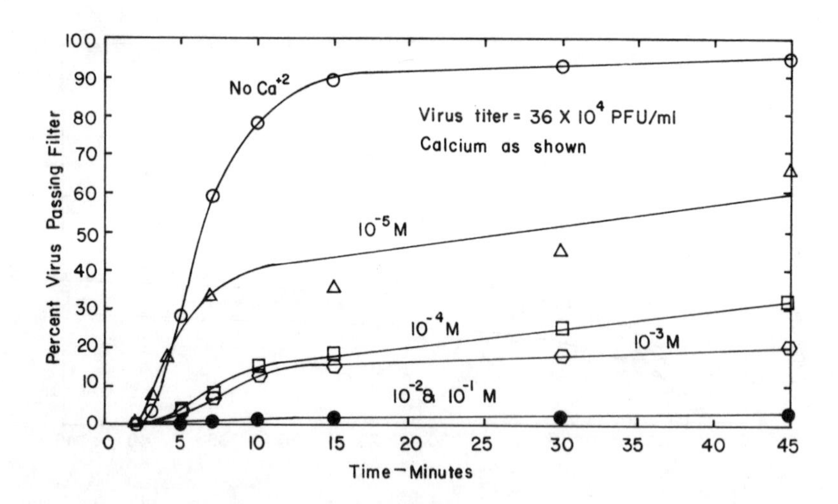

Figure 4. The percentage of virus particles passing the filter virus time as a function of Ca^{2+} concentration.

It is interesting to note that for this virus titer of 36 x 10^4 PFU/ml, a concentration of 10^{-3} *M* Ca^{2+}, which is equivalent to a hardness of 100 mg/l as $CaCO_3$, virus removal rates of about 70 to 80% were achieved. Such a water would be classified as only moderately hard, and there are many waters and wastewaters which have this hardness. Yet, if the other solution variables had similar values, good virus removal could be effected by sand filtration without further addition of Ca^{2+} to the system. However, in natural waters, organic material may interfere with the removal.[15]

Figures 5 and 6 represent data for stated virus titers for calcium concentrations of 10^{-3} *M* and 10^{-4} *M*, respectively. It can be seen from these graphs that the effectiveness of Ca^{2+} for enhancing the removal of virus particles on model sand filters was related to the virus titer. As the virus titer increased, the effectiveness of Ca^{2+} to enhance removal decreased. That is, as the virus titer increased, more Ca^{2+} was needed to achieve the same removal efficiency. Thus, there exists some type of stoichiometric relationship between the amount of Ca^{2+} necessary to effect attachment of virus on sand and the number of virus particles present.

It can be seen in Figure 6 that the percentage of virus particles passing the sand filter reached a plateau and then, after approximately 10 min of filtering time, the percentage passing suddenly increased to a second level. It is believed that this does not reflect the true removal efficiency of the sand filter at a flowrate of 1.358 $1/sec/m^2$ but is the result of inadequate experimental technique. This particular set of data was taken early in the experimentation. Samples were taken at 1-min intervals between time 0 and time 10 min into the filter runs, making it difficult to maintain the flowrate constant. During this time period, as particles were removed, the flowrate decreased and better virus removal was achieved. After 10 min of filter time, fewer samples were taken, thus allowing the flowrate to be monitored more precisely. Later runs, in which the sampling and operating times were better allocated, indicated that with more careful control of the flowrate, this hump did not result.

Data for a Ca^{2+} concentration of 10^{-5} *M* is shown in Figure 7. As with the higher Ca^{2+} concentrations it can be seen, by comparing curves for the virus titers of 52 x 10^4 PFU/ml and 191 x 10^4 PFU/ml, that as the virus titer increased more Ca^{2+} was needed to achieve the same removal efficiency. However, data for a very low virus titer (0.59 x 10^4 PFU/ml) are also shown which are in direct disagreement with this relationship. These data may not be completely representative of actual removal efficiency as the assay technique was not sensitive enough to measure reliably the low virus numbers encountered. Small errors in the total count of virus particles in this titer range would result in large errors on a percentage basis. It could also be true, however,

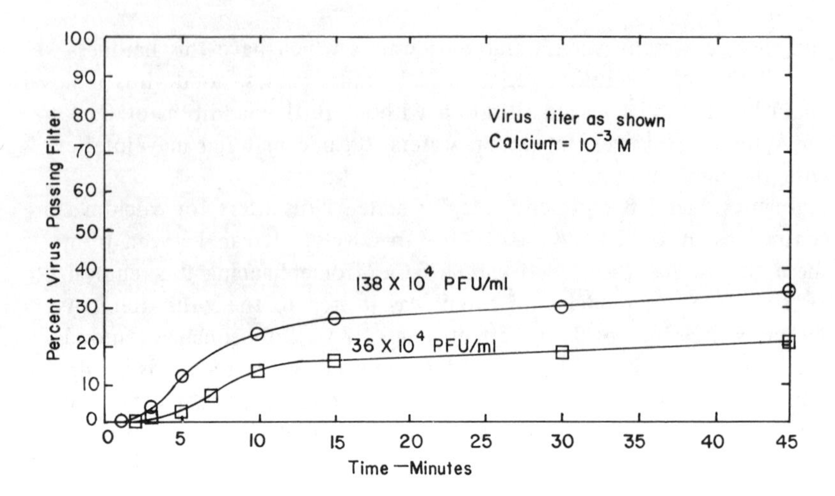

Figure 5. The percentage of virus particles passing the filter virus time as a function of virus titer for a $[Ca^{2+}] = 10^{-3}$ *M*.

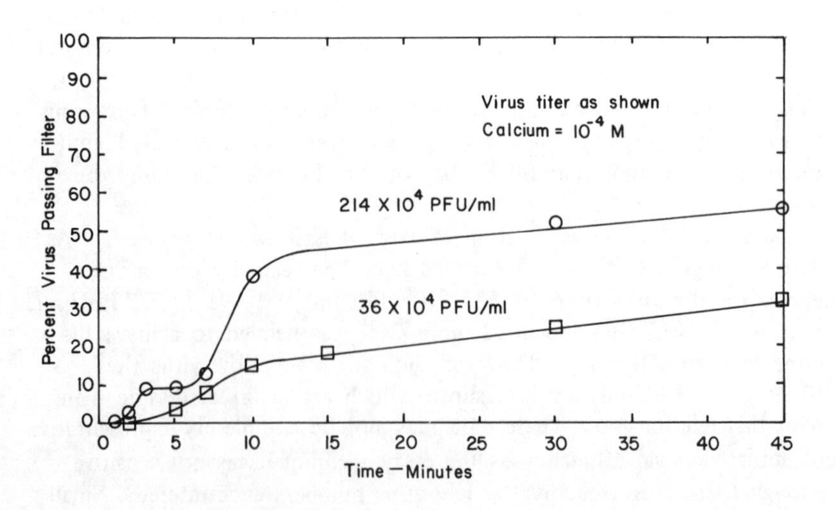

Figure 6. The percentage of virus particles passing the filter virus time as a function of virus titer for a $[Ca^{2+}] = 10^{-4}$ *M*.

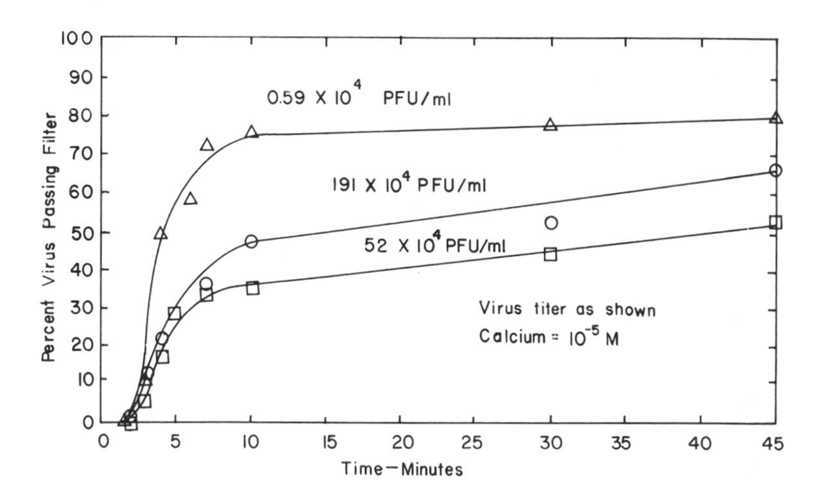

Figure 7. The percentage of virus particles passing the filter virus time as a function of virus titer for a $[Ca^{2+}] = 10^{-5}$ *M.*

that since virus transport to the sand surface is controlled by Brownian motion,[16] the number of particle collisions at this low titer was too small to effect significant removal.

CONCLUSIONS AND SUMMARY

It is believed that there exists an electrostatic repulsive force between the negatively charged filter sand surface and the negatively charged virus that prohibits the filter sand from removing virus particles. With no filter aids present, the sand would electrostatically repel the virus particles; thus, removal would be poor. To obtain significant virus removal, the electrostatic repulsive force must be overcome. When Ca^{2+} is added to the virus suspension, formation of positive sites develop due to the Ca^{2+} complexing with the virus particles, as seen in Equation 5. These positive sites reduce the negative surface of the virus particles to a level such that the electrostatic repulsive force between the sand surface and the virus particles may be overcome by van der Waals forces of attraction. The virus particles are then readily removed by the sand particles.

If complexing between Ca^{2+} and virus particles does reduce the surface charge and cause an increase in filter efficiency, then a relationship between the amount of Ca^{2+} needed and the number of virus particles should exist. As seen in Figures 5, 6 and 7, as the virus titer increased, more Ca^{2+} was needed to achieve the same removal efficiency. Thus, the amount of calcium required to enhance removal was a function of the virus titer, and a stoichiometric relationship does appear to exist.

Subsequent research should include use of alkalimetric titration curves to determine the extent of Ca^{2+} complexing with virus particles, and should consider the effects of pH, organic content, ionic strength and flowrate on removal effectiveness. In addition, removal efficiency for very low virus titers should be investigated.

ACKNOWLEDGMENT

This research was supported in part by the Office of Water Research and Technology, U.S. Department of the Interior, under the Water Resources Research Act of 1964 (PL 88-379) acting through the Auburn University Water Resources Research Institute.

REFERENCES

1. Graun, G. F., and L. J. McCabe. *J. Am. Water Works Assoc.,* 65: 74 (1973).
2. Taylor, F. B. *J. Am. Water Works Assoc.* 66:306 (1974).
3. English, J. N., E. R. Bennett and K. D. Linstedt. *J. Am. Water Works Assoc.,* 69:131 (1977).
4. Robeck, G. G., N. A. Clarke and K. A. Dostal. *J. Am. Water Works Assoc.* 54:1275 (1962).
5. Britton, G. *Water Res.* 9:473 (1975).
6. Stumm, W., C. P. Huang and S. R. Jenkins. *Croat. Chim. Acta.* 42:223 (1970).
7. Jenkins, S. R.. *Environ. Sci. Technol.* 7:43 (1973).
8. Jenkins, S. R., J. Engeset and V. R. Hasfurther. In: *Chemistry of Water Supply, Treatment and Distribution,* A. V. Rubin, Ed. (Ann Arbor, MI: Ann Arbor Science Publishers, Inc. 1974).
9. Cookson, J. T. *J. Am. Water Works Assoc.* 61:52 (1969).
10. Carlson, J.. H., G. H. Ridenour and C. F. McKhann. *Am. J. Public Health* 32:1256 (1942).
11. Lefler, E., and Y. Kott. *Israel J. Technol.* 12:298 (1974).
12. Kohn, A., and P. Fuchs. *Adv. Virus Res.* 18:159 (1973).
13. Vaughn, J. M., and T. G. Metcalf. *Water Res.* 9:197 (1975).
14. Moore, B. E., B. P. Sayik and J. F. Malina. *Water Res.* 9:707 (1975).
15. Oza, P. P., and M. Chandhuri. *Water Res.* 9:707 (1975).
16. Valentine, R. C., and H. C. Allison. *Biochem. Biophys. Acta* 34: 10 (1959).

CHAPTER 7

CHEMICAL TREATMENT FOR NUTRIENTS AND ALGAL REMOVAL FROM OXIDATION POND EFFLUENTS

Adnan Shindala and Marvin T. Bond
Civil Engineering Department
Mississippi State University
Mississippi State, Mississippi

INTRODUCTION

Until recently, waste stabilization ponds have been accepted as satisfying essential waste treatment requirements. However, because of the growing awareness of the pollutional aspects of wastewater and in order to preserve the natural value of streams, higher degrees of treatment must be provided. Currently used waste stabilization ponds do not provide such a high degree of treatment. The poor treatment during the winter months coupled with heavily algae-laden effluents during long hot summer days are two of the main characteristics of waste stabilization ponds. Algae in the pond effluent will eventually die away in the receiving streams and will be available for bacterial degradation, which, besides odor and aesthetic conditions, adds an additional oxygen demand as well as nutrients to the stream. Consequently, posttreatment of oxidation pond effluent to remove the algae is essential to continue the use of ponds as an acceptable means of wastewater treatment.

Information concerning posttreatment methods for waste stabilization pond effluent is scarce. Some of the methods reported in the literature include settling, flotation, microstraining, centrifugation, filtration, bioflocculation, sonic vibration and chemical coagulation. Of all the methods reported, chemical coagulation seems to be most promising. The marked reductions in biochemical oxygen demand (BOD), chemical oxygen demand

(COD), nitrogen and phosphorus, along with the effective removal of algae offered by chemical coagulation, make this method most attractive as a posttreatment process.

In this study, chemical coagulation was investigated as a method to remove algae and to improve the quality of the effluent of a stabilization pond. Several different coagulants were tested and the degree of effectiveness was measured on the basis of reductions in BOD, COD, total phosphates, total nitrogen, coliform bacteria and algal concentrations. A mathematical model relating phosphate removal to coagulant dosage was also developed. Characterizations of the chemical sludges were defined in terms of settleability and dewaterability.

Previous Related Studies

An algal-bacteria symbiotic relationship exists in waste stabilization ponds. Bacteria may degrade the organic matter according to the following simplified relation

$$\text{organics} + O_2 \xrightarrow{\text{bacteria}} CO_2 + H_2O + \text{new bacteria} + \text{degraded organics} \qquad (1)$$

Algae, in turn, reuse the carbon dioxide to form algal biomass:

$$CO_2 + H_2O + \text{energy} + \text{nutrients} \xrightarrow{\text{algae}} O_2 + H_2O + \text{new algae} \qquad (2)$$

Although these relations oversimplify complex biochemical reactions, they do show that a recycling of carbon occurs in the pond. Organic reduction may occur by removal of the algae or removal of the carbon through methane fermentation in an anaerobic sludge layer.[2] The Environmental Protection Agency (EPA) has suggested that physical removal of algae will insure that virtually all of the carbonaceous BOD and nitrogenous BOD in the pond effluent would be removed.[3]

Pavoni[4] has researched algal autoflocculation and flocculation with anionic and monionic synthetic organic polyelectrolytes. One major problem with autoflocculation is that motile, highly charged species of algae and crustaceans are not efficiently removed. McGarry[5] determined that algae removal could be significantly improved with a cationic polyelectrolyte added to a system using alum as a primary coagulant. Anionic and nonionic polyelectrolytes did not aid in the removal of algae. Shindala and Stewart[6] determined the optimum alum dosage for removal of algae to be in the range of 75 to 100 mg/l. An alum dose of 100 mg/l produced a supernatant with a 5-day BOD less than 2.3 mg/l, COD less that 32 mg/l, phosphate less that 1.2 mg/l and total nitrogen less than 6.6 mg/l 90% of the time.

Parker *et al.*[7] investigated the possibility of algal removal by autoflotation and by dissolved air flotation. They have concluded that autoflotation is not a satisfactory method of algae removal when the water is less than saturated with dissolved oxygen. These periods are random in nature and may occur with changes in organic loadings. Furthermore, the dissolved oxygen content may be less than saturated most of the night and early morning hours for much of the year. It would be necessary to resort to other methods of algae removal such as dissolved air flotation or physical-chemical treatment during these times. They reported that chemical coagulation, flocculation and sedimentation removed 60 to 90% of the algae, with the remaining fraction removed by sand filtration.

O'Brien *et al.*[8] suggested that algae may be removed by submerged rock filters or by upflow fly-ash filters. McGhee and Patterson[9] have observed a negative charge on the algal cells and that a number of species are motile. Both factors would tend to limit, or possibly exclude, autoflocculation as a workable process for algal removal. Their results indicate that upflow filtration removed 16.4% of the BOD, 17.9% of the COD and 59.6% of the suspended solids.

McKinney *et al.*[10], after an extensive review of available data, concluded that for small ponds the best method for algae separation was a series arrangement of two or more lagoon cells, with the final pond used for sedimentation. However, Oswald *et al.*[2,3] indicated that although series ponds can be expected to produce a good effluent initially, a deterioration of effluent quality may be expected with prolonged operations.

PROCEDURES AND EQUIPMENT

This chapter summarizes the results of two separate investigations on algal removal through chemical coagulation. The first study evaluated the use of aluminum sulfate (alum), ferric chloride and ferric sulfate; alum, ferric sulfate, lime and various different polyelectrolytes were tested during the second investigation. All polyelectrolytes used were products of Calgon Corporation and are listed in Table I. The polymers were evaluated as primary coagulant as well as coagulant aids in combination with alum. Standard jar tests were conducted through the studies to evaluate the effectiveness of and the optimum conditions (pH and coagulant dosage) for chemical coagulation. The settling and dewatering characteristics of the chemical sludges were evaluated utilizing standard settling tests, Buchner funnel and filter leaf tests.

To widen the application of the results of the laboratory investigation, several ponds receiving different types of wastes and operating under

Table I. List of Polyelectrolytes Tested

Code Name	Ionic State
WT 2690	Nonionic Synthetic Polymer
WT 2870	Cationic Organic Polymer
WT 2630	Cationic Synthetic Polymer
WT 3000	Anionic Synthetic Polymer
WT 2900	Anionic Synthetic Polymer
WT 2700	Anionic Synthetic Polymer
WT Cat-Floc	Cationic Synthetic Polymer

different conditions were sampled. Grab samples were collected from the effluents of three different single-cell lagoons, a two-cell lagoon, and a three-cell lagoon. The samples were collected a random times during daylight hours. Each sample was characterized with respect to PH, alkalinity, algal cell count, suspended solids, BOD, COD, total phosphorus, ammonia nitrogen and total Kjeldahl nitrogen (TKN). All analytical tests were performed according to standard procedures.[1]

RESULTS AND DISCUSSION

Alum

Alum was found to be very effective in separating the algae as well as significantly reducing the pollutional characteristics of the lagoon effluent. Optimum conditions for coagulation were produced at a pH of 5.5 and an alum dosage of 85 mg/l. At the optimum conditions, the clarity of the supernatant (unfiltered) was observed to be that of tap water and with BOD_5 , COD, total phosphate and total nitrogen content equal to or less than 2.3, 3.2, 1.2 and 6.6 mg/l, respectively, 90% of the time.

To facilitate interpretation of the data, the results are reported in terms of frequency of occurrence of any particular characteristic at varying coagulant dosages and are summarized in Table II. The alum dosage was varied from 0 to 200 mg/l. As shown in Table II, the removal efficiencies were observed at dosages greater than 100 mg/l. The clarity of the supernatant was also observed to increase with the increase in coagulant dosage. At a dosage of 100 mg/l alum the clarity of the supernatant was observed to be that of tapwater.

Examination of the different characteristics of the supernatant, as reported in Table II, clearly demonstrates the effectiveness of chemical coagulation as a posttreatment method for oxidation pond effluents. At

Table II. Frequency of Occurrence of Characteristics (Parameters Evaluated for Unfiltered Samples)

	Parameters															
	BOD_5				Total Phosphates				COD				Total Nitrogen			
Alum Dos-age (mg/l)	mg/l Percent Occur*		Percent Remain Percent Occur*		mg/l Percent Occur*		Percent Remain Percent Occur*		mg/l Percent Occur*		Percent Remain Percent Occur*		mg/l Percent Occur*		Percent Remain Percent Occur*	
	50	90	50	90	50	90	50	90	50	90	50	90	50	90	50	90
0	43.0	51.0	100.0	100.0	14.0	19.0	100.0	100.0	142.0	250.0	100.0	100.0	12.0	16.0	100.0	100.0
25	6.0	9.8	15.0	25.0	5.6	11.8	34.0	57.0	42.0	64.0	32.0	47.0	5.0	8.5	43.0	76.0
50	2.3	3.6	7.0	13.5	1.7	3.0	12.0	19.5	27.5	43.0	21.0	29.0	3.7	6.9	32.0	64.0
75	1.6	2.6	4.5	8.0	1.0	1.6	6.4	8.6	23.5	37.0	18.0	29.0	3.5	6.7	30.0	60.0
100	1.5	2.3	3.5	6.0	0.8	1.2	5.0	6.1	21.5	32.0	16.7	26.5	3.4	6.6	29.5	59.0
125	1.4	2.2	3.5	6.0	0.7	1.0	4.5	5.4	20.5	32.0	16.0	26.0	3.4	6.6	29.5	59.0
150	1.4	2:2	3.0	6.0	0.6	0.8	4.0	4.6	20.0	31.0	14.5	24.0	3.4	6.6	29.5	59.0
175	1.2	2.1	3.0	5.5	0.6	0.8	4.0	4.6	20.0	30.0	14.5	24.0	3.4	6.6	29.5	59.0
200	1.2	2.1	3.0	5.5	0.5	0.8	3.5	4.6	20.0	30.0	14.5	23.5	3.4	6.6	29.5	59.0

*Percentage of time value is equal to or less than shown.

an alum dosage of 100 mg/l, for example, the coagulation process resulted in an effluent as clear as tapwater and with characteristics comparable to those expected from tertiary treatment of biologically treated effluents. In addition, chemical coagulation resulted in a marked reduction in the coliform content of the pond effluent with an average value of 5000 coliform/100 ml remaining in the supernatant.

In most cases, the flocs formed were observed to settle rapidly. The size of the flocs was noticed to decrease with increase in dosage, although this seemed to have no effect on the settling characteristics. The problem of algae floating was observed only twice. This was attributed to the release of dissolved oxygen from the supersaturated effluents. An additional rapid mix time of one to two minutes was found to remedy this situation. Typical settling curves are shown in Figure 1.

The sludge produced by alum treatment contained 0.1 to 0.2% solids. The sludge exhibited the properties of hindered settling and was highly compressible. The alum sludge also had the tendency to become septic in a relatively short time. Gravity thickening produced poor results, and the filter leaf test indicated that vacuum filtration would not be applicable for dewatering the sludges.

To estimate the algal contribution to the BOD, COD, total phosphates and total nitrogen, tests were conducted on filtered and unfiltered samples of the pond effluents. Results indicated that, on the average, algae contributed quite heavily (greater than 50%) to the BOD_5, COD and nitrogen, while contributing less than 25% to phosphates.

To enhance future continuous studies, a mathematical model relating influent phosphate concentration, $(PO_4)_e$, to alum dosage, C, was developed. Alum was chosen because of the high cost of the ferric coagulants and the residual iron color. Lime was excluded from the analysis because of the high dosage (500-1000 mg/l) required to provide equivalent treatment, the large volume of sludge produced, and the pH adjustment required prior to effluent discharge. The relationship obtained was as follows

$$C = \frac{(PO_4)_i}{0.06\ (PO_4)_e + 0.132} - 8.0 \qquad (3)$$

This model was applied at various alum dosages, and the predicted effluent phosphate values compared favorably with experimental values as shown in Figures 2 and 3.

Iron Salts

Optimum conditions for chemical treatment with ferric chloride and ferric sulfate included a pH of 4.8 and a chemical dosage of 125 mg/l.

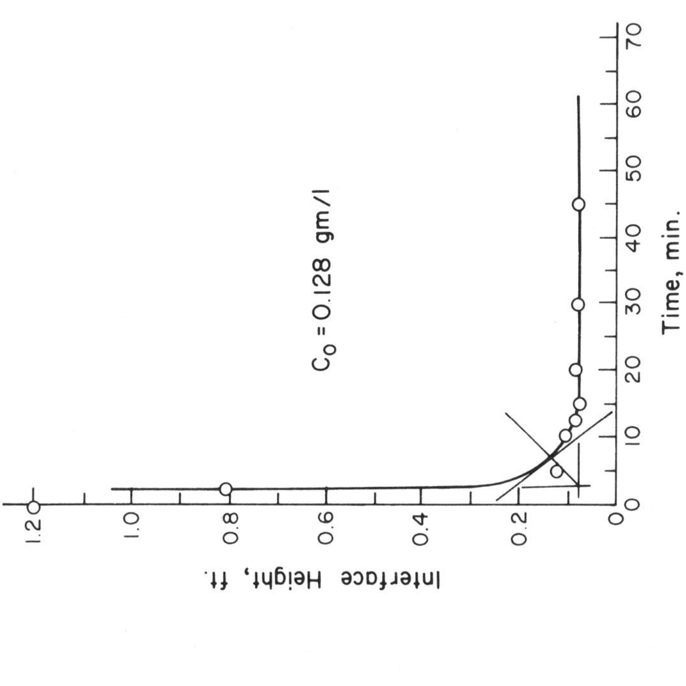

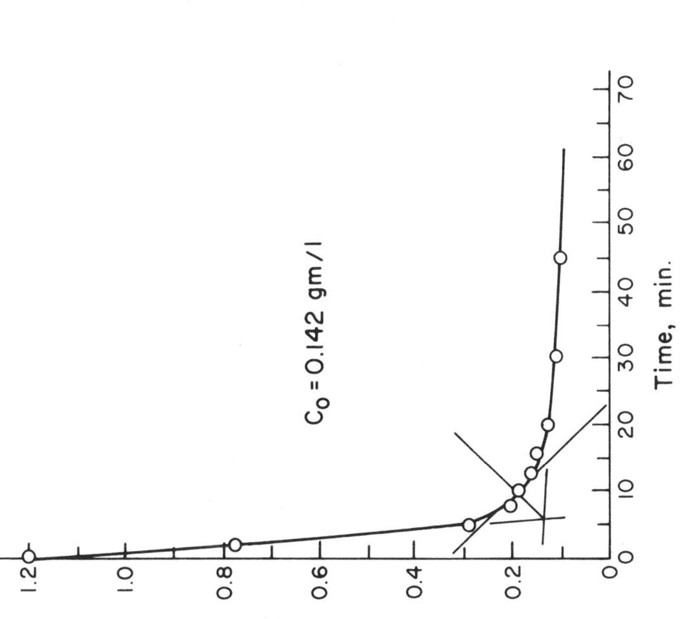

Figure 1. Typical settling curves.

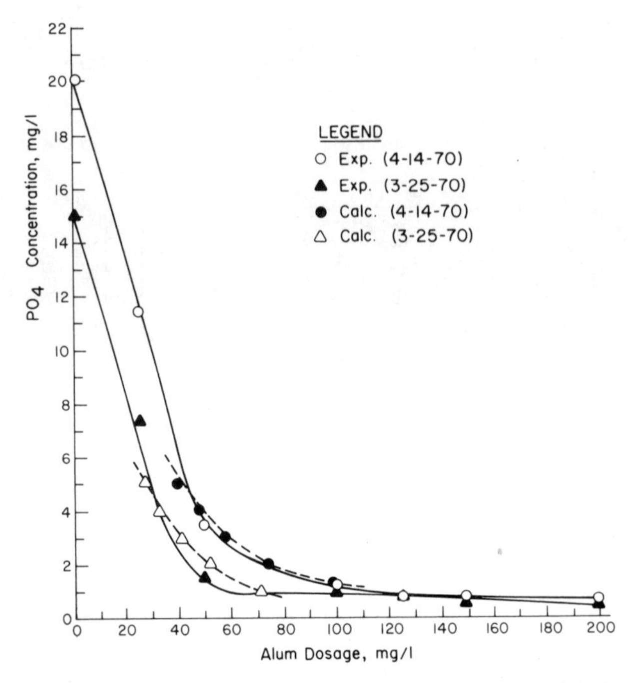

Figure 2. Phosphate concentration after treatment with alum.

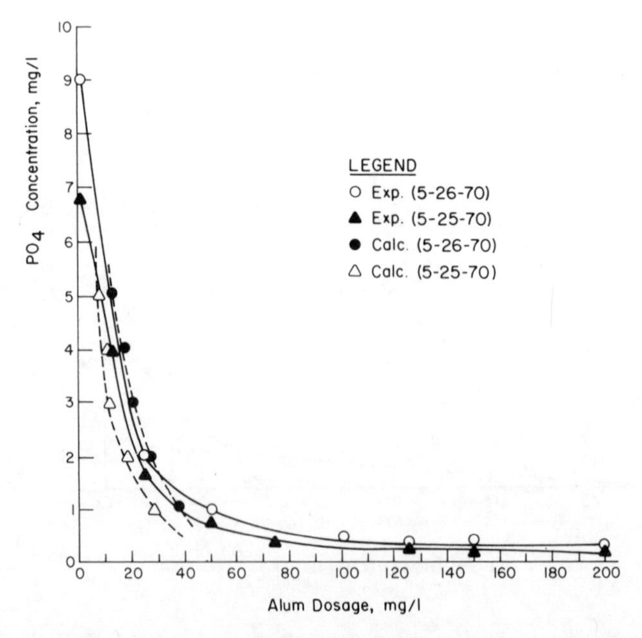

Figure 3. Phosphate concentration after treatment with alum.

Both chemicals produced residual color in the supernatant. At the optimum conditions, the iron salts produced supernatants having qualities comparable to those of alum. The settling characteristics of the chemical sludges were also comparable to those reported for alum. Because of the higher costs of iron salts and the residual iron color observed in the supernatant, no further studies were conducted to characterize the iron sludges.

Lime

Lime requirements were found to vary with the initial pH of the lagoon effluent and appeared to be a function of the ionic strength of the water. Without exception, it was necessary to adjust the pH to a value in excess of 10.5 before a significant reduction of suspended solids was observed. Figures 4 and 5 illustrate the relationship between lime dosage and pH. The optimum for effective treatment was found to be around pH 11.5. At the optimum conditions, it would be necessary to have a lime dosage of approximately 1000 mg/l to approach the efficiency of systems using either alum or iron salts. In addition, pH adjustment through recarbonation or other means must be practiced prior to the final discharge of the effluent into the receiving waters. This, of course, will increase both the capital and operating costs associated with this system.

The use of lime produced ammonia removals superior to those obtained from either alum or iron salts. This could be attributed to ammonia stripping resulting at the high pH.

The lime sludges exhibited excellent settling and thickening characteristics. The solids content ranged from 1.0 to 2.5% by weight. The sludge exhibited the properties of hindered settling and, like alum sludge, was highly compressible. The filter leaf test indicated that lime sludges would be amenable to vacuum filtration providing that chemical conditioning were practiced.

Polyelectrolytes

Although nonionic, cationic and anionic polyelectrolytes were evaluated for their ability to aggregate the algae cells, either most proved to be ineffective or very large doses were required. Poor results were obtained with the nonionic and anionic polyelectrolytes that were listed earlier in this report. They did not produce a floc in concentrations of less than 100 mg/l. It is possible, however, that by controlling the pH with appropriate chemicals, they may find application. It should be pointed out that there are a number of these polymers on the market and others may yield satisfactory results.

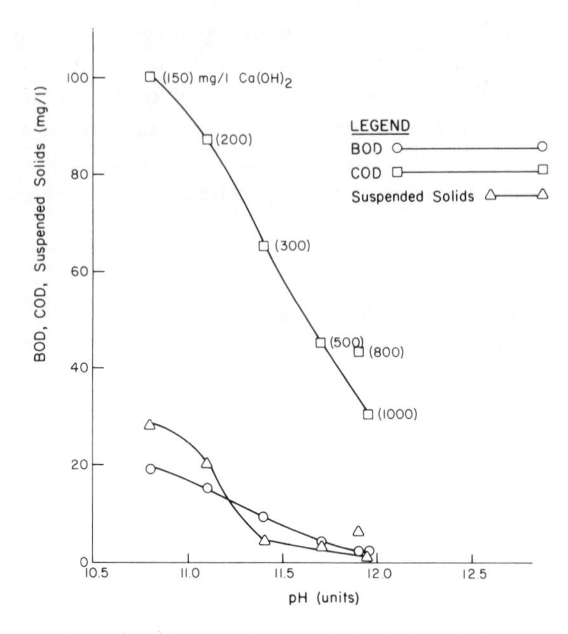

Figure 4. Removal of BOD, COD and suspended solids with lime. Lime dosage as noted ( ) mg/l, initial BOD 30 mg/l, initial COD 142 mg/l, initial SS 52 mg/l.

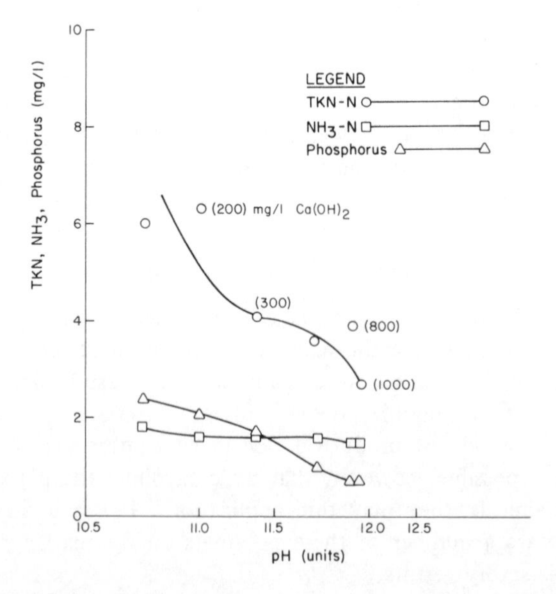

Figure 5. Removal of TKN, ammonia and phosphorus with lime. Lime dosage as noted ( ) mg/l, initial TKN 9.7 mg/l, ammonia 2.1 mg/l, phosphorus 5.0 mg/l.

Cationic polymers produced flocs to some degree. All of the cationics reacted with the algae and gave the desired suspended solids removal. However, they were considered to be inadequate without other chemical aids. The best results were obtained from the polymer Cat-Floc. The removal of suspended solids from a waste stabilization pond effluent with this polymer is shown in Figure 6. The optimum Cat-Floc dosage was found to be between 26 and 30 mg/l and resulted in a decrease in suspended solids from 35 mg/l to 14 mg/l. This is a 60% removal and would normally be considered unacceptable. Furthermore, the cost of the polymer would make this treatment system very expensive to operate. The floc produced was light, fluffy and very easy to shear and its settling properties were poor.

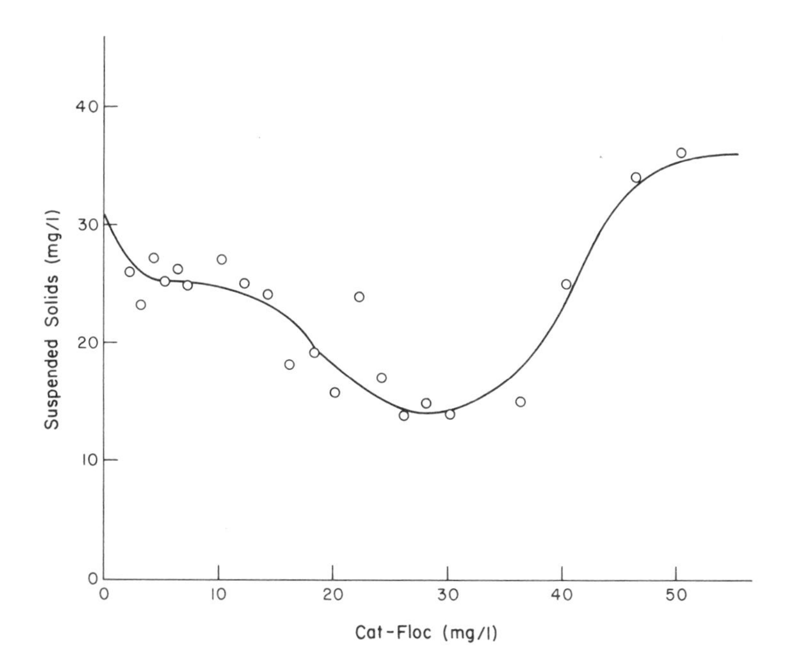

Figure 6. Removal of suspended solids with Cat-Floc; pH 9.0 units, alkalinity 130 mg/l, temperature 13°C, initial SS 35 mg/l.

CONCLUSIONS

From the data collected in this investigation, the following conclusions can be made:

1. Chemical coagulation proved to be an effective posttreatment process for algal removal and for improving the quality of effluents from stabilization ponds.

2. Of the coagulants tested, alum was found to be most favorable. The optimum dosage for best removal of the parameters studied was found to be in the range of 75 to 100 mg/l.

3. A mathematical model (Equation 3) can be used to predict the required dosage of alum to produce an effluent phosphate concentration in the range of 1 to 5 mg/l.

While the flocculation-sedimentation technique has been found to remove algae efficiently from stabilization pond effluent, no study was made of sludge disposal. It is anticipated that a pilot-plant study involving this method of treatment will be made with provisions for recycling the sludge to the stabilization pond.

REFERENCES

1. *Standard Methods for the Examination of Water and Wastewater*, 13th ed. (Washington, D.C.: American Public Health Association, 1971).
2. Oswald, W. J., A. Meron and M. D. Zabat. In: *Second International Symposium for Waste Treatment Lagoons* (Kansas City, Missouri: Missouri Basin Engineering Health Council and Federal Water Quality Administration, 1970), pp. 186-194.
3. Environmental Protection Agency. "Upgrading Lagoons," EPA Technology Transfer (1973).
4. Pavoni, J. L. In: *Water Resources Research Catalog*, Vol. 9, Number 5.0699 (1974).
5. McGarry, M. G. *J. Water Poll. Control Fed.* 42:R191 (1970).
6. Shindala, A., and J. W. Stewart. *Water Sew. Works* 118:100 (1971).
7. Parker, D. S., *et al.* "Algae Removal Improves Pond Effluent," *Water Waste Engineering* 10:26 (1973).
8. O.Brien, W. J., *et al.* *Water Sew.*, *Works* 120:66 (1973).
9. McGhee, T. J., and R. K. Patterson. *Water Sew. Works* 121:82 (1974).
10. McKinney, R. E., J. W. Dornbush and J. W. Venner. "Waste Treatment Lagoons—State of the Art," Missouri Basin Engineering Health Council, EPA WPCRS, 17090 EHX (July 1971).

CHAPTER 8

EVALUATION OF ORGANICS ADSORPTION BY ACTIVATED CARBON

Robert S. Reimers, Andrew J. Englande, Jr., D. Blair Leftwich, Chong-i Paul Lo and Robert J. Kainz*

Department of Environmental Health Sciences
School of Public Health and Tropical Medicine
Tulane University
New Orleans, Louisiana

INTRODUCTION

Objective

Consideration of activated carbon as a wastewater treatment alternative has increased with the enactment of the 1972 Federal Water Pollution Control Act (PL 92-500) and recent concern over trace organic contamination of drinking water supplies. Hence, added impetus has been placed on understanding the potentials and limitations of carbon in adsorbing trace organics. Historically, the prediction of the effectiveness of activated carbon has been empirical. This chapter presents details on a conceptual approach for assessing the utility of activated carbon in conjuction with pertinent wastewater characteristics.

The application of organic waste characteristics to predict affinity for carbon adsorption would be of great value in the determination of waste treatment alternatives. The objective of this investigation was to develop an evaluation technique which could be readily applied to define the relative affinity of selected aqueous organics for activated carbon.

---

*R. J. Kainz, U.S. Army Medical Service Corps, Health and Environmental Activity, Fort Knox, Kentucky.

Background

The application of activated carbon to purify potable water was first reported by Lipscombe[1] in 1863. The first actual study concerned with the effect of molecular structure and solvent pH on adsorption effectiveness took place in England in 1929 when Phelps and Peters[2] observed that the adsorption of the lower fatty acids and simple aliphatic amines depended on the pH of the aqueous solution and the ionic dissociation of acids and bases. They concluded that adsorption occurs only with the undissociated molecules and that the adsorption of organics in aqueous solutions is similar to gases. During the early 1940s, Cheldin and Williams[3] made two important observations: (1) adsorption of 33 amino acids, vitamins, and related substances by activated carbon (Darco 6-60) fit Freundlich adsorption isotherms, and (2) the presence and position of polar groups and the absence of aromatic nuclei are important factors in the aqueous adsorption of organics by activated carbon. Their studies were centered around possible use of carbon for analytical purposes. Organic concentrations, however, were too high to correspond to adsorption situations for water or waste treatment applications.

More recently, with the increased interest in pollution abatement, research efforts directed towards organic adsorption onto carbon were renewed by Weber and Morris[4,5] and Getzen and Ward.[6,7] These investigators observed the steric effect of molecular branching on adsorption, the effects of adsorbate molecular weight on equilibrium capacity and kinetics, and the quantification of ionization constants with respect to aromatic organics and pH. Results included a reduction of capacity with a decrease in molecular weight of organic compounds, an increase in molecular branching and a decrease in the free acid or free base form. The investigation by Weber and Morris also noted an increase in the rate of organic uptake with a decrease in molecular weight of free organics.

Zucherman and Molof[8] also considered the effect of molecular weight of organic constituents in biologically treated effluents. They observed good adsorption with organics of molecular weights around 400. These results, which were determined by employing actual wastestreams, contradicted those observed during the adsorption of synthetic organics.[9] DeWalle and Chian[10] investigated the reasons for significant leakage of organic matter from carbon columns treating secondary effluents. The authors concluded that the major organic fraction adsorbed onto activated carbon was in the weight range of 100 to 10,000. Their results also indicated that the adsorbable organics were fulvic acid-like material and the least adsorbed fractions were low-molecular-weight (less than 100). polar organic molecules and soluble high weight organics (greater than

50,000). Previous work by these investigators[11] showed a decrease in the small polar molecules with an increase in the fulvic and humic-like material during the course of biological degradation. Based on these results, the authors hypothesized that sorption capacity increases with increasing wastewater biodegradation. These results do not agree with those of Weber *et al.*[12] who observed better adsorption with chemically pretreated primary effluent as compared to biologically degraded secondary effluents.

Giust *et al.*[13] investigated the adsorption of 93 petrochemicals by activated carbon. Results of their study indicate increased adsorption with increased molecular weight and decreased molecular polarity, solubility, and branching. Aromatics were observed to exhibit the greatest affinity for adsorption due to low solubilities and/or hydrogen bonding to aromatic surfaces. Molecular functional groups were found to have a substantial effect on adsorbability, especially with respect to solubility and polarity. For straight chained organics the affinity to carbon was found to be in the following order for petrochemicals

Undissociated organic acids > aldehydes > esters > ketones > alcohols > glycols

For compounds greater than four carbons, the alcohol adsorbability is greater than that of esters. In addition, they reported a decrease in adsorption with increasing total surface acidity of the carbon and an increased adsorptive capacity for synthetic organics.

Due to the need for developing a simple analysis, experimental work was undertaken to quantify the affinity of selected organics for activated carbon with specific reference to carbon chain length, aromaticity and functional groups.[14]

Aromatic acids > aliphatic acids and alcohols > aniline > aliphatic amines > phenol

An important observation was that aromatics do not always sorb better than aliphatics, since the affinity for carbon was greatly affected by substituent groups on aromatic rings and by functional groups associated with aliphatics. These findings confirmed literature reports with respect to solubility and acid dissociation of synthetic organics. An extension of the information base was realized by the consideration of acidic and basic aliphatic polymers. Further, a decrease was observed in the adsorptive capacity of at least 1000-fold when pH fluctuations caused charging of the polymer functional groups. The findings recommended a study of the waste characteristics which may be used in the development of a feasibility analysis for the affinity of aqueous organics to carbon. The proposed characteristics include: total organic carbon, total acidity (pH range of 4-7) minus inorganic carbon acidity, organic nitrogen, and aromaticity.

In summary, many factors influence the adsorption of organics from aqueous solutions onto activated carbon; several of these factors are summarized in Table I.[15]

Table I. Influence of Molecular Structure and Other Factors on Adsorbability of Organics by Activated Carbon

1. Increasing solubility decreases its adsorbability.
2. Branched chains are usually more adsorbable than straight chain compounds. Increasing length of the chair decreases solubility.
3. Substituent groups affect adsorbability:

Substituent Group	*Effects*
Hydroxyl	Generally reduces adsorbability depending upon structure.
Amino	Similar to that of hydroxyl but somewhat greater. Many amino acids are not adsorbed to any appreciable extent.
Carbonyl	Varies with molecule; glyoxylic acids are more adsorbable than acetic acid but a similar increase does not occur when higher-molecular-weight fatty acids are introduced.
Double Bonds	Variable as with carbonyl.
Halogens	Variable.
Sulfonic	Usually decreases adsorbability.
Nitro	Often increases adsorbability.
Aromatic Rings	Greatly increases adsorbability.

4. Generally, strongly ionized solutes are not as adsorbable as weakly ionized solutes; *i.e.*, undissociated molecules are in general preferentially adsorbed.
5. The extend of adsorption depends on the ability of hydrolysis to form an adsorbable acid or base.
6. Unless the screening action of the carbon pores intervenes, large molecules are more sorbable than small molecules of similar chemical nature. This is attributed to more solute carbon chemical bonds being formed, making desorption more difficult.
7. Molecules with low polarity are more sorbable than highly polar solutes.

RESEARCH METHODOLOGY

The research was conducted in three phases: experimental, synthesis and application. The experimental phase included the completion of necessary synthetic sorption experiments to fill existing data gaps. Appropriate correlations were developed and applied to results obtained from various field investigations for comparison with accumulated laboratory data.

Selection of Compounds for Study

The following criteria were used to select the specific compounds to be tested:

1. *Effect of Functional Group.* Carboxyl, hydroxyl and amine groups were investigated. In order to demonstrate the effect of the functional group, three, five, seven and nine carbon aliphatic hydrocarbon chains were selected. Longer chain lengths would suppress the effect of the functional groups on adsorption since these compounds are very insoluble.

2. *Effect of Molecular Size.* Aromatic rings with acidic, hydroxyl and amine functional groups attached were investigated. Bulkiness associated with the aromatic ring should enhance adsorptivity. An increase in adsorbability of the aliphatic compounds with increasing carbon content was anticipated and organics of aliphatic and aromatic organics of similar weight were compared.

3. *Effect of pH.* The adsorption of each compound studied was evaluated at two pH values; one above and the other below that associated with the pK_a. One set of isotherm tests exhibited adsorption in the uncharged state, while the other demonstrated adsorption of the dissociated (charged) species. Acidic and basic polymers were also studied to demonstrate the significance of pH.

The waste characteristics analyzed and correlated were total organic carbon, inorganic carbon, organic nitrogen, bicarbonate alkalinity (between pH 8.3 and 4.5) and aromaticity. The analysis for bicarbonate alkalinity and organic nitrogen followed standard procedures.[16] Aromaticity was estimated by absorbance measurements with ultraviolet light at 253.7 nm.

Procedures

Freundlich adsorption isotherms were determined using a Westvaco Nuchar, a widely used carbon type. The carbon was soaked in distilled water for 24 hr, dried at 103°C for 24 hr, and ground to 325 mesh. This mesh size corresponds to a particle diameter of 45 microns.

Carbon dosages for the isothermal experiments normally ranged from 0.1 to 1.2 g. Each weighed sample was carefully placed into a 250-ml ground-glass stoppered erlenmeyer flask. This was followed by the addition by pipet of 100-ml aliquots of the organic stock solution to each flask. At least three carbon dosages were run for each compound evaluated at each adjusted pH level. Stock solutions of the compounds studied were prepared to levels near 100 mg/l.

These prepared samples were hand-shaken briefly to insure complete wetting of powdered carbon present in each flask. Subsequently, each flask was placed on a shaker water bath and moderate agitation was maintained for 4 hr at 25°C. Small samples of about 10 ml of the homogeneous solutions were taken at 24 hr and 48 hr for verification of equilibrium.

Samples were then vacuum filtered through 0.45-μ Millipore filters for activated carbon separation. A Dohrmann model DC-50 Total Organic Carbon analyzer was used to determine the concentration of compounds present in each solution, initially and after equilibrium.[17]

RESULTS AND DISCUSSION

Effect of Acid Groups

Isothermal plots were developed for propionic, valeric, heptanoic, nonanoic and polymethacrylic acids as shown in Figure 1. The slopes and intercepts from the figure are presented in Table II and indicate increased adsorption over the entire concentration range. The ratios of

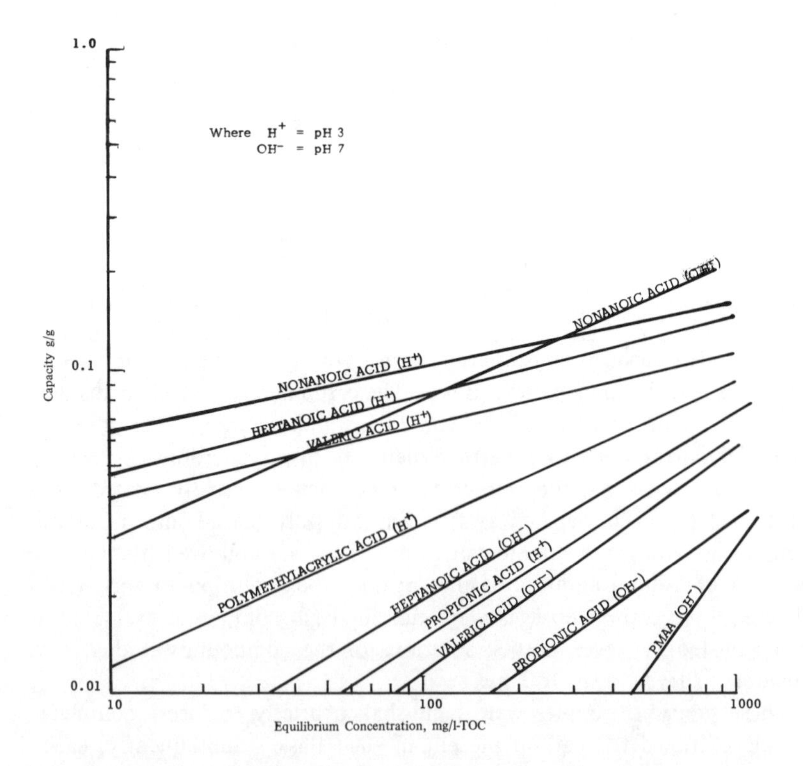

Figure 1. Freundlich adsorption isotherms for carboxyl compounds.

Table II. Extracted Data–Freundlich Isotherms

Compound	pH	pK_a	K (x 10^4)	n
		Organics Containing Carboxyl Groups		
Propionic Acid (H^+)	2.87	4.87	7.5000	1.563
Propionic Acid (OH^-)	6.87	4.87	2.9000	1.447
Valeric Acid (H^+)	2.80	4.80	243.0000	4.545
Valeric Acid (OH^-)	6.80	4.80	4.0000	1.389
Heptanoic Acid (H^+)	2.79	4.79	266.2000	3.950
Heptanoic Acid (OH^-)	6.79	4.79	38.0000	2.813
Nonanoic Acid (H^+)	2.75	4.75	263.3000	2.516
Nonanoic Acid (OH^-)	6.75	4.75	106.7000	2.249
Polymethacrylic Acid (H^+)	3.00		45.6000	2.315
Polymethacrylic Acid (OH^-)	7.00		0.0590	0.819
		Aliphatic Organics Containing Amine Groups		
Pentylamine (OH^-)	12.78	10.78	85.0000	2.800
Polyethylenimine (H^+)	6.00	8.00	0.3000	0.817
Polyethlenimine (OH^-)	10.00	8.00	463.4000	5.917
Nonylamine (H^+)	8.66	10.66	360.0000	0.220
Heptylamine (H^+)	8.64	10.64	198.0000	0.210
		Organics Containing Hydroxyl Groups		
Propanol (H^+)	3.00	18.00	3.5×10^{-8}	0.283
Propanol (OH^-)	10.00	18.00	1.2×10^{-7}	0.299
Pentanol (H^+)	3.00	18.00	261.0000	4.444
Pentanol (OH^-)	10.00	18.00	13.2000	1.362
		Aromatic Organics		
Benzoic Acid (H^+)	2.17	4.17	701.5000	5.128
Phenol (H^+)	7.90	9.90	34.6000	1.567
Pyridine (H^+)	3.23	5.23	2.8×10^{-5}	0.379
Aniline (H^+)	2.00	4.00	7.7000	1.188
Benzoic Acid (OH^-)	6.17	4.17	0.0065	0.536
Phenol (OH^-)	11.90	9.90	50.9000	1.779
Pyridine (OH^-)	7.23	5.23	47.3000	2.088
Aniline (OH^-)	6.00	4.00	377.0000	3.302

organic acidity (mg/l as $CaCO_3$) to organic carbon at equilibrium concentrations of 10, 100 and 1000 mg/l of total organic carbon were determined from the Freundlich isotherms and are presented in Table III. Figures 2 and 3 were developed from Table III and elucidate the influence of organic acidity to total organic carbon ratio on capacity. As the ratio decreased, the capacity increased due to the reduced effect of the uncharged or charged carboxyl group. Also, the uncharging of the carboxyl group under an acidic environment appears to increase the capacity of carbon for organics by one order of magnitude from the

Table III. The Capacity of Carbon for Organic Acids vs Ratio of Organic Acidity to Total Organic Carbon

Compound	Ratio of Acidity to Total Organic Carbon	Capacity[a] at TOC Equilibrium Concentrations of:					
		10 mg/l		100 mg/l		1000 mg/l	
		X	log X	X	log X	X	log X
Propionic Acid (H^+)	1.39	0.003	-2.48	0.015	-1.84	0.064	-1.19
Valeric Acid (H^+)	1.83	0.04	-1.40	0.068	-1.17	0.080	-1.09
Polymethacrylic Acid (H^+)	0.96	0.013	-1.90	0.033	-1.48	0.090	-1.05
Propionic Acid (OH^-)	1.39	0.002	-2.82	0.008	-2.12	0.035	-1.46
Valeric Acid (OH^-)	0.83	0.002	-2.64	0.012	-1.94	0.055	-1.26
Polymethacrylic Acid (OH^-)	0.96	0.000085	-4.07	0.001	-2.85	0.028	-1.56
Heptanoic Acid (H^+)	0.60	0.027	-1.57	0.048	-1.32	0.086	-1.07
Nonanoic Acid (H^+)	0.46	0.027	-1.57	0.065	-1.18	0.105	-1.98
Heptanoic Acid (OH^-)	0.60	0.004	-2.42	0.009	-2.06	0.020	-1.71
Nonanoic Acid (OH^-)	0.46	0.017	-1.78	0.030	-1.57	0.083	-1.08

[a]Capacity in grams of TOC per gram of Carbon, X.

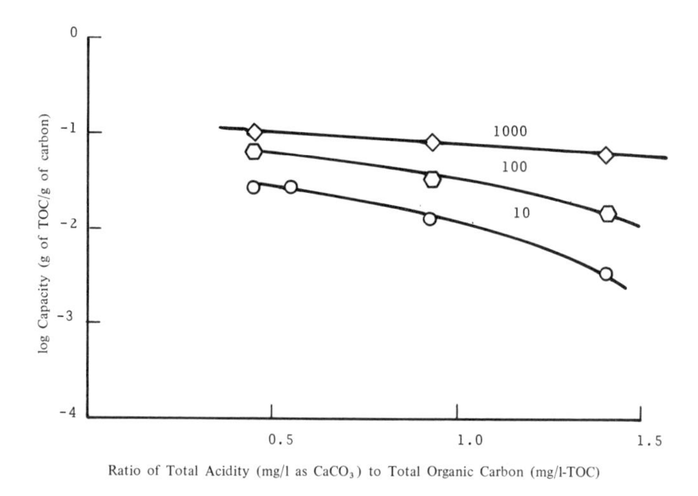

Figure 2. Capacity as a function of the ratio of the acidity; concentrations of 10, 100 and 1000 mg/l-TOC at pH $<$ 4.

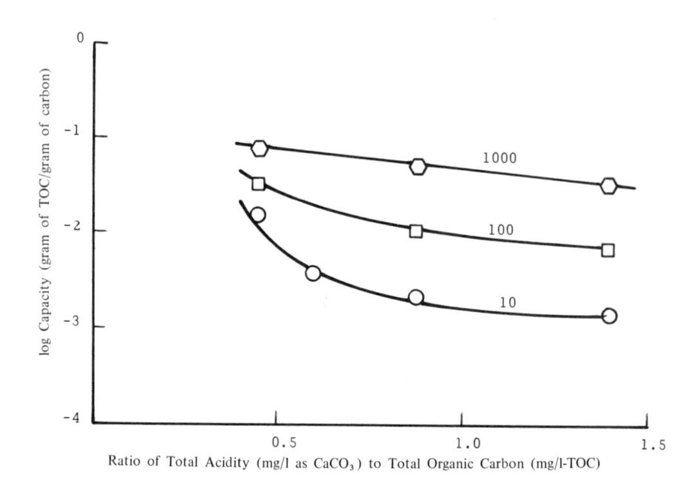

Figure 3. Capacity as a function of the ratio of the acidity; concentrations of 10, 100 and 1000 mg/l-TOC at pH $>$ 6.5.

alkaline environment. These relationships were developed with only alkyl organic acids and do not consider other possible influences from other organic functional groups and modified aspects of molecular structure.

The effect of dissociation on the solubility of aliphatic organic acids is shown in Figure 1. Under neutral or basic conditions, Freundlich isotherms were plotted for the charged species of nonanoic acid and heptanoic acid; under acidic conditions, the isotherms for the uncharged valeric acid and propionic acid were plotted. This figure also illustrates that the uncharging of the carboxylic group (by decreasing the pH) affects sorbability to the same degree as adding four carbons to its chain.

Effect of Amine Groups

The ratio of organic nitrogen to total organic carbon was correlated with the capacity of activated carbon for organic amines. Figure 4 was developed from isothermal tests for pentylamine, heptylamine, nonylamine, and polyethylenimine. The slopes and intercepts from these figures are presented in Table II for each of these compounds, indicating increased adsorption over the entire concentration range.

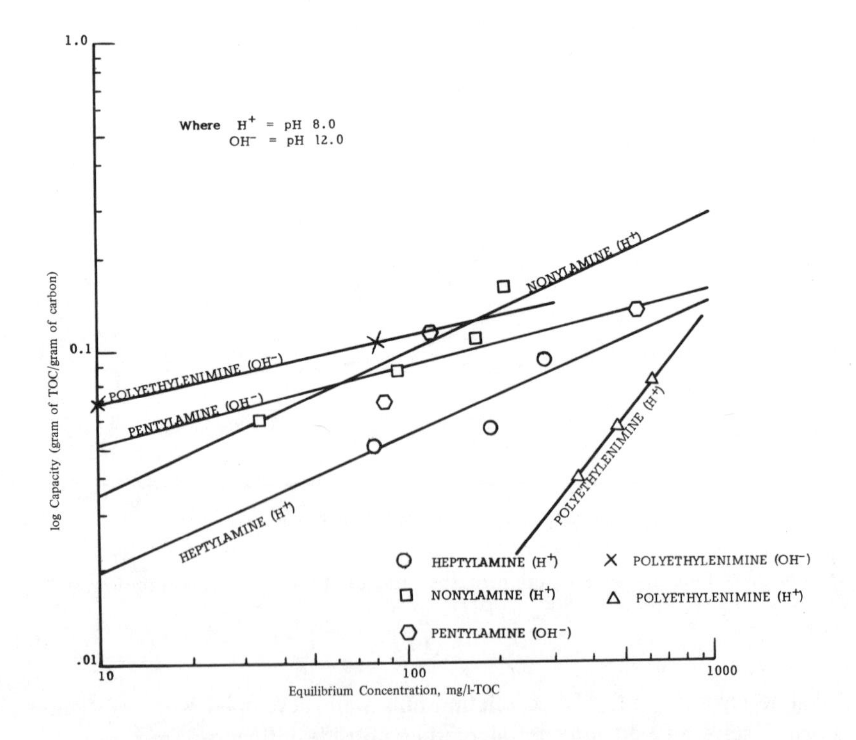

Figure 4. Freundlich isotherms for aliphatic amines.

Table IV was derived from data based on Figure 4. The ratios of organic nitrogen (mg/l-N) to total organic carbon (mg/l-TOC) were calculated and the capacity in grams of total organic carbon at equilibrium concentrations of 10, 100 and 1000 mg of total organic carbon were determined. Employing Table IV as a base, Figures 5 and 6 were constructed to elucidate the effect of the ratio of total organic nitrogen to total organic carbon on the capacity of carbon under neutral or acidic and basic conditions. As the ratio decreased, the capacity increased due to the reduced effect of the uncharged or charged functional amine group. The uncharging of the aliphatic amine under basic conditions appears to increase the capacity by one order of magnitude as compared to the charged species under neutral conditions. Figures 5 and 6 were calculated taking into account alkyl amines; influences resulting from other organic functional groups or various aspects of molecular structure were not considered.

The effect of dissociation on the solubility of aliphatic amines is shown in Figure 4. Freundlich isotherms are plotted for the charged species under neutral conditions of nonylamine and heptylamine, and under basic conditions, the isotherm for the uncharged pentylamine is plotted. Figure 4 illustrates that the unchargin of the amine by increasing the pH affects sorbability to the same degree as adding four carbons to its alkyl chain.

Effect of Alkyl Hydroxyl Compounds

Since the alkyl hydroxyl group is weakly ionized and not measurable for typical waste characterizations, no attempts have been made to develop correlations for organics containing the hydroxyl groups. Isotherms were analyzed, and, as expected, very poor adsorption capacities were observed for highly soluble propanol (see Table II and Figure 7). Pentanol having a reduced solubility did show substantial fluctuations in capacity with varying pH at concentrations less than 300 mg/l-TOC. The adsorption capacity was observed to increase by a factor of 20 concomitant with a corresponding rise in pH from 3 to 10. Only slight variations in the adsorption of phenol was experienced with varying pH. The lower capacity of carbon for the aromatic compound when compared to aliphatic compounds was probably due in part to the large bulkiness of the aromatic ring which in relation to the size of the aliphatic hydroxyl organic reduced the effective adsorption surface area.

Effect of Aromatic Organics

The relative adsorption of aromatic compounds with respect to functional group in both the charged and uncharged state is illustrated

Table IV. The Capacity of Carbon for Organic Nitrogen Compounds for Different Ratios of Organic Nitrogen to Total Organic Carbon

Specific Organic	Ratio of Organic Nitrogen to TOC	Capacity[a] at TOC Equilibrium Concentrations of:					
		10 mg/l		100 mg/l		1000 mg/l	
		X	log X	X	log X	X	log X
Pentylamine (H^+)	0.23	–	–	–	–	–	–
Heptylamine (H^+)	0.17	0.020	-1.71	0.053	-1.28	0.150	-0.82
Nonylamine (H^+)	0.13	0.036	-1.44	0.100	-1.00	0.250	-0.60
Polyethylenimine (H^+)	0.58	0.0005	-3.30	0.009	-2.06	0.114	-0.94
Pentylamine (OH^-)	0.23	0.048	-1.32	0.089	-1.05	0.140	-0.85
Polyethylenimine (OH^-)	0.58	0.069	-1.16	0.110	-0.96	0.160	-0.80
Pyridine (OH^-)	0.23	0.014	-1.85	0.043	-1.37	0.113	-0.95
Aniline (H^+)	0.19	0.006	-2.22	0.021	-1.68	0.225	-0.65
Aniline (OH^-)	0.19	0.069	-1.16	0.125	-0.90	0.220	-0.66
Pyridine (H^+)	0.23	–	–	0.0005	-3.30	0.170	-0.77

[a]Capacity in grams of TOC per gram of Carbon, X.

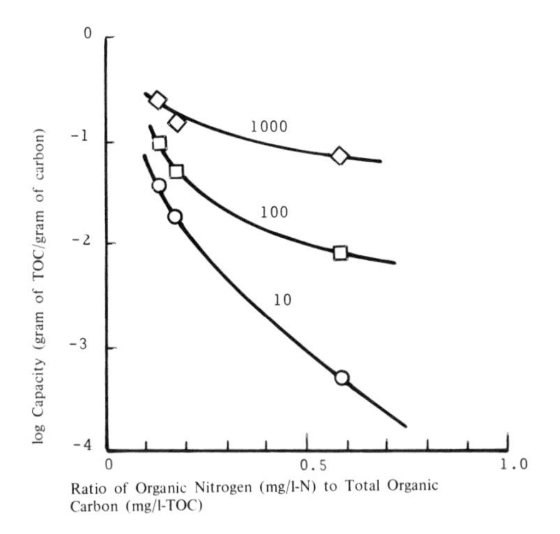

Figure 5. Capacity as a function of ratio of organic nitrogen to total organic carbon; concentrations of 10, 100 and 1000 mg/l-TOC at pH < 8.6 for aliphatic amines.

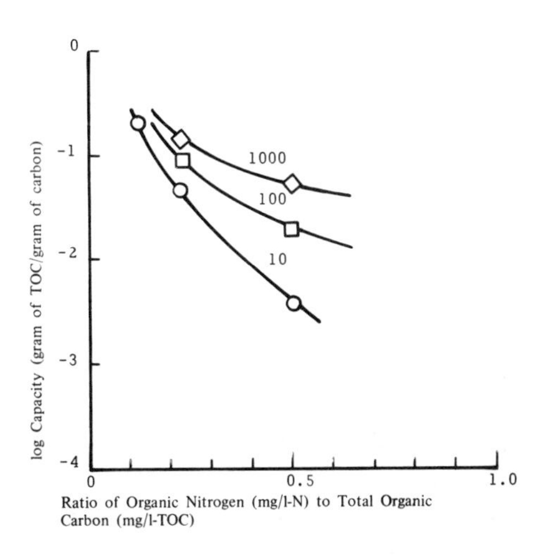

Figure 6. Capacity as a function of ratio of organic nitrogen to total organic carbon; concentrations of 10, 100 and 1000 mg/l-TOC at pH > 12 for aliphatic amines.

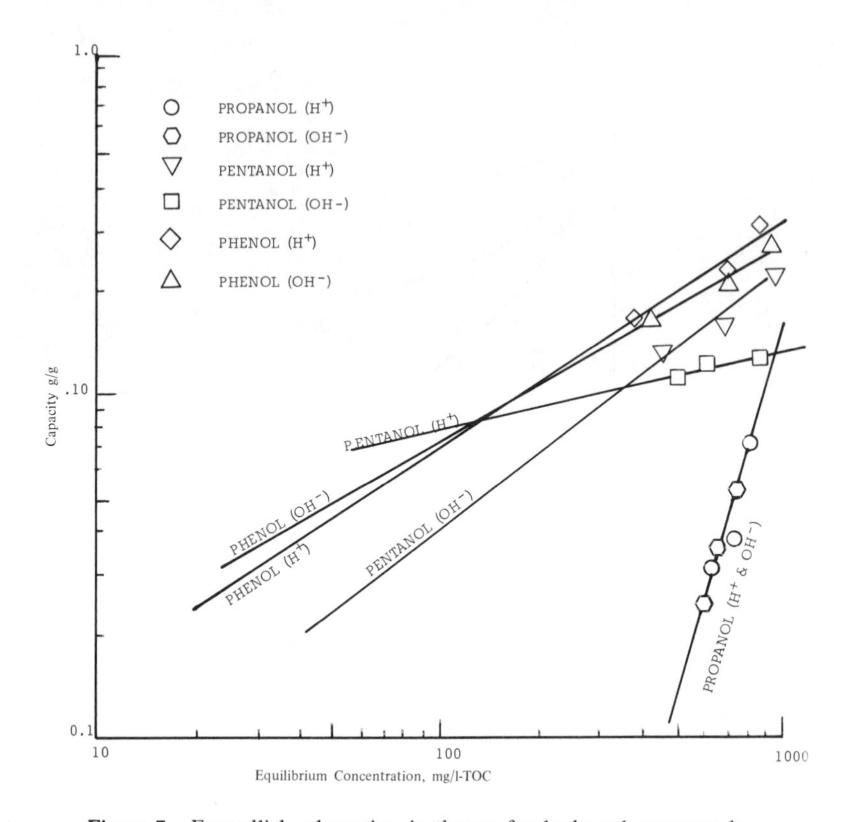

Figure 7. Freundlich adsorption isotherms for hydroxyl compounds.

in Table II and Figure 8. The relative adsorption was shown to follow the sequence

> Benzoic acid (H^+) > aniline (OH^-) > phenol (H^+ + OH^-) > pyridine pyridine (OH^-) > aniline (H^+) > benzoic acid (OH^-) > pyridine (H^+)

It was observed that the adsorption of aromatic organics depended on the substituted group's interaction with the aromatic ring and the charged or uncharged state of the organic compound.

Comparison of aromatic and aliphatic organic compounds at approximately the same molecular weight yielded unexpected results. Figure 9 shows the comparison between benzoic acid (123 g/mol) and heptanoic acid (130 g/mol). Benzoic acid is adsorbed by almost an order of magnitude more than heptanoic acid in an uncharged state (pH 4), but under neutral to alkaline conditions (pH 6) both acids are adsorbed to a considerably lesser extent with heptanoic acid having more affinity for

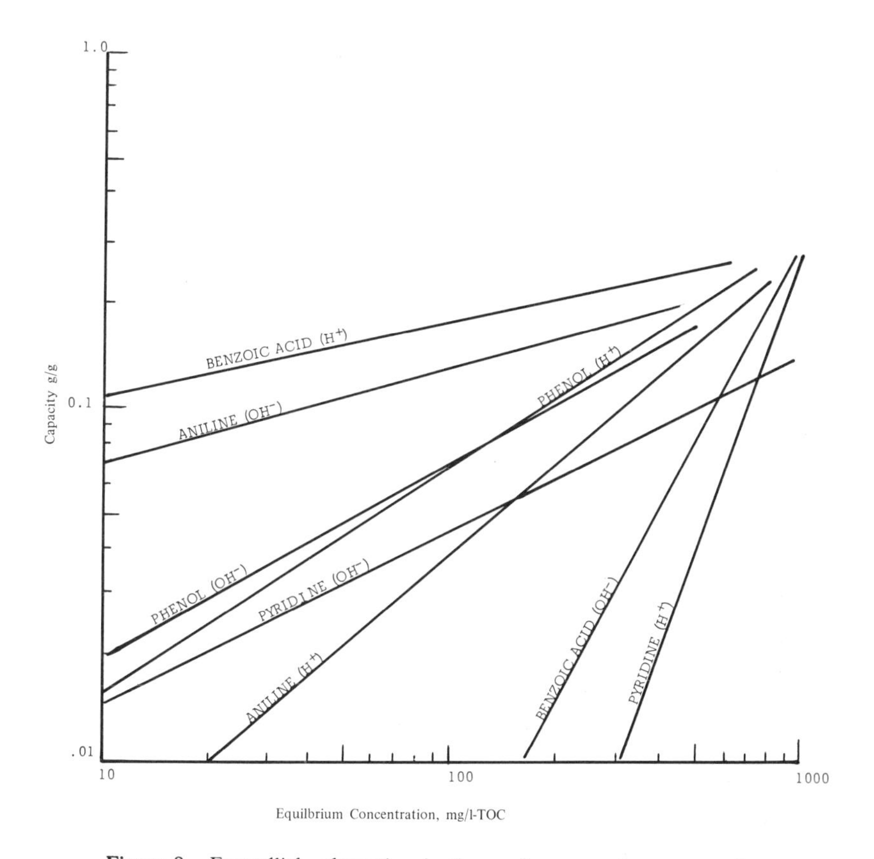

Figure 8. Freundlich adsorption isotherms for aromatic compounds.

the carbon. This may be due to the reduced polarity and stability of the benzene ring of the uncharged benzoic acid. When investigating the difference between hydroxyl aliphatic and aromatic organics there appeared to be a difference between the aromatic and alkyl hydroxyls as shown in Figure 9. The ionization of the alkyl hydroxyl groups, on reducing the pH from 10 to 3, increased the adsorptive capacity of the alkyl organic to above that of phenol. When considering aliphatic and alkyl amines, there appeared to be little difference in sorption. The adsorbability of both aromatic and aliphatic organics, therefore, depends on the functional group of the aliphatic compound or the substituted group of the aromatic compound. When these groups are related to each other, the aromatic compounds do not always adsorb better than their respective aliphatic organics.

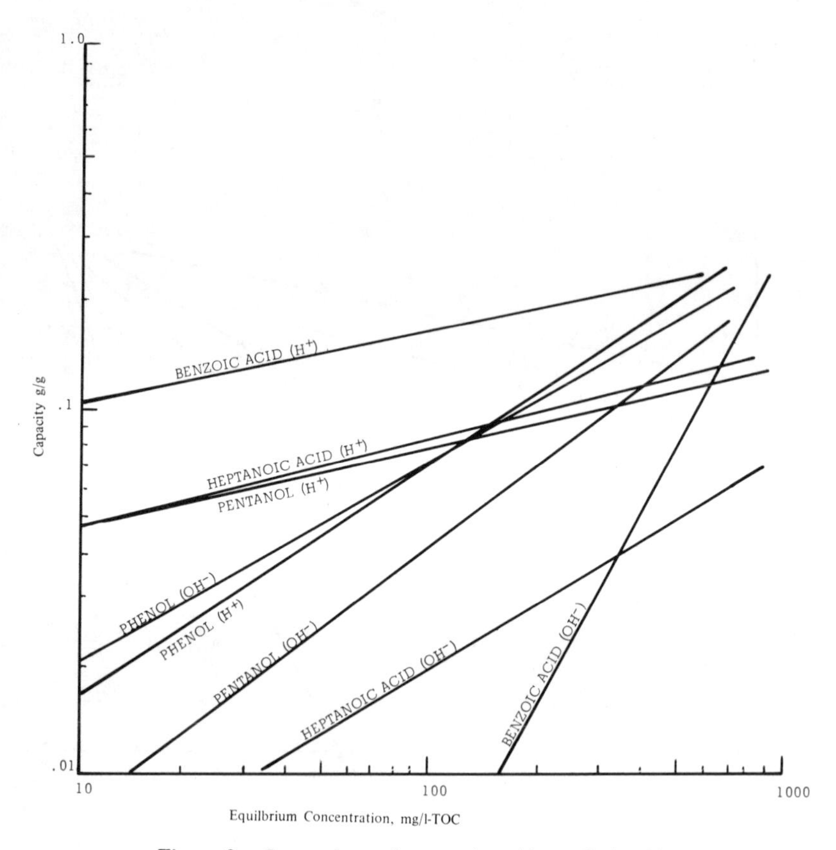

Figure 9. Comparison of aromatic acid to alkyl acids.

APPLICATIONS AND LIMITATIONS

A number of questions relative to the suitability of design alternatives could be resolved by utilizing general waste characteristics in evaluating sorbtion potentials. First, there has been a great deal of conflict over the use of carbon to remove waste organics found in primary, secondary and tertiary effluents. Second, biological treatment causes great fluctuations in the characteristics of the resulting effluent. Thus, by utilizing the proposed criteria, the optimal application of carbon for organics adsorption could be ascertained; the affect on adsorbability of specific coagulants to be used for chemical treatment prior to carbon adsorption may also be determined.

For instance, the South Lake Tahoe facility consists of a chemical mix, flocculation and settling unit, ammonia stripping tower, recarbonation and settling basin, mixed media filters, carbon adsorption and final chlorination. The water quality at various points in the process following 18 months of operation is presented in Table V.[18,19] The ratio of organic nitrogen to total organic carbon was equal to 0.22-0.25, under basic conditions at pH 8, with an equilibrium concentration of 1 to 6 mg/l. Comparing these data to Figures 5 and 6 for the organic nitrogen total organic carbon waste, there appear to be good conditions for removal of organics by carbon adsorption.

In Southwest Africa, the Windhoek Plant was designed for wastewater reuse following physical-chemical polishing of secondary effluent. The influent to the carbon column was similar in quality to that of South Tahoe Plant, and the organic nitrogen was removed predictably, as illustrated in Table V. Other field observations on waste characteristics could not be correlated with organics removal due to the lack of data on chemical oxygen demand and total organic carbon.

For this study, a chemical waste was characterized and batch isotherms were run; the waste characteristics are shown in Table VI. The organic nitrogen/TOC ratio indicates that the waste is relatively sorbable, but inspection of the organic acidity/TOC ratio reveals an expectedly low carbon capacity. Table VI confirms the low affinity of these waste organics for activated carbon.

There are several situations in which the use of the proposed waste characteristics for predicting sorptivity may not be applicable. The first situation is with respect to low-molecular-weight organics. The proposed waste characteristics take into account total organic carbon but do not differentiate between low- and high-molecular-weight organics. The

Table V. Water Quality Before and After Carbon Treatment

Determination	South Lake Tahoe Plant		Windhoek Plant	
	Before	After	Before	After
Organic N (mg/l)	2-4	1-2	0.7	Nil
Total Organic Carbon (mg/l)	8-18	1-6	-	-
BOD (mg/l)	1	1	1	0.3
COD (mg/l)	20-60	1-23	-	-
ABS (mg/l)	0.4-2.9	0.01-0.5	4	0.7
pH	-	-	8.0	8.0
Color (units)	10-30	5	-	-

Table VI. Waste Stream Adsorption Isotherm Data and Waste Characteristics

1. *Waste Characteristics*

pH	=	13.5
Organic N	=	70.0 mg/l-N
COD	=	1330 mg/l-COD = 500 mg/l-TOC
UV	=	1000 absorbance initiates 100 to 200 mg/l of aromatics
Bicarbonate Alk-IOC	=	1400 mg/l-400 = 1000 mg/l as $CaCO_3$

2. *Isotherm Data*

Carbon Dosage (mg/l)	*pH*	*Initial Concentration (mg/l-COD)*	*Equilibrium Concentration (mg/l-COD)*	*Capacity of Carbon (g COD/g carbon)*
2500	7	1290	999	0.117
2500	7	709	630	0.032
1250	7	1160	992	0.134
2500	11	1330	1000	0.132
2500	11	1010	827	0.072
1250	11	1010	913	0.080

Coefficients and Regression Constants

pH 11	*pH 7*
$k = 1.9 \times 10^{-1}$	$k = 20 \times 10^{-3}$
i/n = 307	i/n = 0.95
r = 0.896	r = 0.939

second is with respect to bioactivity on activated carbon as indicated by high BOD/TOC ratios. In this situation, the means of removal is not totally physical-chemical but also biological absorption and degradation. It has been reported that the removal of low-weight organic can be accomplished by absorptive processes.[18,20-23] Initially, adsorption may be predicted by the proposed characterization analysis, but microbes may develop on the carbon and the process change to a bio-adsorption process. This has been observed in both the treatment facilities located at South Lake Tahoe[18] and Cleveland.[23] The proposed method of evaluation of organics adsorption is limited by analytical precision and accuracy of the waste charactertistics. For example, this approach would be limited to total organic carbon concentrations less than 5 mg/l-TOC, organic nitrogen less than 1 mg/l-N and organic acidity less than 8 mg/l as $CaCO_3$.

REFERENCES

1. Lipscombe, F. British Patent 2887, 1862.
2. Phelps, H. J., and R. A. Peters. *Proc. Reg. Soc.* A124:554 (1929).
3. Cheldelin, V. H., and R. J. Williams. *J. Amer. Chem. Soc.* 64:1313 (1942).
4. Weber, W. J., and J. C. Morris. *J. San. Eng. Div., ASCE* 89:31 (1963).
5. Weber, W. J., and J. C. Morris. "Equilibria and Capacities for Adsorption on Carbon" *J. San. Eng. Div., ASCE* 90:79 (1964).
6. Getzen, F. W., and T. M. Ward. *Colloid Interface Sci.* 31:441 (1969).
7. Ward, T. M., and F. W. Getzen. *Environ. Sci. Technol.* 4:64 (1970).
8. Zuckerman, M. M., and A. H. Molof. *J. Water Poll. Control Fed.* 42:437 (1970).
9. Weber, W. J. *J. Water Poll. Control Fed.* 42:456 (1970).
10. DeWalle, F. B., and E. S. K. Chain. *J. Environ. Eng. Div., ASCE* 100:1089 (1974).
11. DeWalle, F. B., and E. S. K. Chain. Paper presented at the 166th Annual Meeting American Chemical Society, Chicago, IL, August 1973.
12. Weber, W. J., C. B. Hopkins and R. Bloom. *J. Water Poll. Control Fed.* 42:82 (1970).
13. Giusti, D. M., R. A. Conway and C. T. Lawson. "Activated Carbon Adsorption of Petrochemicals," *J. Water Poll. Control Fed.* 46:947 (1974).
14. Reimers, R. S., A. J. Englande and H. B. Miles. Paper presented at the 31st annual Industrial Waste Conference, Purdue University, May, 1976.
15. Ford, D. C. Paper presented at the "Open Forum on Management of Petroleum Refinery Wastewaters," Environmental Protection Agency, American Petroleum Institute, National Petroleum Refiners Association, and University of Tulsa, January 1976.
16. *Standard Methods for the Examination of Water and Wastewater*, 14th ed. (Washington, D. C.: American Public Health Association, 1976).

17. Miles, H. B. "The Significance of Physical Chemistry Data of Organic Compounds on Their Adsorption by Activated Carbon," Special Studies for the Department of Environmental Health Sciences, Tulane University (May 1975), pp. 1-36.
18. Slechta, A. F., and G. L. Culp. *J. Water Poll. Control Fed.* 39:787 (1967).
19. DeJohn, P. B. *Chem. Eng.* (April 1975).
20. Directo, L. S., and C. L. Chen. Paper presented at the 47th Annual Water Pollution Control Federation Conference, Denver, CO, October 1974.
21. Hopkins, C. B., W. J. Weber and J. Bloom. "Granular Treatment of Raw Sewage," Water Pollution Series ORD 1705 DAL 05/70 U.S. Department of Interior, Federal Water·Quality Administration, Washington, DC (1970).
22. Guirquis, W., T. Cooper, J. Harris and A. Ungar. Paper presented at the 49th Annual Conference on Water Pollution Control Federation, Minneapolis, MN (1976).
23. Weber, W. J. "An Evaluation of the Effects of Biological Activity in Granular Activated Carbon Adsorption Systems." Unpublished document to Calgon Corporation (June 1975).

CHAPTER 9

EQUILIBRIA OF ADSORPTION OF PHENOLS BY GRANULAR ACTIVATED CARBON

John S. Zogorski

School of Public and Environmental Affairs
Indiana University
Bloomington, Indiana

Samuel D. Faust

Department of Environmental Science
Rutgers, The State University of New Jersey
New Brunswick, New Jersey

INTRODUCTION

The demand for more stringent control and protection of our water resources from pollution has mounted steadily in recent years. While some watersheds in the U.S. will not have enough freshwater of any quality, the majority of water shortages will result from an imminent shortage of clean water. The growth in the demand for clean water has placed a heavy demand on the concept of water reuse. Numerous articles have been reported recently on the engineering feasibility of water reuse systems.[1-5] A major obstacle to water reuse is the potential occurrence in water supplies of dissolved organic compounds that are resistant to conventional biological and physical treatment facilities. Complete removal or in some cases reduction of these persistent organic compounds to an acceptable concentration has become a major concern of advanced water and wastewater treatment technology.

Numerous unit operations have been evaluated for their efficiency in the removal of refractory organics. To date, the most widely utilized treatment process for removing undesirable organic matter from polluted waters has been activated carbon adsorption.

The concept of adsorption of organic materials onto carbon is not new. As early as 1883 charcoal filters were employed in the United States to remove taste and odor from a water supply.[6] Activated carbon systems are currently in use throughout the world in various water and wastewater reclamation applications. While activated carbon systems are in widespread use, very little, if any, consideration to date has been given to the physical and chemical factors involved in adsorption phenomena. Rather, the design of activated carbon units has been based solely upon such criteria as flowrate, head loss, carbon depth and contact time. These factors are important considerations in the design of adsorption units, but their scope falls short of completely describing the physics and chemistry involved in the adsorption process. Other characteristics of the adsorbate, adsorbent and solvent must be considered and evaluated if the most efficient utilization of the adsorptive nature of activated carbon is to be obtained.

The overall objective of this investigation was to evaluate the significance of various physical and chemical parameters on the adsorption of phenols by granular activated carbon. Reported herein are the results of experiments on adsorption equilibria. Other phases of the research have been reported elsewhere.[7-12]

EXPERIMENTAL METHODS AND MATERIALS

Chemicals and Analyses

Chemicals were obtained in the purest form available from commercial supply houses and were used without further purification. Stock solutions of all adsorbates were prepared with distilled water that was passed through a deionizer and an activated carbon filter. These stock solutions were periodically standardized by a Beckman model 915 total organic carbon analyzer. In general, adsorption experiments were initiated within 4 hr after the stock solutions were prepared. In a majority of the experiments, a 0.05 *M* phosphate buffer was incorporated in the stock solutions to minimize pH fluctuations. The procedure for preparing the stock solutions in the presence of the buffer was the same as described previously.

The adsorbate selected was Columbia LCK activated carbon, a product of National Carbon Company, Union Carbide Corporation. This adsorbent was chosen because of its superior resistance to fragmentation. The 12 x 28 mesh provided by the manufacturer was sieved into uniform particle sizes using U.S. Standard Sieves. A household blender was used to provide a range of particle sizes smaller than the manufactured product. The eight uniform fractions prepared were as follows: 12 x 16, 16 x 20, 20 x 30, 30 x 40, 40 x 50, 50 x 60, 60 x 80, and 80 x 100 mesh. The corresponding mean particle sizes were 1.41, 1.00, 0.71, 0.50, 0.36, 0.27, 0.21 and 0.16 mm, respectively. The average N_2-BET surface area was 1,025 m^2/g. Crushing granular Columbia LCK carbon into smaller fragments increased the N_2-BET surface area by about 6% (60 m^2/g).

Considerable care was taken in this investigation to remove as much contamination as possible from the adsorbent before experiments were conducted. After the carbon sample was sifted, the uniform size fractions were washed several times in distilled water. After washing, all carbon samples were placed in glass baking pans, dried in an oven at 105-110°C for 24 hr, and stored in a desiccator at room temperature until used.

Measurement of the concentration of adsorbate in the experimental solutions was made by direct ultraviolet absorbance. The spectrophotometer was a Beckman DK-2A. Preliminary scans were conducted on each adsorbate to determine its wavelength of maximum absorbance. Once this wavelength was established, calibration curves were determined at each experimental pH value studied for path lengths of 1, 5 and 10 cm.

Several equilibrium experiments were conducted using two adsorbates simultaneously. The compounds were phenol and 4-methoxyphenol. The procedure described by Friedel[13] was used to determine the concentration of each phenol in mixtures. Additional information on the procedure used to measure the concentration of phenol and 4-methoxyphenol is reported elsewhere.[12]

Procedures

The experimental system for evaluating the equilibria of adsorption of phenols by granular activated carbon was a batch reactor. The batch technique was selected because of its simplicity and ease of evaluating the parameters which influence the adsorption process. Actual applications of granular carbon will undoubtedly be via column-type operations. However, the evaluation of the fundamentals of adsorption is much simpler in a batch reactor and the basic relationship developed therein can, with due caution, be applied to predict the behavior of continuous adsorption systems.

The reaction vessels were 300-ml round-bottom boiling flasks with ground glass stoppers. Each flask was prewashed with dilute hydrochloric acid, rinsed in distilled water, and air dried at 110°C prior to usage. Five

Burrell wrist-action shakers were used to mix the adsorbate-adsorbent solution. This setup provided the capability to follow the adsorptive characteristics of 80 adsorption systems simultaneously. The shakers were housed in a dark room, which was maintained at ± 2°C of the desired temperature.

Adsorption isotherms were developed to describe the adsorptive properties of Columbia LCK carbon at equilibrium. The procedure consisted of maintaining a constant carbon concentration and varying the initial adsorbate concentration from flask to flask. The activated carbon added to each reactor was 0.2 g, except for experiments with 2,4-dinitrophenol in which only 0.1 g was added.

Large volumes of adsorbate solution and appropriate dilution water were then prepared and 200 ml were placed accurately into a series of 250-ml Erlenmeyer flasks. The concentration of the adsorbate was varied to cover the range of equilibrium solute concentrations of interest. After equilibration to the proper temperature, the solutions were transferred from the Erlenmeyer flasks into the reaction vessels. The reaction vessels were sealed and shaken until equilibrium was obtained. Preliminary experiments indicated that no measurable change in the amount of adsorption occurred after a contact time of 5 days.[7] Unless otherwise noted, the reaction vessels were agitated continuously for 12 to 14 days.

Immediately after the shakers were turned off, each reaction vessel was opened and samples were removed. Prior to analysis all samples were centrifuged at 8,000 rpm for 5 min to remove colloidal matter. Preliminary studies showed that essentially all of the interference in the analysis of the adsorbate was eliminated by this procedure. The amount of adsorption was computed from the difference between the initial concentration and the equilibrium concentration of the adsorbate.

Attempts were made to fit the experimental data to Langmuir and Freundlich adsorption isotherms. The purpose was to provide a means by which the influence which various systematic parameters exhibited on the adsorption process could be evaluated. Plots for the seven different adsorbates studied indicated that the experimental data did not adhere to either the Freundlich or Langmuir model completely. However, it was observed that the experimental data could be fitted to either model provided one was selective of the range plotted. For example, experimental data describing the adsorption isotherm of 4-nitrophenol were fitted extremely well by the Freundlich adsorption isotherm in the equilibrium concentration range from 50 to 3,500 μmol/l. However, if the equilibrium concentration range of 4-nitrophenol was changed to 1 to 200 μmol/l or from 1 to 3,500 μmol/l, the isotherm failed to define the results satisfactorily.

Since the entire set of equilibrium data could not be properly described by either the Freundlich or the Langmuir equation, the approach taken was to use the equation that was most applicable for the segment of experimental data being evaluated. For the situation where the experimental data was not described properly by either the Langmuir or Freundlich model, the data were compared directly without the aid of the equations. Additional discussion on the use of these equations to describe the equilibria of adsorption of phenols has been reported previously.[12]

RESULTS AND DISCUSSION

Phosphate Buffer

The influence of a 0.05 *M* phosphate buffer on the adsorption of 2,4-dichlorophenol (2,4-DCP) and 2,4-dinitrophenol (2,4-DNP) was evaluated and has been reported elsewhere.[9] In summary, 0.05 *M* phosphate buffer had no measurable influence on the equilibria of adsorption of 2,4-DCP and 2,4-DNP at pH values where these adsorbates were in an undissociated form. However, the presence of a 0.05 *M* phosphate buffer affected the amount of adsorption of the anionic form of these two adsorbates. The adsorption of the dissociated forms of 2,4-DCP and 2,4-DNP was increased in the presence of the buffer, although the magnitude of the influence varied from adsorbate to adsorbate. These findings concur with results from similar studies by Snoeyink *et al.*[14] and DiGiano and Weber.[15]

Adsorption Isotherms

Isotherms were developed for the seven different phenolic compounds studied over a wide range of equilibrium concentrations, C_e. The pH values in these experiments were adjusted such that the adsorbates were predominantly in an undissociated form. Typical plots are shown in Figures 1, 2 and 3. The adsorption isotherms all exhibited first-order behavior in the low equilibrium concentration range and tended to plateau at high bulk solution concentrations.

The data for each adsorbate were analyzed for adherence to the Langmuir and Freundlich adsorption equations. Neither equation was suitable over the entire C_e range. Therefore, a smooth curve was drawn through the data plotted in the figures and the amount of the adsorption was read directly from the graphs. A comparison of the adsorption of the seven compounds is presented in Table I. Columbia LCK carbon

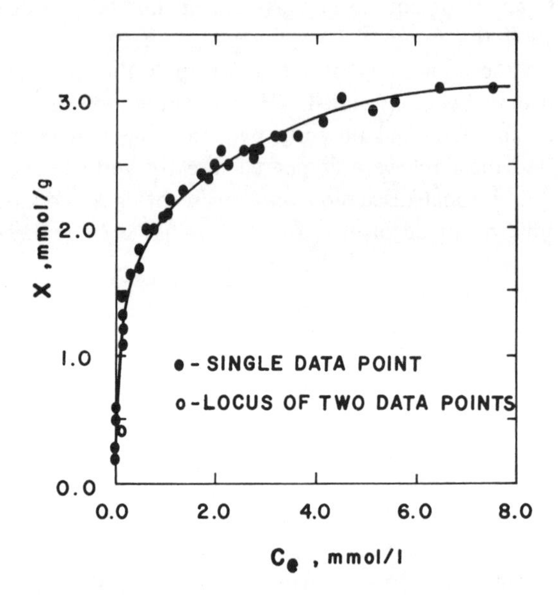

Figure 1. Adsorption isotherm for phenol at pH 6.3; 20°C; 16 x 20 mesh adsorbent.

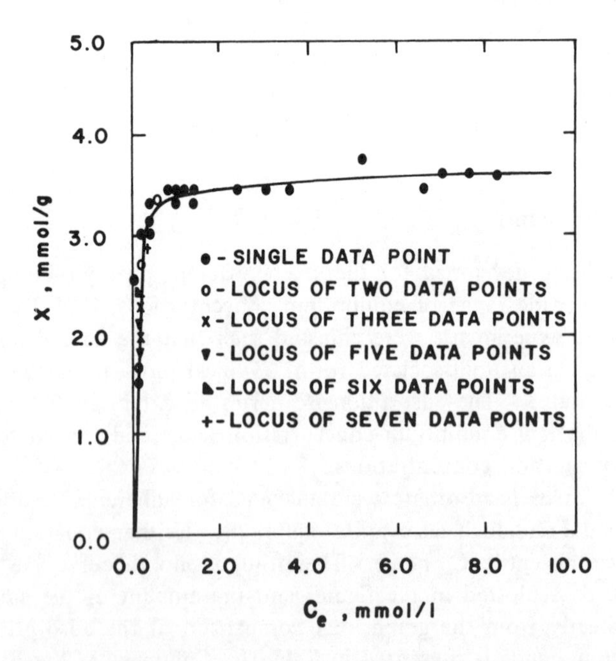

Figure 2. Adsorption isotherm for 2,4-dichlorophenol at pH 6.3; 20°C; 16 x 20 mesh adsorbent.

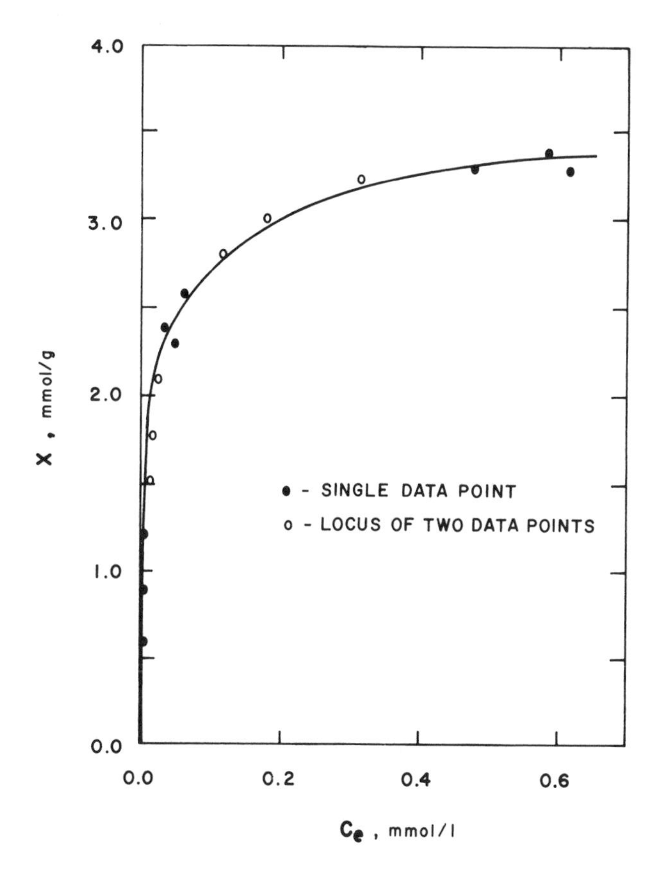

Figure 3. Adsorption isotherm for 2,4-dinitrophenol at pH 2.1 or 3.05; 20°C; 30 x 40 mesh adsorbent.

exhibited approximately the same affinity for the first five compounds listed in the table, while the adsorption of phenol and 4-hydroxyphenol was considerably less.

The amount adsorbed of the seven phenolic compounds was plotted as a function of the solubility of each adsorbate in 0.05 *M* phosphate buffer solution. The amount adsorbed was found to increase with decreased solubility. The excellent fit of the experimental data to a smooth curve was unexpected since phenols with various functional groups were studied rather than a homologous series.

Carbon Particle Size

Four preliminary experiments were conducted to determine the influence which the size of the adsorbent has on the adsorption of phenols. In

Table I. Summary of Adsorption Data for Several Adsorbates[a]

Compound	Amount Adsorbed in mmol/g at the Following Equilibrium Concentrations					
	0.1 mmol/l	0.2 mmol/l	0.4 mmol/l	0.6 mmol/l	1.0 mmol/l	4.0 mmol/l
2,4-dinitrophenol[b]	2.7	3.0	3.3	3.4	–	–
2,4-dichlorophenol	2.8	2.9	3.1	3.3	3.4	3.5
4-nitrophenol	2.2	2.4	2.6	2.8	3.0	3.3
4-chlorophenol	2.1	2.3	2.6	2.7	2.7	3.5
4-methoxyphenol	2.3	2.5	2.8	2.9	3.0	3.3
4-hydroxyphenol	1.1	1.4	1.5	1.6	1.7	2.1
phenol	1.1	1.4	1.7	1.9	2.1	2.8

[a] 20°C, pH = 6.3 except for 2,4-dinitrophenol and 0.05 *M* buffer concentration.
[b] pH = 2.1 or 3.0, and contact time = 8 days.

these experiments the initial adsorbate and adsorbent concentrations were kept constant, while the carbon size was varied from flask to flask. Both 2,4-DCP and 2,4-DNP were evaluated, each at two different initial adsorbate concentrations. The initial concentrations of 2,4-DCP evaluated were 2.45 and 3.07 mmol/l, while initial 2,4-DNP levels of 0.54 and 0.89 mmol/l were tested. Results of these studies are presented in Figure 4. As shown, the particle size of the adsorbent had essentially no influence on the amount of 2,4-DCP and 2,4-DNP adsorbed. To verify that this finding was valid over a much wider initial concentration range, adsorption isotherms for 2,4-DCP were developed for each of the eight

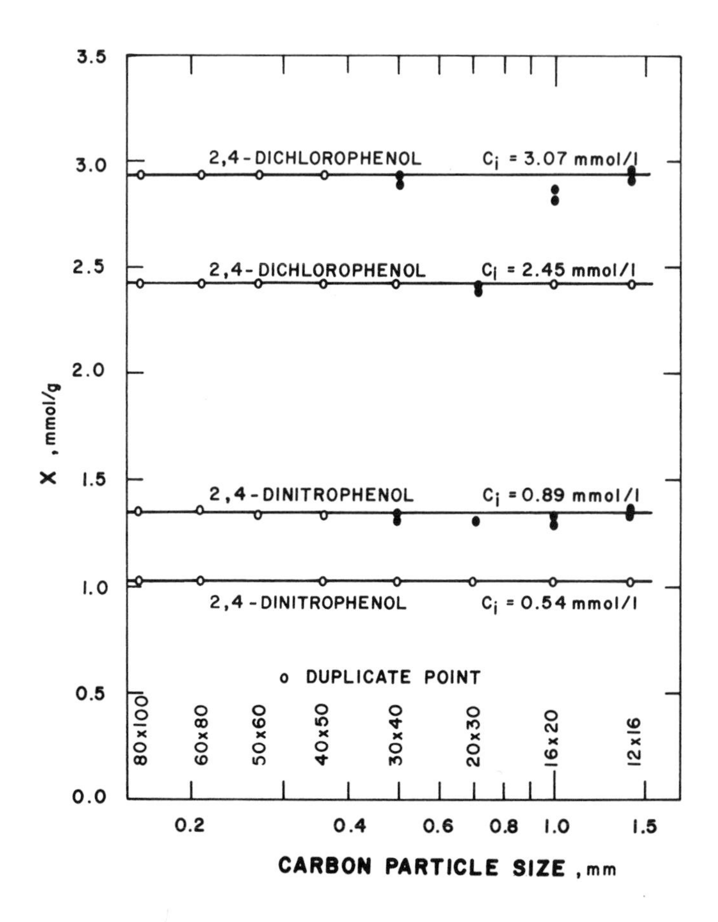

Figure 4. Influence of carbon particle size on the adsorption of 2,4-dinitrophenol and 2,4-dichlorophenol at 20°C.

sizes of adsorbent. The range in the equilibrium concentrations of the adsorbate for these experiments was 0.01 to 2.0 mmol/l. Over this concentration range, the experimental data were plotted as a straight line on logarithmic paper and the constants for the Freundlich isotherm were determined by least square analysis. Table II lists these constants for each mesh size tested. As shown, the constants k and 1/n are approximately the same for each mesh size. The average values were 0.348 and 0.10, respectively, where C_e is expressed in mg/l and X in grams of 2,4-dichlorophenol per gram of carbon.

Table II. Freundlich Constants for 2,4-Dichlorophenol with Various Sizes of Adsorbent[a]

Carbon Particle Size		Number of	Range in C_e	Freundlich Constants[c]	
(mesh size[b])	(mm)	Data Points	(mmol/l)	(k)	(1/n)
12 x 16	1.41	10	0.012-1.4	0.354	0.095
16 x 20	1.00	29	0.031-1.4	0.334	0.100
20 x 30	0.71	14	0.009-1.7	0.335	0.113
30 x 40	0.50	13	0.014-2.0	0.343	0.088
40 x 50	0.36	12	0.012-1.8	0.351	0.110
50 x 60	0.27	12	0.011-1.8	0.358	0.096
60 x 80	0.21	12	0.008-1.6	0.359	0.102
80 x 100	0.16	12	0.012-1.7	0.355	0.099

[a]20°C, pH 6.3, and 0.05 *M* phosphate buffer.
[b]U.S. Standard Sieve Series.
[c]Where C_e is expressed in mg/l and k in grams of 2,4-dichlorophenol per gram of activated carbon.

Temperature

Adsorption isotherms were developed for phenol at temperatures of 8, 20 and 29°C. The results are shown in Figure 5 and illustrate that the adsorption of phenol is an exothermic process. The amount of phenol adsorbed was decreased with higher temperatures. Temperature exhibited more of an influence on the adsorption process at lower equilibrium concentrations of phenol. This is illustrated in Table III, which lists the amount of phenol adsorbed as a function of temperature and equilibrium concentration. The amount of adsorption was influenced significantly by temperature at phenol concentrations less than 200 μmol/l. For example, when the concentration of phenol was 30 μmol/l, an increase in

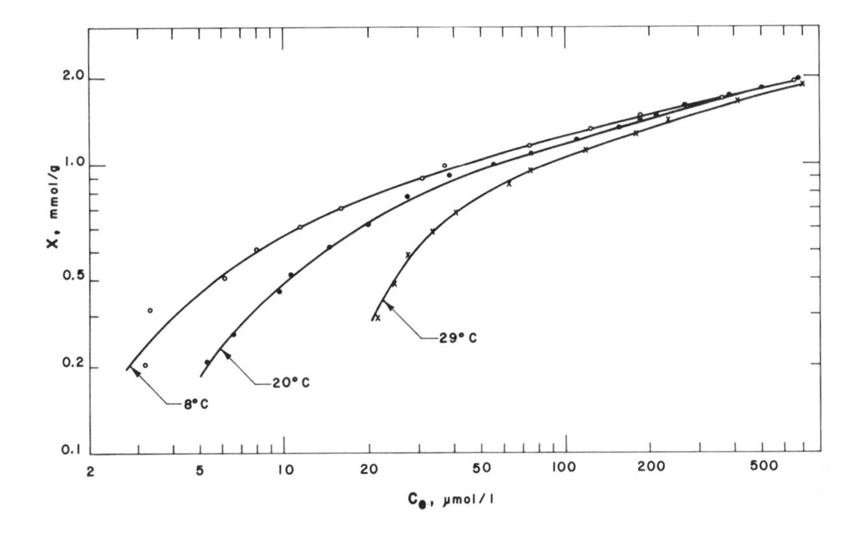

Figure 5. Influence of temperature on the adsorption of phenol at pH 6.3; 16 x 20 mesh adsorbent.

Table III. Amount of Phenol Adsorbed as a Function of Temperature and Adsorbate Concentration

Concentration of Phenol[a] (µmol/l)	$X_{8°C}$[b] (µM/g)	$X_{29°C}$[b] (µM/g)	Decrease in Adsorptive Capacity of Adsorbent (percent)
30	895	520	42
50	1040	780	25
100	1250	1050	16
200	1480	1320	11
300	1620	1500	7
400	1710	1625	5
500	1800	1725	4

[a]In the solution phase at equilibrium.
[b]Amount of phenol adsorbed per unit weight of activated carbon at the temperature indicated.

temperature from 8 to 29°C decreased the amount of adsorption from 895 µmol/g to 520 µmol/g. This represents a decrease of 42% in the efficiency of the adsorbent.

Hydronium Ion Concentration

Several experiments were completed to describe the influence of pH on the adsorption of phenols by activated carbon. Adsorption isotherms were developed at various pH values for both 2,4-DCP and 2,4-DNP. The acidic dissociation constants (pK_a values) for these two compounds are 7.8 and 4.0, respectively.

A 0.05 *M* phosphate buffer was included in these studies to minimize variation in the pH within an experiment. Previous studies illustrated that the inclusion of the buffer only influenced the adsorption of the dissociated form of the adsorbate; increased adsorption of the dissociated form would be expected in the presence of the buffer.[9]

The experimental data for 2,4-DCP were described very well by the Freundlich adsorption isotherm. Regression analyses were completed on the data to determine the equation constants and the results are presented in Table IV. The values of k and 1/n are nearly constant over the pH range of 2 to 8.3. However, a marked increase in 1/n and a decrease in k occurred over the pH range of 8.3 to 11.7. The dependence of the adsorption of 2,4-dichlorophenol upon the pH of the solution is presented more comprehensively in Figure 6. The amount adsorbed was calculated by substituting selected C_e values into the derived Freundlich equations for each pH. Maximum adsorption of 2,4-DCP occurred over the pH range of 5.0 to 7.5 and dropped off sharply at pH values greater than the pK_a value of the adsorbate.

Table IV. Freundlich Constants for 2,4-Dichlorophenol at Several pH Values[a]

Number of Data Points	C_e Range (mmol/l)	pH Value	Freundlich Constants[b]	
			k	1/n
12	0.080-1.6	2.0	0.275	0.114
10	0.048-1.3	2.6	0.262	0.124
16	0.056-1.5	3.1	0.286	0.123
29	0.031-1.4	6.3	0.322	0.100
16	0.046-1.4	7.9	0.326	0.099
12	0.009-0.94	8.3	0.306	0.117
13	0.057-1.4	10.9	0.109	0.207
9	0.032-0.94	11.5	0.079	0.214
13	0.069-1.7	11.7	0.072	0.202

[a] 21-24°C, and 0.05 *M* phosphate buffer.
[b] Where C_e is expressed in mg/l and k in grams of 2,4-dichlorophenol per gram of activated carbon.

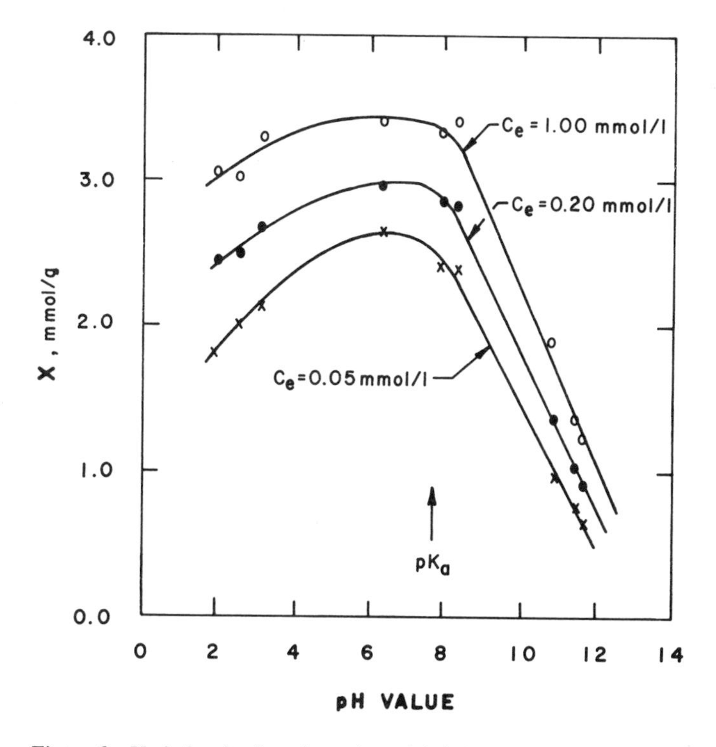

Figure 6. Variation in the adsorption of 2,4-dichlorophenol with pH.

Adsorption isotherms were developed also for 2,4-DNP between pH 2.1 and 11.9. Maximum adsorption occurred near pH 3. The isotherms for 2,4-DNP were also subjected to statistical analysis. The data fit the Freundlich equation fairly well over the C_e range 5 to 600 μmol/l. The equation constants for each pH evaluated are listed in Table V. Values of 1/n vary only slightly with pH whereas k values are definitely decreased at values greater than pH 3.0.

The decrease in adsorption of 2,4-DNP at values greater than pH 3.0 is illustrated in Figure 7, which was developed in a manner similar to Figure 6. The figure also shows clearly that the adsorption of 2,4-DNP drops off rapidly at pH values greater than the pK_a of this phenol. It is also evident from the figure that the hydroxide ion apparently does not compete significantly with the 2,4-DNP anion for adsorption sites. If hydroxide ion were sorbed at sites suitable for phenols, the adsorption of 2,4-dinitrophenolate would be expected to decrease with increasing pH. Such a relationship was not observed.

Table V. Freundlich Constants for 2,4-Dinitrophenol at Several pH Values[a]

Number of Data Points	C_e Range (mmol/l)	pH Value	Freundlich Constants[b] k	1/n
7	0.008-0.62	2.1	0.321	0.145
7	0.033-0.58	3.0	0.351	0.124
9	0.005-0.34	4.3	0.295	0.135
9	0.010-0.53	5.1	0.232	0.139
7	0.006-0.61	6.0	0.166	0.156
5	0.054-0.77	7.0	0.153	0.121
5	0.009-0.86	8.0	0.129	0.128
10	0.003-0.53	10.4	0.116	0.147
3	0.015-0.61	11.0	0.117	0.139
5	0.005-0.62	11.9	0.113	0.141

[a]21-24°C, and 0.05 *M* phosphate buffer.
[b]Where C_e is expressed in mg/l and k in grams of 2,4-dinitrophenol per gram of activated carbon.

Dissociation of Adsorbate

A large decrease in the amount adsorbed occurs at pH values greater than the pK_a of the adsorbate. This suggests that the amount of dissociation of the adsorbate influences the extent of adsorption. The variation of adsorption with degree of dissociation is not totally unexpected since it is well known that the dissociated form is more soluble in solution and, therefore, is less adsorbable.[17,18] The dissociated components are adsorbed less because: (1) stronger adsorbate-solvent bonds must be broken before adsorption can occur; and (2) repulsive forces between the anionic sorbate and the adsorbent can become significant at high pH values, especially when the carbon surface has a net negative charge.

Experimental data were collected and evaluated to determine if 2,4-DCP and 2,4-DNP behaved similarly with respect to extent of dissociation. The results of this analysis are shown in Figure 8. The relative amount adsorbed was defined arbitrarily as $X_{(pH)}$ divided by $X_{(max)}$; that is, the amount of adsorption which occurred at any pH value was divided by the maximum amount measured. The term pH-pK_a was selected as a means of comparing the two compounds on the same basis. Both phenols exhibited similar adsorptive characteristics when plotted in this fashion. This suggests that different phenols behave similarly with respect to the extent of dissociation. Additional studies with other adsorbates would be required to confirm this contention.

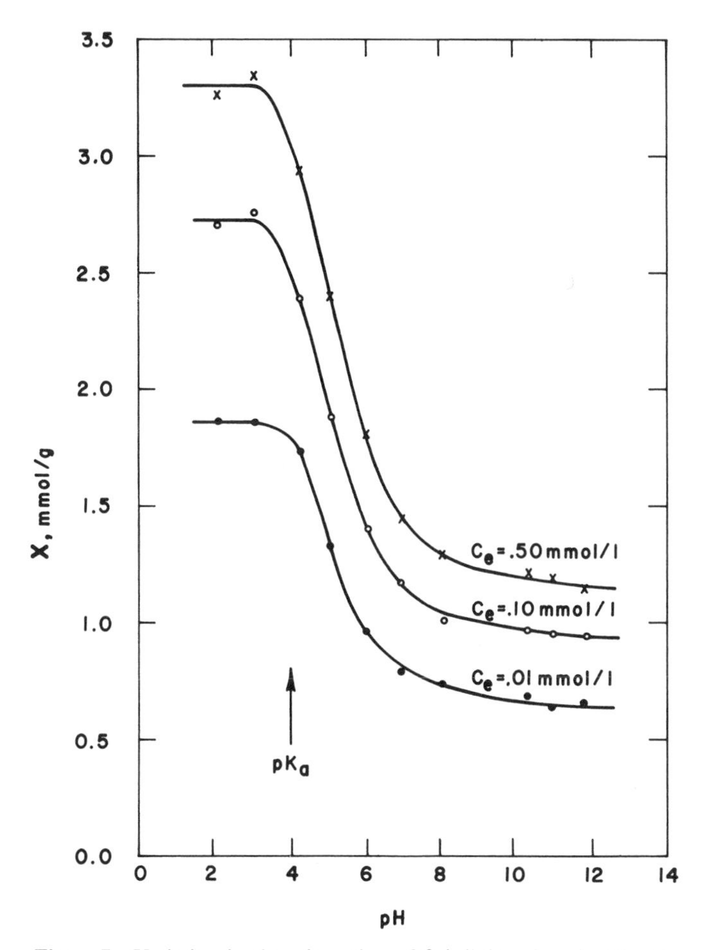

Figure 7. Variation in the adsorption of 2,4-dinitrophenol with pH.

Adsorption-Desorption Isotherms

Adsorption and desorption isotherms were developed for 2,4-DCP. An adsorption isotherm was developed in the usual fashion, while the desorption isotherm was developed by decanting the supernatant from each reaction vessel (at the end of the adsorption part of the experiment) and then reinserting 200 ml of distilled water into each flask. The reaction flasks were then resealed and placed back on the Burrell shaker and allowed to reequilibrate for a period of ten days. At the end of this period, the concentration of adsorbate was determined in each flask. These equilibrium concentrations in conjunction with corresponding adsorption phase data were used to compute the desorption isotherm.

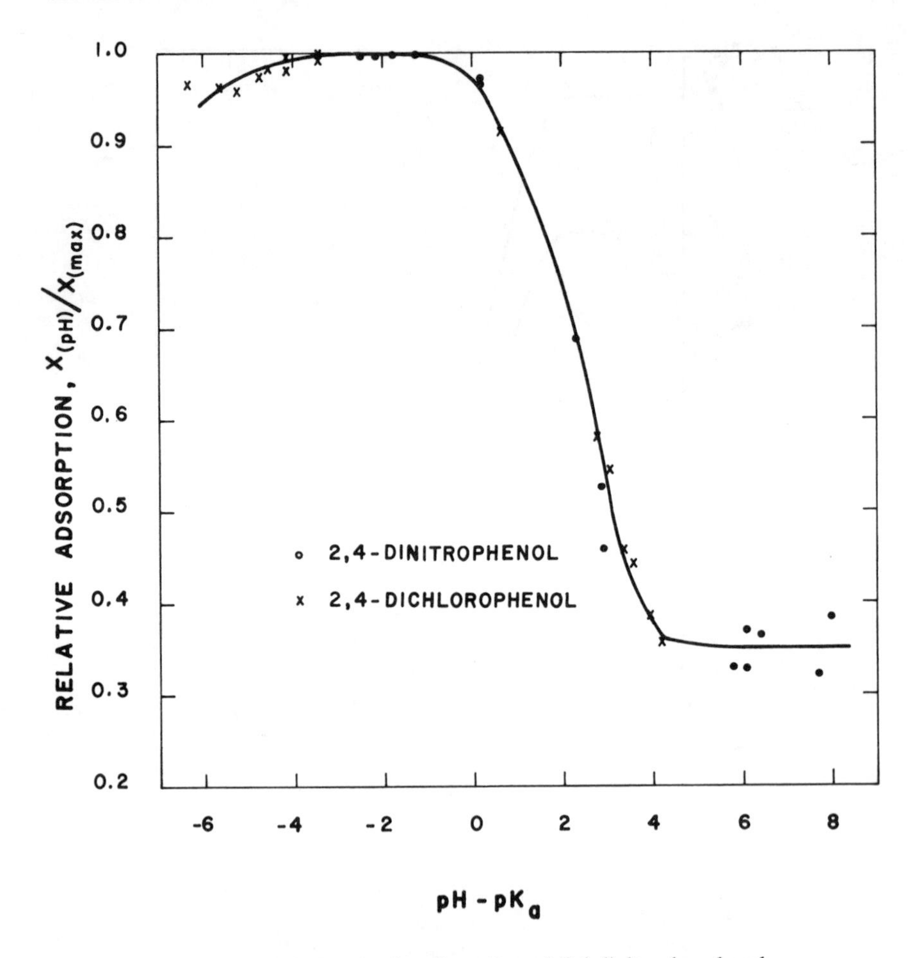

Figure 8. Variation in the adsorption of 2,4-dinitrophenol and 2,4-dichlorophenol with extent of dissociation.

The results of the adsorption-desorption isotherms are presented in Figure 9. Significant hysteresis of the 2,4-dichlorophenol molecules did not occur. The adsorption isotherm falls slightly above that for the desorption experiments, but the difference shown in the figure is considered insignificant because of the inherent difficulties in developing desorption isotherms. The near coincidence of the adsorption and desorption isotherms shows that very little, if any, of the adsorbate undergoes an irreversible chemical reaction at the carbon surface or is entrapped in "ink bottle" pores.

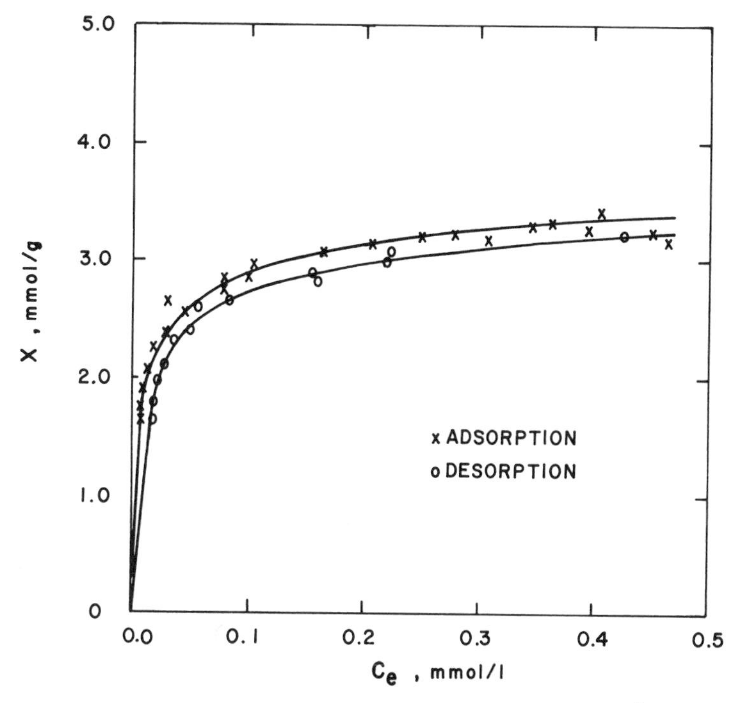

Figure 9. Adsorption-desorption isotherm for 2.4-dichlorophenol at 20°C and pH 6.3.

CONCLUSION

Results from the experiments indicate that the extent of adsorption of phenols at equilibrium is highly dependent upon the solubility of the adsorbate, the extent of protolysis of the adsorbate, and the concentration of adsorbate equilibrated with the absorbent. The equilibria of adsorption of phenols were found to be independent of the mesh size of the adsorbent. Temperature had only a minor influence on the adsorption process, especially at high surface coverages. Also, the adsorption of phenols appears to be a reversible process.

In summary, the technical feasibility of removing phenolic compounds from aqueous solution via granular activated carbon adsorption is excellent. High adsorptive capacities are the rule, rather than the exception, for phenols. However, a clear understanding of the many physical and chemical factors influencing the extent of phenol adsorption is a prerequisite before the adsorptive capacity of activated carbon can be exploited to its fullest extent.

ACKNOWLEDGMENTS

The authors express their sincere gratitude to John H. Haas, Jr. for his assistance in completing the laboratory experiment.

This research was made possible through a research grant awarded by the Office of Water Resources Research, U.S. Department of the Interior, through the Water Resources Research Institute at Rutgers University.

REFERENCES

1. Cecil, L. K., Ed. *Complete Water Reuse* (New York: American Institute of Chemical Engineers, (1973).
2. Cecil, L. K., Ed. *Water Reuse,* Symposium Series 78 (New York: American Institute of Chemical Engineers, 1967).
3. *J. Amer. Water Works Assoc.* 63:609 (1971).
4. Horsefield, D. R. *J. Amer. Water Works Assoc.* 66:238 (1974).
5. Phillips, W. J. *J. Amer. Water Works Assoc.* 66:231 (1974).
6. American Water Works Association. *Water Quality and Treatment* (New York: McGraw-Hill, 1971).
7. Zogorski, J. S., and S. D. Faust. "Removal of Phenols from Polluted Waters," final report submitted to New Jersey Water Resources Research Institute, 1974.
8. Zogorski, J. S., and S. D. Faust. Proceedings of Second World Congress, International Water Resources Assoc., New Delhi, India, Vol. II, (1975), pp. 89-97.
9. Zogorski, J. S., and S. D. Faust. *J. Environ. Sci. Health* 11:501
10. Zogorski, J. S., S. D. Faust and J. H. Haas, Jr. *J. Colloid Interface Sci.* 55:329 (1976).
11. Zogorski, J. S., and S. D. Faust. in *Water–1976: I. Physical, Chemical Wastewater Treatment,* Symposium Series 166, 73:54, (New York: American Institute of Chemical Engineers, 1977).
12. Zogorski, J. S. Ph.D. Thesis, Rutgers University (1975).
13. Friedel, R. A., and M. Orchin. *Ultraviolet Spectra of Aromatic Compounds* (New York: Wiley, 1951).
14. Snoeyink, V. L., W. J. Weber, Jr. and H. B. Mark, Jr. *Environ. Sci. Technol.* 3:918 (1969).
15. DiGiano, F. J., and W. J. Weber, Jr. Technical report T-69-1, Department of Civil Engineering, University of Michigan (1969).
16. Snoeyink, V. L. Ph.D. Thesis, University of Michigan (1968).
17. Hassler, J. W. *Activated Carbon* (New York: Chemical Publishing Co., 1963).
18. Mattson, J. S., and H. B. Mark, Jr. *Activated Carbon* (New York: Marcel Dekker, Inc., 1971).

CHAPTER 10

INFLUENCE OF PORE SIZE DISTRIBUTION ON THE HOCl-ACTIVATED CARBON REACTION

Makram T. Suidan

School of Civil Engineering
Georgia Institute of Technology
Atlanta, Georgia

Vernon L. Snoeyink, William E. Thacker and Dennis W. Dreher

Department of Civil Engineering
University of Illinois
Urbana, Illinois

INTRODUCTION

Free chlorine ($HOCl + OCl^-$) is commonly used in water and wastewater treatment for disinfection, ammonia removal, and taste and odor control. In addition to its desirable effects, free chlorine residual has a number of undesirable characteristics such as toxicity to aquatic species in receiving waters, taste and odor in drinking water, and interference with some industrial processes, and thus it may need to be removed or at least reduced in concentration. Dechlorination is also of interest since minimizing the time of contact of chlorine with certain waters should reduce the extent of formation of chlorinated organic compounds. These and other reasons illustrate the importance of having a well understood process for the removal of chlorine from water.

A number of processes have been developed for reducing chlorine levels in water, and the use of activated carbon in a fixed bed reactor is one of the more common ones. A description of these processes may be found in a recent literature review of the subject of dechlorination.[1]

A mathematical model for the reaction between aqueous HOCl and activated carbon has been developed.[2] It is based on both surface reaction and pore diffusion as rate limiting steps and accounts for the reduction in reaction rate owing to the buildup of reaction products on the carbon surface. Constants for the model were determined by analysis of data obtained from small-scale batch reactors, and the model was then used to predict chlorine concentration as a function of time for various carbon sizes, chlorine doses and carbon doses in batch reactors. It was further employed to predict chlorine effluent concentration as a function of time for packed bed reactors at different detention times, particle sizes and chlorine concentrations. The applicability of the model to both types of reactors was then verified by laboratory experimentation. A parallel study[3] was carried out to investigate the effects of pH and temperature on the reactions of HOCl and OCl^- with carbon, and to predict the behavior of packed beds of carbon for various values of pH and temperature. The model thus developed can be used to design packed bed activated carbon dechlorination reactors.

The objective of the research reported herein was to determine the effect of the pore volume distribution on the dechlorination efficiency of seven commercially available carbons. The investigation was carried out at pH 4 where HOCl is the predominant free chlorine species.

MATHEMATICAL MODEL

The reaction between HOCl and activated carbon can be described by[4,5]

$$C^* + HOCl \rightarrow CO^* + H^+ + Cl^- \tag{1}$$

where C* represents a reactive site on the carbon surface and CO* represents a surface oxide. This reaction is heterogeneous; it is assumed to proceed by a reversible adsorption step, followed by irreversible dissociation[2]

$$C^* + HOCl \underset{k_2}{\overset{k_1}{\rightleftharpoons}} CHOCl^* \tag{2}$$

$$CHOCl^* \overset{k_3}{\rightarrow} CO^* + H^+ + Cl^- \tag{3}$$

where CHOCl* represents an adsorbed molecule and k_1, k_2 and k_3 are rate constants. The reaction rate decreases as oxides build up on the carbon surface, but not all oxides remain on the surface. The fraction of active sites occupied by the oxides, f_o, can be given by[6]

$$f_o = \frac{Q}{k_5 + Q} \tag{4}$$

where Q represents the amount of chlorine reacted per gram of carbon. Assuming the reaction in Equation 3 to be rate limiting and the reaction in Equation 2 to be at equilibrium, the following surface reaction rate expression can be developed:[2]

$$R_c(C, f_o) = \frac{k_3 SC}{k_4 + C}(1-f_o) \qquad (5)$$

where $k_4 = k_2/k_1$
S = the total number of reactive sites
C = the concentration of HOCl

The pores of an activated carbon granule are a series of tortuous, interconnected paths of varying cross-sectional areas. Consequently, it would not be feasible to describe diffusion within each or any of the pores. Suidan *et al.*[2] assumed that the pore volume of a carbon granule was made up of straight, cylindrical, open-ended pores of diameter d_p and length 2 L_p. The two ends of each pore are exposed to the concentration of the free chlorine species at the external surface of the carbon particle. Such a model is roughly equivalent to a spherical particle of diameter 6 L_p.[1] Surface film diffusion resistance was shown to be negligible relative to pore diffusion under the conditions employed.[2] Incorporating the surface reaction rate expression, Equation 5, into a material balance in the pore, the following expressions are obtained

$$\frac{\partial C}{\partial t} = [\frac{D_c}{L_p^2}]\frac{\partial^2 C}{\partial \eta^2} - [\frac{4k_3 S}{d_p}]\frac{C}{C + k_4}(1-f_o) \qquad (6)$$

$$\frac{\partial f_o}{\partial t} = [\frac{k_3 S}{k_5}]\frac{C}{C + k_4}(1-f_o)^3 \qquad (7)$$

where η = a reduced position variable measured from the mouth of the pore
D_c = the diffusivity of the chlorine species

the boundary conditions are $C = C_o$ at $\eta = 0$ and $\partial C/\partial \eta = 0$. The initial conditions are taken to be the state of fresh carbon with $f_o = 0$.

The bulk concentration, C_B, in the boundary condition in Equation 7 is obtained through a material balance on free chlorine in the bulk fluid phase of the reactor. For a closed batch reactor in which the concentration is initially C_o and is allowed to decrease with time, the material balance results in the expression[2]

$$\frac{dC_B}{dt} = -\left(\frac{A_p d_p}{4}\right)\left(\frac{m}{\epsilon}\right)\left(\frac{D_c}{L_p^2}\right)\frac{\partial C}{\partial \eta}\bigg|_{\eta = 0} - k_d C_B \tag{8}$$

where A_p = the pore surface area per unit weight of carbon
m = the concentration of carbon in the reactor
ϵ = the porosity of the reactor
k_d = the first order self-decomposition rate constant which is experimentally determined from the rate of decrease of the chlorine concentration in a comparable "blank" reactor containing no carbon.

For a packed bed reactor, a material balance on a packed bed with axial dispersion yields[2]

$$\frac{\partial C_B}{\partial t} + V\frac{\partial C_B}{\partial Z} = D_A\frac{\partial^2 C_B}{\partial Z^2} + \left(\frac{A_p d_p}{4}\right)\left(\frac{m}{\epsilon}\right)\left(\frac{D_C}{L_p^2}\right)\frac{\partial C}{\partial \eta}\bigg|_{\eta = 0} \tag{9}$$

where V = the interstitial liquid velocity in the bed
m = the mass of carbon per unit bed volume
ϵ = the porosity of the packed bed
Z = the distance dimension as measured from the entrance end of the bed
D_A = is the axial dispersion coefficient. The boundary conditions are $VC_{O_o} = VC_B - D_A\, \partial C_B/\partial Z$ at $Z = 0$ and $\partial C_B/\partial Z = 0$ at $Z = L_b$[8] where L_b is the bed length. The models were solved numerically as outlined by Suidan.[9]

The quantity (0.25 Apdp) appearing in Equations 8 and 9 represents the total pore volume per unit weight of carbon if the pore volume truly consists of straight cylindrical pores of diameter d_p. In actuality, however, the pore volume is distributed over a wide range of pore diameters, as discussed later, and not all the pore volume is accessible to the free chlorine species with their waters of hydration. A more appropriate description of the quantity 0.25 Apdp is to regard it as a measure of the effective or accessible pore volume. The objective of this research is to evaluate the importance of this quantity and to attempt to relate it to the observed performance characteristics of several commercially available carbons.

EXPERIMENTAL MATERIALS AND METHODS

Activated Carbon

Seven different carbons were used including four bituminous coal base carbons, hereafter referred to as carbons A, B, C and D. Three other carbons were used: a lignite base carbon, a petroleum base carbon and a charred resin carbon.

The activated carbons were prepared by mechanical grinding followed by sieving into a number of particle size ranges. They were then washed with deionized water to remove fines and were subsequently dried at 105-110°C.

Chlorine Measurement and Solutions

The DPD ferrous titrimetric procedure[10] was used to determine the concentration of aqueous HOCl. Measurements were found to be accurate within 0.01 mg/1 in the concentration range of 0.2-4.0 mg/1 as Cl_2. Samples of higher concentrations were diluted to give concentrations within this range.

Stock free chlorine solutions were prepared by making appropriate dilutions of the household bleach, Clorox.® The free chlorine solutions used in the experiments were prepared by adding measured volumes of the stock solution to deionized water, and adding to these dilutions 5 ml of 0.5 *M* phosphate buffer per liter. In this study all experiments were carried out at pH 4.0 with HCl and NaOH solutions being used for pH adjustment.

Closed Batch Experiments

All experiments carried out in this study were of the closed batch, completely mixed reactor type. The solution volume in every experiment was maintained at approximately four liters. A "blank" reactor with all reagents added except activated carbon was included in each experiment to determine the rate of disappearance of free chlorine owing to factors other than reaction with activated carbon. The reactors were mixed using T-line laboratory stirrers (Talboys Engineering Corporation, Emerson, NJ) and a Teflon®* stirring rod. The speed of mixing was maintained between 764 and 917 rpm. The mixing speed was high enough so that an increase in mixing rate did not result in an increased chlorine reduction rate. This was done in order to eliminate any major film diffusion resistance.

An automatic titrator was used to maintain the pH in the test reactor at 4.0. A 0.5 *M* NaOH solution was used for pH control. The temperature of the solutions was maintained between 22.5° and 23.5°C with 23°C as a mean. The reactors were covered with foil to prevent the photodecomposition of the chlorine.

*Registered trademark of E. I. du Pont de Nemours and Company, Inc., Wilmington, Delaware.

MATHEMATICAL MODEL CONSTANTS

Previous Determination of Constants

The mathematical model, Equations 6 through 10, contains six groups of constants ($4\ k_3S/d_p$), (k_3S/k_5), k_4, (D_c/L_p^2), (0.25 Apdp) and k_d. The value of k_d was obtained from the first-order rate of disappearance of HOCl from the blank reactor. The group (D_c/L_p^2) has a value[2] of $36 \times 10^{-5}/L_m^2\ sec^{-1}$ where L_p is considered to be one-sixth of the arithmetic mean particle diameter, L_m, in centimeters, and D_c has a value of $10^{-5}\ cm^2/sec$. This value for D_c was based on the Wilke-Chang approximation[11] at 23°C. The remaining four groups of constants were determined by Suidan *et al.*[2] for the bituminous coal base carbon A by a trial-and-error fit of the mathematical model to the experimental data obtained from semibatch reactors. The model was subsequently tested against data obtained from closed batch and packed bed reactors and the resulting correlations were very satisfactory. The model was further tested for particle size effect with excellent correlations. Table I contains the values of the six groups of constants for bituminous base activated carbon A at 23°C and pH 4.

Table I. Values of Constants in Mathematical Model

$4k_3S6d_p$	$6.08 \times 10^3\ g/cm^3$-sec
k_3S/k_5	$2.7\ sec^{-1}$
k_4	$1.81 \times 10^{-5}\ g/cm^3$
k_d	$2.23 \times 10^{-7}\ sec^{-1}$
D_c/L_p^2	$36 \times 10^{-5}/L_m^2\ sec^{-1}$
$0.25\ A_pd_p$	$0.94\ cm^3/g$

Dependence of Constants on Carbon Type

In order to simplify the mathematical model and facilitate its use in comparing dechlorination efficiencies of different activated carbons, the quantity 0.25 Apdp, which describes the effective or accessible pore volume within the activated carbon, was assumed to be the only variant with the type of carbon. Although it may be an oversimplification, this assumption was made to provide a means for describing relative efficiencies.

The first order HOCl decay rate constant, k_d, is likely to be independent of the type of activated carbon. The surface dissociation rate constant, k_3, and the reversible adsorption equilibrium constant, k_4, may be functions of the chemistry of the reacting system. However, to simplify calculations, the different carbons were assumed to possess the same surface chemistry. The values of the above two constants evaluated for carbon A were assumed to apply to all seven carbons studied. Similarly, the constants, S and k_5, which respectively represent the total reactive sites concentration per unit carbon surface area and the denominator constant in the expression for f_o, were assumed to be independent of the type of activated carbon used.

The HOCl pore diffusivity, D_c, and the cylindrical pore diameter, d_p, are interrelated. The results of several studies[1 2] have indicated that the smaller the pore diameter, the smaller the diffusivity. A typical activated carbon, however, possesses pores that vary in diameter over a very wide range (see Figures 1 and 2), and thus it would be impossible to describe the heterogeneous reactions in each pore. The carbon surface associated with pore diameters below a certain value would not be readily accessible to the HOCl molecules, and pores possessing comparatively very large diameters would not have any appreciable surface area associated with them. Therefore, there is a certain effective pore diameter range within which most of the dechlorination reaction occurs. It is practically impossible to determine a representative value of D_c and d_p for each of the carbons used; consequently, it was assumed in this study that the values of D_c and d_p were the same for all the carbons studied.

As stated previously, if all of the carbons possessed straight cylindrical pores of one diameter, d_p, then the quantity 0.25 Apdp would represent the total pore volume per unit weight of carbon. This, however, is not actually the case, since the pore surface area A_p is not entirely accessible to HOCl and the pore diameter values vary over a very wide range of sizes. Consequently, with d_p assumed constant, the quantity 0.25 Apdp is no longer equal to the total pore volume but can be regarded as a measure of the effective or accessible pore volume within the activated carbon. This group of constants was assumed to be the only group that varied for the different carbons.

PORE VOLUME DISTRIBUTION ANALYSIS

The seven activated carbons used in this investigation were each analyzed for their pore volume distribution. Nitrogen adsorption isotherms were used to determine the distribution in the pore radius range of 10 Å to 60 Å, and a mercury porosimeter was used to analyze the distribution for the larger pores. The results of these analyses are presented in Figures 1 and 2. Figure 1 contains the pore volume distributions of the four

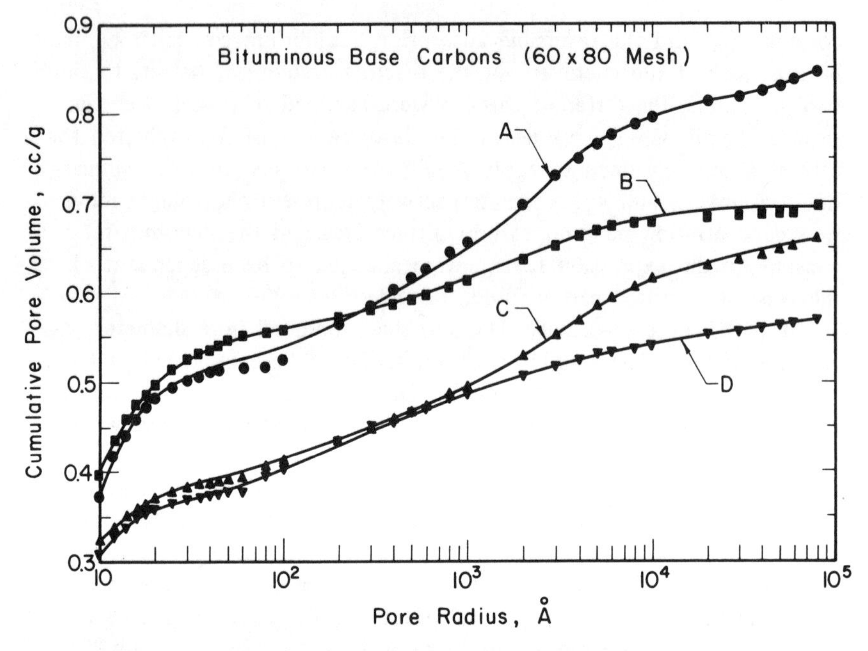

Figure 1. Pore size distribution.

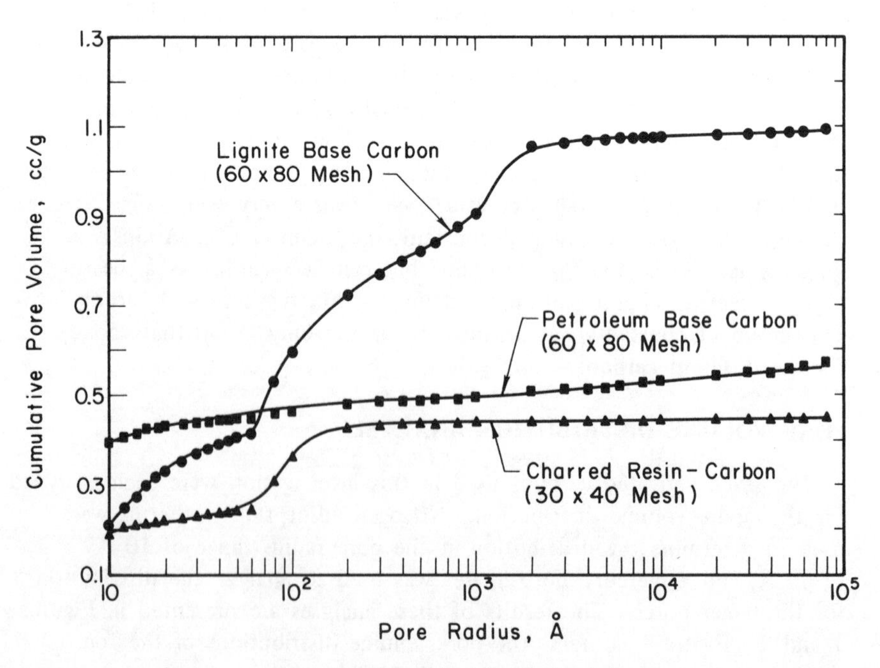

Figure 2. Pore size distribution.

bituminous base carbons A, B, C and D. Although the pore volume distributions for the four carbons vary over the whole range of pore radii, one major difference between these carbons is the pore volume associated with pore radii smaller than 10 Å. The diameter of the HOCl molecule and its waters of hydration was estimated a 11.5 Å.

Figure 2 represents the pore volume-pore radius distribution for the lignite base carbon, the petroleum base carbon and the charred resin carbon. The petroleum base carbon contains very little pore volume associated with pore radii larger than 10 Å. For the charred resin carbon, on the other hand, approximately one third of the total pore volume lies below a pore radius of 10 Å, while the remaining two thirds of the pore volume appear to be in the pore radius range between 70 Å and 120 Å. The pore volume of the lignite base carbon appears to be more uniformly distributed with approximately 90% of the volume falling within the pore radius range of 10 Å to 2000 Å. The remaining 10% is associated with pore radii below 10 Å.

COMPARATIVE BATCH STUDIES

The data in Figure 3, from earlier work,[2] represent the concentration-time results as obtained from two closed batch experiments carried out at pH 4.0 and temperature 23°C. In both experiments, the initial concentrations of HOCl and activated carbon were 30 mg/l and 10 mg/l, respectively. The carbon used in both experiments was 60 x 80 mesh bituminous base activated carbon A. As can be noted from the figure, the reproducibility of the results is excellent. The continuous curve in Figure 3 represents the mathematical model prediction using the constants given in Table I.

Similar closed batch experiments were conducted on bituminous base carbons B, C and D. The pH was maintained at 4.0 and the temperature was controlled at 23°C. In all three experiments the carbon used was 60 x 80 mesh. The initial concentrations of HOCl activated carbon were, respectively, 30.7 mg/l and 15 mg/l for the carbon B test, 31.2 mg/l and 15 mg/l for the carbon C test, and 29.6 mg/l and 10 mg/l for the carbon D test. The HOCl concentration-time data from the three batch experiments are shown in Figure 4.

The three continuous curves in Figure 4 represent the curves for the trial-and-error fit of the mathematical model to the experimental data. The three curves were obtained using the same model constants given in Table I except for k_d and 0.25 Apdp. The decay rate constant for each experiment was determined from a parallel blank run. The value of the group of constants 0.25 Apdp was estimated by a trial-and-error fit of the model to the experimental data. The values of the two groups

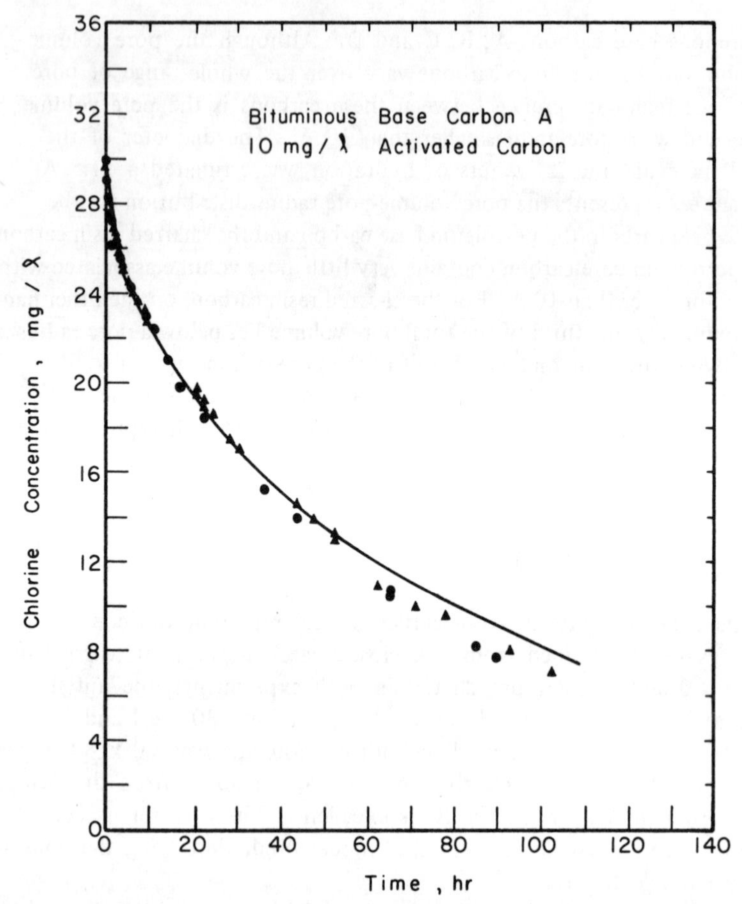

Figure 3. Chlorine carbon (60 x 80 mesh) reaction in a closed batch reactor, pH = 4, 23°C.

of constants k_d and 0.25 Apdp for each of the three experiments are given in Table II.

Similar closed batch experiments were conducted on the three other types of carbon. The pH was maintained at 4.0 and the temperature was controlled at 23°C. In each of the lignite base and the petroleum base carbon experiments, 15 mg/l of 60 x 80 mesh carbon were used, while 50 mg/l of 30 x 40 mesh carbon were used in the charred resin carbon experiment. The initial concentration of HOCl in the lignite base, petroleum base and the charred resin carbon experiments were 29.8 mg/l, 31.5 mg/l and 30.7 mg/l, respectively. The HOCl concentration vs time data for the three experiments are shown in Figure 5. The three continuous

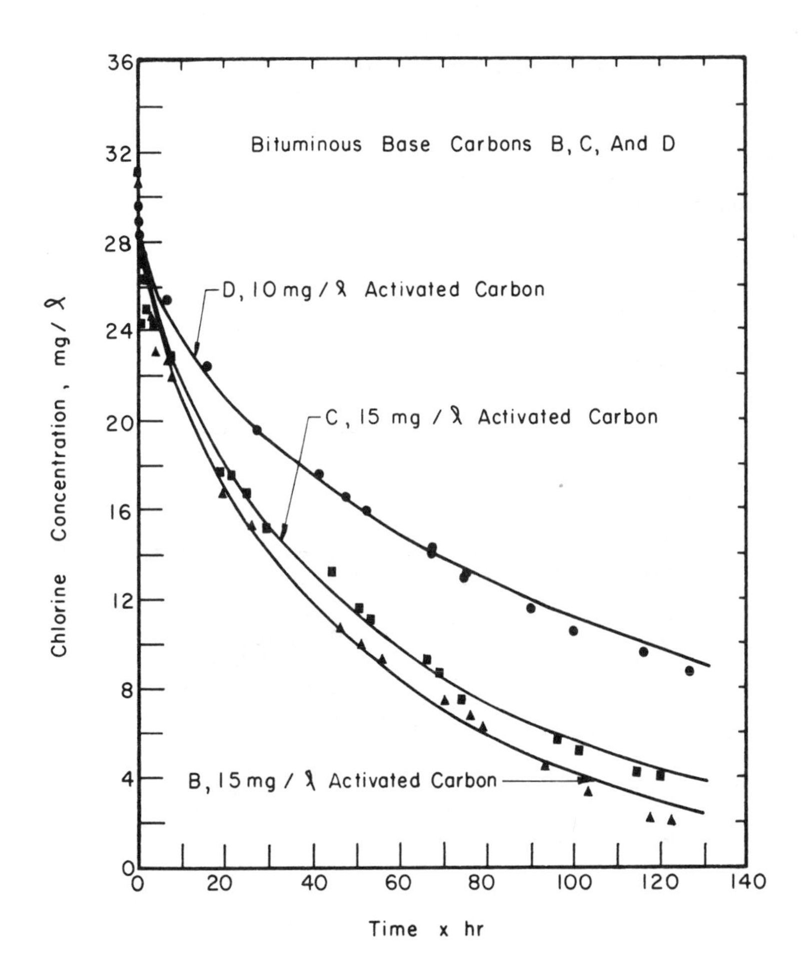

Figure 4. Chlorine carbon (60 x 80 mesh) reaction in a closed batch reactor, pH = 4, 23°C.

curves in the figure represent the mathematical model trial-and-error fit. The values of the two groups of constants k_d and 0.25 Apdp were obtained for each experiment in a similar manner as with the previous set and are given in Table II.

RELATIONSHIP WITH PORE VOLUME

The overall rate of HOCl reduction by activated carbon is directly proportional to the quantity 0.25 Apdp, provided all other parameters

Table II. Values of k_d and 0.25 Apdp for the Seven Carbons

Carbon Type	0.25 Apdp/cm^3/g	k_d min^{-1}
Bituminous Base Carbon A	0.94	1.34×10^{-5}
Bituminous Base Carbon B	0.80	1.3×10^{-5}
Bituminous Base Carbon C	0.79	8.2×10^{-6}
Bituminous Base Carbon D	0.74	9.7×10^{-6}
Lignite Base Carbon	0.91	1.2×10^{-5}
Petroleum Base Carbon	0.55	1.3×10^{-5}
Charred Resin Carbon	0.09	1.3×10^{-5}

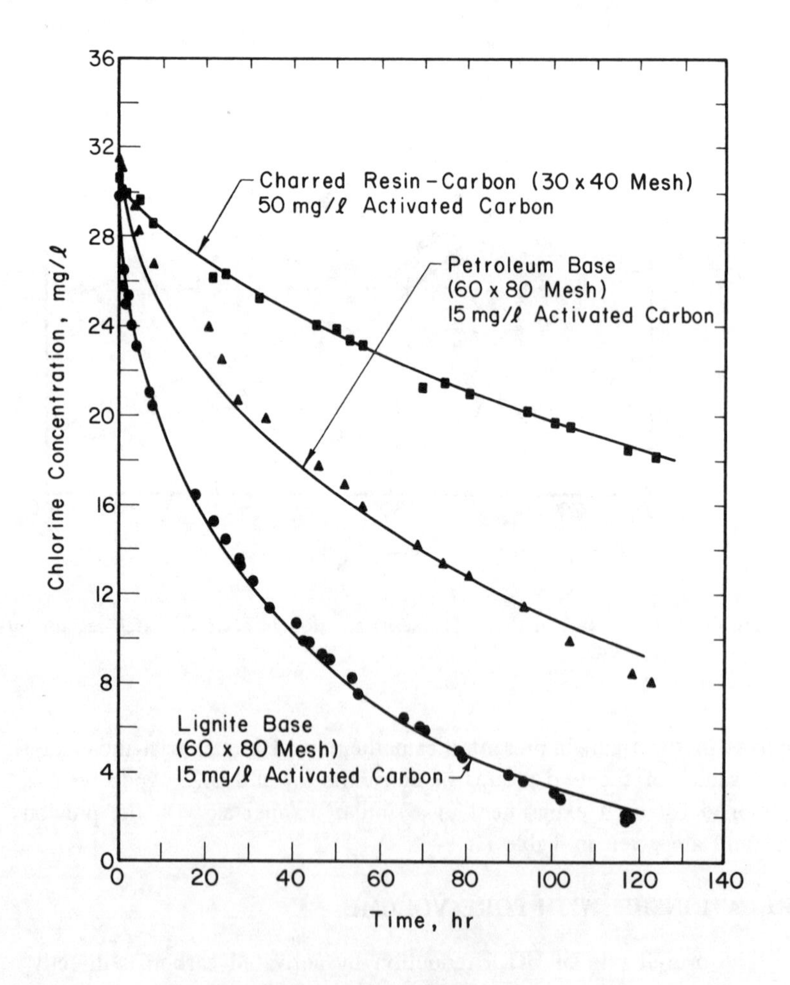

Figure 5. Chlorine-carbon reaction in a closed batch reactor, pH = 4, 23°C.

remain constant, in accordance with Equation 8. The results of the batch experiments reproduced in Table II indicate that bituminous base carbon A possesses the highest reactivity with HOCl among the seven activated carbons studied. This carbon is followed by the remaining six carbons based on their reactivity according to the following order: (2) lignite base carbon, (3) bituminous base carbon B, (4) bituminous base carbon C, (5) bituminous base carbon D, (6) petroleum base carbon, and (7) charred resin carbon.

If the carbons are truly similar in structure and surface chemistry, one would expect a linear relationship between the fitted quantity 0.25 Apdp, *i.e.*, reactivity, and pore volume. Therefore, attempts were made to establish a linear relationship between 0.25 Apdp and certain fractions of the measured pore volume distributions.

Figure 6 presents a plot of 0.25 Apdp vs the total pore volume of all seven carbons. The four bituminous base carbons disply a roughly linear relationship. The other three carbons, however, show considerable scatter. Also in this figure is a plot of 0.25 Apdp against the pore volume fraction falling between pore radii of 10 Å and 6000 Å. This latter plot produces

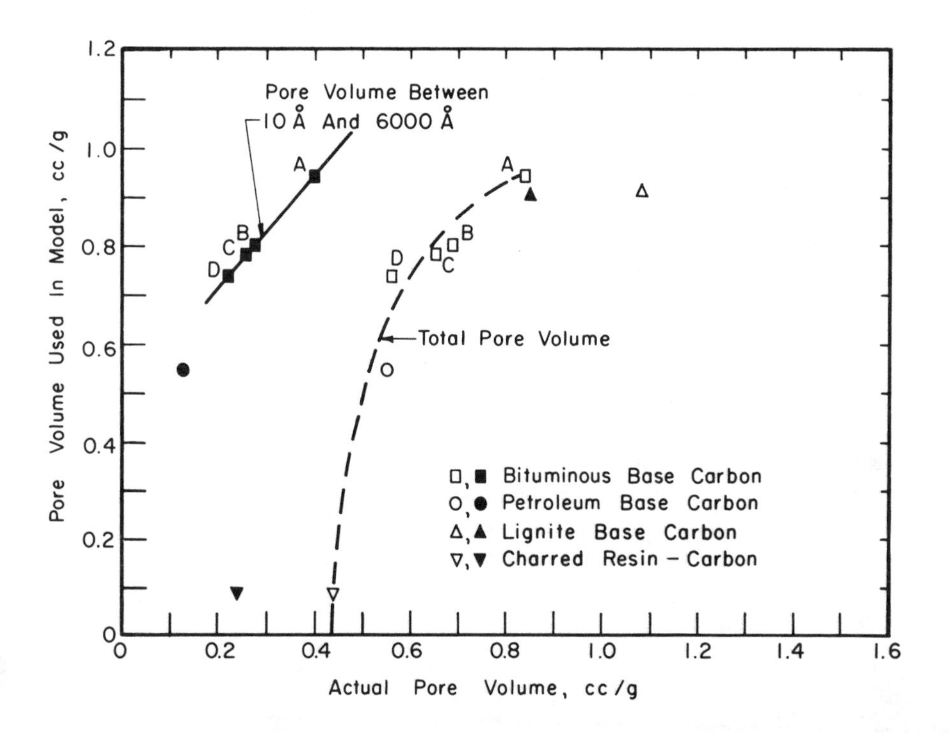

Figure 6. Pore volume relationship.

a linear relationship with excellent fit for the bituminous carbons, but the straight line does not pass through the origin. The other three carbons do not follow the linear trend.

The two plots suggest that the bituminous carbons have surface chemistries that are quite similar, at least with respect to the reaction with HOCl. Furthermore, the other three carbons differ appreciably from the bituminous carbons in surface chemistry and structure.

The plot for the 10 Å to 6000 Å pore radius fraction for the bituminous base carbons additionally suggests that it is this portion of the pore volume which is primarily involved in the dechlorination reaction, although penetration of some pores less than 10 Å radius cannot be ruled out. Pore radii above 6000 Å probably possess relatively little surface area, and, as a result, also are not important.

PACKED BED STUDIES

The ability of the packed bed mathematical model to predict the performance of a packed bed reactor has been demonstrated using constants obtained from a batch reactor. The quality of the model prediction was excellent when compared to experimental chlorine breakthrough curves obtained from columns packed with small size granular carbon (60 x 80 mesh) and from a packed bed experiment where a larger carbon granule (18 x 20 mesh) was used.[2]

Predicted Performance

Using the constants obtained in this study, the predicted performance of the seven carbons in packed beds was compared. Figure 7 shows the predictions for the seven carbons. In each case, the influent concentration, temperature and pH were 5 mg/l, 23°C, and 4.0, respectively. The carbon was 18 x 20 mesh, the length of each column was 10 cm, and the flowrate was 4 gal/min/ft^2. As is apparent from the figure, the breakthrough curves from the bituminous base carbon A and the lignite base carbon were almost indistinguishable. A similar statement could be made about the bituminous base carbon B and C runs. The results indicate that the efficiency is not very sensitive to the quantity 0.25 Apdp when the value of this quantity varied between 0.94 and 0.74 cm^3/g. Below that value the efficiency dropped rapidly, but the value for 0.25 Apdp dropped rapidly also.

CONCLUSIONS

The main accomplishment of this study was to show that a simple closed batch reactor experiment is sufficient to provide the information needed for the design of a packed bed activated carbon dechlorination reactor. The effect of particle size[2] and pH and temperature[3] on the dechlorination efficiency of bituminous base activated carbon A have already been investigated. The results from the previous studies, combined with the information provided in this paper provide sufficient information to estimate the performance of packed bed reactors using these seven carbons. For additional reliability, the column predictions should be checked with laboratory column studies using the water to be treated. This is especially true because organics may cause some interference by reacting with HOCl and/or adsorbing onto the carbon.

In the case where a new carbon is to be investigated, the procedure recommended by the authors is as follows: (1) a closed batch experiment should be conducted on the carbon in question; (2) the closed batch reactor mathematical model should then be used to estimate the value of the quantity 0.25 Apdp; (3) once the value of the quantity 0.25 Apdp has been estimated, then sufficient information is available to use the packed bed mathematical model to predict and design such a reactor; and (4) check design with lab test using water to be treated.

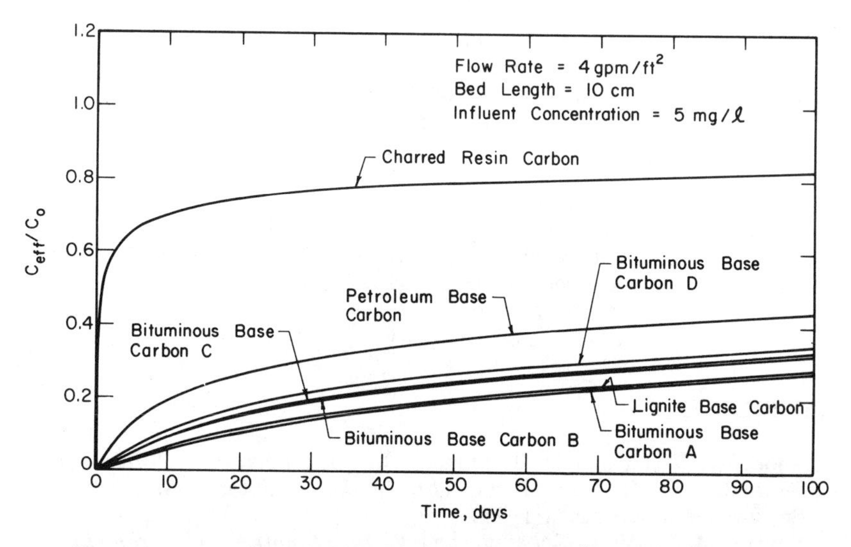

Figure 7. Chlorine breakthrough curves, pH = 4, 23°C, 18 x 20 mesh.

ACKNOWLEDGMENT

The analysis of pore volume distribution by Westvaco is gratefully acknowledged. Dennis W. Dreher was supported by an Environmental Protection Agency Traineeship.

NOTATION

A_p	pore surface area per unit weight of carbon, cm^2/g
C	concentration of HOCl, g/cm^3
C_B	bulk HOCl concentration, g/cm^3
C_o	initial or influent bulk HOCl concentration, g/cm^3
C*	reactive sites concentration per unit surface area of carbon, $sites/cm^2$
CHOCl*	HOCl surface complex, $sites/cm^2$
CO*	oxidized sites concentration per unit surface area of carbon, $sites/cm^2$
D_A	packed bed axial dispersion coefficient, cm^2/sec
D_c	free chlorine pore diffusion coefficient, cm^2/sec
d_p	diameter of straight cylindrical pore, cm
f_o	fraction of active sites occupied by surface oxides
k_1	adsorption rate constant, $cm^3/g\text{-}sec$
k_2	desorption rate constant, sec^{-1}
k_3	surface dissociation rate constant, g/sec-site
k_4	adsorption-desorption equilibrium constant, g/cm
k_5	denominator constant in expression for f_o, g/cm^2
k_d	first order rate constant for HOCl disappearance, sec^{-1}
L_b	packed bed length, cm
L_m	mean granule diameter, cm
L_p	pore half length, cm
m	mass of carbon per unit reactor volume, g/cm^3
Q	total amount of HOCl, as Cl, that has reacted per unit area of carbon, g/cm^2
$R_c(C,f_o)$	the rate of conversion of HOCl by the carbon, $g/cm^2\text{-}sec$
s	total reactive sites concentration per unit surface area of carbon, $sites/cm^2$
t	time, sec
v	interstitial velocity in packed bed reactor, cm/sec
z	position variable in packed bed model, cm
ϵ	reactor porosity
η	reduced position variable in the pore

REFERENCES

1. Snoeyink, V. L., and M. T. Suidan. In: *Disinfection of Water and Wastewater,* J. D. Johnson, Ed. (Ann Arbor, Michigan: Ann Arbor Science Publishers, Inc., 1975).
2. Suidan, M. T., V. L. Snoeyink and R. A. Schmitz. *J. Environ. Eng. Div. ASCE* (in press).

3. Suidan, M. T., V. L. Snoeyink and R. A. Schmitz. *Environ. Sci. Technol.* (in press).
4. Magee, V. *Proc. Soc. Water Treat. Exam.* 5:17 (1956).
5. Puri, B. R. In: *Chemistry and Physics of Carbon*, P. L. Walker, Editor, Vol. VI (New York: Marcel Dekker, 1970).
6. Snoeyink, V. L., H. T. Lai, J. H. Johnson and J. F. Young. In: *Chemistry of Water Supply, Treatment and Distribution,* A. Rubin, Editor (Ann Arbor, Michigan: Ann Arbor Science Publishers, 1974).
7. Levenspiel, O. *Chemical Reaction Engineering* (New York: Wiley, 1972).
8. Bischoff, K. B. *Chem. Eng. Sci.* 16:131 (1961).
9. Suidan, M. T. Ph.D. thesis, University of Illinois, Urbana, Illinois (1975).
10. *Standard Methods for the Examination of Water and Wastewater,* 13th ed. (New York: American Public Health Assoc., 1971).
11. Bird, R. B., W. E. Stewart and E. N. Lightfoot. *Transport Phenomena* (New York: Wiley, 1960).
12. Satterfield, C. N. *Mass Transfer in Heterogeneous Catalysis* (Cambridge, Massachusetts: MIT Press, 1970).

CHAPTER 11

NONVOLATILE ORGANICS IN DISINFECTED WASTEWATER EFFLUENTS: CHARACTERIZATION BY HPLC AND GC/MS

Robert L. Jolley and W. Wilson Pitt, Jr.
Oak Ridge National Laboratory
Oak Ridge, Tennessee

INTRODUCTION

Increasing awareness of the relationship between environmental pollutants and human disease has focused national attention on the necessity for identification and measurement of organic constituents in water and, also, for determination of possible reaction by-products of disinfection with chlorine and ozone. Although the major fraction of the organics in natural waters and wastewater effluents has not been identified, it probably consists of compounds with relatively low volatility. These compounds are considered to be mostly humic materials.[1,2]

Analytical methodology for determination of volatile constituents has been developed by many investigators and is comparatively advanced.[3] A principal reason for this progress is the relative ease with which volatile organics are separated from the aqueous matrix; *e.g.*, this can usually be done with simple solvent extraction or gas sparging.[4] Concentration and separation of nonvolatile organic constituents present at μg/l concentrations as complex mixtures in aqueous solutions are more difficult. Adsorption,[5] differential freezing,[6] reverse osmosis[7] and vacuum distillation[8] are several methods that have been used for concentrating nonvolatile organics in water samples. Because of its success in separating complex mixtures, liquid chromatography is a principal method for analyzing nonvolatile organics.[3,4,8]

This chapter reports the progress achieved at the Oak Ridge National Laboratory in the identification and quantification of organic constituents that are considered to have low volatility (or be nonvolatile) and are present in μg/l concentrations in disinfected effluents from wastewater treatment plants.

METHODS

Identification and measurement of micropollutants, *i.e.*, constituents present at ppb or less concentration, in water required three essential steps: concentration of constituents to detectable levels, separation and isolation of constituents, and identification of separated constituents. At the Oak Ridge National Laboratory a variety of polluted waters of environmental concern have been examined using the following experimental regimen.[8-10]

1. *Concentration.* Low-temperature (< 35°C) vacuum distillation is routinely used to concentrate water samples 500- to 5000-fold. Wastewater effluents from the primary stage are customarily concentrated 1000-fold. Secondary effluents must be concentrated to a greater degree, usually 3000-fold. After vacuum concentration, the residue is treated with acetic acid to destroy the carbonate precipitate and to redissolve occluded organics. The solution is then centrifuged to remove inorganic salts. The supernatant is frozen and lyophilized. The freeze-dried residue is taken up in chromatographic buffer and the resulting mixture is centrifuged to remove solids; finally, the volume of the supernatant is adjusted to give the desired concentration factor.

2. *Chromatography.* The organic constituents in the concentrate are detected and separated using high-pressure liquid chromatography (HPLC). The principal component of the HPLC system (see Figure 1) is a 316 stainless steel column, 1.5 m x 2.2-3.0 mm i.d., that is packed with strongly basic anion exchange resin (Bio-Rad Aminex A-27, nominal diameter of 8 to 12 μ). The chromatographic buffer used as eluent is an ammonium acetate-acetic acid (pH 4.4) solution whose acetate concentration increases from 0.015 to 6.0 *M* over a period of 24 to 36 hr at a flowrate of 10 ml/hr. Separated constituents are usually detected by UV absorbance at 254 and 280 nm and collected in fractions for subsequent identification and characterization. Alternative detection modes may also be used. For example, the eluate can be mixed with a Ce(IV)-sulfuric acid reagent and the Ce(III) fluorescence measured (see Figure 1). The cerate oxidative detector indicates the presence of oxidizable constituents.[4] In radioactivity tagging experiments, the separated tagged constituents may be detected by monitoring the eluate for radioactivity.[12]

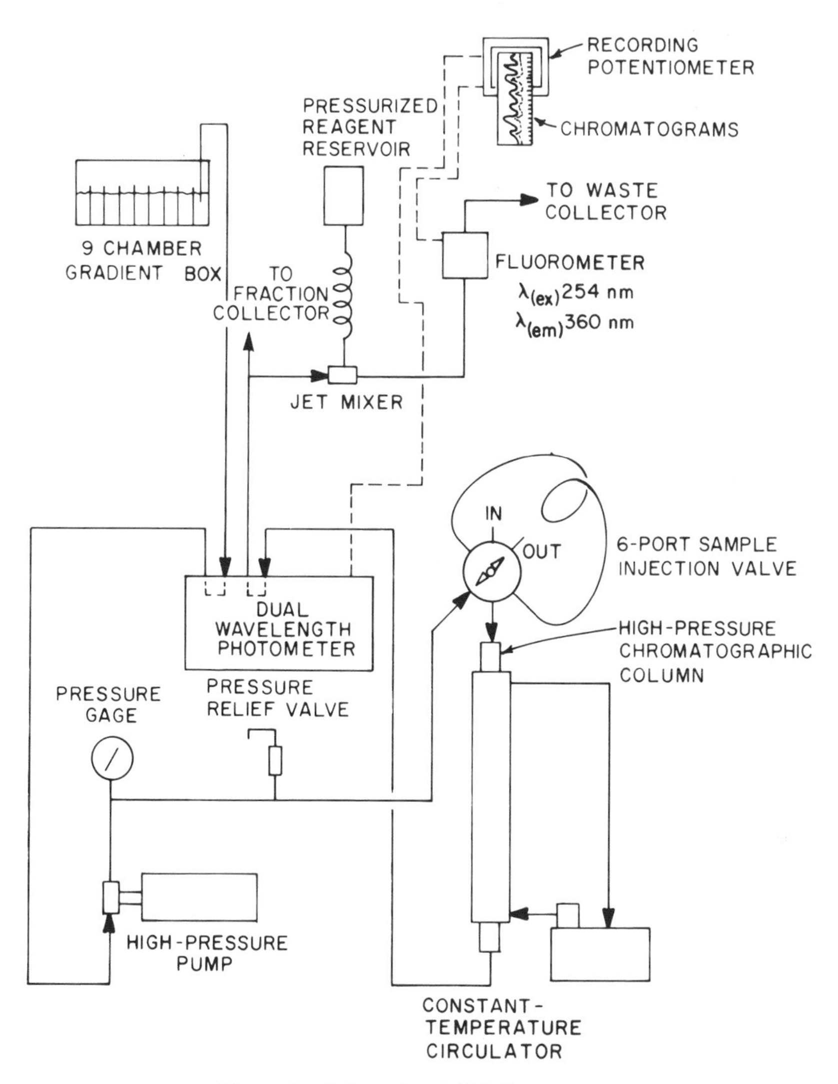

Figure 1. Schematic of HPLC system.

3. *Identification.* Characterization and identification of the separated constituents are accomplished by using several analytical methods (see Figure 2). After HPLC (anion exchange) separation, compounds eluting in the first portion of the chromatographic run (12 hr) are rechromatographed on cation exchange resin. The ammonium acetate buffer salts and water are removed from the eluate fractions by freeze-drying. The lyophilized residues are dissolved in methanol for storage. The nonvolatile

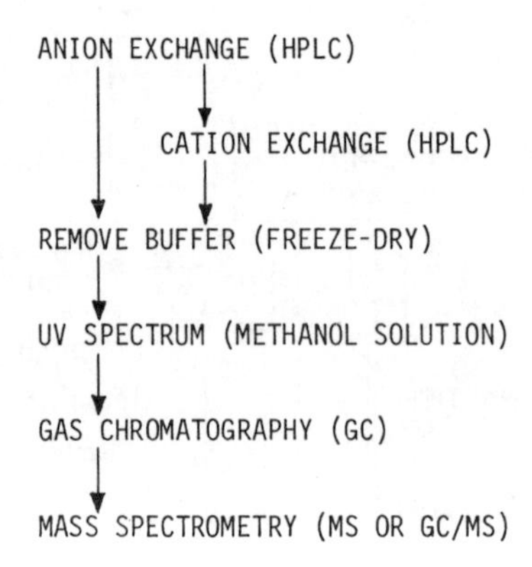

Figure 2. Identification methodology.

organic compounds in aliquots of the methanol solution are methylated or trimethylsilylated for gas chromatography or combined gas chromatography-mass spectrometry.[9,10] Tentative identifications of separated constituents are made based on liquid chromatographic elution position or retention volumes, UV absorbance and/or cerate oxidizability. Positive identification requires corroborative information from two or more analytical methods.

Wastewater effluent samples were grab samples taken from wastewater treatment plants operated by the city of Oak Ridge, Tennessee (courtesy of J. Robinson and H. E. Sturgill, Supervisors of the Oak Ridge Wastewater Treatment Plants). The wastewater of Oak Ridge is essentially the product of domestic sources with relatively little industrial contribution. Details concerning the capacity and operation of these wastewater treatment plants have been previously presented.[13]

RESULTS

Seventy-four molecular constituents that have relatively low volatility have been identified in effluents from domestic wastewater treatment plants. Most of these were isolated from effluent disinfected with chlorine to residuals (orthotolidine) of 0.5 to 1 mg/l. A typical HPLC chromatogram of a primary wastewater effluent is shown in Figure 3. Organic compounds identified to date in primary effluent are labeled in this chromatogram at their respective elution positions. Table I lists the compounds, identification methods and concentrations in the wastewater effluent. Many other constituents, not yet identified, were characterized with respect to gas chromatographic and mass spectral properties.[8]

DISCUSSION

The chemical nature and the concentrations of organic constituents in effluents from wastewater treatment plants are of considerable national interest, particularly with respect to the increasing need for water reuse and possible buildup of concentrations of relatively biorefractory constituents in receiving waters which are sources of municipal water supplies. Cause-and-effect relationships between human disease and environmental pollutants cannot be determined without quantitative data concerning these organic contaminants. Furthermore, cost-benefit analyses of disinfection processes cannot be determined without quantitative data concerning organic constituents before and after disinfection, in addition to knowledge concerning public health aspects of waterborne diseases and the relationship of such diseases to the disinfection process of wastewater effluents.

The data summarized in this chapter represent a first step toward developing the information needed for evaluating disinfection processes. Many simple chlorinated organics are formed during the disinfection of wastewater with chlorine. The relative toxicity of these compounds at ppb concentrations, their chemical stability and transport in the aquatic environment, and their significance in sources of potable water supplies remain to be established. Similar HPLC separations and GC/MS characterizations are planned for wastewater effluents disinfected with ozone and UV irradiation. The data determined in these projected studies should be helpful in evaluating these alternative disinfection methods.

ACKNOWLEDGMENTS

The authors wish to express their gratitude to G. Jones, Jr., and J. E. Thompson for their technical assistance, and to J. E. Attrill, C. W. Hancher, and M. G. Stewart for critically reading the manuscript.

This research was supported by the Energy Research and Development Administration and the U.S. Environmental Protection Agency. Oak Ridge National Laboratory is operated by Union Carbide Corporation for the Energy Research and Development Administration.

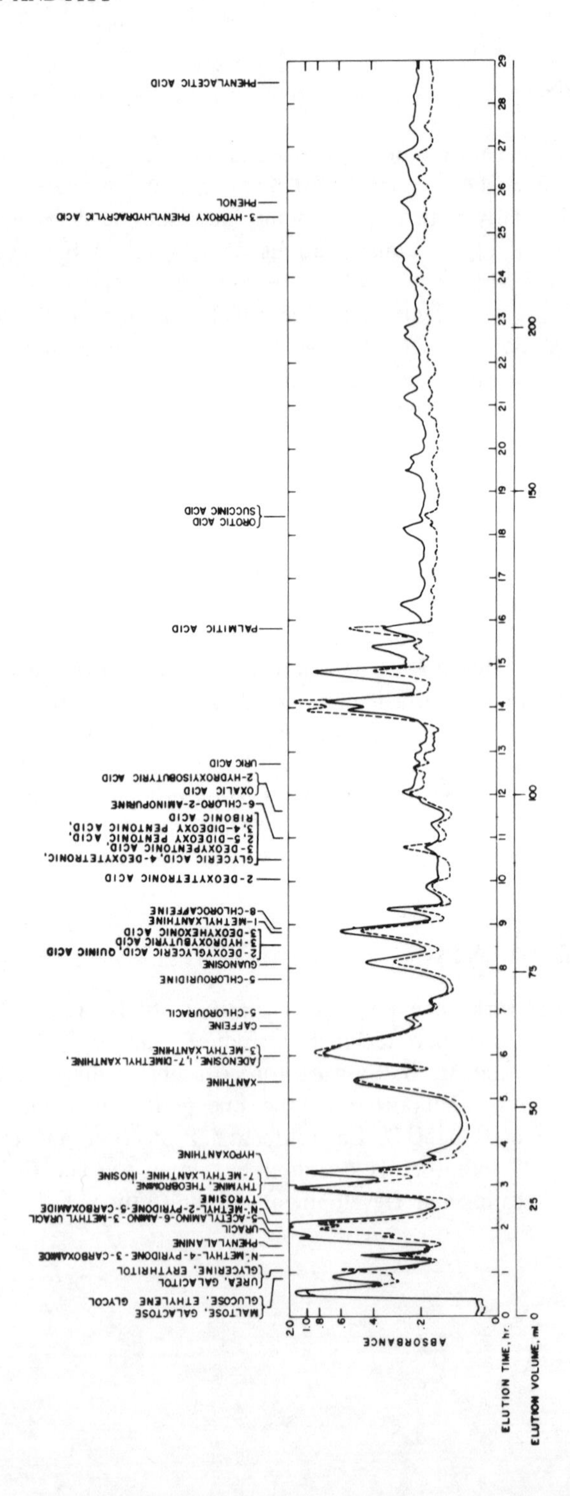
PHENYLACETIC ACID
PHENOL
3-HYDROXY PHENYLHYDRACRYLIC ACID
SUCCINIC ACID
OROTIC ACID
PALMITIC ACID
URIC ACID
2-HYDROXYISOBUTYRIC ACID
OXALIC ACID
6-CHLORO-2-AMINOPURINE
RIBONIC ACID
3,4-DIDEOXY PENTONIC ACID,
2,5-DIDEOXY PENTONIC ACID,
3-DEOXYPENTONIC ACID,
GLYCERIC ACID, 4-DEOXYTETRONIC,
2-DEOXYTETRONIC ACID
8-CHLOROCAFFEINE
1-METHYLXANTHINE
3-DEOXYHEXONIC ACID
3-HYDROXYBUTYRIC ACID
2-DEOXYGLYCERIC ACID, QUINIC ACID
GUANOSINE
5-CHLOROURIDINE
5-CHLOROURACIL
CAFFEINE
3-METHYLXANTHINE
ADENOSINE, 1,7-DIMETHYLXANTHINE,
XANTHINE
HYPOXANTHINE
7-METHYLXANTHINE, INOSINE
THYMINE, THEOBROMINE,
TYROSINE
N-METHYL-2-PYRIDONE-5-CARBOXAMIDE
5-ACETYLAMINO-6-AMINO-3-METHYL URACIL
URACIL
PHENYLALANINE
N-METHYL-4-PYRIDONE-3-CARBOXAMIDE
GLYCERINE, ERYTHRITOL
UREA, GALACTITOL
GLUCOSE, ETHYLENE GLYCOL
MALTOSE, GALACTOSE
ABSORBANCE
ELUTION TIME, hr.
ELUTION VOLUME, ml

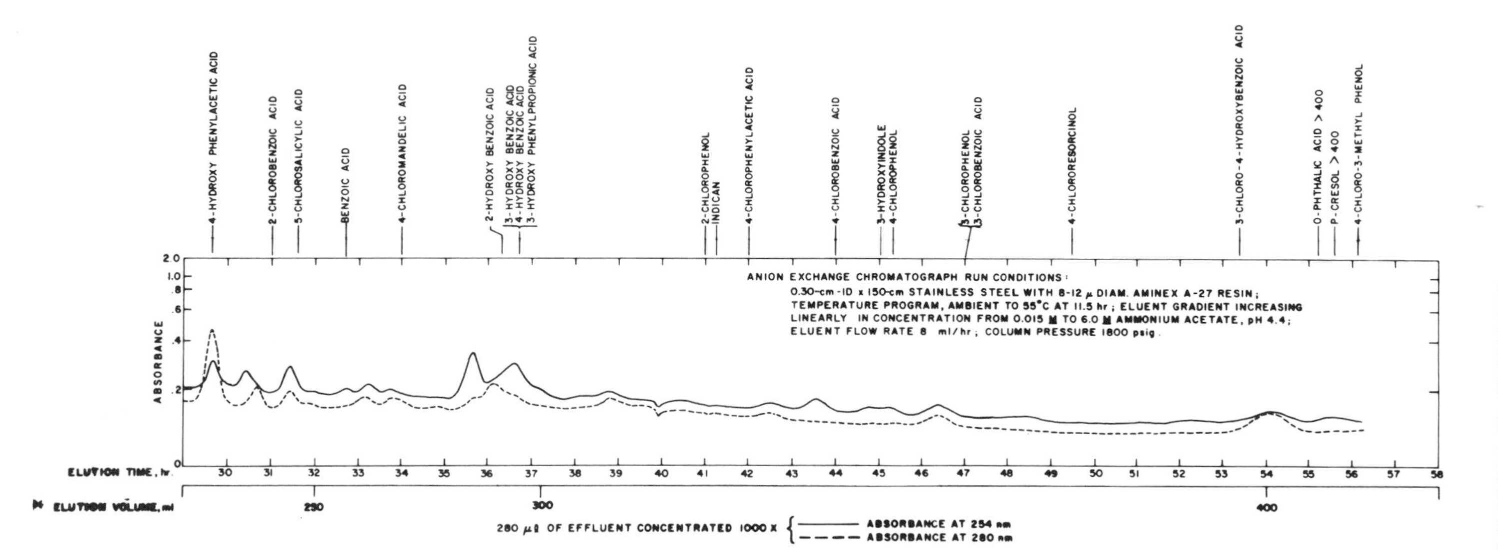

Figure 3. HPLC chromatogram of nonvolatile organics in effluent from the primary stage of a wastewater treatment plant. Composite identifications from several chromatograms are labeled at their respective elution positions.

Table I. Nonvolatile Molecular Constituents in Effluents from Municipal Wastewater Treatment Plants

Constituent[a]	Identification Method[b]	Concentration (μg/l)
Carbohydrates and Polyols		
Erythritol	AC,GC,MS	5
Ethylene Glycol	AC,GC,MS	3
Galacitol	AC,GC,MS	2
Galactose*	AC,GC	
Glucose*	AC,GC	
Glycerine	AC,GC,MS	4-20
Maltose	AC,GC	0.5
Aliphatic Organic Acids		
3-Deoxy-*arabino*-hexonic Acid	MS	7
3-Deoxy-*erythro*-pentonic Acid	MS	4
2-Deoxyglyceric Acid	MS	7
2,5-Dideoxypentonic Acid	MS	6
3,4-Dideoxypentonic Acid	MS	10
2-Deoxytetronic Acid	MS	6
4-Deoxytetronic Acid	MS	6
Glyceric Acid	MS	5
4-Hydroxybutyric Acid	GC,MS	6
2-Hydroxyisobutyric Acid	GC,MS	4
Oxalic Acid	AC,GC,MS	2
Quinic Acid	MS	50
Ribonic Acid	MS	4
Succinic Acid	AC,GC,MS	20
Aromatic Organic Acids		
Benzoic Acid	AC,GC,MS	3
2-Hydroxybenzoic Acid	AC,GC,MS	2-7
2-Chlorobenzoic Acid	AC	0.3
3-Chlorobenzoic Acid	AC	0.6
4-Chlorobenzoic Acid	AC	1
3-Chloro-4-hydroxybenzoic Acid	AC	1
4-Chloromandelic Acid	AC	1
4-Chlorophenylacetic Acid	AC	0.4
5-Chloro-2-hydroxybenzoic Acid	AC	0.2
3-Hydroxybenzoic Acid*	AC,GC,MS	7-40
4-Hydroxybenzoic Acid*	AC,GC	1
4-Hydroxyphenylacetic Acid	AC,GC,MS	20-190
3-Hydroxyphenylhydracrylic Acid	AC,UV,GC,MS	10-20
3-Hydroxyphenylpropionic Acid*	AC,GC,MS	6-20
Phenylacetic Acid	AC,GC	10
o-Phthalic Acid	AC,UV,MS	200
Fatty Acids		
Palmitic Acid	GC,MS	6-10
Amino Acids		
Phenylalanine	AC,GC,MS	50-90
Tyrosine	AC,GC,MS	30

Table I., continued

Constituent[a]	Identification Method[b]	Concentration (μg/l)
Amides		
Urea	AC,GC,MS	20-40
Phenolic Compounds		
Catechol*	MS	1
4-Chloro-3-methylphenol	AC	2
2-Chlorophenol	AC	2
3-Chlorophenol	AC	2
4-Chlorophenol	AC	0.7
4-Chlororesorcinol	AC	1
p-Cresol	AC,GC,MS	20-30
Phenol	AC,GC,MS	6-10
Indoles		
3-Hydroxyindole	MS	2
Indican	AC,GC,F	1-2
Indole-3-acetic Acid*	MS	10
Purine Derivatives		
Adenosine	AC,CC,UV,GC,MS	10
Caffeine	AC,CC,GC,MS	10-50
8-Chlorocaffeine	AC	2
6-Chloro-2-aminopurine	AC	0.9
8-Chloroxanthine	AC	2
1,7-Dimethylxanthine*	AC,CC	6
Guanosine	AC,CC,UV,GC,MS	4-50
Hypoxanthine	AC,GC,MS	10-40
Inosine	AC,CC,UV,GC,MS	10-50
1-Methylinosine*	AC,CC,UV	80
1-Methylxanthine*	AC,CC,UV	6-70
3-Methylxanthine*	AC,CC	
7-Methylxanthine	AC,CC,GC	2-90
Theobromine*	AC,CC	
Uric Acid*	AC,GC,MS	20
Xanthine	AC,CC,UV,GC,MS	2-70
Pyrimidine Derivatives		
5-Acetylamino-6-amino-3-methyluracil*	AC,CC,UV,GC	30-140
5-Chlorouracil	AC,CC	4
5-Chlorouridine	AC	2
Orotic Acid	AC,UV,GC,MS	2-5
Thymine	AC,CC,GC,MS	7-30
Uracil	AC,CC,UV,GC,MS	20-60

[a]All constituents were identified in chlorinated effluents except those designated with an asterisk.

[b]AC–anion exchange chromatography; CC–cation exchange chromatography; UV–ultraviolet spectroscopy; GC–gas chromatography on two columns; MS–mass spectrometry; F–fluorometry.

REFERENCES

1. Christman, R. F., and R. A. Minear. In: *Organic Compounds in Aquatic Environments,* S. D. Faust and J. V. Hunter, Eds. (New York: Marcel Dekker, Inc., 1971).
2. Rebhun, M., and J. Manka, *Environ. Sci. Technol.*, 5:606 (1971).
3. Keith, L. H., Ed. *Identification and Analysis of Organic Pollutants in Water* (Ann Arbor, MI: Ann Arbor Science Publishers, Inc., 1976).
4. Garrison, A. W., J. D. Pope and F. R. Allen. In: *Identification and Analysis of Organic Pollutants in Water,* L. H. Keith, Ed. (Ann Arbor, MI: Ann Arbor Science Publishers, Inc., 1976).
5. Junk, G. A., J. J. Richard, M. D. Grieser, D. Witiak, J. L. Witiak, M. D. Arguello, R. Vick, H. J. Svec, J. S. Fritz and G. V. Calder. *J. Chromatogr.* 99:745 (1974).
6. Baker, R. A. *Water Res.* 1:97 (1967).
7. Kopfler, F. C. "Extraction and Identification of Organic Micropollutants–Reverse Osmosis Methods," in *Proc. Conf. Aquatic Pollutants and Biological Effects with Emphasis on Neoplasia, Ann. N.Y. Acad. Sci.* (in press).
8. Pitt, W. W., Jr., R. L. Jolley and S. Katz. "Automated Analysis of Individual Refractory Organics in Polluted Water," EPA-660/2-74-076, U.S. Environmental Protection Agency, Washington, D.C. (August 1974).
9. Jolley, R. L., S. Katz, J. E. Mrochek, W. W. Pitt, Jr., and W. T. Rainey. *Chem. Technol.* 1975:312 (1975).
10. Pitt, W. W., Jr., R. L. Jolley and C. D. Scott. *Environ. Sci. Technol.* 9:1068 (1975).
11. Katz, S., and W. W. Pitt, Jr. *Anal. Lett.* 5:177 (1972).
12. Jolley, R. L. *Environ. Lett.* 7:321 (1974).
13. Jolley, R. L. *J. Water Poll. Control Fed.* 47:601 (1975).

CHAPTER 12

EFFECT OF CHLORINATION ON DIFFERENTIATED COLIFORM GROUPS AFTER DILUTION OF A TREATED EFFLUENT

E. Crispin Kinney

United States Public Health Service
Office of Environmental Health
Anchorage, Alaska

David W. Drummond and N. Bruce Hanes

Civil Engineering Department
Tufts University
Medford, Massachusetts

INTRODUCTION

The primary objective of chlorination is disinfection. Research has shown that although short-term disinfection can be achieved with chlorine, chlorination may have questionable long-term effects. The practice of chlorination of wastewater has been shown to produce chlorinated compounds,[1-3] some of which are believed to be carcinogenic to humans. It has also been demonstrated that chlorination of wastewater produces an effluent with increased toxicity to aquatic life.[1,4,5] On the other hand, apparent bacterial regrowth has been observed in chlorinated wastewaters.[6-10]

For the most part, these studies of bacterial regrowth have been concerned with the total and fecal coliform groups of bacteria. Various conclusions can be drawn concerning the significance of this regrowth to public health, depending on the assumptions which are made concerning the relationship between the coliform group of bacteria and pathogens. It has been pointed out in the literature that this relationship is ambiguous because both the total and fecal coliform groups are heterogeneous with respect to genera of organisms.[11] In addition, neither the total nor the fecal coliform groups necessarily simulate the behavior of microbial pathogens in aquatic environments.[11]

Given these shortcomings, it is difficult to interpret the significance of coliform bacterial regrowth following chlorination. The membrane filter method (mC) developed by Dufour and Cabelli[11] enables the rapid differentiation of three major subgroups of the total coliform group of bacteria. These groupings are *Escherichia coli*, *Klebsiella* sp. and *Enterobacter-Citrobacter* sp. The information obtained by the mC procedure should be helpful in determining the behavior of the various genera of coliform bacteria and assists in the interpretation of the coliform test.

The purpose of this study was to investigate the phenomenon of apparent regrowth of these three major subgroups of the total coliform group of bacteria in a chlorinated wastewater effluent. A comparison was also made between the apparent regrowth of these bacteria in the chlorinated effluent and their survival in a nonchlorinated effluent. If the long-term effect of chlorination on the death of bacteria is not significantly different from the effect of natural dieoff, the practice of chlorination must be questioned, especially in cases where a wastewater will not come into human contact for a considerable time period after discharge.

MATERIALS AND METHODS

All laboratory procedures, unless otherwise stated, were performed in accordance with those recommended in *Standard Methods.*[12] Analyses were performed in random order to facilitate statistical tests.

Sampling

The wastewater effluent used in this study was obtained at the Marlborough Easterly Wastewater Treatment Plant, Marlborough, Massachusetts. This plant consists of a two-stage activated sludge system with alum addition for phosphorus removal. The effluent was collected from the tertiary clarifier overflow prior to chlorination. The diluting streamwater used in this study was obtained from Hop Brook at a location above the point of discharge of the treatment plant effluent. Flow of the wastewater during the study averaged 3.24 mgd (12,300 m^3/day). Sampling was performed on weekdays at approximately 10:00 a.m. Samples were brought to the laboratory in less than one hour and were stored at 4°C until used. In no case did the storage period exceed six hours.

Chlorine Determinations

Sodium hypochlorite was added to effluent samples in an amount necessary to achieve a 1 mg/l total chlorine residual after a 15-minute contact period. The chlorinated samples were mixed during the 15-minute contact time. Chlorine determinations were performed by the amperometric method, using the back titration modification.[12] Chlorine residuals were divided into free available, combined available and total available chlorine. The amount of hypochlorite actually applied and the resulting residuals are given in Table I.

Table I. Chlorine Concentrations in Marlborough Easterly Wastewater Treatment Plant Effluent

Experiment Number	Dosage Applied (mg/l Cl_2)	Chlorine Concentrations (mg/l Cl_2) Residuals After 15 Minute Contact		
		Free Available	Combined Available	Total Available
1	2.35	0.40	0.55	0.95
2	3.55	0.65	0.45	1.05

Experimental Procedure

The experimental procedure is presented in Figure 1. The effluent sample was split into two halves. One-half of the effluent sample was chlorinated to approximately 1 mg/l total available residual after a 15-minute contact time. The actual residual value was measured immediately (Table I). The other half of the effluent sample was not chlorinated. The unchlorinated and chlorinated portions of the effluent were combined with the Hop Brook sample in the ratio of three parts effluent and one part receiving stream. This ratio was proportional to the flows of the effluent and receiving stream at Marlborough. The resulting mixtures were placed in four-liter flasks. Three or four replicate samples were prepared for each of the two conditions. The mixtures were placed on a shaker table and were kept at 20°C in a constant temperature room. The samples were maintained in the dark for the five to seven day period of the study. Five-day biochemical oxygen demand (BOD_5) values were determined for the unchlorinated and chlorinated effluent, Hop Brook, and the effluent-stream mixtures (Table II).

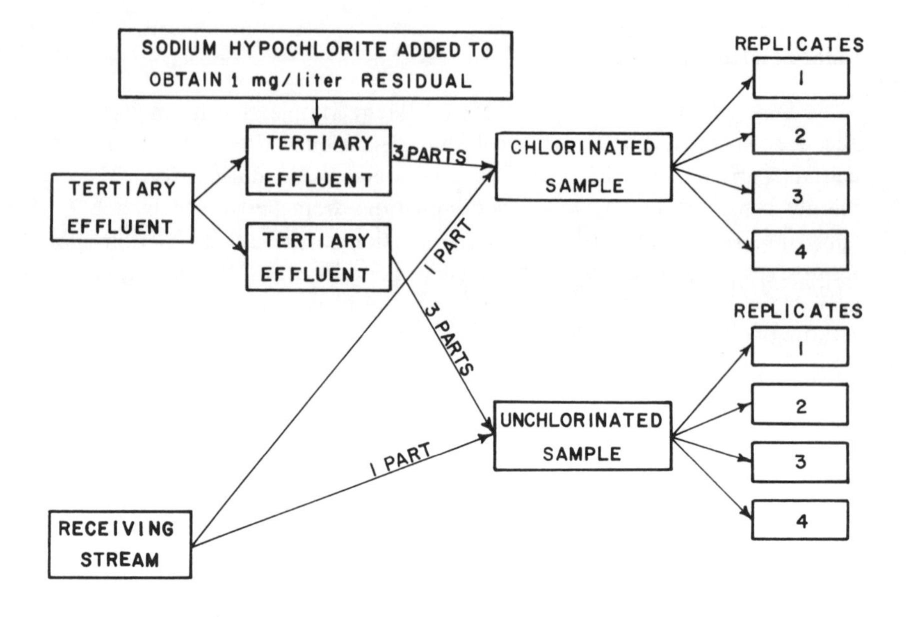

Figure 1. Summary of experimental procedure for unchlorinated and chlorinated effluent.

Table II. Biochemical Oxygen Demand Values of Various Samples

Sample	BOD_5 (mg/l O_2)	
	Experiment 1	Experiment 2
1 Marlborough Effluent; Unchlorinated	11.5	7.2
2 Marlborough Effluent; Chlorinated	2.0	4.1
3 Hop Brook	2.2	2.6
4 Mixture of 1 and 3	10.5	6.9
5 Mixture of 2 and 3	2.2	3.4

Bacteriological Methods

Total coliform bacteria were measured by the membrane filter method (mC) developed by Dufour and Cabelli.[11] This procedure enabled the quantitative differentiation of the total coliforms into *Klebsiella, E. coli* and *Enterobacter-Citrobacter.* The mC medium of Dufour and Cabelli was used with Millipore-type HA membrane filters. The oxidase step suggested by Dufour and Cabelli was not used after preliminary work indicated it was not necessary for the samples. During the study, crowding of the mC bacterial plates by background colonies occurred, particularly in the samples from chlorinated waters. To alleviate this problem, the minimum recommended incubation time was used. Buffered saline was used as dilution water.

Bacteria were measured daily in the mixtures of effluent and streamwater. Initial concentrations of bacteria were also determined in the Marlborough effluent, both chlorinated and unchlorinated, as well as in the Hop Brook samples (Table III).

Data Analysis

Logarithmic values of the bacterial concentrations were used throughout the analysis of the data. The apparent growth and death appeared to be logarithmic in nature and the error appeared to be a constant percentage of the bacterial densities.

The data analysis was divided into three parts. First it was desired to determine if bacterial concentrations were similar in the unchlorinated and chlorinated samples after a period of time. The Smith-Satterthwaite modification[13] of the independent t-test (one-tail) was used because it was determined that the assumption of equal variances in the two samples was invalid. Second, the initial bacterial concentrations in the chlorinated samples were compared to the peak values for each of the coliform groups

Table III. Initial Concentrations of Differentiated Coliform Groups

Differentiated Coliform Group	Experiment No.	Bacterial Concentrations (cells/100 ml)		
		Hop Brook	Unchlorinated Effluent	Chlorinated Effluent
Total Coliforms	1 1	190	72,000	56
	2 2	25	15,000	180
Klebsiella	1 1	53	18,000	6[a]
	2 2	6	8,100	130
Enterobacter-Citrobacter	1 1	27	32,000	56
	2 2	6	5,000	10
E. coli	1 1	110	22,000	6[a]
	2 2	13	1,900	6

[a]Maximum estimates based on volume filtered.

to determine if the apparent regrowth was significant. A one-tail independent t-test was used for this analysis. Third, relative growths of the different bacterial groups in the chlorinated samples were compared by computing the area under their growth curves. The area was corrected for the initial concentrations by computing only the area above a horizontal line representing the initial value. The computed area was termed "log cell-days." In all tests of significance the 95% confidence level was used.

RESULTS AND DISCUSSION

General

Two experiments were performed with similar results. All of the various coliform groups increased in number after chlorination. Peak bacterial densities were achieved in five to seven days. The bacteria died rapidly in the samples which had not been chlorinated. Data describing these trends for Experiment 1 are presented in Figure 2.

The increase in bacterial numbers that occurred in the chlorinated samples agrees with earlier studies which dealt with sewage,[6,7] stormwater[8,9] and a secondary effluent.[10] However, this study is the first to demonstrate the apparent regrowth phenomena in an advanced wastewater treatment plant effluent following chlorination. Inspection of the trends in Figure 2 suggests that the mechanism causing the increasing bacterial numbers following chlorination is recovery of damaged cells rather than growth of surviving cells. It is unlikely that growth of the same microorganisms would occur in the chlorinated effluent-river mixtures while death of these organisms is simultaneously occurring in the matched unchlorinated effluent-river mixtures.

Equivalency

Initially the densities of all the coliform groups were found to be significantly different in the chlorinated and unchlorinated samples. After a period of one to four days, however, these differences became statistically insignificant as shown in Table IV. This equivalency of bacterial numbers was the result of both increasing numbers in the chlorinated samples and decreasing numbers in the unchlorinated samples. Thus, it appears that chlorination has little effect on the numbers of coliform bacteria after a certain time interval. The bacterial behavior observed suggests that the need for chlorination of a well-treated effluent may be seriously questioned in some cases. Chlorination may not be needed where a receiving water is not used for water supply, recreation, irrigation or shellfish harvesting in the vicinity of an effluent outfall. In

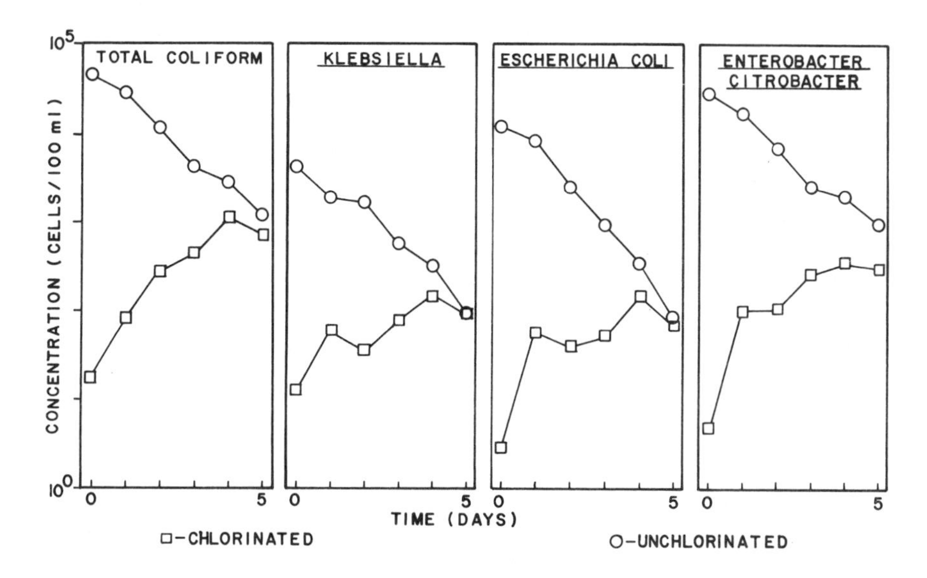

Figure 2. Log mean concentrations for coliform bacteria in unchlorinated and chlorinated samples.

Table IV. Comparison of Unchlorinated and Chlorinated Samples at Various Times Using Smith-Satterthwaite Test (One-Tail) at 95% Confidence Level[a]

Differentiated Coliform Groups	Experiment No.	Time (Days)							
		0	1	2	3	4	5	6	7
Total	1	+	+	+	+	-	-		
Coliform	2	+	-	-	-	-	-	-	-
Klebsiella	1	+	+	+	+	-	-		
	2	+	+	+	-	-	-	-	-
Enterobacter-	1	+	+	+	+	-	-		
Citrobacter	2	+	-	-	-	-	-	-	-
E. coli	1	+	+	+	+	-	-		
	2	+	+	+	-	-	-	-	-

[a]"+" Numbers are significantly different. "-" Numbers are not significantly different.

these cases, more effective disinfection of a well-treated effluent may be achieved by allowing bacteria to die naturally in a receiving stream rather than by chlorinating, which also produces potentially carcinogenic compounds and increased toxicity of the effluent to aquatic life.

It is difficult to evaluate the meaning of coliform concentrations obtained either at the outfall of a chlorinated effluent or farther downstream until the behavior of pathogens following chlorination is known. If it is assumed that pathogenic bacteria similarly increase in number, then the current practice of monitoring coliform concentrations immediately after chlorination is potentially dangerous because higher concentrations of pathogenic organisms will occur downstream. If it is assumed that pathogenic bacteria do not regrow following chlorination, then although coliform concentrations at an outfall probably would be related to the pathogenic concentrations, the coliform concentrations downstream would be unrealistic indicators of pathogens and of the safety of water. In either case a more rational approach to evaluating the effectiveness of chlorination is required. Determination of the bacterial concentrations of a sample immediately after chlorination and after a period of storage would provide valuable information concerning the true numbers of viable coliform bacteria surviving chlorination.

Coliform Regrowth in Chlorinated Samples

The data were analyzed to determine if the apparent regrowth of the coliform bacteria was statistically significant. It was found that all of the differentiated coliform groups exhibited a significant increase in numbers in both experiments reaching peak values in four to five days (Table V).

The areas under the apparent growth curves were computed to determine the duration and magnitude of the increased bacterial numbers and the results are given in Table VI.

The *Enterobacter-Citrobacter* coliform group regrew to the greatest extent in both experiments, achieving the largest peak values (Table V). This same group also maintained the highest numbers of organisms for the longest period of time as indicated by the areas under the various growth curves (Table VI).

The initial numbers of *Klebsiella* in the chlorinated samples were the highest of the subgroups of total coliforms (Table V). This finding agreed with those of Ptak *et al.*[14] who found *Klebsiella* to account for 59% of the total coliforms found in chlorinated water samples. *Klebsiella* also demonstrated the least regrowth of the coliform groups (Table VI). The resistance to chlorination and apparent regrowth of *Klebsiella* may be important to public health since certain species of this genus are considered to be pathogenic.[15]

CONCLUSIONS

The bacterial behavior observed in this study questions the need for chlorination of a well-treated effluent in which bacterial death occurs without disinfection. In this case, more effective and safer disinfection of

Table V. Comparison of Initial and Peak Bacterial Concentrations in Chlorinated Samples; One-Tailed Paired T-Test Performed at 95% Confidence Level on Logs of Data

Differentiated Coliform Group		Log Mean Value (cells/100 ml)		Significance	Peak Time (days)
		Initial	Peak		
Total	1	17	1,100	+	4
Coliform	2	81	14,000	+	5
Klebsiella	1	13	160	+	4
	2	78	1,900	+	5
Enterobacter-	1	5	360	+	4
Citrobacter	2	3	7,500	+	5
E. coli	1	3	150	+	4
	2	2	290	+	5

Table VI. Area Under Growth Curves (Log Cell-Days) of Chlorinated Samples; Corrected for Initial Values

Differentiated Coliform Group	Log Cell-Days (Mean Value)	
	Experiment 1 (4 replicates)	Experiment 2 (3 replicates)
Total Coliforms	6.0	7.3
Klebsiella	3.5	3.9
Enterobacter-Citrobacter	6.9	11.8
E. coli	4.9	7.3

a well-treated effluent may be achieved by allowing bacteria to die naturally in a receiving stream which does not come into human contact near the plant outfall.

The results of this study also suggest that a different technique, a delayed bacterial analysis of stored samples, should be used for the evaluation of the efficiency of the chlorination procedure.

Finally these data provide circumstantial evidence that the increase in numbers of coliform bacteria following chlorination is the result of the recovery of damaged cells rather than bacterial growth.

REFERENCES

1. "Disinfection of Wastewater," USEPA, Office of Research and Development, Washington, DC (1975).
2. Murphy, K. L., R. Zaloum, and D. Fulford. *Water Res.* 9:389 (1975).
3. Jolley, R. L., and W. W. Pitt, Jr. "Chloro-Organic By-Products from Sewage Effluent Chlorination," Proc. ASCE National Conf. Environmental Engineering Research Development and Design, Seattle, Washington (1976).
4. Bradley, R. M. *Effluent Water Treat. J.* 683 (1973).
5. Esvelt, L. A., W. S. Kaufman and R. E. Selleck. *J. Water Poll. Control Fed.* 45:1558 (1973).
6. Rudolfs, W., and H. W. Gehm. "Sewage Chlorination Studies," New Jersey Agricultural Experimental Station, New Brunswick, New Jersey, Bulletin 601 (1936).
7. Heukelekian, H. *Sew. Ind. Wastes* 23:273 (1951).
8. Eliassen, R. *J. San. Eng., ASCE* 94:371 (1968).
9. Evans, F. L., E. E. Geldreich, S. R. Weibel and G. G. Roebeck. *J. Water Poll. Control Fed.* 40:R162 (1968).
10. Shuval, H. I., J. Cohen and R. Kolodney. *Water Res.* 7:537 (1973).
11. Dufour, A. P., and V. J. Cabelli. *Appl. Microbiol.* 29:826 (1975).
12. *Standard Methods for the Examination of Water and Wastewater*, 13th ed. (Washington, DC: American Public Health Association, 1971).
13. Miller, I., and J. E. Freund. *Probability and Statistics for Engineers* (Englewood Cliffs, NJ: Prentice-Hall, 1965).
14. Ptak, D. J., W. Ginsburg and B. F. Willey. *J. Am. Water Works Assoc.* 65:604 (1973).
15. "*Klebsiella pneumoniae* Infection: A Review with Reference to the Waterborne Epidemiological Significance of *K. pneumonia* Presence in the Natural Environment," Technical Bulletin No. 254, National Council of the Paper Industry for Air and Stream Improvement, Inc., New York (1972).

CHAPTER 13
CHEMISTRY AND DISINFECTANT PROPERTIES OF BROMINE CHLORIDE

Jack F. Mills
Halogens Research Laboratory
Dow Chemical USA
Midland, Michigan

INTRODUCTION

The disinfection of wastewater is of obvious public health importance since pathogenic organisms, if not destroyed, could be transmitted to man through sewage contamination of water for drinking, food processing, irrigation, shellfish culture or recreational purposes. A variety of infectious microorganisms are generally found in municipal wastewaters.[1] As a result, outbreaks of gastroenteritis, typhoid, shigellosis, salmonellosis, ear infections and infectious hepatitis have been reported among individuals drinking or swimming in sewage-contaminated waters or consuming raw shellfish harvested therefrom.[2,3] The most serious of these outbreaks seems to be due to the enteric viruses. There are over 100 viruses shed from the enteric and urinary tracts of humans. These viruses may pass through waste treatment plants, persist in the receiving waters and enter potable water supply facilities. Using specially developed sampling techniques in Houston, TX, nearly 100% of the samples collected five miles downriver from the nearest sewage outfall were positive for enteric viruses.[4]

Chemical disinfection of wastewater is now the most effective and practical means of protecting the public health from contagious bacteria and viruses. However, to meet more strict disinfection requirements, a waste treatment plant may be required to maintain or increase chlorine dosages, but this practice may result in the addition of an excessive amount of toxicity to the aquatic environment. Chlorination has recently come under fire as a potential threat to marine life downstream from wastewater treatment plant outfalls. Studies with caged fish held in streams below treatment plants in Michigan have demonstrated that waters in these areas were toxic to fish.[5] Also, the toxicity of residual chlorine to various forms of aquatic life has been reviewed and well documented by Brungs.[6]

The concern over toxicity of trace chlorine residuals has led states like California and Maryland to adopt waste discharge requirements specifying a maximum chlorine residual in the undiluted effluent of 0.1 mg/l or below. Since much higher chlorine residuals are needed to obtain adequate disinfection, the waste discharger using present chlorination practices has a dilemma wherein he cannot meet strict disinfection requirements and at the same time maintain safe chlorine residuals in his waste treatment plant effluent. The solution to this problem may be the use of an alternative disinfection practice which is both efficient and environmentally safe. Among several of the alternatives proposed are bromine chloride, dechlorination, ozonation and chlorine dioxide.

THE ROLE OF CHLORINE AND ITS PROBLEMS

Although for more than 75 years chlorine has been the exclusive chemical disinfectant used to control spread of diseases, it is now being challenged by more efficient and less hazardous disinfectants. White has given an excellent review of the chemistry of chlorination and the germicidal significance of chlorine residuals.[7] Perhaps the most important reaction in the chemistry of chlorination is the reaction between chlorine and the ammonia present in wastewater to form chloramines. These compounds play a dominant role in disinfection and toxicity of waste treatment plant effluents. Several studies have reported that chloramines are both poor disinfectants and toxic to fish at very low concentrations. There is also concern that chlorine can form chlorinated organics which are potentially toxic to man.

It is also generally conceded that present chlorination practices are not adequately destroying viruses.[8,9] Many viral disease epidemics have been suspected, and some have been proven, to be attributable to the failure of chlorination procedures in waste treatment plants.[10,11] Furthermore, because of the greater resistance of viruses and other pathogenic organisms, it is generally recognized that coliform indices and chlorine residuals are not accurate indications of safe water.[12,13] Indeed, if we desire to control viruses in wastewater, chlorination is not the process of choice. The relatively poor disinfectant properties of chloramines have been reported by Kabler and others.[14] Chloramines required approximately a hundredfold increase in contact time over free available chlorine to obtain the same amount of disinfection. It has also been shown that about 25 mg/l of chloramine is required for adequate viral inactivation of the most resistant enteric viruses, and that probably more than 100 mg/l of hypochlorite (OCl^-) is required for similar inactivation. In contrast, only 0.5-1.0 mg/l of hypochlorous acid (HOCl) will give satisfactory viral inactivation in 30 minutes. Unfortunately, very little HOCl exists in the presence of ammonia and the normal pH range of wastewater.[15]

The disinfection efficiency of chlorine is helped significantly by rapid mixing at the point of application.[16] Rapid mixing breaks up solid clumps and permits better distribution and use of free chlorine as HOCl in the first few seconds of contact before the chloramines are produced.

A number of studies have reported the toxic effects of chlorinated wastewater on marine life.[6,17,18] The Michigan Department of Natural Resources[5] reported the effects on fish in several receiving streams below wastewater discharges. Within 96 hours, 50% of the rainbow trout died at TRC concentrations of 0.014 to 0.029 mg/l as far as 0.8 mile below the outfall. Under actual stream conditions, it has been observed that fish populations tend to avoid toxic materials; thus, broad reaches of receiving water may become unavailable to many fish. A side effect is the blockage of upstream migrations of certain fish during the spawning season. Recent increases in chlorination usage along these waterways complicate the problem.

CHLORINATED ORGANICS FROM CHLORINATION OF WASTEWATER

It has been reported recently that chlorination of water and wastewater results in the formation of halogenated organic compounds that are suspected of being toxic to man. The formation of chloroform and other similar volatile chlorinated organics identified in water samples from 60 major cities in the United States has been attributed mainly to the chlorination of naturally occurring fulvic acids.[19] However, in the chlorination of wastewater, only small differences in the concentration levels of volatile chlorinated organics were observed before and after chlorination.[20]

R. L. Jolley[21] has shown that the chlorination of municipal wastewater results in the formation of numerous other chlorinated organic compounds, seventeen of which were identified. The yield of these chlorinated organic compounds comprised in total only about 1% of the chlorine dose. All of the compounds identified by Jolley contained only a single chlorine atom. In general, such chlorinated compounds are much more readily degraded biochemically than are the polychlorinated materials such as the PCBs that have been found to be so persistent in the environment. Therefore, it seems unlikely that any of these compounds will persist in receiving waters long enough to cause major problems at downstream water treatment plants or that they will accumulate biologically in organisms to any great extent.[22] Several properties of wastewater, including higher ammonia levels, can lead to results far different from those observed when drinking water is chlorinated.

EPA TASK FORCE RECOMMENDATIONS

An EPA Task Force was formed in early 1974 to develop the necessary background information for consideration of agency policy on wastewater disinfection requirements and the use of chlorine. Its main objective was to provide information in the form of guidance on public health and water quality requirements, the potential toxic effects of chlorination to both the aquatic and human environments and alternate methods for disinfection. After careful study, the EPA Task Force recommended the following[23]:

1. Disinfection of wastewater is needed to protect public health where receiving waters are used for purposes such as downstream water supply, recreation, irrigation, and shellfish harvesting.
2. Modify the present standards and regulations for disinfection in order to allow flexibility in regard to year-round requirements and where protection of public health is not involved.
3. The exclusive use of chlorine for disinfection should not be continued where protection of aquatic life is of primary consideration. However, when chlorine is used, the total residual chlorine in the receiving waters should not exceed the recommended levels outside a described mixing zone. *Use of alternate processes should be encouraged by the agency through a vigorous promotion of the new alternatives.*
4. The use of alternate disinfectants should be further pursued because of recent findings of the potentially hazardous halogenated organics in drinking water.

Prior to the issuance of this Task Force report, top priority was given to the need for developing new alternatives to chlorination. An EPA grant with the City of Wyoming, MI, was a major part of this program to evaluate other possibilities. Although the bioassay project at Wyoming was only 50% completed at the time of this report, the study had produced significant results and had shown that dechlorination, ozonation and bromine chloride are effective processes with lesser toxic effects than chlorine.

BROMINE CHLORIDE AS AN ALTERNATIVE

Bromine chloride is the newest candidate in the search for an alternate disinfectant. It is a heavy, fuming dark red liquid with a sharp penetrating odor. Bromine chloride exists in equilibrium with bromine and chlorine in both the gas and liquid phases

$$2\ BrCl \rightleftharpoons Br_2 + Cl_2 \qquad (1)$$

Most of the physical properties of the mixture are intermediate to those of bromine and chlorine as shown in Table I.

Table I. Properties of Bromine Chloride

Property	Br_2	BrCl	Cl_2
FP (°C)	-- 7.3	-66	-101
BP (°C)	58.8	5	- 34
Density (20°C)	3.12	2.34	1.40
Vapor Pressure psig (20°C)	0	30	100
Solubility (g/100 g water)	3.3	8.5	0.75

For example, cylinders containing BrCl are under less pressure (30 psig) than those containing chlorine (100 psig). Also, the lower freezing point of BrCl eliminates winter freezing problems like those experienced with liquid bromine.

The chemistry of bromine chloride is similar to chlorine except for some important differences.[24] Chemically it is a very active oxidizing agent, both as a liquid and as a vapor. Liquid BrCl rapidly attacks the skin and other tissues, producing irritation and burns which heal very slowly, and, as with chlorine, even comparatively low concentrations of vapor are highly irritating and painful to the respiratory tract. The warning properties of BrCl are such that a person will avoid gross overexposure if he is capable of getting to fresh air.

However, compared to chlorine, the results of competitive reactions of bromine chloride with water, ammonia and various reducing agents (including organics) are more complementary to disinfection.[25] Bromine chloride appears to hydrolyze exclusively to hypobromous acid, an active disinfectant which is more active at higher pH than hypochlorous acid, the product of chlorine hydrolysis

$$BrCl + H_2O \rightarrow HOBr + HCl \quad (2)$$

The hydrolysis rate for bromine chloride is estimated at less than 0.35 milliseconds, which is faster than the hydrolysis of either bromine or chlorine. Its higher rate of hydrolysis as well as its higher solubility in water (8.5 g/100 g at 20°C–11 times that of chlorine) can be attributed to the polarity of the BrCl molecule.

In wastewater, bromine chloride and its hydrolysis products react rapidly with ammonia to form bromamines. The major products are mono- and dibromamine. Both are more chemically active, but less stable, than chloramines.[25]

$$3\ BrCl + 2\ NH_3 \rightarrow NH_2Br + NHBr_2 + 3\ HCl \quad (3)$$

An important property of bromamine residuals is their relatively high biocidal activity and low stability compared to chloramine residuals. These properties may be attributed to the lower bond strengths of bromine compounds compared to their chlorine analogs. The effect of ammonia concentrations on the germicidal efficacy of BrCl compared to chlorine using *Escherichia coli* are shown in Table II. Although the efficiency of BrCl residuals was essentially unaffected, even low concentrations of ammonia showed a significant decrease in the rate of kill by chlorine residuals.

Bromamines have been demonstrated to be far superior to chloramines in both bactericidal and virucidal activity, almost equal to free bromine

Table II. Disinfection Efficiency of BrCl on *Escherichia coli* Compared to Chlorine[a]

Contact Time (min)	Survival (%)					
	0 ppm NH_3		0.5 ppm NH_3		10 ppm NH_3	
	Cl_2	BrCl	Cl_2	BrCl	Cl_2	BrCl
0.5	>40	23.7	>72	2.6	>56	7.9
5	>18	0.0007	>55	0.0001	>55	0.016
15	1.8	0.0001	28.5	0.0001	26.6	0.0001
30	0.015	0.0001	1.6	0.0001	18.6	0.0001

[a]A.O.A.C. procedure was followed using 1 x 10^6 organisms per ml and 0.5 ppm halogen as initial concentrations.

in the pH range of wastewater.[25,26] Bromamines are also superior disinfectants over a more practical pH range than chloramines. According to Johannesson,[26] the very strong bactericidal properties of bromamines can be exploited by using a small combined bromine residual since there is no need to reach breakpoint conditions to achieve maximum disinfectancy as is the case for chlorine. In breakpoint chlorination, it becomes necessary to add nine- to tenfold weight excess chlorine to ammonia to obtain the more efficient free chlorine residuals.

Bromamines also show a faster and more complete destruction of virus than chloramines[27,28] Polio II virus in the presence of 10 mg/l ammonia has been sterilized in less than 5 min with bromine chloride, while equivalent chlorine concentrations failed to kill all the virus in over 60 min. In the absence of ammonia, BrCl was also found to be significantly better in killing Polio I virus in the pH 7-9 range.[29] Bromine chloride was shown to be only slightly more efficient than chlorine in killing f^2 bacterial virus.[30]

In a test of sewage disinfection effectiveness, total bacteria counts for secondary effluent treated with BrCl were consistently lower than the same effluent treated with equal weight dosages of chlorine.[24] The fast rate of kill and short half-life of BrCl residuals make longer contact times required by chlorine unnecessary in disinfecting wastewater with bromine chloride. Good operational control with efficient disinfection was obtained using 5 min contact time with residuals measuring 0.5-0.8 mg/l (total BrCl) at the end of 5 min. Data showing the higher efficiency and faster rate of kill for BrCl compared to chlorine are given in Table III. In secondary effluent a dosage of 1 mg/l of BrCl gave at the end of 5 min comparable results to chlorine at 2.5 mg/l after 15 min contact time.[31] In view of these data, the construction and use of contact tanks would not improve the efficiency of BrCl residuals and thus may not be necessary.

The EPA-funded Wyoming (Michigan) study on alternative disinfectants concluded that bromine chloride is an effective disinfectant on secondary effluent with less toxic effects than chlorine to aquatic life.[32] This study was conducted to determine the comparative effectiveness of chlorine, bromine chloride and ozone as wastewater disinfectants, and to determine the comparative effectiveness of chlorine, bromine chloride and ozone as wastewater disinfectants, and to determine any residual toxicity associated with wastewater disinfection with these agents or with chlorinated wastewater which had been dechlorinated with sulfur dioxide. Each of the five wastewater streams was used in acute toxicity tests with several species of fishes and macroinvertebrates, in addition to a life cycle study with the fathead minnow. For the most part, the respective fecal and

Table III. Comparing Disinfection Efficiency of BrCl to Chlorine in Secondary Effluent

Disinfectant Dosage (mg/l)	Contact Time (min)	Total Coliforms[a] (Average No./100 ml)	
		BrCl	Cl_2
1.0	5	<100	TNT
	15	12	>5,000
1.5	5	<100	TNT
	15	24	>4,000
2.0	5	25	∼8,000
	15	20	280
2.5	5	20	3,000
	15	8	20

[a]Total coliforms of untreated effluent were 6 x 10^5/100 ml TNT (too numerous to count), suspended solids 18 mg/l, pH 7.3, temperature 57°F.

total coliform densities in chlorinated, dechlorinated and chlorobrominated effluents did not differ significantly. The ozonated effluent gave significantly higher fecal and total coliform densities than other disinfected effluents unless there was multimedia pressure filtration applied to the effluent before ozonation. Aftergrowth of microorganisms was considerably less apparent in the chlorobrominated effluent stream than in the dechlorinated and ozonated effluent streams.

Since 100% nondisinfected effluent was lethal to fathead minnows, no conclusions could be drawn on the toxicity of the 100% disinfected effluents. Continuous exposure to chlorinated effluent concentrations of 50% or less containing mean residual BrCl levels of 0.043 mg/l or less had no effect on the growth, reproduction, or survival of fathead minnows. In comparison, 14% and 20% chlorinated effluent concentrations with mean total chlorine residuals of 0.045 mg/l or more caused growth retardation and mortality of continuously exposed fathead minnows.

FORMATION OF BROMINATED ORGANICS

The dominant reactions involving BrCl in wastewater are oxidation-reduction reactions producing innoxious halide salts. Like chlorine, the major portion of the available halogen during the disinfection of wastewater effluents was consumed in oxidation reactions. The total halogenation yield was estimated at less than 1% of the halogen dosage applied. In addition to inorganic reducing agents, such as sulfides and nitrites, there are many organic reducing agents. Organic alcohols, aldehydes, amines and mercaptans are oxidized by bromine chloride reaction species resulting in harmless chloride and bromide salts. For example,

$$CH_3SH + BrCl \xrightarrow{H_2O} CH_3SO_3H + HBr + HCl \qquad (4)$$

The chemistry and reactivity of bromine species in aqueous solutions are much different from what is observed in nonaqueous solvents. In water, bromine species are much better oxidizing agents than analogous chlorine species, whereas the latter are generally more active halogenating species. For example, the oxidation of cellulose by hypobromous acid is much faster than by hypochlorous acid.[33] Also, the oxidation of glucose to gluconic acid using hypobromite is 1,360 times faster than hypochlorite.[34] Because the effective oxidation potentials of the bromine systems are lower than those of the analogous chlorine systems, the relative oxidizing strength of these halogen species as measured by their oxidation potentials are not true indicators of their chemical reactivity.[35] This anomaly, where bromine species unexpectedly show higher rates of oxidation, indicates that different factors such as reaction mechanisms, bond strengths and steric hindrance influence the reaction rates more than oxidation potentials.

It has also been found that, in general, compounds which are more readily brominated are more susceptible to either hydrolytic or photochemical degradation. Accordingly, one would not expect that any appreciable quantities of these materials would persist in receiving waters long enough to cause major problems at downstream water treatment plants, or that they would accumulate biologically in organisms to any great extent. However, more work in this area is warranted.

FEEDING AND ANALYZING BROMINE CHLORIDE

While application of bromine chloride in wastewater treatment plants is similar to that of chlorine, there are some important differences. Some of these differences along with specific details describing the properties and handling of bromine chloride have been published by the chemical suppliers.[36,37]

The corrosivity of BrCl to steel is sufficiently low (0.93 mils/yr) to allow storage in steel constructed cylinders and tank cars for extended periods of time.[38]

Special care should be taken in avoiding plastic materials which may be attacked by liquid BrCl or its concentrated vapors. Polyvinyl chloride (PVC) and ABS plastic which are common to chlorine feeding systems should be avoided. However, plastic pipes made of PVC can be used to carry a water stream containing bromine chloride from the point of injection. Kynar®*, Teflon®*, and Viton®* and other equivalent, highly resistant plastics are recommended as replacements for rubber, PVC or ABS plastics in the feed system. Bromine chloride should always be removed from a cylinder or tank car as a liquid because the dissociation above the liquid produces a chlorine-rich vapor. Therefore, cylinders loaded with BrCl contain dip tubes for easy removal of the liquid under its own pressure. When liquid is withdrawn by the prescribed method, its composition will remain substantially constant throughout the removal process. In contrast to chlorine feed systems, which require an evaporator only at high feed rates, BrCl vapor feed systems always include an evaporator. The evaporator consists of a chamber which is heated to increase the temperature of the BrCl liquid, causing it to boil. Bromine chloride liquid enters the chamber and maintains the required liquid level and pressure necessary to meet the vaporization rate for the gas demand. The gas released passes through the chamber and is superheated as it leaves on the way to a vacuum regulator. The vacuum regulator reduces the pressure to a negative (vacuum) pressure and is connected to a remote flow meter. The metered gas is directed to the ejector which creates the vacuum and mixes the gas with a water system. The BrCl solution is piped from the ejector to the point of application or diffuser. If the vacuum is lost, the vacuum regulator will shut off, stopping BrCl flow.

The analysis of BrCl or bromamine residuals in wastewater uses essentially the same standard methods for total chlorine except for the final calculation. Bromine chloride residuals are 1.6 times as great as the residuals measured and calculated as total chlorine. A measured quantity of solution aliquot containing BrCl when added to *neutral* iodide solution will oxidize the iodide to yield two equivalents of iodine per BrCl molecule. The iodine is then titrated with standard sodium thiosulfate (hypo) solutions. In measuring low residual concentrations typically found in wastewater, the standard method using an amperometric titration with phenyl arsine oxide is recommended.[39] Less interference by other oxidizing agents is obtained using the back titration procedure with a standardized iodine solution. Other methods useful in determining total bromine chloride residuals are the DPD total chlorine analysis[39] and a spectrophotometric method of analysis.[40]

COST COMPARISONS

The total treatment costs for some of the alternate methods of disinfection have been tabulated in Table IV. Because of the difficulty in

*Registered trademark of E. I. du Pont de Nemours and Company, Inc., Wilmington, DE.

Table IV. Total Treatment Cost for Disinfection Alternatives [Cost Expressed as Cents/1,000 gal Treated (Capital Costs Included)]

	Plant Size (mgd)		
Disinfection Alternatives	1	10	100
1. Chlorination	3.6	1.5	0.77
2. Chlorination-Dechlorination with SO_2	4.5	1.9	0.96
3. Chlorination-Dechlorination with SO_2-Post Aeration	7.8	2.5	1.3
4. Chlorination-Dechlorination with Activated Carbon[41]	19.1	8.7	3.4
5. Chlorobromination (BrCl)	4.4	2.4	1.6
6. Ozonation (from oxygen)	8.0 (air)	4.2	3.1

determining the relative efficiencies for disinfecting various quality effluents, these costs were estimated using 8 ppm dosage for each disinfectant. The total treatment costs in ¢/1,000 gal include amortization of capital costs (@ 5-7/8% for 20 yr), operation and maintenance costs, including materials and supplies, and chemical costs. The cost estimates were made for three plant sizes. The effect of the plant size on total unit cost is probably more pronounced for systems 3, 4 and 6 in Table IV because of their high capital costs. The chemical costs were adjusted to June 1977 using chlorine prices (22, 12.5 and 7¢/lb) and projected BrCl prices (35, 27 and 20¢/lb) in 150-lb cylinders, ton cylinders and tank car quantities, respectively. Power costs were estimated at 2.0¢/kWh. Total costs for the above systems using aqueous sodium hypochlorite (15%) instead of liquid chlorine can be estimated by adding on 1.6, 0.9 and 0.5¢/lb to the 1-, 10- and 100-mgd plant, respectively.

The results indicate that liquid chlorine is the most economical among all the alternative disinfectants studied on an equal performance basis. Among the alternatives to chlorination, bromine chloride appears to be competitive without taking any advantage of its higher efficiency as a disinfectant.

ENERGY REQUIREMENTS

It is important to point out that the energy requirements for the manufacture and use of various chemical disinfectants more than any other cost factors may affect the future choice of which alternative to use. The energy requirements for some of the alternative wastewater disinfectants are given in Table V. Calculations were made of the energy required to disinfect 1,000 gal of effluent using 8 ppm dosage of each disinfectant chemical. Although the halogens are somewhat energy intensive, the energy required to produce ozone, including recycling and drying requirements, is the major operational cost for ozonation.

Table V. Estimated Energy Requirements of Alternative Wastewater Disinfectants

Disinfectant	Chemicals Used[a] (kWh/lb)	Effluent Treated (kWh/1,000 gal)
Chlorination	0.70	0.074
Chlorination-Dechlorination with SO_2[b]	1.5	0.158
Chlorobromination (BrCl)	2.0	0.213
Ozonation (oxygen)[42]	7.0	0.742
(air)	11.0	1.18

[a]Energy requirements include pumping and drying energies.
[b]Estimated using 2:1 weight ratio of Cl_2/SO_2.

REFERENCES

1. Geldreich, E. E. In: *Water Borne Pathogens in Water Pollution Microbiology,* R. Michel, Ed. (New York: Wiley-Interscience, 1972).
2. Mosely, J. W. "Epidemiological Aspects of Microbial Standards for Bathing Beaches," Paper No. 9, International Symposium on Discharge of Sewage from Sea Outfalls, London, England, August 1974.
3. Fisher, L. M. *Am. J. Public Health* 27:180 (1973).
4. Grinstein, S., *et al. Bull. World Health Org.* 42:291 (1970).
5. Basch, R. E., M. R. Newton, J. G. Truchau and C. M. Fetterolf. *Water Pollution Control Research Series EPA,* WQO Grant No. 18050 G22 (October 1971).
6. Brungs, W. H. *J. Water Poll. Control Fed.* 45:2180 (1973).
7. White, G. C. *Handbook of Chlorination,* Chapter 4 (New York: Van Nostrand Reinhold Company, 1972).
8. Kelly, S., and W. W. Sanderson. *Sew. Ind. Wastes* 31(6):683 (1959).
9. Carlson, H. J., *et al. Am. J. Public Health* 33:1083 (1943).
10. Cox, C. R. "Gastroenteritis and Public Water Supplies," ***A Symposium** on Hydrobiology* (Madison, WI: University of Wisconsin Press, 1941), p. 260.
11. Dennis, J. M. *J. Am. Water Works Assoc.* 51:1288 (1959).
12. Clarke, N. A., *et al. Adv. Water Poll. Res.* 2:523 (1964).
13. Baumann, E. R. *Water Sew. Works* 109:21 (1962).

14. Symposium on Transmission of Viruses by Water, Cincinnati, Ohio, December 6-8, 1965.
15. Morris, J. C. "Chlorination and Disinfection–State of the Art," *J. Am. Water Works Assoc.* 63:769 (1971).
16. White, G. C. *Handbook of Chlorination* (New York: Van Nostrand Reinhold Company, 1972), p. 444.
17. Zillich, J. A. *J. Water Poll. Control Fed.* 44:212 (1972).
18. Brungs, W. A. "Effects of Waste Water and Cooling Water Chlorination on Aquatic Life," EPA 600/3-76-098 (August 1976).
19. Rook, J. J. *Environ. Sci. Technol.* 11(5):478 (1977).
20. Bellar, T. A., *et al.* "The Occurrence of Organohalides in Chlorinated Drinking Waters," EPA 670/4-74-008 (November 1974).
21. Jolley, R. L. "Chlorination Effects on Organic Constituents in Effluents from Domestic Sanitary Sewage Treatment Plants," Oak Ridge Nat. Laboratory Publication ORN-TM-4290 (October 1973).
22. Morris, J. C. "Formation of Halogenated Organics by Chlorination of Water Supplies," EPA-600/1-75-002 (March 1975).
23. "Disinfection of Waste Water." Task Force Report, USEPA Office of Research & Development, Washington, DC (July 1975).
24. Mills, J. F. "The Disinfection of Sewage by Chlorobromination," paper presented at the 165th National Meeting of the Division of Water, Air and Waste Chemistry, American Chemical Society, Dallas, TX, April 1973.
25. Sollo, F. W., *et al.* In: *Disinfection–Water and Waste Water,* J. D. Johnson, Ed. (Ann Arbor, MI: Ann Arbor Science Publishers, Inc., 1975).
26. Johannesson, J. D. *Am. J. Public Health* 50:1731 (1960).
27. Mills, J. F. In: *Disinfection–Water and Waste Water,* J. D. Johnson, Ed. (Ann Arbor, MI: Ann Arbor Science Publishers, Inc., 1975).
28. Krusé, C. W. "Mode of Action of Halogens on Bacteria, Viruses and Protozoa in Water Systems," U.S. Army Medical R&D Contract No. DA-99-193-MD-2314, John Hopkins University, Baltimore, MD (1968).
29. Haufler, K. Z. "A Comparison of Bromine Chloride Inactivation of Poliovirus Type I," Master's Thesis, Virginia Polytechnical Institute, Blacksburg, VA (1976).
30. Yin, R. L. "A Comparison of the Bactericidal and Viricidal Activity of Bromine Chloride and Chlorine in Sewage," Master's Thesis, John Hopkins University, Baltimore, MD (1976).
31. Jackson, S. C. "Chlorobromination of Secondary Sewage Effluent," Proc. Wyoming Workshop on Disinfection of Waste Water and Its Effect on Aquatic Life, Wyoming, MI, October 1974.
32. "Disinfection Efficiency and Residual Toxicity of Several Waste Water Disinfectants–Volume I–Grandville, Michigan," EPA 600/2-76-156 (October 1976).
33. Giertz, H. W. *Tappi* 34:209 (1951).
34. Lewin, M. Ph.D. Thesis, Hebrew University, Jerusalem (1947).
35. Mills, J. F. "Competitive Oxidation and Halogenation Reactions in Disinfection of Waste Water," International Ozone Institute Meeting, Cincinnati, OH (November 1976).

36. Mills, J. F., and J. A. Schneider. *Ind. Eng. Chem. Prod. Res. Develop.* 12:160 (1973).
37. Dow Chemical USA. *Bromine Chloride Handbook*, Publication No. 101-15, Inorganic Chemicals Department, Midland, MI (1972).
38. Mills, J. F., and B. D. Oakes. *Chem. Eng.*, pp. 102-106 (August 6, 1973).
39. *Standard Methods for the Examination of Water and Wastewater,* 13th ed. (Washington, DC: American Public Health Association, 1971).
40. Custer, J. J., and S. Natelson. *Anal. Chem.* 21(8):1005 (1949).
41. Smith, R., R. G. Eilers and W. F. McMichael. "Cost of Alternative Processes for Waste Water Disinfection," Proc. Wyoming Workshop on Disinfection of Waste Water and Its Effect on Aquatic Life, Wyoming, MI, October 1974.
42. Rosen, H. M., F. E. Lowther and R. G. Clark. *Water Wastes Eng.* 8:25 (1974).

CHAPTER 14

DIBROMAMINE DECOMPOSITION KINETICS

John L. Cromer, Guy W. Inman, Jr., and J. Donald Johnson
Department of Environmental Sciences
and Engineering
School of Public Health
University of North Carolina
Chapel Hill, North Carolina

INTRODUCTION

Recent concern over the high fish toxicity[1] and low disinfection efficiency of chlorine in wastewater[2] has led to a search for alternative disinfectants. When bromine is used as a disinfectant the 10-20 mg/l of NH_3-N typical of wastewater reacts with either BrCl or Br_2 to form the bromamines.[3,4]

$$HOBr + NH_3 \rightarrow \underset{\text{monobromamine}}{NH_2Br} \tag{1}$$

$$NHBr_2 + HOBr \rightarrow \underset{\text{dibromamine}}{NHBr_2} \tag{2}$$

$$NHBr_2 + HOBr \rightarrow \underset{\text{tribromamine}}{NBr_3}. \tag{3}$$

Unlike chlorine, which forms only monochloramine, NH_2Cl, with the high ammonia present under sewage conditions,[5] bromine forms primarily dibromamine and only small amounts of monobromamine.[4]

Dibromamine was first prepared in low-temperature ether solutions by Coleman *et al.*[6] and was later shown to exist in aqueous solutions by Johanneson.[7] He showed[8] that the bromamines are much better bactericides than the chloramines with a colicidal efficiency approaching that of free bromine. The bromamines also have been found to be superior to the chloramines as disinfectants with viruses.[9,10] Galal-Gorchev and Morris[11] confirmed dibromamine's existence in aqueous solution and showed that the distribution of bromamines was a function of both pH and the ratio of ammonia to bromine concentration. They identified the bromamines by their characteristic UV spectra and performed some stability studies. Later studies showed that the decomposition kinetics of dibromamine were approximately second order with respect to time.[4]

Since the major bromamine in wastewater disinfected with bromine is dibromamine, its stability is intimately related to the disinfecting ability of bromine in wastewater. The toxicity of the discharge of a brominated wastewater is also directly related to its stability. Ward *et al.*[12] have shown that if bromamines are present the discharge is toxic. The low toxicity found by Mills[13] can be attributed to the rapid decomposition of dibromamine. The bromamines are not only bacteriologically active but are also chemically reactive as substituting and oxidizing agents. Unlike monochloramine, monobromamine reacts rapidly with itself at neutral pH to form dibromamine.

In addition to substitution reactions which do not decrease the oxidizing equivalents or total residual oxidant present, the bromamines act as oxidizing agents and are reduced to bromide. Dibromamine is unstable in the absence of reducing agents and decomposes directly to nitrogen gas and bromide according to the following overall reaction when excess ammonia is present.

$$3\ NHBr_2 + NH_3 \rightarrow 3\ N_2 + 6\ HBr \qquad (4)$$

The purpose of this study was to determine rate constants and develop the underlying mechanism for this reaction under typical ammonia and pH conditions encountered in wastewater.

EXPERIMENTAL

Most of the reagents and procedures used in this study were identical to those used in a previous study of the decomposition kinetics of tribromamine.[14] The preparation of all reagents and buffers, the use of the thiosulfate-iodine amperometric titration procedure for the determination of total reducible bromine, the determination of bromamine concentrations by UV spectrophotometry, the use of the thermostated aluminum syringe holder with T connector for the quick mixing of initial hypobromous acid and ammonium chloride solutions directly in the quartz cell of the spectrophotometer, and also the use of a FORTRAN IV computer program for the calculation of initial rates are all described in detail by Inman *et al.*[14]

At pH 7.0, the reaction mixture was monitored at a wavelength of 232 nm (λ_{max} for dibromamine) and its concentrations were calculated from Beer's law using the molar absorptivity of 1900 l-mol $^{-1}$ cm $^{-1}$ determined by Galal-Gorchev and Morris.[11] In order to resolve the overlapping spectra of dibromamine and tribromamine at pH 6.0, parallel kinetic runs of the same initial conditions were made with UV data taken at both 232 nm and 258 nm (λ_{max} for tribromamine). The concentrations of both dibromamine and tribromamine were calculated by solving simultaneous equations derived from Beer's law. Similarly, at pH 8.0 the concentrations of dibromamine and monobromamine were calculated from data taken at 232 nm and 278 nm (λ_{max} for monobromamine). Values for the molar absorptivities used in these calculations are given in Table I. Those not taken from Galal-Gorchev and Morris[11] were determined in this laboratory from UV scans and extrapolated kinetic data.

Table I. Molar Absorptivities of the Bromamines at Selected Wavelengths

	Molar Absorptivities, M^{-1} cm^{-1}		
Compound	λ = 232 nm	λ = 259 nm	λ = 278 nm
NH_2Br	190		390[11]
$NHBr_2$	1,900[11]	900	780
NBr_3	3,810	5,000	

In studying the effect of ammonia on the decomposition rate, concern was focused upon the total excess ammonia, NH_3 and NH_4^+, present in the system as dibromamine decomposes. This quantity was calculated by taking the initial added ammonium chloride concentration less the stoichiometric amounts needed to form the bromamines and the amount oxidized by bromine. The fractions of the total excess ammonia present as NH_3 and as NH_4^+ were calculated using K_b equal to 2.28×10^{-5}, the ionization constant at 20.0°C and a solution ionic strength of 0.10.

Table II gives the initial conditions under which the decomposition rates were studied together with the estimated errors of each parameter. All experiments were performed at a temperature of 20.0 ± 0.5°C and with a constant solution ionic strength of 0.10 ± 0.002. Replicate kinetic experiments included as many as five to a series. A 6% error in the instantaneous rate was found to be typical.

Table II. Range of Initial Conditions for the Rate Study

pH	$[HOBr]_o$ M		$[NH_4Cl]_o$ M	
	Maximum	Minimum	Maximum	Minimum
6.00 ± 0.04	$4.07 \pm 0.05 \times 10^{-4}$	$1.10 \pm 0.02 \times 10^{-4}$	$2.00 \pm 0.01 \times 10^{-2}$	$3.00 \pm 0.02 \times 10^{-4}$
7.00 ± 0.04	$8.02 \pm 0.29 \times 10^{-4}$	$0.60 \pm 0.08 \times 10^{-4}$	$4.00 \pm 0.02 \times 10^{-2}$	$3.00 \pm 0.02 \times 10^{-4}$
8.00 ± 0.03	$4.03 \pm 0.05 \times 10^{-4}$	$1.00 \pm 0.03 \times 10^{-4}$	$8.00 \pm 0.04 \times 10^{-3}$	$3.00 \pm 0.02 \times 10^{-4}$

RESULTS

Figures 1, 2 and 3 show typical concentration-time data obtained at pH 6.0, 7.0 and 8.0. As the pH is increased, tribromamine concentration decreased while monobromamine increased. Also the formation and decomposition rate of dibromamine apparently became slower. These trends were also observed with an increase of the ammonia-to-bromine molar ratio. Instantaneous rates were calculated at points on the concentration-time curve located sufficiently beyond the formation peak so that the calculated rates were not obviously affected by the formation reaction, yet not so far out on the curve as to defeat the purpose of using initial rates. In practice this meant examining each concentration-time curve to decide at which point to calculate the instantaneous rate referred to as an initial rate. Generally the points were chosen at about 45 sec at pH 6.0, from 1.0 to 1.5 min at pH 7.0 and in the range of 2.0 to 3.0 min at pH 8.0.

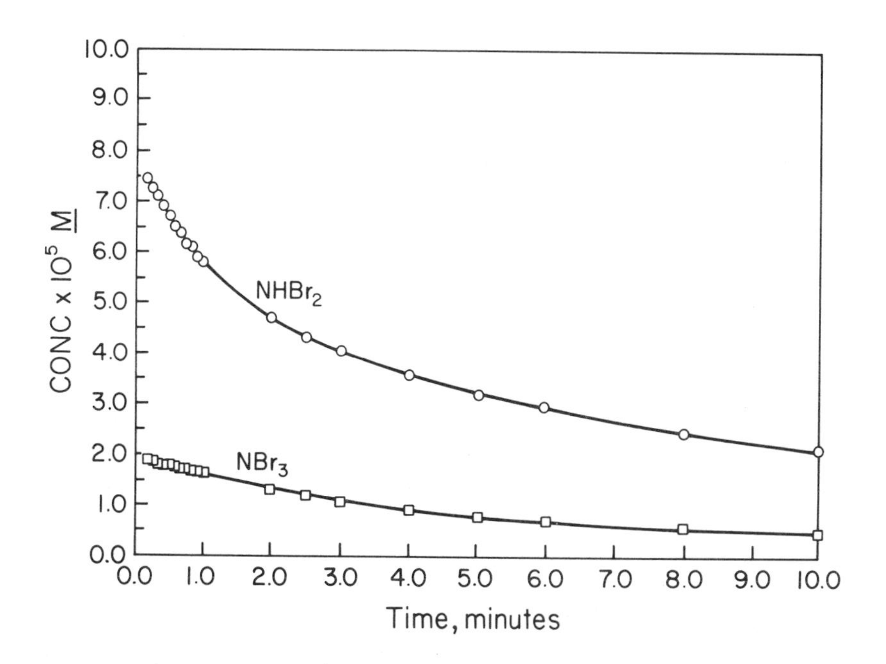

Figure 1. Bromamine concentrations as a function of time, at pH 6.0. $[HOBr]_o = 2.0 \times 10^{-4} M$; $[NH_4Cl]_o = 3.0 \times 10^{-4} M$; I = 0.10; 20°C.

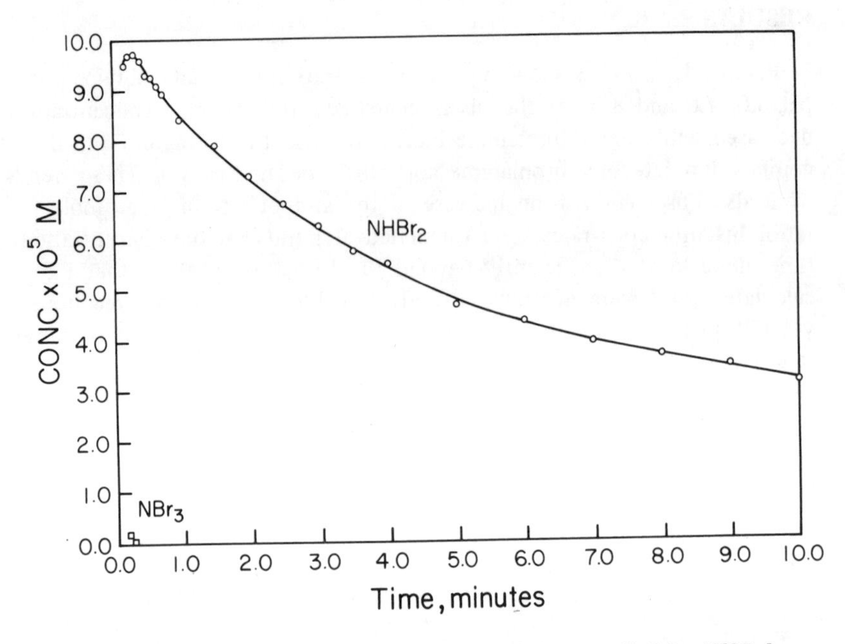

Figure 2. Bromamine concentrations as a function of time at pH 7.0. $[HOBr]_o = 2.0 \times 10^{-4}$ *M*; $[NH_4Cl]_o = 5.0 \times 10^{-4}$ *M*; I = 0.10; 20°C.

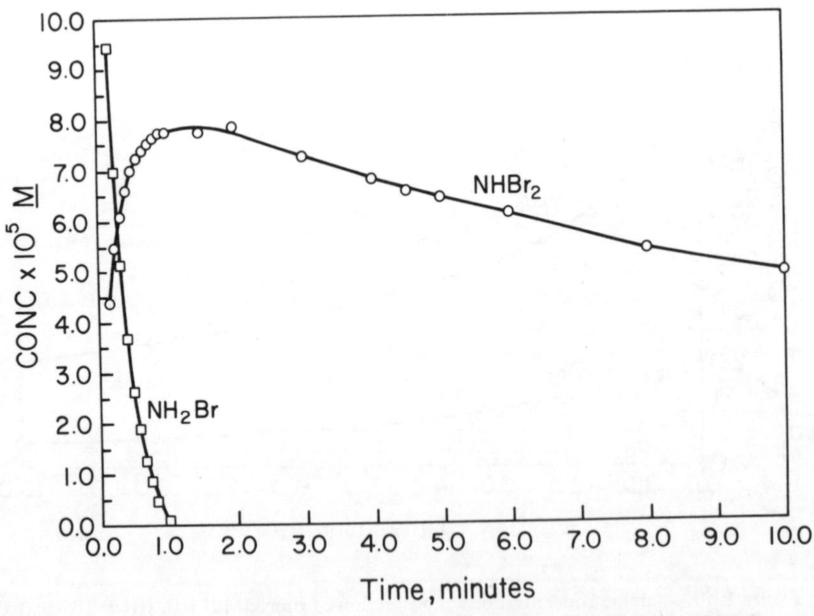

Figure 3. Bromamine concentrations as a function of time at pH 8.0. $[HOBr]_o = 2.0 \times 10^{-4}$ *M*; $[NH_4Cl]_o = 1.0 \times 10^{-3}$ *M*; I = 0.10; 20°C.

The results of the initial rate studies for dibromamine and ammonia appear in Figures 4,5 and 6. Slopes of 1.97 ± 0.03 at pH 7.0 and 2.14 ± 0.09 at pH 8.0 indicate that the decomposition is second order with respect to dibromamine. But at pH 6.0 the slope of Figure 5, 2.64 ± 0.13, gives the order with respect to dibromamine near 2.5. The slope of Figure 6, -0.56 ± 0.01, gives the order with respect to ammonia at pH 6.0 as -0.5. Studies of the decomposition rate with variable ammonia concentration at pH 7.0 and 8.0 failed to give linear results with log rate against log concentration plots, but a similar inhibition of the rate by ammonia was observed and slopes were found to be in the range of -0.3 to -0.4.

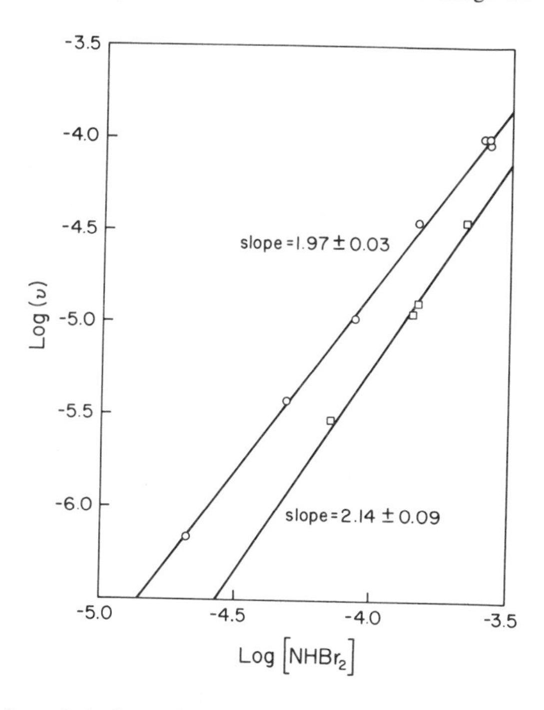

Figure 4. Log of the instantaneous rate of decomposition, υ, as a function of the log of dibromamine concentration. Lines represent a least-squares fit to the data. $[NH_4^+]$ = 1.9×10^{-3} *M*; I = 0.10; 20°C; ○ pH 7.0; □ pH 8.0.

Figure 7 shows the second-order rate constant, $\upsilon/[NHBr_2]^2$, as a function of the square root of the reciprocal of the ammonia concentration at pH 7.0 and 8.0. The linearity of this plot shows that the order with respect to ammonia is -0.5. However, the non-zero intercept indicates the existence of a two-term rate equation with one term being independent of the pH and the ammonia concentration.

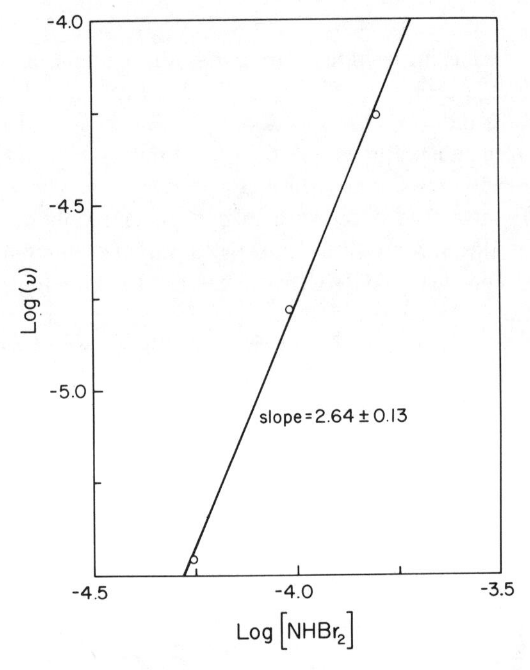

Figure 5. Log of instantaneous rate, υ, as a function of log dibromamine at pH 6.0. The line represents a least-squares fit to the data. $[NH_4^+] = 1.9 \times 10^{-3}$ *M*; I = 0.10; 20°C.

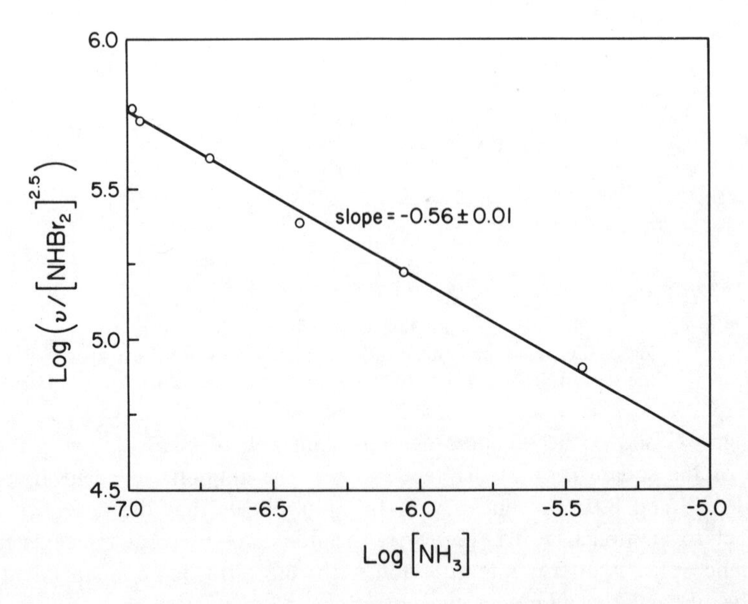

Figure 6. Log $(\upsilon/[NHBr_2]^{2.5})$ as a function of log ammonia at pH 6.0. $[HOBr]_o$ = 4. 2.0×10^{-4} *M*; I = 0.10; 20°C.

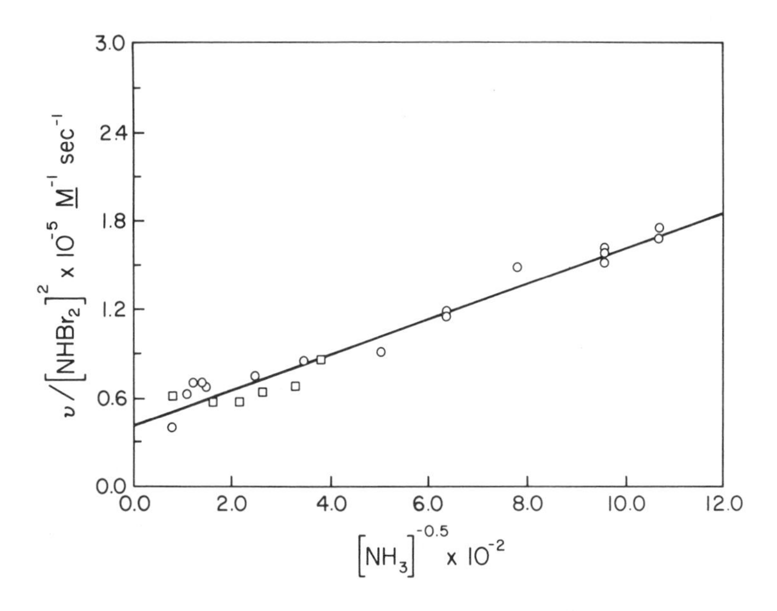

Figure 7. Observed second-order rate constant, $v/[NHBr_2]^2$, as a function of $[NH_3]^{-0.5}$. The line is a least-squares fit to the data. $[HOBr]_o = 2.0 \times 10^{-4} M$; I = 0.10; 20°C. ○ pH 7; □ pH 8.0.

DISCUSSION

The observed reaction orders at pH 6.0 are most consistent with the rate equation

$$v = \frac{-d[NHBr_2]}{dt} = k_a \frac{[NHBr_2]^{2.5}}{[NH_3]^{0.5}} \tag{5}$$

which may also be derived from the following mechanism

$$3\ NHBr_2 \overset{K}{\leftrightharpoons} 2\ NBr_3 + NH_3 \tag{6}$$

$$NHBr_2 + NBr_3 \overset{k_1}{\rightarrow} N_2Br_4 + H^+ + Br^- \tag{7}$$

$$N_2Br_4 + H_2O \overset{k_2}{\rightarrow} N_2 + 2\ H^+ + 2\ Br^- + 2\ HOBr \tag{8}$$

The rate-determining step, Reaction 7, accounts for the observation that mixtures of dibromamine and tribromamine are most unstable when equivalent amounts of each are present.[14] The reactive intermediate decomposes rapidly to end products in Reaction 8. The equilibrium expression for Reaction 6 is obtained if the following equilibria are assumed for each of the bromamines[4]

$$NH_3 + HOBr \overset{K_1}{\rightleftharpoons} NH_2Br + H_2O \tag{9}$$

$$NH_2Br + HOBr \overset{K_2}{\rightleftharpoons} NHBr_2 + H_2O \tag{10}$$

$$NHBr_2 + HOBr \overset{K_3}{\rightleftharpoons} NBr_3 + H_2O \tag{11}$$

Reaction 6 is merely a way of expressing the equilibrium between dibromamine and tribromamine when ammonia rather than HOBr is present in excess.

If $K_1 = \dfrac{[NH_2Br]}{[NH_3]\ [HOBr]}$, $K_2 = \dfrac{[NHBr_2]}{[NH_2Br]\ [HOBr]}$, and $K_3 = \dfrac{[NBr_3]}{[NHBr_2]\ [HOBr]}$

then an equilibrium expression for ammonia, dibromamine and tribromamine is

$$\frac{K_3{}^2}{K_1\ K_2} = K \tag{12}$$

According to Equation 7 the rate of dibromamine disappearance is given by

$$\upsilon = k_1\ [NHBr_2]\ [NBr_3], \tag{13}$$

which is similar to the hydroxide catalyzed equation proposed in the tribromamine decomposition study.[14] When the equilibrium expression from Equation 6 is solved for tribromamine and substituted into Equation 13, the rate law becomes

$$\upsilon = k_1 K^{0.5} \frac{[NHBr_2]^{2.5}}{[NH_3]^{0.5}} \tag{14}$$

where $k_1 K^{0.5} = k_a$

To explain the data at pH 7.0 and 8.0, a second decomposition pathway consisting of a bimolecular rate-limiting reaction of dibromamine with itself can be postulated:

$$NHBr_2 + NHBr_2 \overset{k_3}{\rightarrow} N_2HBr_3 + H^+ + Br^- \tag{15}$$

$$N_2HBr_3 + H_2O \overset{fast}{\rightarrow} N_2 + 2\ Br^- + 2\ H^+ + HOBr \tag{16}$$

The additional term from Reaction 15 added to Equation 14 gives the following rate expression:

$$v = k_1 K^{0.5} \frac{[NHBr_2]^{2.5}}{[NH_3]^{0.5}} + k_3 [NHBr_2]^2 \quad (17)$$

This equation satisfies the requirements of the data taken at pH 7.0 and 8.0. The only pH effect predicted by Equation 17 is that associated with the equilibrium shift of ammonium ion to ammonia as pH increases. Equation 17 can also be written with ammonium ion in the denominator of the first term, in which case the hydrogen ion must appear in the numerator with an order of 0.5.

Figure 8 represents a test of Equation 17. Lines are least-squares fits to the data. The slopes 1.22 ± 0.06 and 1.13 ± 0.06 x 10^4 are equal within experimental error. The slopes, with the y intercept of the pH 7.0 and 8.0 data, give values for the rate constants: $(k_1 K^{0.5})$ = 1.18 ± 0.06 x 10^4 M^{-1} sec^{-1} and k_3 = 4.79 ± 1.06 x 10^4 M^{-1} sec^{-1}. The initial rate order determinations and the y intercepts of Figure 8 show that the pH 6.0 data are most consistent with a single-term rate equation, while the pH 7.0 and 8.0 data fit a two-term rate equation.

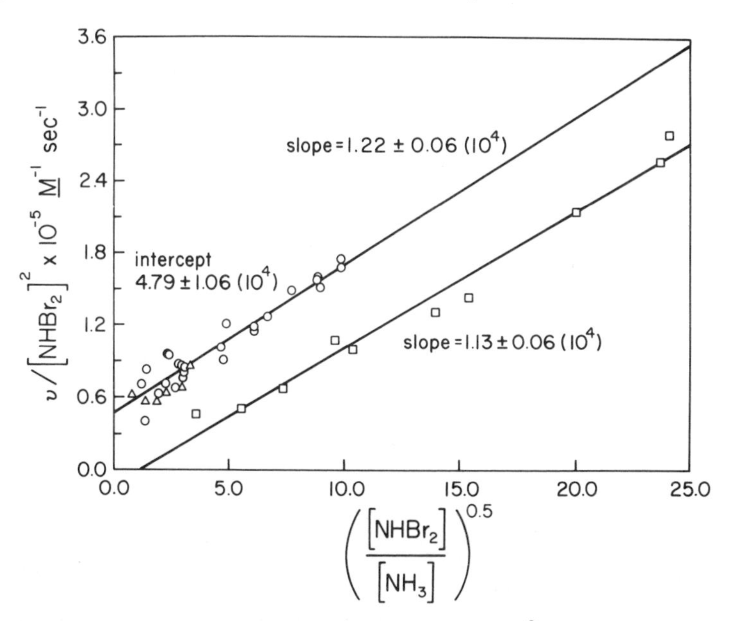

Figure 8. Observed second-order rate constant, $v/[NHBr_2]^2$, as a function of $([NHBr_2]/[NH_3])^{0.5}$ using full range of the data. Lines represent least-squares fit to the data. [HOBr] = 0.6 to 8.0 x 10^{-4} M; $[NH_3]$ = 0.011 to 1.6 x 10^{-4} M; I = 0.10; 20°C. □ pH 6.0; ○ pH 7.0; △ pH 8.0.

Although the rate expressions are not consistent between pH 6 and pH 7 and 8, the mechanism proposed describes the 5/2 order with respect to dibromamine observed at pH 6.0 and explains both the pH effect associated with the ammonia-ammonium ion equilibrium and why the fastest decomposition occurs when tribromamine and dibromamine are present in equivalent amounts. Estimates from the data indicate that k_1 is about three orders of magnitude greater than k_3, suggesting that the dibromamine-tribromamine reaction is the preferred pathway for decomposition. This also explains the slower rates observed at higher pH values and ammonia-to-bromine ratios where there is less tribromamine. However, just how the equation is transformed between pH 6.0 and 7.0 from a one-term equation is not understood. The slowing of the dibromamine formation rate observed at pH 7.0 and 8.0 might indicate that the assumed equilibria become kinetically limited. A study of the monobromamine and dibromamine formation kinetics using a stopped-flow apparatus might yield valuable results.

Practically, the increased stability of dibromamine and the possible formation of monobromamine at high pH and high ammonia concentrations have serious implications in wastewater disinfection and for situations in which brominated effluent is discharged into waters of high salinity and pH, such as estuaries. Because of the second order and higher dependence on dibromamine concentration, the rate of decomposition of dibromamine becomes quite slow at high dilution, high pH and high ammonia levels. At pH 8, for a 100-fold molar excess of ammonia at 20°C and 1 mg/l Br_2, the half-line of dibromamine is nearly three days. Thus previous studies which show that bromine in wastewater is quite unstable[12] are influenced by the loss of bromine because of its reactivity as an oxidizing agent, *i.e.* demand reactions, as well as the inherent instability of dibromamine.

ACKNOWLEDGMENT

This work was supported in part by a U.S. Army Medical Research and Development Command contract number DADA 17-72-C-2053.

REFERENCES

1. Brungs, W.A. "Effects of Wastewater and Cooling Water Chlorination on Aquatic Life," EPA Report No. 600/3-76-098 (August 1976).
2. White, G.C. *Handbook of Chlorination* (New York: Van Nostrand-Reinhold, 1972) p. 379.
3. Mills, J. F. in: *Disinfection–Water and Wastewater,* J. D. Johnson, Ed. (Ann Arbor, MI: Ann Arbor Science Publishers, Inc., 1975), p. 122.
4. Johnson, J.D., and R. Overby. *J. San. Eng. Div.,* ASCE 97:617, (1971) and 99:371 (1973).

5. Weil, I., and J.C. Morris, *J. Amer. Chem. Soc.,* 71, 1664 (1948).
6. Colman, G.H., H. Soroos and C.B. Yager. *J. Amer. Chem. Soc.* 56: 965 (1934).
7. Johannesson, J.K. *J. Chem. Soc.* 1959:2998 (1959).
8. Johannesson, J.K. *Nature* 181:1799 (1958).
9. Taylor, D.G., and J.D. Johnson. In: *Chemistry of Water Supply, Treatment and Distribution,"* A.J. Rubin, Ed., (Ann Arbor, MI: Ann Arbor Science Publishers, Inc., 1974), p. 369.
10. Floyd, R., D.G. Sharp and J.D. Johnson. *Environ. Sci. Technol.* In press.
11. Galal-Gorchev, H.A., and J.C. Morris. *Inorg. Chem.* 4:899 (1965).
12. Ward, R.W., R.D. Giffen, G.M. DeGraeve and R.A. Stone. "Disinfection Efficiency and Residual Toxicity of Several Wastewater Disinfectants," Vol. I, EPA Report No. 600/2-76-156 (October 1976).
13. Mills, J.F. In: *Disinfection–Water and Wastewater,* J.D. Johnson, Ed. (Ann Arbor, MI: Ann Arbor Science Publishers, Inc. 1975), p. 113.
14. Inman, G.W., T.F. LaPointe and J.D. Johnson. *Inorg. Chem.* 15:3037 (1976).

CHAPTER 15

CHEMICAL INTERACTIONS IN A RAPID INFILTRATION SYSTEM ACCOMPLISHING TERTIARY TREATMENT OF WASTEWATERS

Donald B. Aulenbach

Department of Chemical and Environmental Engineering
Rensselaer Polytechnic Institute
Troy, New York

INTRODUCTION

Lake George is a beautiful recreational lake location in the eastern portion of the Adirondack Park of New York State. The lake is noted for its fine recreational facilities, its beautiful tree covered shoreline and the clarity of its waters. As an indication of the quality of its waters, the lake has been given a special class AA designation. This means that the water may be used for direct consumption with only chlorination required for public water supplies.[1] Many residents who live in cottages surrounding the lake secure their drinking water directly from the lake and consume it with no treatment whatsoever.

In order to maintain the high quality of the lake water, regulations were passed[2] restricting the discharge of any wastewaters into the lake directly or into any stream which flows into the lake. Properly operating subsurface disposal systems such as septic tanks with seepage fields have been considered as adequate treatment provided there is no surface runoff into the lake or into any stream which flows into the lake. With the concentration of population at the southern tip of the lake, in 1936 plans were made to construct a waste treatment plant for the Village of Lake George. The treatment plant, consisting of secondary treatment by means of trickling filters and application of the final effluent onto natural delta sand beds, was put into operation in 1939 and has been operating continuously since that time.

The sewage from the village is collected by gravity to a central location in a park adjacent to the lake from which it is raised approximately 55 m (108 ft) through a 1.6-km (1-mi) force main to the treatment plant. The initial design of the plant was for a wintertime flow of 0.15 mgd (600 m^3/day) and a summertime flow of approximately three times the winter flow, or 0.5 mgd (1,900 m^3/day). The plant itself, as shown in Figure 1, was built in triplicate so that one-third could be used for winter flows and the entire plant for summer flows. The treatment plant consisted of Imhoff tanks, dosing siphons, trickling filters, final clarification in 2 circular secondary tanks and the discharge of the final effluent without chlorination onto one of 6 sand beds. Since then two of the three Imhoff tanks have been replaced by Clarigesters which still provide a separate compartment for sludge digestion. One of the trickling filters is covered with boards on sawhorses and is used exclusively during the winter. Two rectangular secondary settling tanks with continuous sludge scrapers have been added. The original 6 sand infiltration beds have now been expanded to 21 with a total filter area of 2.15 ha (5.3 ac). The preliminary treatment portion of the plant is considered to be designed for a flow of 1.75 mgd (6,600 m^3/day). Maximum flows observed on a summer weekend have reached 1.25 mgd (4,700 m^3/day).

The sand beds are normally operated by dosing one north and one south bed during the period from approximately 8 a.m. to 4 p.m. and dosing another pair of north and south beds for the remaining 16 hours of the day. During a weekend two north and two south beds are dosed for a 24-hr period. During extremely high summer flows, additional beds are dosed as needed. Under normal operating conditions the beds drain dry in approximately 1 to 3 days. The beds are allowed to remain dry as long as possible for aeration. With frequent dosing, the beds slowly clog and the infiltration rate is decreased. Periodically, in the range of twice a year, the surface of the beds is scraped. A small amount of the sand is removed and then the beds are raked and releveled prior to being put back into service. The first few dosings after this cleaning procedure result in very rapid infiltration rates. In 1973 approximately 1 ft of the surface sand was removed from each of the sand beds to remove any fine sand grains, any surface clogging material and the sand which most likely has its phosphate removal capacity expended.

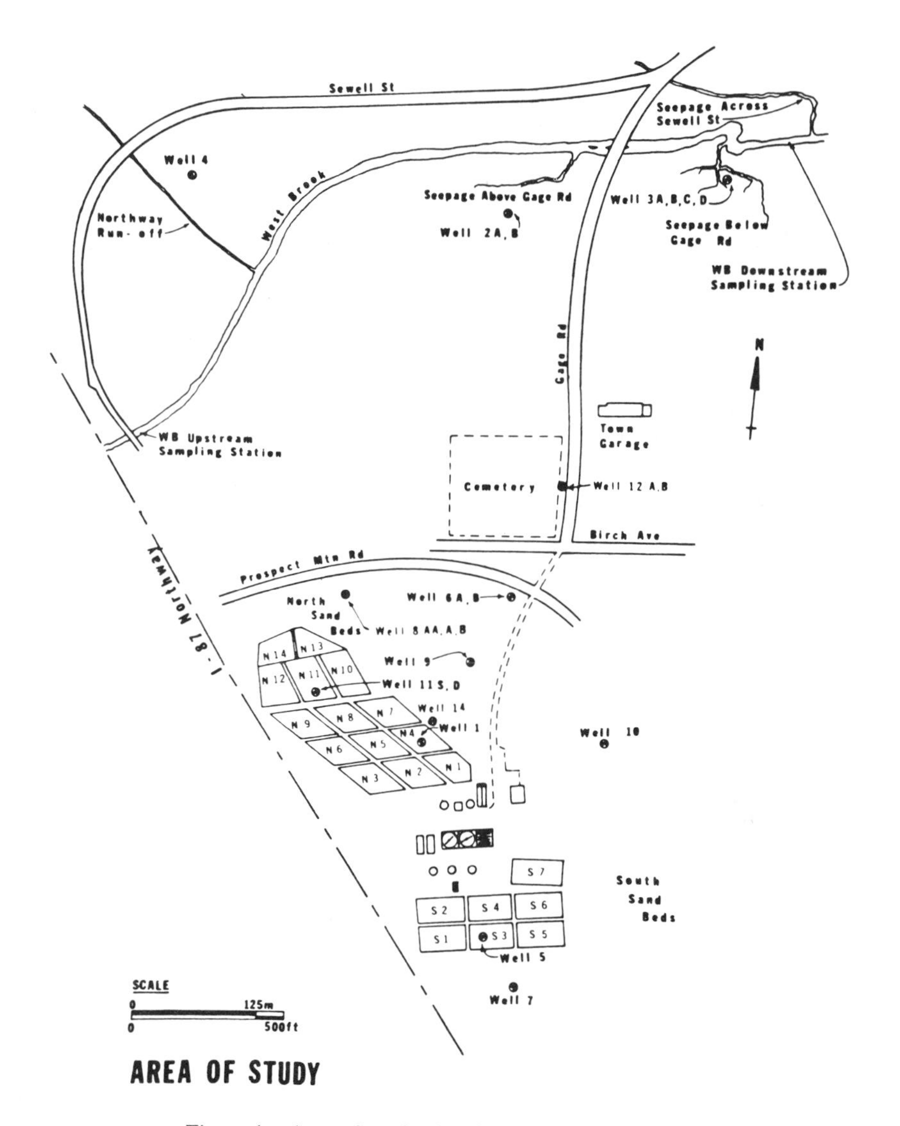

Figure 1. Area of study showing sampling locations.

INFILTRATION RATES

Infiltration rates were estimated based on the amount of sewage applied to each bed, the time it takes for the sewage to drain through a bed and the frequency of dosing. Since precise flow data to each sand bed are not available, it was assumed that half of the flow reaches the treatment plant between 8 a.m. and 4 p.m. and the other half of the daily flow occurs during the 16-hr nighttime period. Since two beds are being dosed simultaneously during each period, it is assumed that each bed receives approximately one-quarter of the daily flow.

The estimated monthly loading rates over a full year are shown in Table I.[3] It may be seen that the maximum loading rate occurred during the month of August with a loading of 4.83 gal/ft^2-day or 0.65 ft/day (1.37 m^3/ha-min or 0.2 m/day). This is not an infiltration rate. It represents the amount of liquid applied to the bed during this period including the resting period, that is, the volume that can be safely applied without exceeding the total infiltration capacity. During the latter part of August in both 1975 and 1976 a period existed when the sand beds were all completely loaded and the normal drying time between dosing was either very short or nonexistent.

Table I. Wastewater Loading Rates, Lake George, New York

	Flow (mgd)	Loading Rate	
		(gal/ft^2-day)	(ft/day)
1974			
Sep	0.740	3.17	0.42
Oct	0.588	2.52	0.34
Nov	0.414	1.77	0.24
Dec	0.491	2.10	0.28
1975			
Jan	0.499	2.14	0.29
Feb	0.513	2.20	0.29
Mar	0.568	2.43	0.32
Apr	0.699	2.99	0.40
May	0.661	2.83	0.38
Jun	0.779	3.33	0.45
Jul	0.970	4.15	0.55
Aug	1.128	4.83	0.65
Average	0.67	2.87	0.38

The actual infiltration rate was measured in several of the sand beds by installing a water level recorder. The rate of infiltration increased with the head of liquid on the sand bed. The lowest rates recorded with less than 1 ft of liquid on the sand bed were in the range of 0.25 to 0.6 ft/day (0.08-0.18 m/day) under normal operating conditions. Bed S7 (see Figure 1) had a rate exceeding 1 ft/day (0.3 m/day) with a water depth on the bed of 2 ft (0.6 m). An infiltration rate exceeding 2 ft/day (0.6 m/day) was measured on a freshly scraped bed with a water depth of 1 ft. With continued intermittent operation the flowrate decreased gradually to the values previously stated.

PURIFICATION OF WASTEWATERS IN THE SOIL

Studies were made of the changes in water quality of the applied sewage effluent in both the vertical and horizontal transport through the soil. The soil in this case consists exclusively of delta sand deposits containing no observed quantities of clay. The quality changes during vertical transport were measured in north sand bed 11 (Bed N11), and the changes during horizontal transport were measured in a series of observation wells installed between the infiltration beds and West Brook, a small tributary to Lake George approximately 2,000 ft north of the treatment plant. The groundwater, including the applied sewage effluent, emerges from the ground as seepage along the south bank of the flood plain of the stream. The sampling locations are shown in Figure 1.

Purification With Depth

Bed N11 (see Figure 1) was chosen for the quality changes with depth because it was one of the older beds and one in which earlier studies had been conducted to determine changes in water quality with depth.[4] A series of well points was driven into the sand of the bed at intervals of 2 ft from 2 to 14 ft (0.6 to 4.28 m). Porous cup lysimeters were also installed at 5-ft intervals from 5 to 65 ft. Only four of these lysimeters at depths of 3, 7, 11 and 18 m proved to be functional. In addition, two wells with 6-in. casing were equipped with submersible pumps to obtain samples directly from the aquifer under Bed N11.

Samples for the study were secured on an approximately biweekly basis. Fewer results were obtained during the winter due to difficulties caused by freezing of the sampling equipment and the sampling wells. Thus, the average data for the winter season are not so reliable as for the other three seasons. In the figures depicting the results of quality changes with depth in Bed N11, the influent to the treatment plant is indicated at the top of the graph and the values shown at the zero depth represent the effluent from the treatment plant applied to the sand beds. The difference or change between these two values represents the degree of treatment or change in passing through the conventional portion of the sewage treatment plant. The level of water in the saturated aquifer varied between

20 and 22.5 m from the surface. Thus, the (normally) two data points below this depth represent the quality of water within the saturated aquifer.

In general, the temperature within Bed N11 increased with depth during the fall and the winter and decreased with depth during the spring and summer. The temperature of the water in the saturated aquifer was fairly constant throughout the year. The pH, as shown in Figure 2, varied from a low of 6.5 at the 11-m depth in fall to a high of 7.4 in the shallower pumped well during summer. There were no consistent trends in pH with depth, although the spring and summer curves were similar. The fall results were similar to the spring and summer curves below the 11-m depth. The total dissolved solids showed a slight increase in depth in the unsaturated zone of the sand bed. Values within the saturated aquifer were consistent throughout the year and were much lower than the values in the unsaturated zone. This suggests a considerable dilution of the high dissolved solids in the sewage effluent by the lower dissolved solids in the natural water in the aquifer.

There was an increase in the dissolved oxygen (DO) content of the waste as it passed through the treatment plant as indicated in Figure 3. The DO values obtained within the sand bed may be somewhat in error due to aeration caused by the sampling techniques used. Well 11S was monitored for DO in the fall, winter and spring by inserting the DO probe directly into the well. During the summer the DO was measured in samples pumped from this well to the surface. The values measured within the well were lower than the pumped values. Thus, the DO values for the samples from the wells within the sand bed, with the exception of the fall, winter and spring samples in Well 11S, may be slightly high due to aeration of the samples during pumping. The redox potential was determined only during the spring of 1976. The trickling filter treatment increased the redox potential markedly, resulting in effluent values slightly on the positive side which continued in all samples except at the 18-m depth. There appeared to be significant reduction at this depth within the sand bed, at least during spring. This low value correlates fairly well with a consistently low DO value at the 18-m depth.

There was little significant variation in the chloride content with depth in any of the seasons. In the shallow pumped well the chloride levels were higher during the spring and summer than at the deeper pumped well, whereas in the fall the chloride levels were higher in the deeper pumped wells.

In order to compare the changes in the various forms of nitrogen, all three forms measured (nitrate, ammonia and total Kjeldahl nitrogen) are shown in the same figure. The values for summer and fall (Figure 4) show

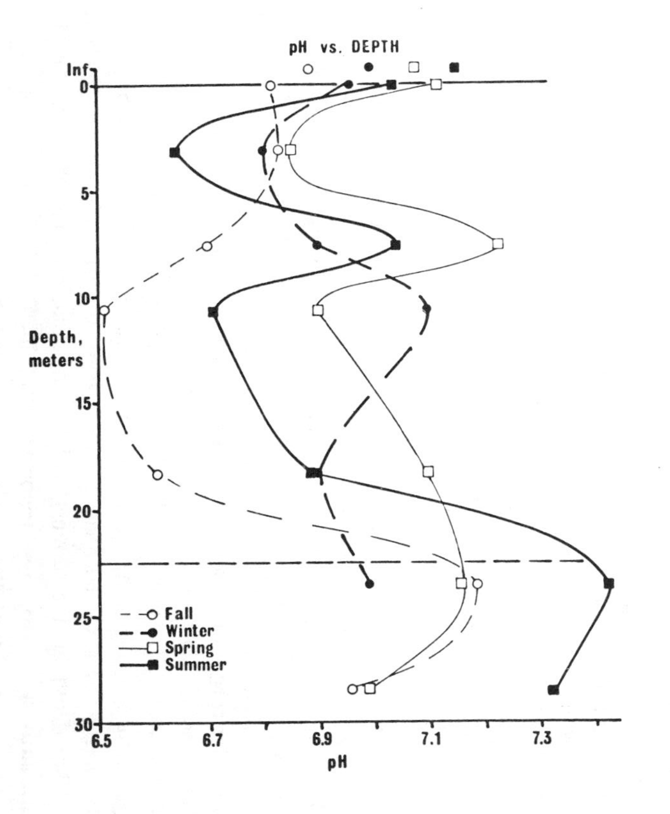

Figure 2. Seasonal pH variations within sand bed N11.

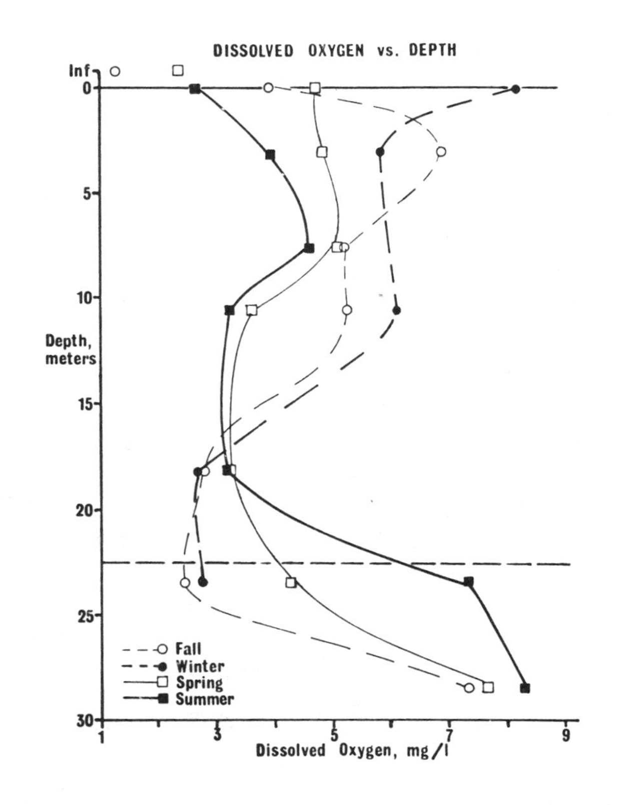

Figure 3. Seasonal dissolved oxygen variations within sand bed N11.

similar trends. In both cases there was a decrease in the ammonia and Kjeldahl nitrogen content with a corresponding increase in the nitrate content at the 3-m depth. At slightly greater depths there was a reduction in nitrate with a significant increase in the ammonia and Kjeldahl nitrogen as the liquid approached the 18-m depth. In both cases the nitrate content at the 18-m depth was less than 1 mg/1-N. During the summer a relatively high nitrate content was observed in the shallow pumped well, but all other forms of nitrogen were low in concentration in both of the pumped wells. During the fall all forms of nitrogen were low in the two pumped wells. The data suggest an oxidation of the reduced nitrogen compounds in the upper aerated portion of the sand bed, with reduction of this nitrogen to ammonia or organic nitrogen at a slightly greater depth. Since all forms of nitrogen showed significant reduction by the 18-m depth, there was apparently some loss of total nitrogen from the aqueous system. During the spring, the results (Figure 5) showed somewhat different trends. There was an increase in the nitrate content in the upper 3 m of the sand bed with a subsequent gradual reduction to less than 1 mg/1-N at the 18-m depth. The Kjeldahl and ammonia nitrogen contents dropped significantly to less than 1 mg/1-N at all depths of 3 m and greater. There was approximately 2 mg/1-N of nitrate in the shallow pumped well, but all other values in both the shallow and deep pumped wells were less than 1 mg/1-N. The results for spring indicate that there was an initial oxidation of the reduced nitrogen compounds to nitrate followed at greater depths by reduction of the nitrate to nitrogen gas, which escaped the aquatic system. It is not clear why, during the summer and fall, the reduction in nitrate resulted in a corresponding increase in the ammonia and Kjeldhal nitrogen. The spring nitrate and oxidation-reduction potential curves were nearly identical in shape with the possible exception of slightly higher values of redox potential in the saturated portion of the aquifer. This supports the conclusion that oxidation to nitrate occurs in the upper 3 m of the sand bed with subsequent reduction of the nitrate to nitrogen gas in the lower portions of the sand bed.

The total and orthophosphate phosphorus are compared for the fall and spring in Figures 6 and 7, respectively. During both seasons, the sewage treatment plant accomplished a significant reduction in total phosphorus. During the spring there was also a reduction in orthophosphate in passing through the treatment plant, but during the fall there was a slight increase in orthophosphate, possibly indicating the conversion of polyphosphates to orthophosphates in the treatment system. In all cases, the orthophosphate was reduced to less than 0.1 mg/1-P by the time the effluent reached the 10-m depth. In the summer and fall the total phosphorus was reduced significantly in the top 3 m followed by an increase in concentration at the 8-m depth. During the spring the concentration of total phosphorus

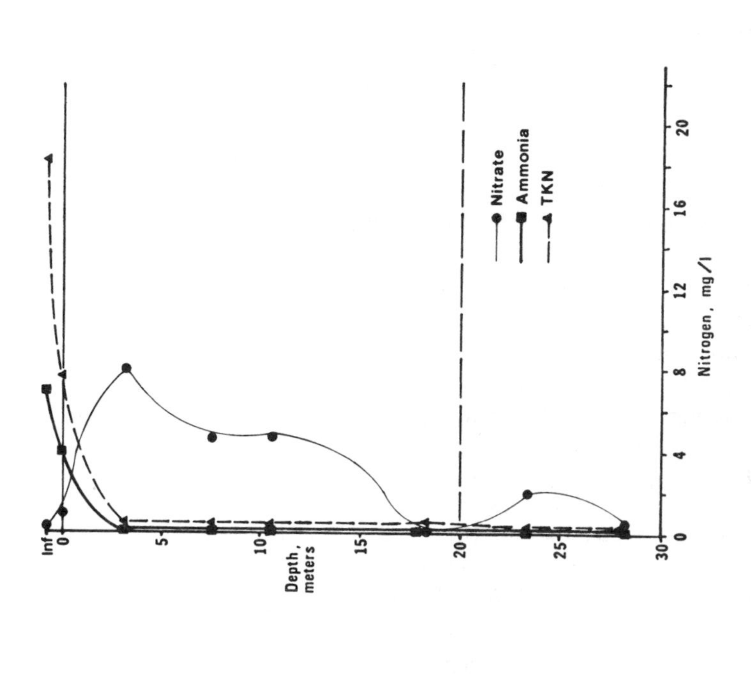

Figure 5. Variations in nitrogen content in sand bed N11 during spring.

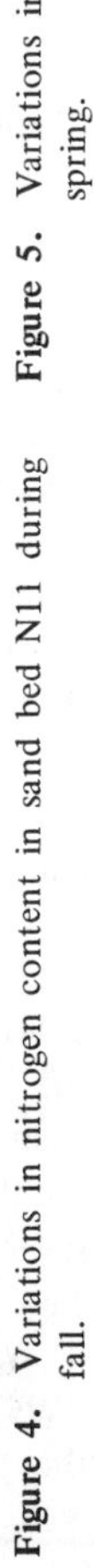

Figure 4. Variations in nitrogen content in sand bed N11 during fall.

was very similar to that of the orthophosphate. Slight amounts of total phosphorus were observed in the shallow pumped well during the summer and fall, but during the spring and at the deeper pumped well the levels were consistently less than 0.1 mg/l-P. The orthophosphate was reduced to levels lower than could be achieved by conventional physical-chemical treatment methods of phosphate removal by the time the liquid reached the 11-m depth sampling location. Approximately 0.4 mg/l of total phosphate was observed in the shallow pumped well during the summer and fall.

Additional determinations were made for calcium, magnesium, alkalinity, iron, sodium, potassium, potassium to sodium ratio and copper. The concentrations of copper observed were always below the detectable limit of 0.05 mg/l by the atomic absorption technique. BOD, COD and coliforms were essentially completely removed in the top 3 m of the sand bed.

Change in Quality With Distance

After the applied sewage effluent travels vertically to the aquifer, the groundwater including the sewage effluent travels in an approximately northerly direction toward West Brook. Tracer studies using tritium and rhodamine WT are presently being concluded confirming this direction of flow. The effluent recurs as seepage along the south bank of the flood plain of West Brook.[5] Observation wells were driven into the aquifer between the infiltration beds and the seepage areas. Wherever possible, well points were driven to at least two different depths within the aquifer at the location of that well. Wells designated as A are the shallow well points, that is, located near the top of the aquifer, and subsequent alphabetical letters indicate well points progressively deeper into the aquifer. The relationship between the treatment plant, the observation wells, the seepage and West Brook are shown in Figure 1.

Figures 8, 9 and 10, showing changes in quality with distance, represent typical data secured from September 1975 to August 1976.[6] The figures show data for the influent and effluent of the conventional portion of the sewage treatment plant to compare the changes in quality in the treatment plant with those in the soil. Where not specifically designated on the figures, the open symbols indicate the shallower wells, whereas the solid symbols represent the deeper wells. Where appropriate, separate lines were drawn to connect the shallower wells and the deeper wells. In Figure 8, Well 2 was connected to seepage above Gage Road and the values from Well 3 were connected to seepage below Gage Road (indicated as S on the figures), since the seepage above Gage Road is located adjacent to well site 2 and the seepage below Gage Road is located adjacent to well

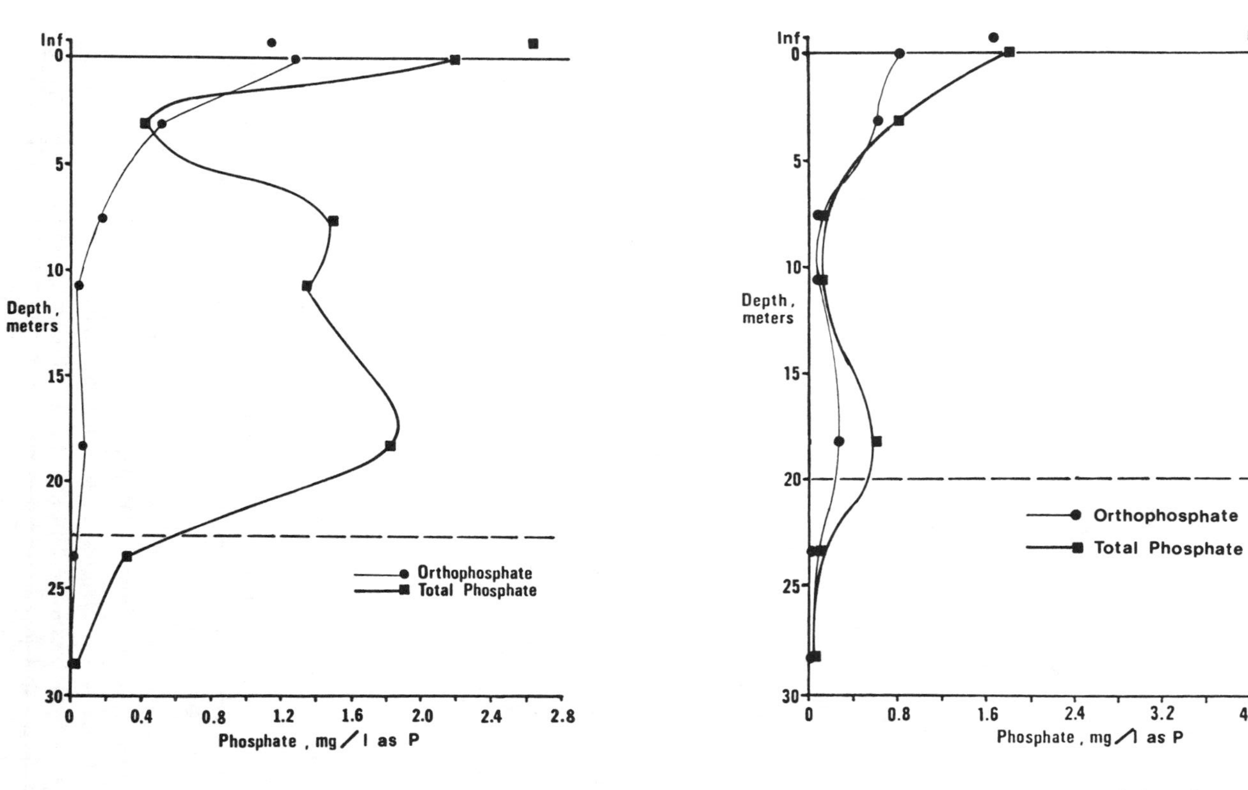

Figure 6. Total and orthophosphate variations in sand bed N11 during fall.

Figure 7. Total and orthophosphate variations in sand bed N11 during fall.

site 3. In addition, the values for West Brook upstream (WBUS) and West Brook downstream (WBDS) show the effect of the seepage upon the quality of the water within West Brook.

There was little change in pH in the aquifer during the spring. However, during the summer there was a marked increase at Well 6 in both the shallow and the deeper sampling points. The highest seasonal value obtained was pH 8.8 in Well 6A during the summer. Thereafter, there was a gradual decrease in the pH in all the wells, but it remained higher than pH 7.0 which was the average value of the sewage treatment plant influent. In both the spring and the summer there was a slight decrease in the pH as West Brook passed the location of the inlet of the two seepages. During the fall there was a trend similar to that during the summer, but the values reached a maximum of only pH 7.67. There was little significant change in the pH with distance during the winter. During both fall and winter there was a slight increase in the pH in West Brook in passing the area of the seepages. Quite consistently during all seasons the shallower wells indicated higher values of dissolved solids than the corresponding deeper samples.[7] This indicates the potential for the sewage containing higher dissolved solids to remain nearer the surface of the aquifer, with the lower sampling points more representative of the normal groundwater. The control for the dissolved solids analyses was Well 7, which showed values of 100 mg/l during the fall and winter and 68 mg/l during the spring. It must be pointed out here that, as with chloride analyses, above well site 3, which also represents seepage below Gage Road, there is a highway department garage which formerly stored highway deicing salt in the open at this location. It becomes quite obvious that the salt has leached in the ground and affects the dissolved solids and chloride determinations at well site 3 and seepage below. Thus, these two locations are not representative of the effects of the sewage treatment plant effluent upon the groundwater. Of particular note are the increases in dissolved solids during the spring and summer in both Wells 8 and 6 at the shallower depths. Well 6 also indicated somewhat higher values during the fall and winter. A possible explanation could be runoff from highways containing salt used for deicing during the winter.

The shallower wells consistently maintained a higher DO content than the deeper wells as shown in Figure 8 for the spring and summer seasons. DO was measured at all times; the lowest values occurred during the summer, with 0.5 mg/l being measured in Wells 9 and 3D. There was a slight trend toward increasing DO levels with increasing distance from the sand infiltration beds. During all seasons there was a decrease in the DO content in West Brook as it picked up the discharges of the two seepages. This could be due in part to increase in the temperature, resulting in a

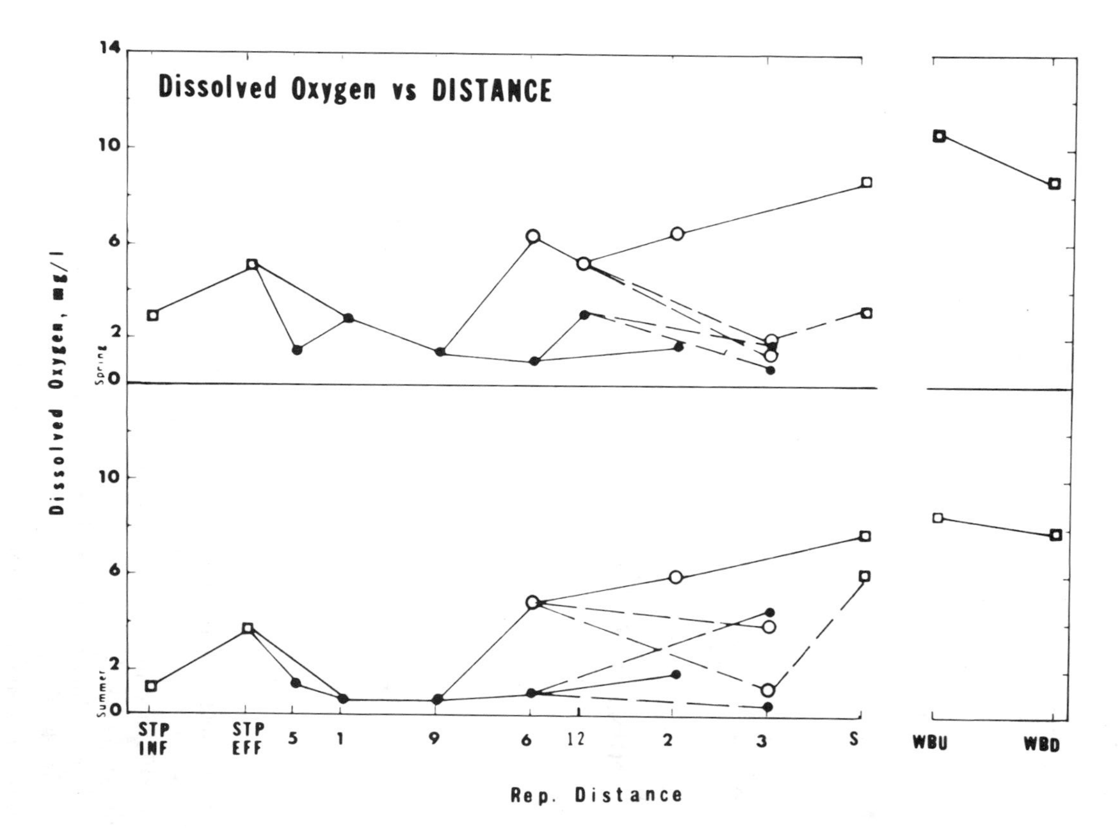

Figure 8. Variations in dissolved oxygen with distance from the Lake George Village sewage treatment plant during spring and summer.

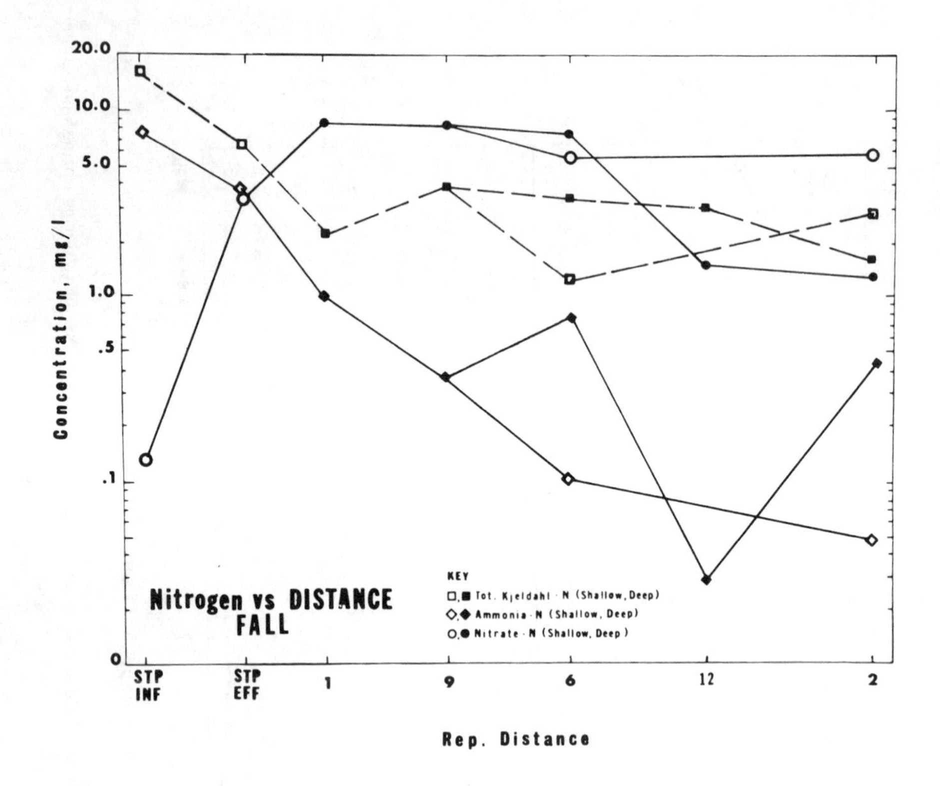

Figure 9. Variations in nitrogen compounds with distance from the Lake George Village sewage treatment plant during fall.

lower DO saturation value. The average seasonal increases in temperature in this stretch of West Brook were a minimum of 0.4°C in spring and fall and a maximum of 1.4°C in winter.

Redox potential data are available for only the spring of 1976. In general, the values varied between +100 and +150 mV with the highest value being +170 mV in Well 12B. Surprisingly, the lowest values were observed in the control wells with a value of +55 mV at Well 10 and +65 mV in Well 4. The other control, Well 7, had a redox potential of +105 mV. There was an increase in West Brook from +90 mV at the upstream sampling to +107 mV at the downstream station. With the exception of the samples from Well 6A during the summer and fall, there was a general trend of decreasing chloride with distance from the treatment plant.[6] Well 3 and seepage below were high in chloride, reflecting the highway deicing salt which was stored at the local highway garage immediately above this

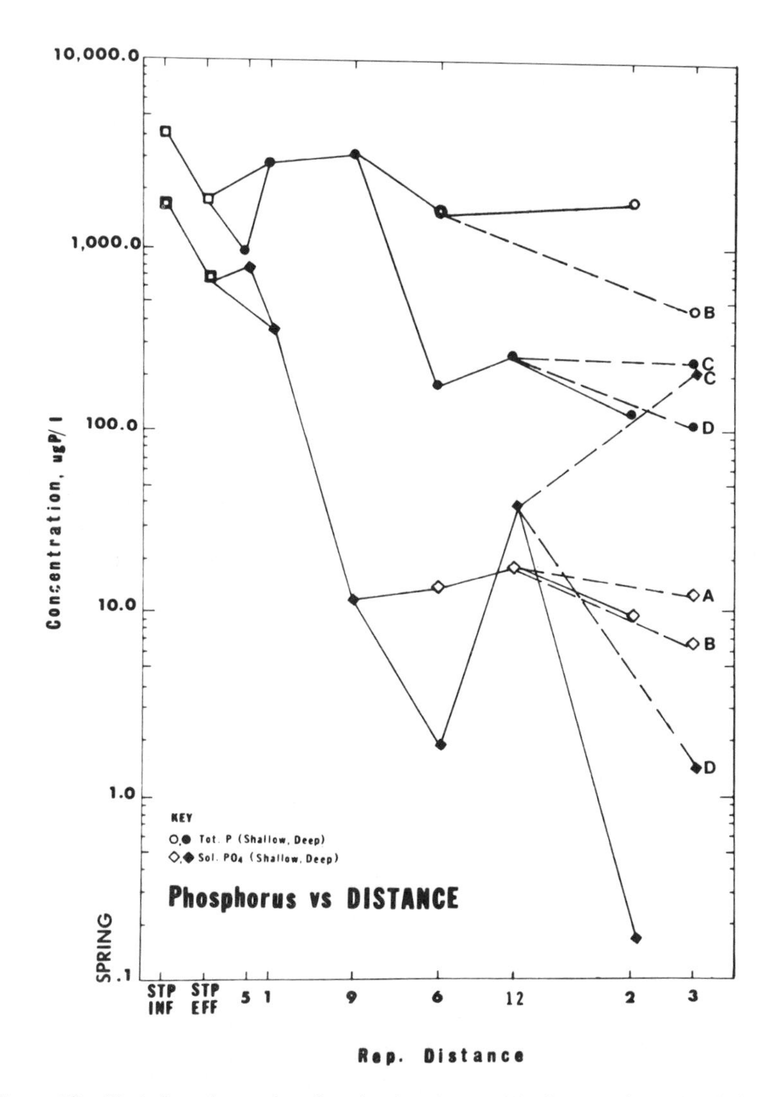

Figure 10. Variations in total and orthophosphate with distance from the Lake George sewage treatment plant during spring.

location, as mentioned previously. In all cases there was a slight increase in the chloride content of West Brook due to the seepage flows.

In order to show the interrelationship between the various oxidized and reduced forms of nitrogen, these were plotted together. Typical data for the fall are shown in Figure 9. In general, there was a gradual reduction in the total Kjeldahl nitrogen (TKN) from about 6 mg/l in the applied sewage effluent to approximately 2 mg/l at Well 2. There was a much more significant reduction in the ammonium nitrogen from an average value of about 4 mg/l in the applied sewage effluent to values less than 0.1 mg/l at Well 2A. During the fall and spring there were higher values of ammonia nitrogen in Well 2B than in Well 12B, with values ranging between 0.3 and 0.5 mg/l. The nitrate values were approximately the inverse of the TKN and ammonia values. During the summer there was a marked reduction in the nitrate between Well 6 and Well 2, with values in Well 2B of approximately 0.6 mg/l. During the fall, winter and spring, the nitrate values at Well 2A ranged between 5 and 7 mg/l. This represents nearly a quantitative conversion of ammonia and TKN to nitrate within the system. During the summer, however, there appears to have been some loss of total nitrogen through the system. This is possibly due to the higher temperatures which enhance the denitrification reaction in which nitrate is reduced to nitrogen gas, which then escapes from the aqueous system. During all seasons the nitrate content was lower in Well 2B than in Well 2A. The high values of nitrate in Well 12 correlate with the highest redox potential of any of the observation wells.

Changes in total and soluble reactive phosphate as P with distance are shown in Figure 10 for the spring. In general, the sand system lowered the total phosphorus concentration in the applied sewage effluent from an average of 3 mg/l to less than 0.2 mg/l. The only significant exception was in Well 3B during the spring, at which time the average concentration was 0.5 mg/l. The total phosphate at Well 2A was consistently nearly the same as in the applied sewage effluent; however, at Well 2B the values were reduced to 0.1 mg/l or slightly above. The soluble reactive phosphate (SRP) showed even more dramatic reductions. The poorest reductions were observed in Wells 5 and 1, both of which are located within the sand infiltration beds. All of the other observation wells with the exception of Well 3C showed SRP values on a seasonal basis of 0.01 mg/l or less. The lowest values of less than 0.2 ug/l (the minimal detectable limit of the method utilized) were found in Well 2B. The soil application system is doing an excellent job of removing the soluble reactive phosphate.

CONCLUSIONS

More chemical interrelationships were apparent in the vertical transport through the unsaturated aquifer than in the horizontal transport through the saturated aquifer. This correlates with the fact that the majority of the removal of the pollutional substances took place within the top 10 m of the sand beds in the vertical transport through the unsaturated aquifer.

Fewer reactions remained to take place at greater depths and within the saturated zone.

It did not appear that pH was a major controlling factor in any of the chemical reactions observed. The changes in pH in vertical transport were minimal; although a high value of pH 8.8 was observed in well 6A during the summer, this did not correlate with any other significant chemical change. Phosphorus also did not appear to be affected by any other chemical reaction within the soil. The major phosphorus removal was within the top 8 m or less in the vertical transport through the unsaturated aquifer.

There appeared to be a direct relationship between DO and redox potential. There was also a direct relationship between these two parameters and the production of nitrate in the vertical transport through the unsaturated aquifer. At the 3-m depth, where the DO and redox potential were high, there was great production of nitrate with a corresponding decrease in organic and ammonia nitrogen. However, at the 8- and 11-m depths there was a decrease in nitrate during the fall with a corresponding increase in ammonia and organic nitrogen. Unfortunately, redox potential data are not available for the fall, but it is assumed that the values would be similar to the DO values which were low at these depths. Thus, where oxidizing conditions were present, such as in the top 3 m of the sand bed, ammonia and organic nitrogen were oxidized to nitrate. Whenever reducing conditions appeared at a greater depth, there was the possibility or reduction of nitrate apparently to ammonia and organic nitrogen. Ultimately, there must have been some reduction to nitrogen gas, as at the 18-m depth all three forms of nitrogen were low in concentration. This appears to present one problem in that (as seen in Figures 4 and 5) the nitrate content at the 18-m depth in both fall and spring was less than 1 mg/l, whereas the nitrate content in the seepage was in the order of 7 mg/l as N. The changes in quality with depth were studied in bed N11, in which there is approximately 22 m of vertical flow in the unsaturated zone; other beds, particularly the newer south beds, are in the order of only 5 m deep. It may be observed from Figures 4 and 5 that at the 5-m depth the nitrate concentration was quite high. If it is assumed that the same oxidation of organic and ammonia nitrogen to nitrate occurs in all of the beds, then the quality of the water which enters the saturated aquifer from the shallower beds would reflect the conditions at the 5-m depth in bed N11. This condition is a high nitrate content with low organic and ammonia nitrogen. This, then, would provide the source of the high nitrate content in the seepage.

For oxidation of reduced nitrogen compounds to nitrate, aerobic conditions are important. Thus, the need for allowing the beds to rest between dosing in order to allow air to enter the beds becomes obvious. Furthermore, total nitrogen removal can be obtained if greater depths of sand are available in which reducing conditions can occur.

The use of deep rapid sand infiltration techniques for the further treatment of secondary treated domestic sewage can be relied upon to produce an effluent which meets the recommended drinking water quality standards.[8] This has been accomplished at the Lake George Village Sewage Treatment Plant. Total nitrogen removal may be accomplished if sufficient depth and reducing conditions are available. Aeration of the sand beds is important in oxidizing reduced nitrogen compounds. The appropriate chemical interactions within the soil achieve the desired treatment.

REFERENCES

1. "Official Code Rules and Regulations of the State of New York," Part 830, Title 6, Item 48, Water Index No. C-101-T367, Class AA.
2. "The Environmental Conservation Law," Chapter 664, Sec. 17, Title 17, Art. 1709.
3. Beyer, S.M. M.S. thesis, Rensselaer Polytechnic Institute, Troy, NY (1976).
4. Aulenbach, D.B., T.P. Glavin and J.A. Romero-Rojas. *Ground Water* 12:161 (1974), FWI Report 74-1.
5. Chiaro, P.S. M.E. thesis, Rensselaer Polytechnic Institute, Troy, NY (1976).
6. Hajas, L. M.S. thesis, Rensselaer Polytechnic Institute, Troy NY (1975).
7. Middlesworth, B.C. M.E. thesis, Rensselaer Polytechnic Institute, Troy, NY (1976).
8. Environmental Protection Agency. *Federal Register* 40(51):11990 (1975).

CHAPTER 16

COMPARISON OF WASTEWATER FLOCCULATION IN JAR TEST EXPERIMENTS, CONTINUOUS-FLOW REACTORS AND LARGE-SCALE PLANTS

Hermann H. Hahn, Birgit Eppler and Rudolf Klute

University of Karlsruhe
Karlsruhe, Germany

INTRODUCTION

Coagulation and flocculation have become important unit processes in the treatment of domestic and industrial sewage in conjunction with conventional mechanical and biological plants as well as physico-chemical plants.[1,2] Traditionally, the so-called jar test as described in the literature[3] has been used to determine in general whether certain wastewaters can be treated by this unit process and, if so, what specific conditions with respect to coagulant/flocculant type and concentration, ionic medium and pH, in particular, must be met in order to attain high suspended solids removal. Furthermore, the jar test is used as a convenient routine analysis for the on-line control and adjustment of process efficacy.

In a jar test, coagulation and flocculation are accomplished under batch conditions in a cylindrical reactor. The reagent dosing and mixing (destabilization) and the subsequent sedimentation (phase separation) take place in the same reaction chamber. In a large-scale coagulation and flocculation plant, a specific reactor or part of a reactor (in compact plants) is assigned to each step of the overall process, *i.e.*, a mixing chamber, a flocculation chamber and a sedimentation tank. The reactors are operated under continuous-flow condition; frequently complete mixing is assumed. Applying the results of jar test studies to the dimensioning and operation of large-scale plants, it is assumed that significant process variables can be kept comparable and that nonquantifiable effects are negligible.

Factors that are controlled and therefore in first approximation comparable are reaction time (detention time), energy input and average solution conditions (such as ionic medium and reagent concentration). However, since a batch system is compared with a continuous-flow system, the comparison of reaction time is not unambiguous. Likewise, the average energy dissipation per unit reactor volume presents problems in identifying process-determining flow conditions (velocity gradients). Effects that are difficult to quantify encompass wall effects, concentration fluctuations and concentration gradients. Whether these phenomena can be neglected in all instances or not can be determined only after careful evaluation of the specific situation. Thus, in order to use results of jar tests for plant design and operation, the following questions should be investigated in detail: (1) the effect of different residence time distributions predominantly on the destabilization and aggregation phase; (2) the relationship between geometry of reaction vessels, energy dissipation and process-controlling flow conditions (measured as effective particle aggregation); (3) the general effect of varying geometry and scale of the reaction vessels on overall process efficiency; and (4) the different dampening effects of different size and shape reactors on concentration fluctuations and the concomitant effect of concentration gradients on overall particle removal.

A review of reported data on coagulation and flocculation studies indicated that these questions have received relatively little attention, both from the point of view of a basic clarification and from the practical viewpoint. An early study of aggregation of kaolinite coagulated with aluminum in batch- and continuous-flow reactors[4] indicated that rate constants and process efficiency were comparable under identical conditions. Another recent investigation also reports direct compatibility between data from flocculation in jars and from flocculation in water purification plants.[5] Other investigations focused on various phases of the process. A study of colloid destabilization revealed the necessity to keep chemical conditions identical in particular with respect to gradients in the coagulant dosing,[6] and to reproduce hydrodynamic conditions as closely as possible by keeping energy input, Reynolds number and stirring characteristics identical.[7,8] Detailed experimentation on the transport phase gave clear evidence that the originally proposed global G-value[9] was not a sufficient criterion of equivalence; the results showed that the absolute value of energy dissipated, the distribution of energy dissipation and the spectrum of this distribution are of significance.[10,11] A study on the sedimentation of suspensa in columns comparable to jar test reactors seemed to indicate wall effects.[12]

Summarizing the results of this brief review, one can state that the significance of each individual phenomenon is established. In most instances the effects have been demonstrated qualitatively. A few parameters have been merely identified and quantified. Furthermore, the overall effect of these individually investigated phenomena is not known.

It was the purpose of this investigation to study the general relationship of data on aggregation of suspended solids in a jar test apparatus and in a large-scale wastewater treatment plant when operated under identical conditions. In order to have a basis for the discussion of the observed relationships, additional investigations were made: (1) on the effect of detention time on colloid destabilization and on particle aggregation, (2) on the effect of reactor geometry on flow conditions in the reactor and on particle aggregation and (3) on the effect of changes of reactor scale on overall particle aggregation. All studies on the composite effects were done as field studies with wastewater from different sources, while all other studies on specific phenomena were bench-scale experiments with clearly defined conditions of soluble and insoluble matter. The aim was to justify the applicability of the convenient and inexpensive jar test and to establish the limits of the validity of these data.

EXPERIMENTAL

Suspensions

For the comparison of global effects, sewage was flocculated in a jar, in a continuous-flow reactor (Figure 1) and in a large-scale plant. (Figure 2 summarizes the general approach.) In one series of experiments at Edenkoben, the wastewater source was a middle-size community with a moderate component of small-industry wastes; flocculation followed primary and secondary treatment. The characteristics of the secondary effluent corresponded to known average values and are listed in Table I.

In a second set of experiments at Ludwigshafen, the wastewater source was a large diversified chemical plant; flocculation followed neutralization and crude pretreatment in a grit chamber. The characteristics of this wastewater are very high absolute COD values, a high COD/BOD ratio and a high suspended solids content (Table II).

All *in vitro* experiments on aggregation were done with suspensions of silica (Min-U-Sil 30-Pennsylvania Glass Sand Corp.) with an average particle diameter of 4.2 μ suspended in 10^{-3} mol/l reagent-grade NaCl.

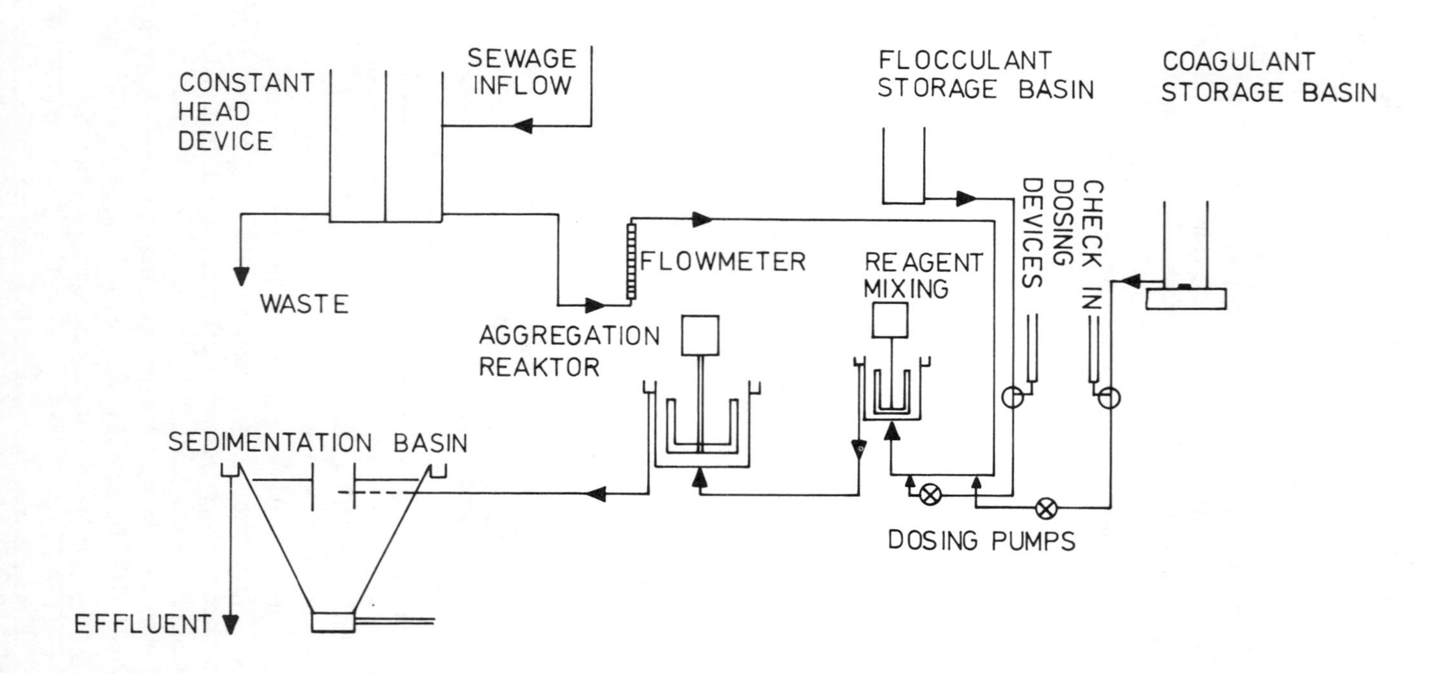

Figure 1. Scheme of the continuous-flow reactor (CFR), a scaled-down replica of the large-scale plant.

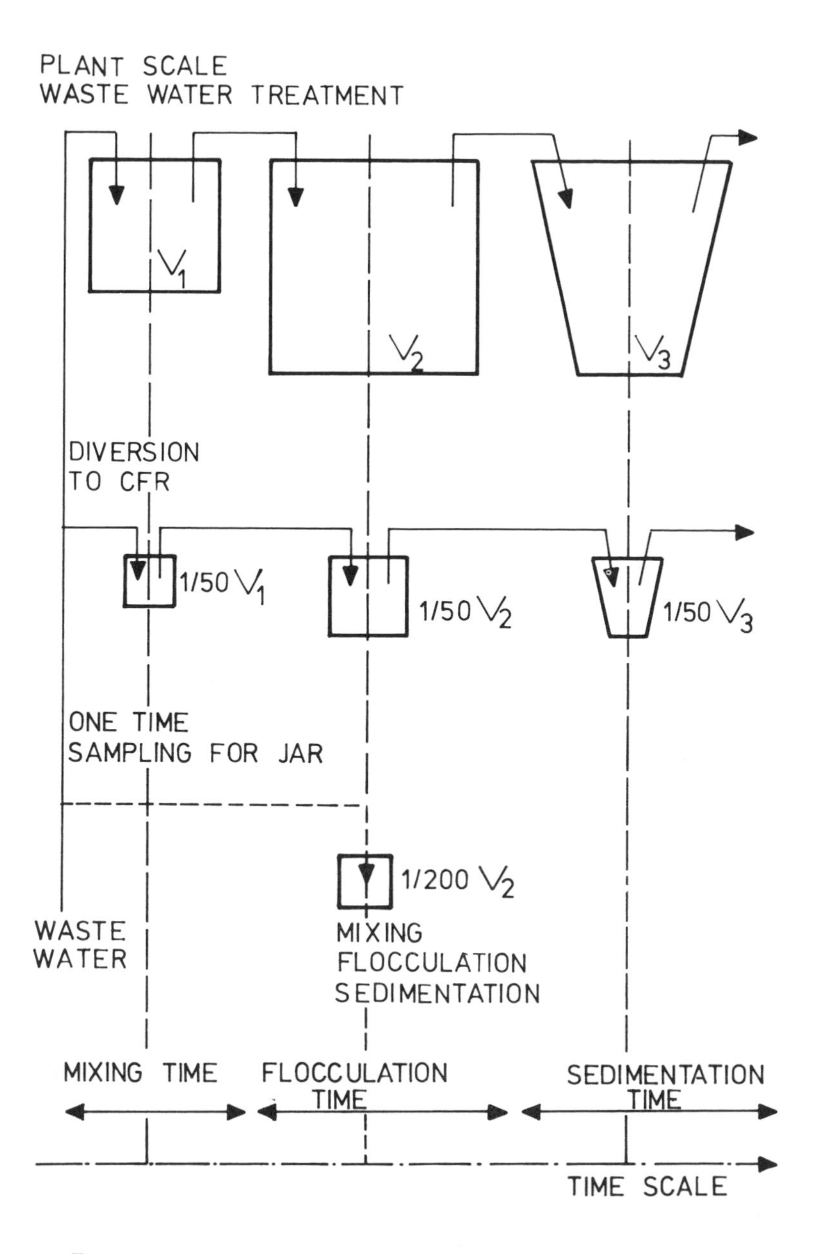

Figure 2. Schematic flow diagram of experimental procedures used.

Table I. Routine Analysis of the Effluent of the Edenkoben Two-Stage Plant (throughput 33 liter/sec)

Parameter		Treatment Plant Inflow	Plant Effluent
Temperature	(°C)	20	19
Settleable Material	(ml/l)	12	1.1
pH	(mg/l)	7.4	6.4
COD (permanganate)	(mg/l)	575	44
BOD_5	(mg/l)	360	15
NH_4^+	(mg/l)	50	44
NO_2^-	(mg/l)	0	0.3
Cl^-	(mg/l)	84	81

Table II. Routine Analysis of the Influent to the Ludwigshafen Treatment Plant

BOD	500-700 mg/l
COD	1200-1800 mg/l
pH Before Neutralization	pH_{min} = 2 pH_{max} = 10
pH After Neutralization	6-8.5
Conductivity	3000-5000 μ-s-cm^{-1}
TOC	250-300 mg/l

Chemicals

All chemicals used in *in vitro* studies were reagent grade (Merck). Flocculants used were as available from commercial suppliers (Brothers Giulini GmbH, Boliden-Knapsack GmbH). Characteristics as given by the supplier are listed in Table III. Adjustment and control of pH as required for *in vitro* studies was effected with 10^{-1} *M* HCl and NaOH, respectively.

Methods

Hydraulic flowrates were measured with ball flowmeters (Fischer-Porter). Dissipated energy was determined by measuring the torque at the stirrer shaft (Rotationsviskosimeter RV3 Haake) or by estimating energy input as hydraulic head loss or from stirrer motor specifications for large-scale plants. Electrophoretic mobility of the particles was determined using a particle electrophoresis instrument (Zetameter) with a specially constructed flow-through cell. Particle aggregation was determined in two ways: (1) immediate counting of total number of particles with a particle analyzer (Coulter Counter ZB) and (2) indirect evaluation by measuring the turbidity after a 30-min sedimentation phase (Hach Turbidimeter). Other wastewater characteristics, such as permanganate COD, total phosphorus and orthophosphate, were determined according to "Deutsche Einheitsverfahren."[13]

Results

In order to establish the limits of the applicability of jar test studies for plant design and operation, the efficiency of reduction of pertinent wastewater constituents in a large-scale plant and in a standard jar test arrangement was observed. Since the plant is operated under continuous-flow conditions and the jar test under batch-flow conditions, an additional series of comparative experiments was performed in reactors operated under continous-flow conditions.

Comparison of Plant and Jar Test Data

Flocculation as a unit process (in conjunction with a phase separation process) specifically reduces suspended solids. The efficiency of the flocculation process in plant and jar was, therefore, evaluated on the basis of turbidity before and after treatment. Conventional wastewater treatment plant design and operation in Western Germany concentrates on the reduction of oxygen-consuming substances; this parameter was used in addition to determine process efficiency. In some instances changes in the concentration of phosphorus components were also recorded since it is well known that some of these components adsorb onto suspended solids and are removed in part by this unit process.

Figure 3 summarizes all data on the reduction of turbidity, COD and phosphorus fractions through flocculation in a jar and in a plant. Comparing mean values of dimensionless concentrations it is seen that in all instances the efficiency of reduction in the plant is as high or higher compared with the jar. It is never lower in the plant. This is the case

Table III. Summary of Flocculants Used (Supplier, Brand Name, Specification)

Supplier	Brand Name	Active Compound	Al_2O_3	Fe_2O_3	pH	Non-Dissolved
Gebr. Giulini GmbH Ludwigshafen/Rh	ferri-FLOC	Iron(III) sulfate	4-5%	13-15%	1.5 5% solution	10-12%
Boliden-Knapsack GmbH Hurth-Knapsack	AVR	Aluminum sulfate	13-14%	4-5%	2.6 1% solution	2-3%

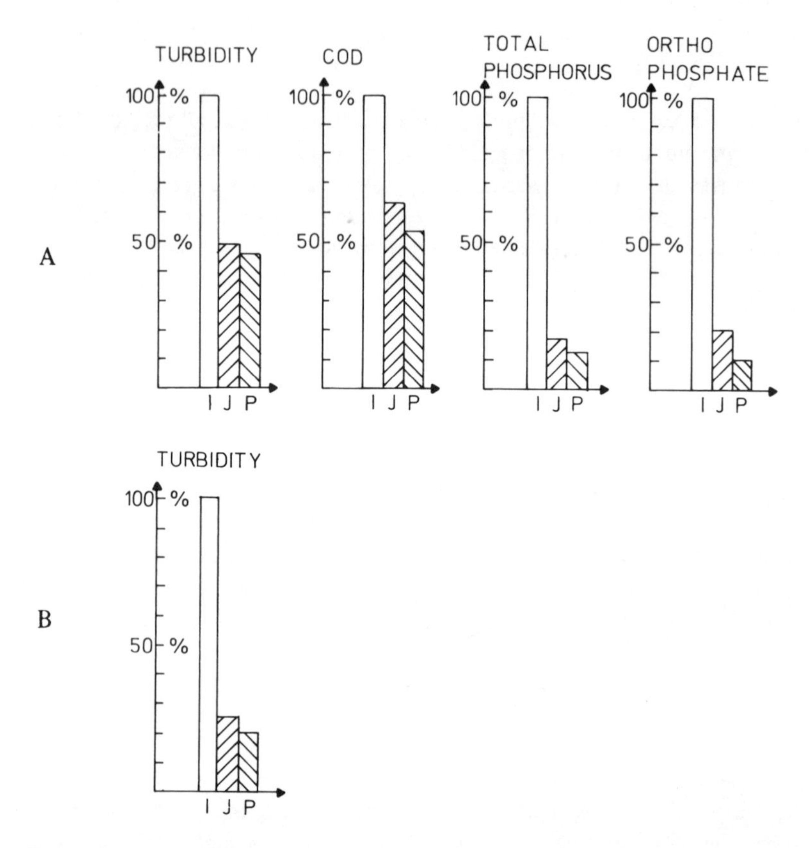

Figure 3. Comparison of jar test data and plant performance data. A. Plant at Edenkoben–municipal wastewater. B. Plant at Ludwigshafen–industrial wastewater.

for flocculation of secondary effluent (Figure 3A: Edenkoben) and for flocculation of raw sewage from a chemical plant (Figure 3B: Ludwigshafen). Statistical analysis of the data material on COD reduction from Edenkoben, tested for the hypothesis that both (not necessarily Gaussian-distributed) series of observations belong to the same statistical population, show that data coincide with a relatively high level of confidence; about 92.5% confidence in a X^2 test (see Table IV). This means that the two series are similar in their characteristics.

Table IV. Summary of Statistical Analysis and Evaluation of Significance of Agreement of Plant, Jar and CFR Data (COD Reduction at Edenkoben)

	Plant	CFR
Jar	90-95%	80-90%

Comparison of CFR and Jar Test Data

Parallel investigations on the reduction of turbidity in a continuous-flow reactor of a size relatively comparable to the jar test reactor should yield information on the effect of operating conditions (flow conditions). Figure 4 summarizes all data from experiments with Edenkoben and Ludwigshafen sewage, as well as with sewage from two other locations. In all instances the average values of dimensionless concentration of suspended solids/turbidity in the CFR are slightly lower than in the jar test. The phosphate reduction at Edenkoben represents the only exception; the relatively numerous results, however, still do not allow a statistical analysis on the significance of this difference. Again, the hypothesis was tried that both series of observations result from the same statistical population (Table IV); here the hypothesis only holds with a lower degree of confidence or must be rejected on the basis of the usual confidence interval (*i.e.,* the differences are significant). This means that despite small difference in the average values, the series of observations are less comparable than those for plant and jar test, where average values were less comparable.

Summarizing these results one can state that for the investigated situation, in particular with respect to the wastewater characteristics, plant performance data are better or at least as good as jar test data in view of removal efficiencies. Continuous-flow reactor data are equivalent to jar test data only on a lower level of significance; flocculation conditions in a jar test and in a CFR arrangement of comparable size and comparable intensity of parameter control are appreciably different. An explanation and, if possible, a quantification of the differences might therefore concentrate in first approximation on a comparison of these two systems.

DISCUSSION

Comparison of process efficiency data from the three types of reactors investigated has led to the conclusion that jar test results will always show lower, or at best equally high, efficiency than the results from the large-scale plant. The differences in performance between the intermediary

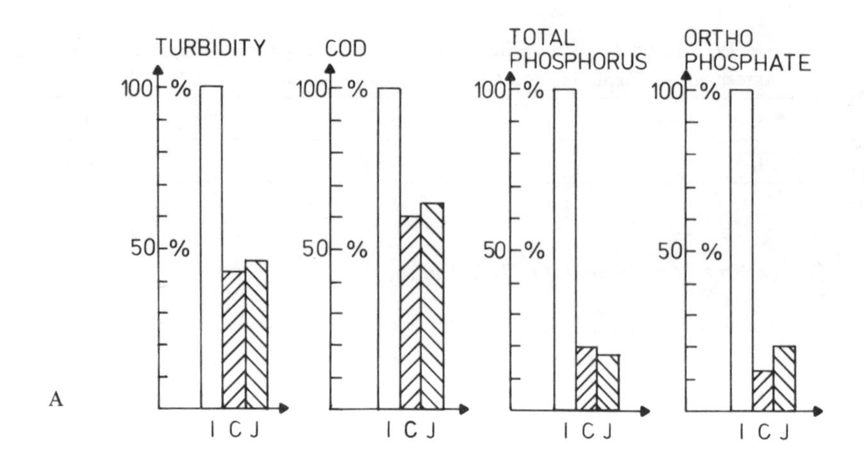

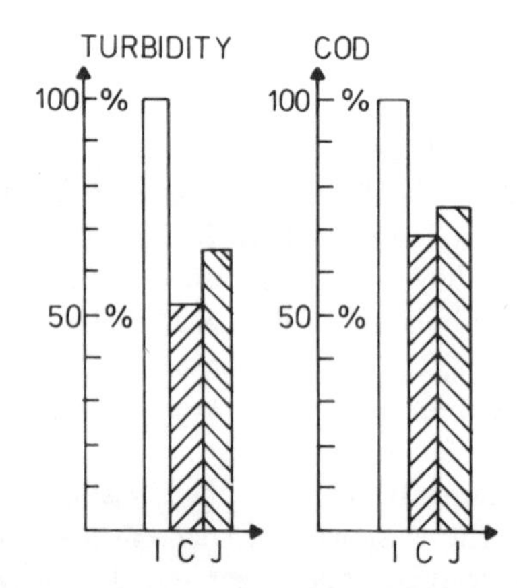

reactor, the bench-scale continuous-flow reactor, to the jar test arrangement are significant. Consequently, a discussion of these differences might yield a basis for the explanation of differences between jar test and plant. Three phenomena have been investigated in more detail and found to explain differences between jar test system and continuous-flow reactor.

Effect of Residence Time Distribution

In a batch system, reaction time and residence time are clearly defined by the conduction of the experiment, *i.e.*, the commencement (sampling) and the halting of the process (analysis). In a continuous-flow system, individual reactant particles are retained longer or shorter than a theoretical residence time, *i.e.*, reactor volume divided by the flowrate, depending on the hydraulic characteristics of the reactor. There are definite indications

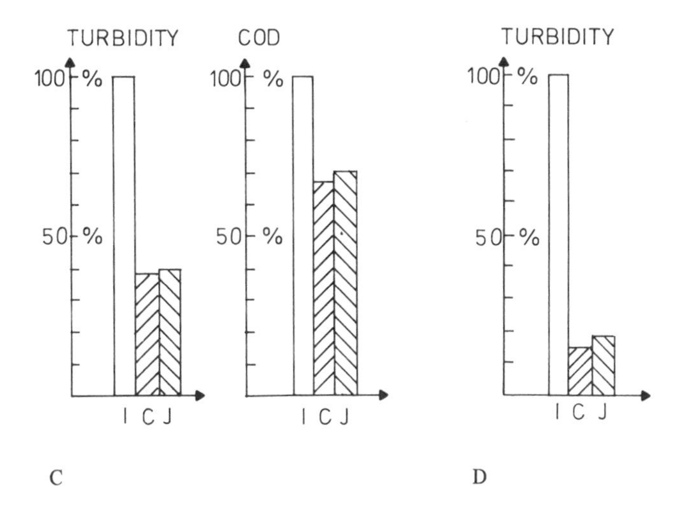

Figure 4. Comparison of jar test data and continuous-flow reactor data. A. Wastewater from Edenkoben. B. Wastewater from Berghausen–medium-size municipal waste source. C. Wastewater from Karlsruhe–large municipal/industrial waste source. D. Wastewater from Ludwigshafen.

that the actual residence time will affect particle destabilization, measured as electrophoretic mobility (Figure 5A).[14] Juxtaposing this effect to the actual residence time in an equivalent batch and continuous-flow system (Figure 5B), it is seen that differences in particle destabilization and/or aggregation will arise. It appears that wider distributions (CFR) affect destabilization positively. The differences in resulting effluent turbidity* from a jar and two differently constructed continuous-flow reactors (with identical mean residence time but different geometry and thus different residence time distribution are shown in Figure 5C. Reactor geometry affects not only residence time distribution but also the pattern of energy dissipation.

Effect of Energy Dissipation and Flow Pattern

In a number of studies the effect of energy dissipation on the transport phase has been shown.[10,11] Presently, there are indications that even the

*Turbidity of the supernatant after sedimentation as a measure of particle aggregation is not always unambiguous. In this instance turbidity analyses were used mainly for the field studies. However, they were periodically compared with particle counting analyses and found to be satisfactory.

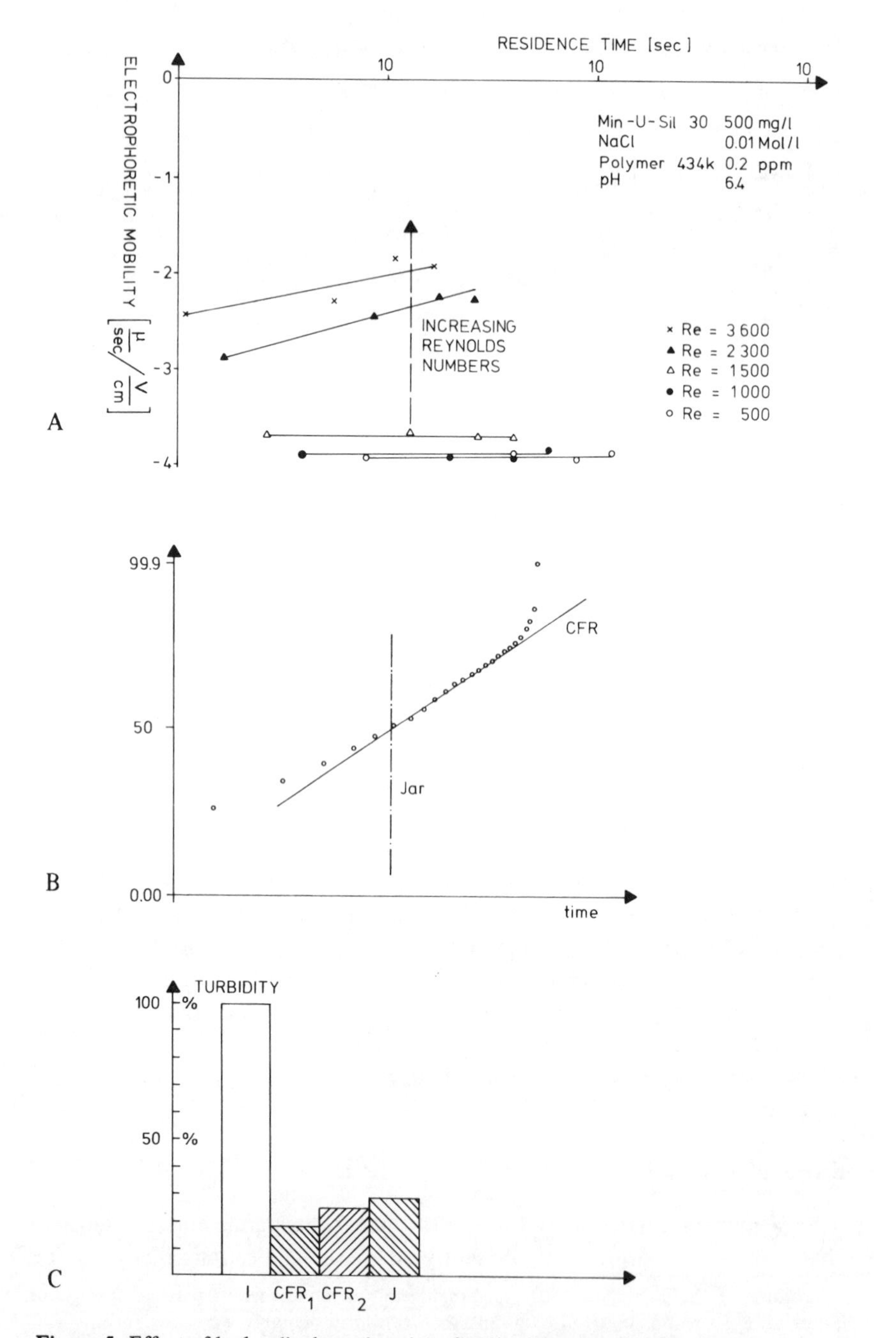

Figure 5. Effect of hydraulic detention time distribution upon floc formation in jar test and CFR and large-scale plant. A. Effect on destabilization. B. Differences in detention time distribution. C. Effect on actual turbidity removal.

destabilization phase is affected by the pattern of energy dissipation in the reactor.[15] Differences in performance of cylindrical batch reactors and noncylindrical continuous-flow reactors might therefore also arise from differences in the pattern of energy dissipation. In a detailed hydraulic investigation of two reactor systems, it was found that Reynolds numbers, as defined for such systems, differ significantly for a continuous-flow reactor as compared to a jar for comparable energy dissipation (Figure 6). In all instances the Reynolds number was higher in the CFR system. The results in Figure 6 furthermore demonstrate the significant differences

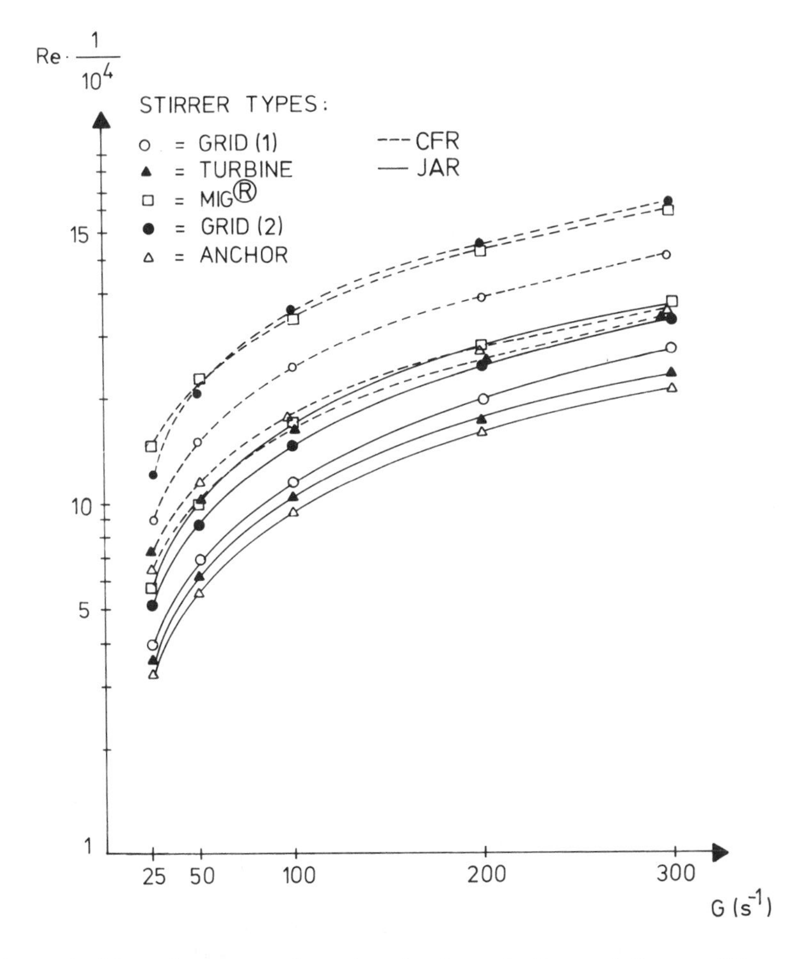

Figure 6. Effect of geometry of reaction chamber and stirrer and flow conditions (batch, continuous flow) on the energy dissipation pattern.

between differently shaped stirrers, attesting the importance of individual flow patterns.[8]

For most systems at the reaction times under consideration, higher Reynolds numbers in this order of magnitude may lead to more efficient particle aggregation. This is confirmed by measuring the change in total particle number due to aggregation in a jar and in a continuous-flow reactor (Figure 7A). Only at very high levels of energy dissipation, rarely encountered in practice, is the reverse observed. Even a slight change in the energy dissipation pattern already results in a change in effluent turbidity/particle aggregation (Figure 7B).

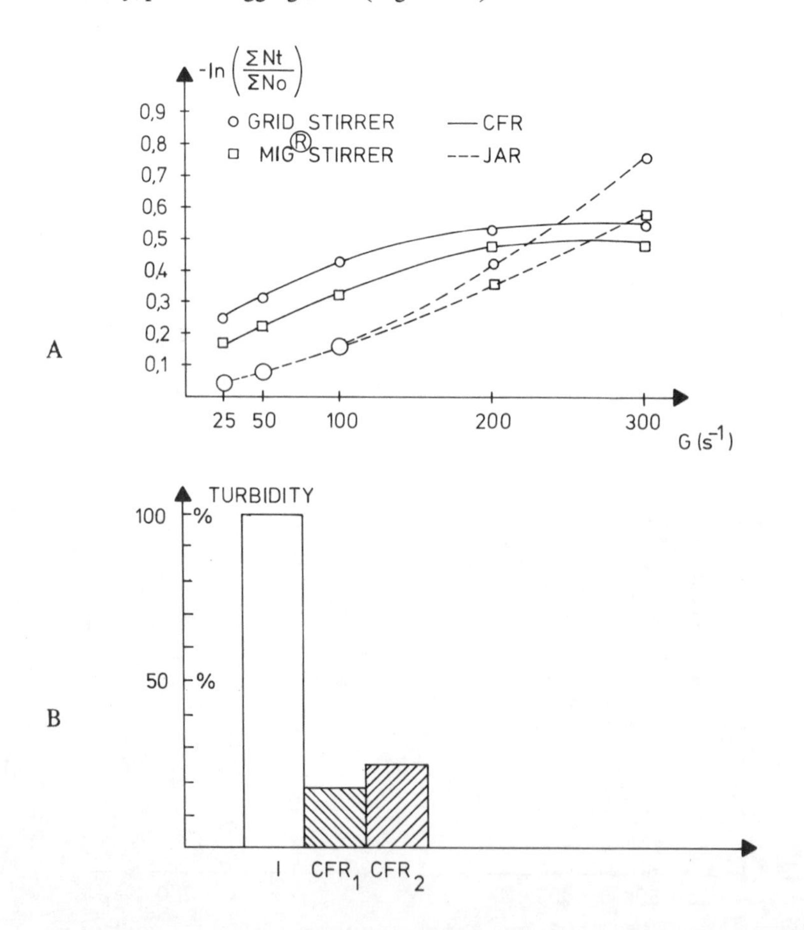

Figure 7. Effect of energy dissipation upon particle aggregation for jar test and rectangular continuous-flow reactor. A. Reduction in total particle number (Min-U-Sil). B. Comparison of two differently constructed CFR (wastewater at Ludwigshafen).

Effect of Concentration Fluctuations

Fluctuations and gradients in the concentration of constituents that control the aggregation process may arise from varying inflow, from changes in the composition of the inflow, from nonsynchronous dosing of reagents and from noncomplete mixing. It has been indicated that slight changes in the absolute and relative size of chemical parameters may result in significant changes in particle aggregation.[16] Furthermore, even different gradients at relatively comparable average solution conditions may affect particle destabilization through the formation of different flocculant complexes or flocculant colloid complexes.[16] It is, however, difficult to determine such fluctuations and gradients. Thus, in this study, the distribution of the measured effluent concentration was taken as being indicative of the extent of fluctuations and gradients in the system under consideration.[17] Figure 8 shows, for two different wastewaters, the respective distribution of performance data. It is seen that the distributions are significantly more narrow in the case of the CFR, indicating more stable conditions with respect to the concentration of flocculation-affecting constituents. This, again, might lead to more effective particle aggregation in a continuous-flow reactor as compared to a jar test system.

SUMMARY

Performance data from jar tests and from large-scale plants on suspended solids removal differ. The plant shows in most instances a higher removal efficiency. Comparing a jar reactor and a continuous-flow reactor of similar size, it is found that a significant difference in performance arises from this change in form and flow pattern. In general, three different factors have been found to cause more efficient particle aggregation in the continuous-flow reactor, thus indicating at the same time the reasons for different removal efficiencies in jar and plant: a difference in residence time/reaction time distribution favoring particle destabilization and/or aggregation in the CFR, an effect of reactor geometry on the energy dissipation pattern leading again to higher aggregation rates in the CFR, and a more narrow distribution of observed concentration values in the CFR indicating reduced concentration fluctuations and gradients and favorable conditions for particle aggregation. At the same time, the results should be taken only as indications of possible effects that might explain differences in performance data. More detailed analysis of residence time distribution, of velocity gradients and their distribution, and of actual reactor concentrations as function of time and location is needed to clarify basic relationships. It seems possible, however, to state that for reasons indicated, plant performance data will never fall below jar test data if operating conditions are kept comparable.

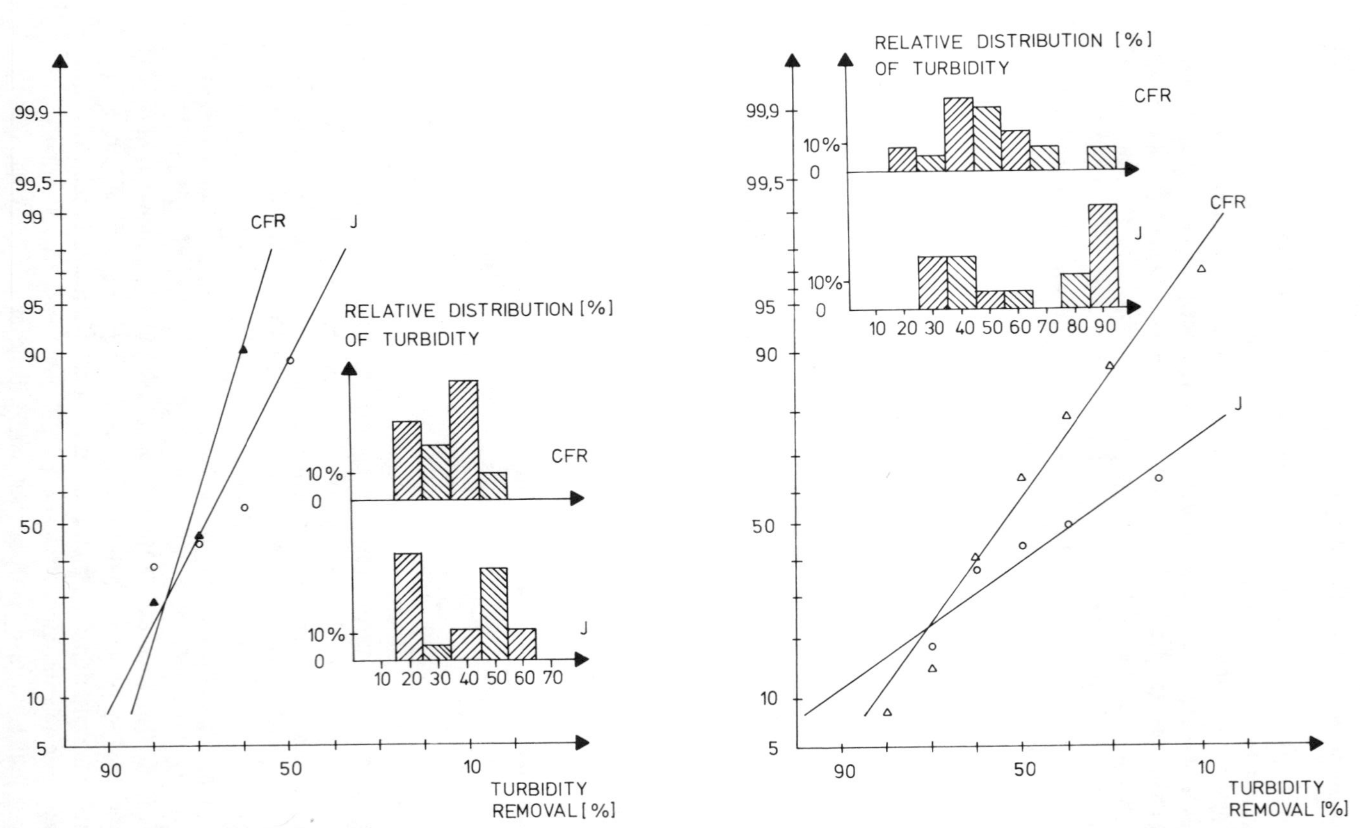

Figure 8. Comparison of distribution of effluent turbidity concentrations in a CFR and a jar for two different wastewaters. A. Turbidity from Edenkoben. B. Turbidity from Berghausen.

REFERENCES

1. Culp, W., and R. Culp. *Physico-Chemical Wastewater Treatment* (New York: Van Nostrand Reinhold Company, 1972).
2. Weber, W. J. *Physico-Chemical Processes for Water Quality Control* (New York: Wiley-Interscience, 1972).
3. Jenkins, D., *et al.* "Water Chemistry Laboratory Manual," Assoc. of Environm. Eng. Prof. (1973).
4. Harris, H. S., W. J. Kaufman and R. B. Krone. *J. San. Eng. Div., ASCE* 92:95 (1966).
5. AWWA Research Foundation Report, *J. Am. Water Works Assoc.* 68:46 (1976).
6. Hahn, H. H. Ph.D. Thesis, Harvard University, Cambridge, MA (1968).
7. Klute, R., and H. H. Hahn. Technical Report, Institut für Siedlungswasserwirtschaft, University of Karlsruhe, Germany (1976).
8. Mertsch, V. Master's Thesis, University of Karlsruhe, Germany (1976). (1976).
9. Camp, T. R., and P. C. Stein. *J. Boston Soc. Civ. Eng.* 30:219 (1943).
10. Argaman, Y., and W. J. Kaufman. *J. San. Eng. Div., ASCE* 96:223 (1970).
11. Klute, R., and H. H. Hahn. *Vom Wasser* 43:215 (1975).
12. Koglin, B. *Chem. Ing. Techn.* 44:515 (1972).
13. Deutsche Einheitsverfahren zur Wasser-, Abwasser- und Schlammuntersuchung. Verlag Chemie, Weinheim, Germany (1971).
14. Bernhardt, H., H. H. Hahn, R. Klute and H. Schell. "Standardisierung einer Flockungstestapparatur," Bericht an das KfW (1976).
15. Klute, R. Ph.D. Thesis, University of Karlsruhe, Germany (1977).
16. Hahn, H. H., and W. Stumm. Chapter in *Advances in Chemistry Series* Vol. 79 (Washington, DC: American Chemical Society, 1968).
17. Eppler, B., and G. Hinrichs. "Abwasserflockung," Bericht an das BMI, Bonn, Germany (1977).

17

APPLICATION OF POLYELECTROLYTES FOR INDUSTRIAL SLUDGE DEWATERING

Francis J. Mangravite, Jr., Christopher R. Leitz, and Donald J. Juvan

Betz Laboratories, Inc.
Trevose, Pennsylvania

INTRODUCTION

Continued emphasis on environmental protection has resulted in state and federal timetables for reducing the discharge of pollutants. To meet these requirements, industry is finding it necessary to upgrade its waste treatment facilities from primary to secondary treatment. Waste biological sludge generated from secondary treatment is often less than 2% suspended solids. Such low solids content coupled with an increasing difficulty in obtaining permits for land disposal is forcing some industrial plants to dewater their biological sludge prior to disposal. The latter may take the form of incineration, landfill, recycle or sale as a soil conditioner.

Mechanical dewatering of biological sludges by vacuum filters, belt filters, centrifuges and sometimes filter presses cannot be accomplished without first chemically conditioning the sludge. Inorganic coagulants such as ferric chloride and lime and high-molecular-weight cationic polymers are the most effective destabilizing agents for biological sludges.

The cationic polymers used vary in their physical and chemical properties. Two important variables are molecular weight and charge density. This chapter discusses the relative effects of these two variables in dewatering several industrial secondary sludges as observed in laboratory tests.

Laboratory measurements of the effectiveness of sludge dewatering aids usually involve one of the following methods: pressure filter test (PFT), filter leaf test (FLT), Buchner funnel test (BFT) or capillary suction time (CST) test. This study employed the last three tests and, in the case of one sludge, compared the three methods to each other. By applying filtration theory, the average specific resistance of a sludge can be determined. The value of this parameter in predicting sludge treatability has not been fully demonstrated. This study compares the specific resistance of untreated sludges with their ability to be dewatered by several different polymers.

Extrapolation of laboratory dewatering test results to plant performance is not always successful. In each case, mixing conditions, the nature of the sludge and the type of dewatering equipment used in the plant should be taken into consideration.

EXPERIMENTAL

The five polymers used in this study are described in Table I. Each is essentially linear in structure. Polymers A through D are copolymers of acrylamide and a quaternary nitrogen-containing monomer; thus, the charge densities are constant over a broad pH range. Polymer E is also linear but derives its cationic charge from the protonation of unquaternized nitrogen groups such that its charge should increase with decreasing pH over the range investigated in this study, 5.7 to 7.7. Molecular-weight rankings were based on viscosity, and charge density rankings were based on the monomer charge composition with the understanding that conversion was high.

Fresh sludge was sampled from each plant at a point before the addition of any chemical conditioning agents for dewatering or secondary clarification. Testing was usually complete in 25 to 48 hours after sampling. During this time the effects of aging were observed to increase somewhat the optimum dosage of a polymer, but not to affect significantly the filterability at the optimum or the order of performance of the different polymers. Generally, the five polymers were all evaluated within a period of five hours, during which aging effects were negligible.

For CST and BF tests, a 200-ml sludge sample was stirred with the designated quantity of 0.5% polymer solution at about 500 rpm for 20 sec in a 250-ml beaker with a 3-bladed (2-in. diam) propeller-type stirrer. Testing commenced immediately after mixing.

BF tests used a 12-cm funnel, 11-cm Whatman No. 4 filter paper, 15 in. of Hg vacuum and 200 ml of sludge. The CST test employed either the 1.0- or 1.8-cm cells and Whatman No. 17 chromatographic grade filter paper. The CST for deionized water was about 5 to 6 sec for the

Table I. Polymer Identification

Polymer	Cationic Charge Density	Relative Molecular Weight
A	same as B	high
B	same as A	higher than A
C	higher than B	not reported
D	higher than C	not reported
E	high	high

1.8-cm cell and 9 to 10 sec for the 1.0-cm cell. For the BFT, 200 ml of water passed through the filter in about 2 to 3 sec.

FL tests were conducted on a 0.1-ft^2 round leaf, with a 60 x 40 count, 7/1 satin fil-warp yarn, 12 oz/yd^2-weight Eimco nylon cloth. A 1300-ml sample of stirred conditioned sludge was subjected to a 1-min pickup time and a 3-min drying time at 15 in. Hg vacuum.

All polymer solutions were freshly prepared every two days with deionized water.

RESULTS

Effect of Polymer Structure

Secondary sludge was obtained from several industrial plants. The filterability of the sludges when treated with the various cationic polymers listed in Table I was measured by BF or CST tests. The optimum dosage and filterability at the optimum dosage was used to determine the order of performance of the polymers. In this manner relative conclusions were drawn concerning the effects of increasing molecular weight and charge.

Paper Mill A (Figure 1)

This treatment facility consisted of primary clarification, aeration basin and secondary clarification. A fresh sample was taken from the secondary sludge recycle line. Of the five polymers tested, the three with the highest charge density (C, D and E) gave the best filterability. Increasing the molecular weight reduced the optimum dosage significantly (B vs A) but also tended to narrow the dosage range of performance. Increasing molecular weight did not significantly affect filterability.

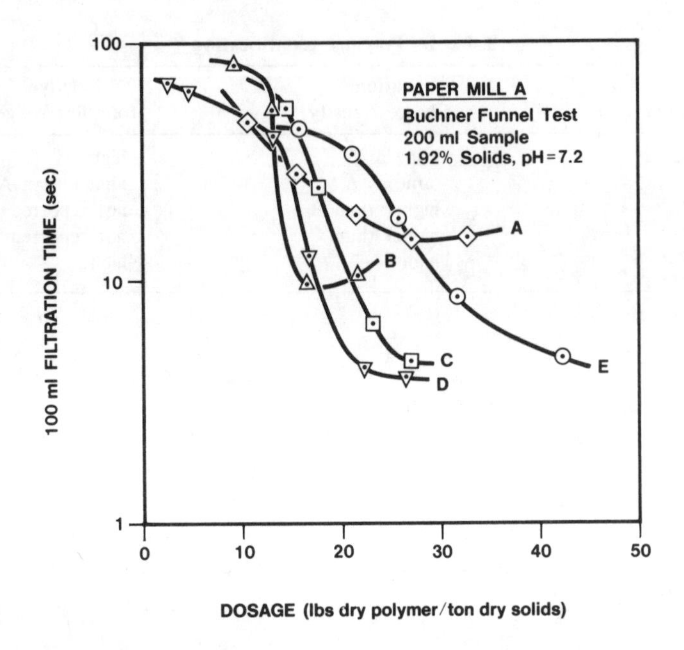

Figure 1. Effects of polymers A, B, C, D and E on the filtration times of biological sludge from paper mill A.

Refinery A (Figure 2)

Refinery A's waste was successively treated by an API separator, an aeration basin and a secondary clarifier. The excess secondary sludge was further concentrated by flotation units from which the sample was obtained. Polymer A was essentially inactive in this system. Comparing the results of polymers A and B indicates that increasing the molecular weight improved the filterability of the conditioned sludge. Increasing the charge up to that of polymer D also improved the rate of dewatering. Polyelectrolyte E gave a very broad performance curve.

Refinery B (Figure 3)

This refinery treated its waste by an API separator, an aeration basin and a secondary clarifier. The excess secondary sludge was thickened before dewatering. A sample of the thickened sludge was used for this evaluation. As in the case of the paper mill sludge, increasing molecular weight narrowed the performance curve while shifting it to lower dosages. Increased molecular weight improved the rate of filtration only slightly

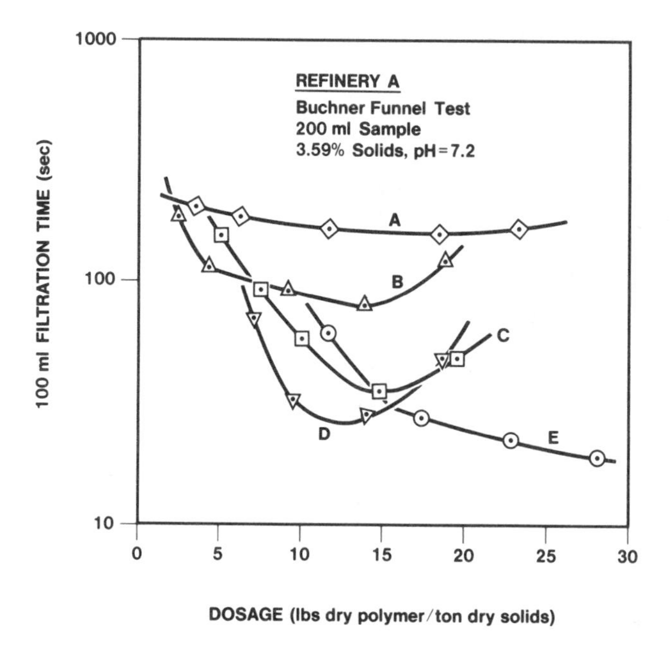

Figure 2. Effects of polymers A, B, C, D and E on the filtration times of biological sludge from refinery A.

at the optimum dosage. Increasing the cationic character of the polymer improved the rate of filtration up to polymer D.

Pet Food Processing Plant (Figure 4)

The effluent from this plant was treated by centrifugation to remove tallow, followed by flotation, extended aeration and secondary clarification. A sample of the aeration pond effluent was settled overnight and the supernatant decanted before the evaluation was started. Increasing the polymer charge increased the rate of filtration.

Fiber Plant (Figure 5)

This nylon plant's waste treatment consisted of biodiscs and clarification. The settled biodisc sludge was further treated in an aerobic digester from which the sample was obtained. As observed previously, increasing molecular weight reduced the optimum dosage while increasing the cationic charge density improved the dewaterability up to polymer D. Generally, this sludge was quite diffcult to dewater with any of the polymers. The performance curves of polymers B, C and D were relatively narrow.

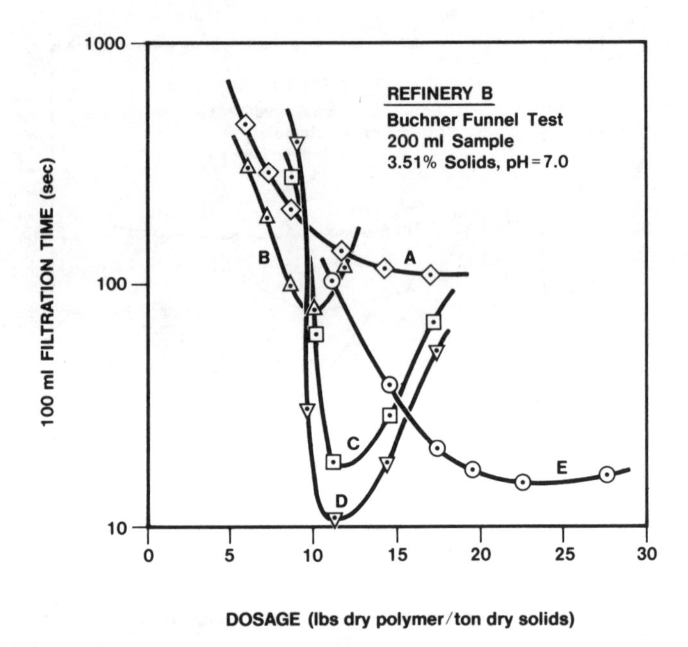

Figure 3. Effects of polymers A, B, C, D and E on the filtration times of biological sludge from refinery B.

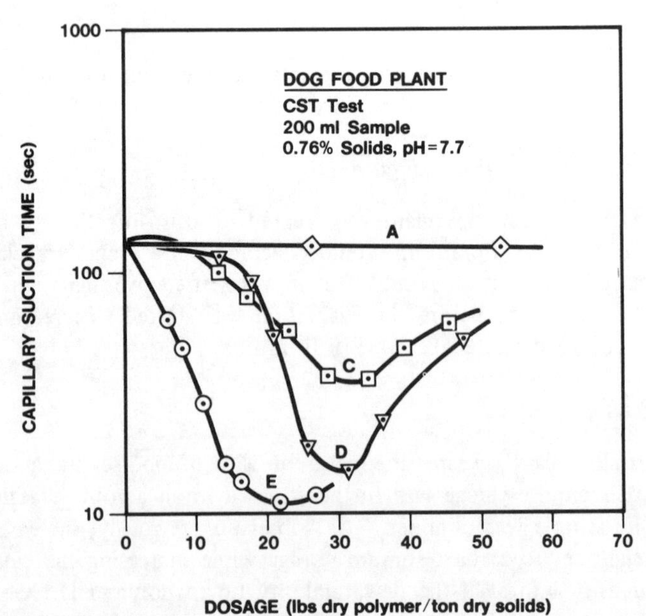

Figure 4. Effects of polymers A, C, D and E on the capillary suction times of biological sludge from a pet food processing plant.

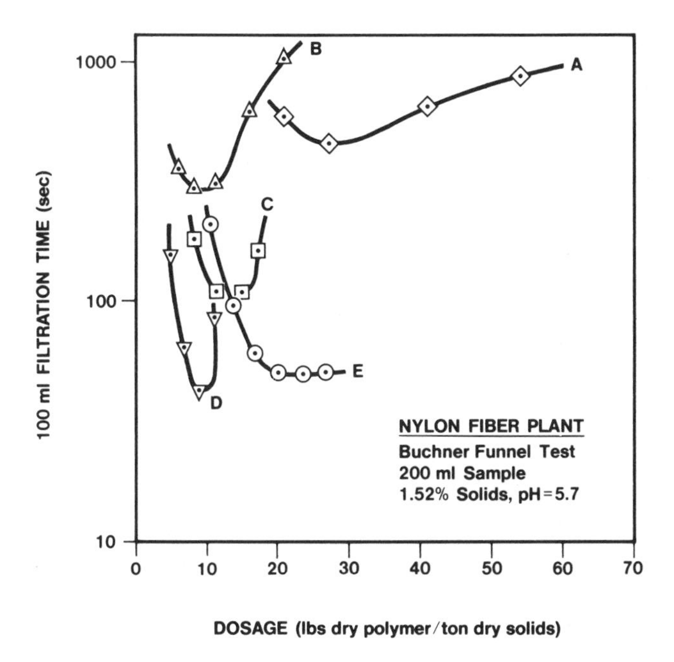

Figure 5. Effects of polymers A, B, C, D and E on the filtration times of biological sludge from a nylon fiber plant.

Comparison of Test Methods

As mentioned above, three of the most common laboratory methods used to evaluate sludge dewatering aids are the filter leaf test (FLT), the Buchner funnel test (BFT) and the capillary suction time test (CST). The last of these was developed as a rapid, portable method for screening dewatering aids.[1,2] It relies on gravity and the capillary suction of a piece of thick filter paper to draw out the water from a small sample of conditioned sludge. The sample is placed in a cylindrical cell on top of the chromatography grade filter paper. The time for the wetted area to enlarge between two fixed points is recorded electronically as the capillary suction time. In effect, it is the time for a specific volume of water to be drained from the sample.

The CST test is closely related to the BFT, which relies on mechanically induced vacuum to draw water from the sludge through a thin piece of filter paper. A correlation between CST values and average specific resistances determined by the BFT has been demonstrated. However, it should be realized that unless the BF tests are conducted at a vacuum equivalent to that of the capillary pressure of the CST filter paper, error is introduced into the correlation. For the commonly employed Whatman #17 chromatographic grade CST paper, an equivalent pressure of 100 g/cm^2 (2.9 in. Hg) has been estimated.[1]

The CST test loses differentiation when the sludge solids are low or the sample well flocculated. In these cases the CST value approaches that of water. When the untreated sludge gives a CST of less than about 25-50 sec with the small diameter cell, greater differentiation can be achieved by concentrating the sludge solids or using two pieces of filter paper.

The BFT permits the recording of the time to filter any number of filtrate volumes. This can be done as a function of pressure (vacuum) in order to determine sludge compressibility.[3] Most biological sludges are relatively compressible. The suspended solids of the filtrate can also be recorded.

According to the Carman-Kozeny equation, a plot of time over volume against volume should yield a straight line during cake filtration.[4] Curvature is normally observed initially due to cake formation and after most of the liquid has been filtered due to cake drainage. In between, the slope of the linear portion of the line is directly proportional to the average specific resistance. In turn, the average specific resistance is related to the specific surface area of the solids and porosity.[5]

Average specific resistance values have been widely employed as a measure of filterability and in a few cases as a means of characterizing sludges.[6,7] However, when determining the effectiveness of various polymers as conditioning agents, it is not necessary to determine resistances. Simply measuring the time to filter some unit volume of filtrate will provide the same information without tedious calculations or curve plotting. For rapidly filtering samples this simple method is much more advisable since the time over volume against volume plots becomes almost horizontal, resulting in large errors when calculating the slope.

The FLT has traditionally been used to predict vacuum filter performance. The round leaf covered with an appropriate filter cloth is inverted into a stirred sludge sample for a specific length of time while vacuum is applied. The leaf with the resulting cake is withdrawn and the vacuum continued to simulate the vacuum filter's drying cycle. With this procedure, it is possible to determine yield (weight dry solids/unit area/unit time) as well as filtrate suspended solids. The effect of different filter cloths on cake pickup, blinding, cake release and filtrate quality can also be observed. This can be quite important since some well flocculated sludges will not adhere to certain filter cloths. Disadvantages of the method are the need for large samples, greater operator error, the less portable nature of the equipment and the longer time required for each test.

In order to compare the three methods, a sample of secondary sludge was obtained and separately conditioned with two different cationic polymers. As rapidly as possible, filterability was determined simultaneously by CST, BF and FL tests. The two polymers were polymer A (see Table I) and polymer F, which is a medium-charge density, high-molecular-weight cationic (quaternary) polymer. The purpose was not to compare the polymers, but rather to compare the three test methods on the same sludge treated with two different polymers. For this study, biological sludge was obtained from the waste secondary sludge thickener of a refinery.

The filtration rates of the sludge conditioned with various amounts of polymers A and F as determined by the BFT and CST tests are given in Figure 6. Because the sludge dewatered slowly, the time to filter 25 ml from the 200-ml sample was recorded during the BFT. The change in filtration rate with increasing dosage measured by the two methods closely parallel one another. This was true for both polymers even though they differed significantly in their ability to condition this sludge.

Both the CST test and the BFT showed the minimum polymer F dosage necessary for near maximum filterability to be about 8.5 lb/ton. At higher dosages, the BFT showed a decrease in filtration times up to 13 lb/ton. The CST test, however, showed a slight increase in filtration times between 8.5 and 13 lb/ton. It is believed that at dosages above the optimum, unadsorbed polymer in the filtrate retards the rate of wetting of the CST filter paper.

Results with the FLT are shown in Figures 7 and 8. The optimum cake yield for polymer A (Figure 7) occurred at 9 lb/ton. For this polymer the BFT and CST test filtration rates were less definitive in indicating an optimum dosage. Maximum filtration rates did not occur until about 11 lb/ton. For polymer F (Figure 8) the FLT showed an optimum yield around 8.5 lb/ton. This agrees with the CST data but is at the lower end of the optimum dosage range (8.5-13 lb/ton) determined by the BF test.

One advantage of the filter leaf test is its ability to monitor cake solids (moisture). Polymer A (Figure 7) had little effect on the cake solids. However, polymer F increased the cake solids from 13 to 17% when concentrations of 11 or more lb/ton were added The minimum dosage necessary to achieve maximum cake solids was higher than for maximum yield and qualitatively agreed more with the BFT data than the capillary suction times.

Usually, filtrate quality is not measured with the BFT because of the use of filter paper rather than a cloth. However, for the purposes of conducting a more thorough comparison of the test methods, the absorbance

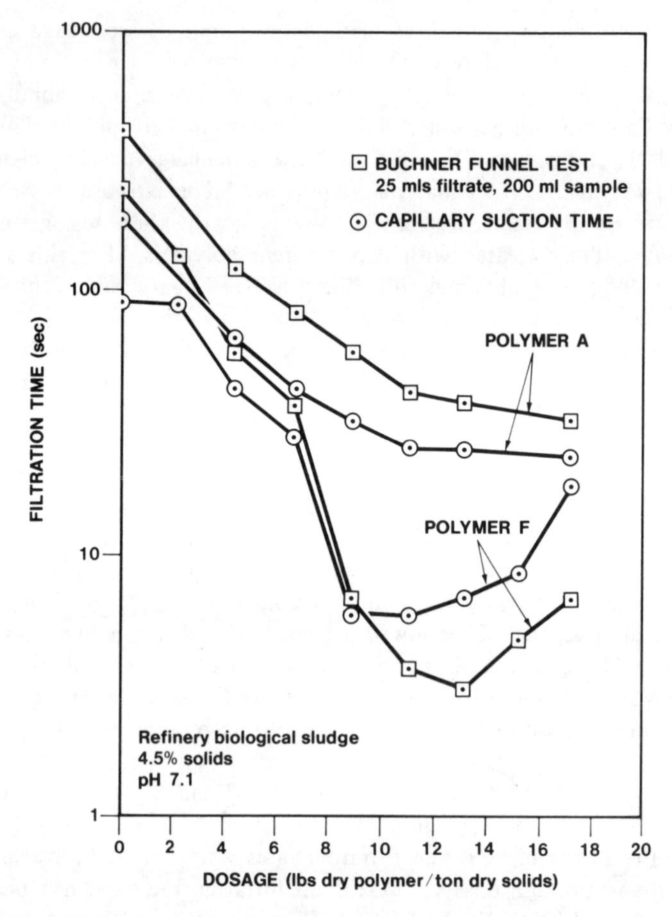

Figure 6. Comparison of the Buchner funnel test with capillary suction time for polymers A and F on refinery biological sludge.

and turbidity of the BFT filtrates were recorded. Figures 9 and 10 compare these values as a function of dosage for polymers A and F with the filtrate suspended solids found with the FLT. Evaluation of filtrate quality is not possible with the CST test.

For both polymers it is obvious that there is little correlation between the filtrate qualities of the two methods. Only unflocculated colloidal particles can be expected to penetrate the filter paper in the BFT. Use of a filter cloth in the FLT allows the passage of unflocculated colloids as well as some small flocs. The size distribution, shear strength and compressibility of the flocs could play an important role in determining

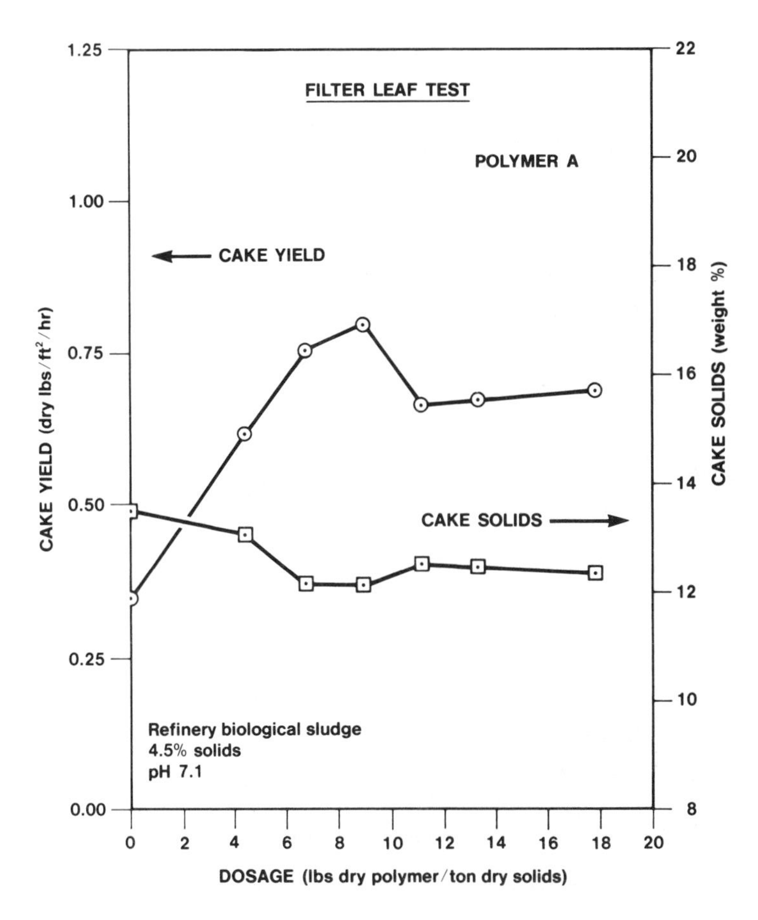

Figure 7. Cake yield and cake solids vs dosage of polymer A on refinery biological sludge.

penetration of the cloth. With both tests the porosity of the filter cake is expected to govern the amount of unflocculated colloids which reach the filter media. This is believed to be the reason for the relatively low BFT filtrate turbidities of the untreated sludge. Generally, the turbidity and absorbance curves paralleled one another and neither measurement appeared to offer an advantage over the other.

For polymer A (Figure 9), increasing dosage beyond 5 lb/ton gradually improved filtrate quality. No dramatic change occurred around 8-11 lb/ton where optimum yield and filtration rates were observed. However, for polymer F (Figure 10) the BFT filtrate quality deteriorated with increasing polymer concentration up to the optimum dosage (8-11 lb/ton)

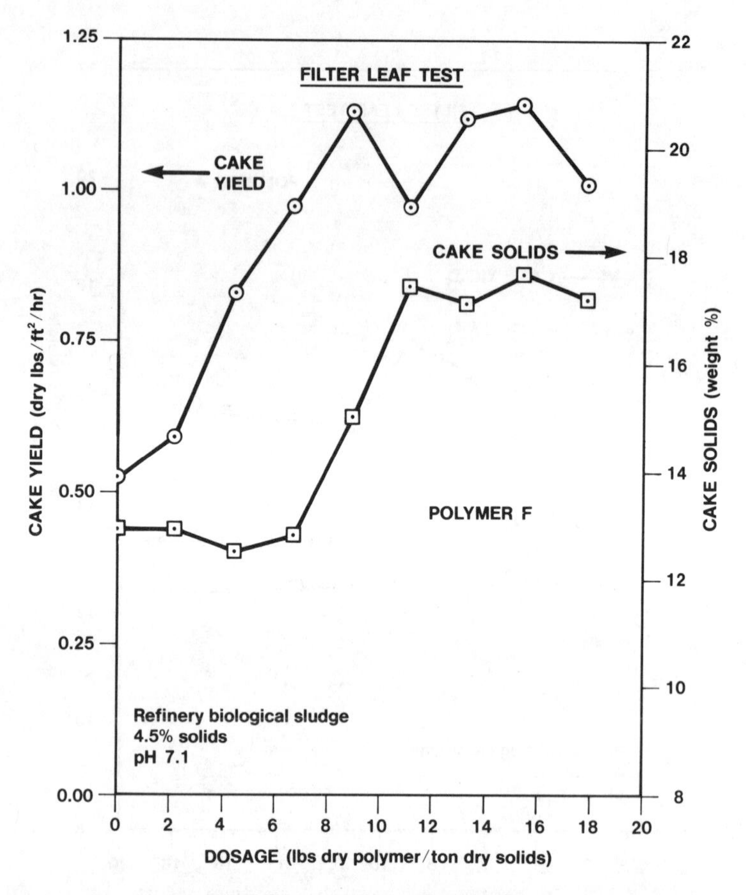

Figure 8. Cake yield and cake solids vs dosage of polymer F on refinery biological sludge.

after which it rapidly improved. FLT filtrate suspended solids decreased up to the optimum dosage after which they leveled off. Obviously, the mechanisms whereby particles enter the filtrate during Buchner funnel and filter leaf tests are quite different.

With the BFT it is possible to determine the average specific resistance of sludge samples. As mentioned before, for the purpose of comparing chemical dewatering aids, it is not necessary to actually determine resistances. Usually, biological sludges have a complex chemical and physical composition that changes with time. Detailed characterization is not feasible. Thus, observation of specific resistance would appear to offer a practical method of characterization.

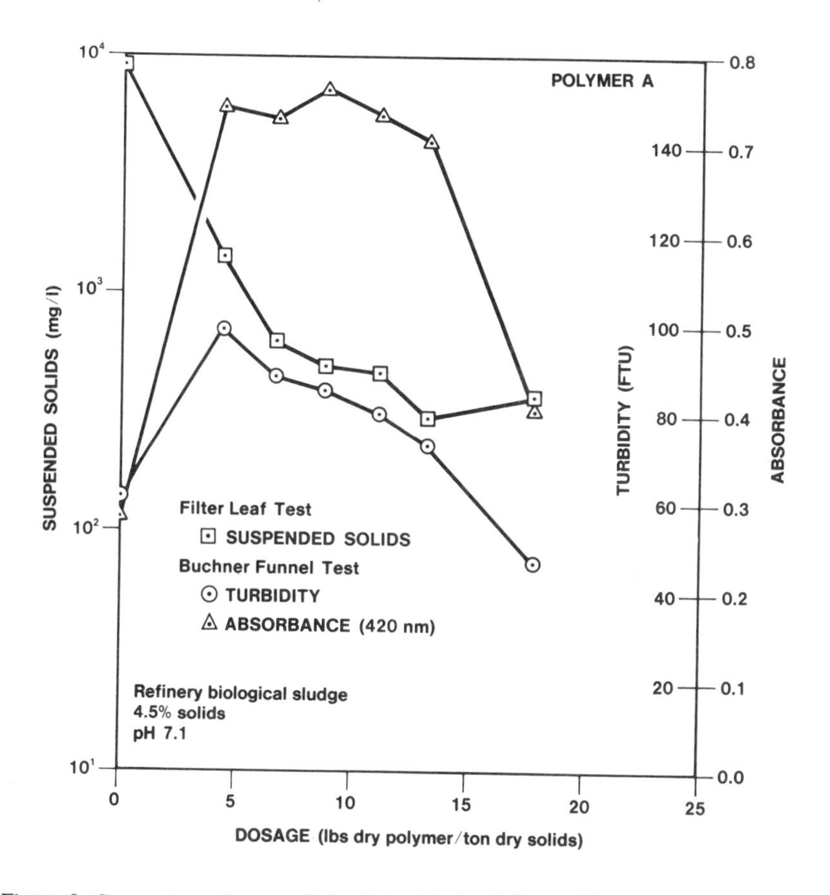

Figure 9. Comparison of suspended solids of filter leaf test filtrate with turbidity and absorbance of Buchner funnel test filtrate for various dosages of polymer A.

To define the relationship between the average specific resistance of the untreated industrial biological sludges and their response to conditioning by several polymeric dewatering aids, the data in Table II were assembled. The optimum dosage and the filtration rate (time) at the optimum dosage for five different polymers are tabulated along with the untreated specific resistances. Four of the seven sludges are the same as those in Figures 1, 2, 3 and 5. Table II clearly indicates that there is no obvious relationship between untreated specific resistance and the activity of polymeric dewatering aids. For instance, the dosage of polymer A is the same for paper mill A and the fiber plant even though their specific resistance differs by a factor of 32. Filtration rates for all five polymers at the

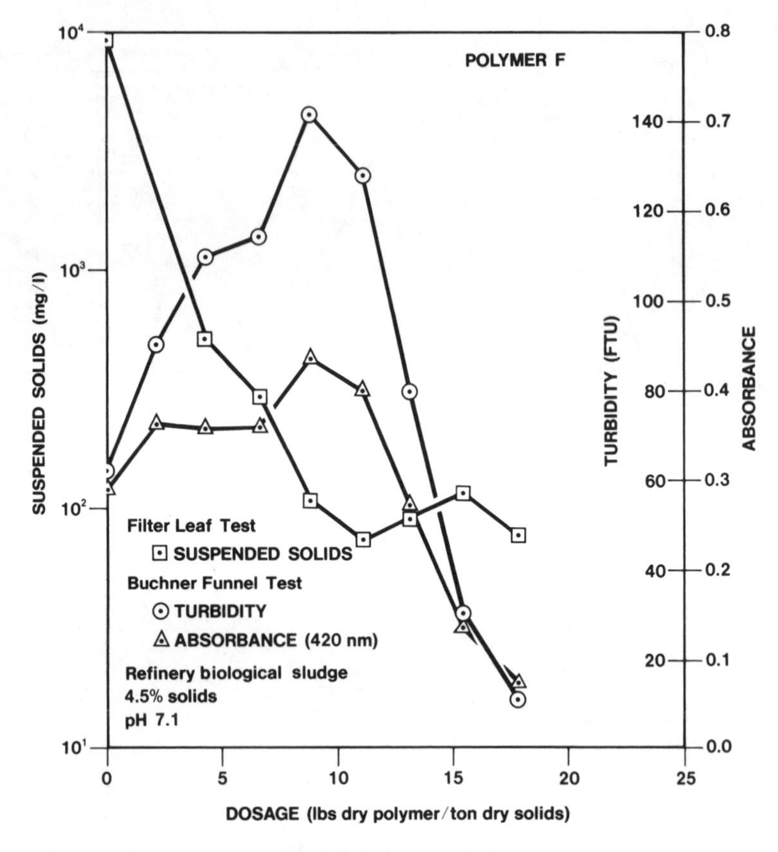

Figure 10. Comparison of suspended solids of filter leaf tests filtrate and turbidity and absorbance of Buchner funnel test filtrate for various dosages of polymer F.

optimum dosage were faster for the sludge with an average specific resistance of 4.1 x 10^{12} cm/g than the two sludges with resistances of 3.1 and 9.5 x 10^{12} cm/g.

DISCUSSION

There are many commercial polyelectrolytes employed for biological sludge dewatering. While dual polymer treatments and combination inorganic coagulant polymer treatments may utilize anionic polymers, most single polymer treatments are cationic. These commercial cationic polymers differ widely in their chemical and physical properties such as average molecular weight, molecular weight distribution, type of cationic

Table II. Comparison of Specific Resistance and Polymer Performance

Biological Sludge Source	Untreated Specific Resistance X 10^{-12} cm/g	Polymer									
		Optimum Dosage (lb/tds)					Filtration Time[a] (sec)				
		A	B	C	D	E	A	B	C	D	E
Paper Mill A	0.30	27	17	30	22	42	15	10	5	4	6
Refinery A	1.0	c	13	14	10	17	160	82	36	29	20
Paper Mill B	1.7[b]	16	12	19	-	32	6.5	7	5	-	4
Municipality	1.8	11	7	9	9	6	15	8	6	4	38
Refinery C	3.1	c	-	12	13	19	300	-	32	20	38
Refinery B	4.1	17	10	11	11	20	100	80	19	11	16
Fiber Plant	9.5	26	9	13	8	20	440	300	110	40	50

[a]Time to filter 100 ml from a 200-ml sample.
[b]40% secondary/60% primary sludge.
[c]Polymer showed no optimum dosage.

charge group, degree of branching, charge location relative to the backbone, chain stiffness, charge density, charge distribution, counterion type and chemical structure. For example, the charge density of polymers with amine groups varies with pH and combines with chlorine atoms while those with quaternary nitrogen groups do not.

Comparison of the appropriate values for each polymer in Table II illustrates the general effects of molecular weight and charge density on the dewatering activity of polymers for several industrial biological sludges. Increasing the molecular weight from polymer A to polymer B (the Brookfield viscosity of a 1% solution of polymer B was twice that of A) reduced the optimum dosage by about 40% and the filtration time by roughly 30%. Obviously, molecular weight has an important effect on activity.[8]

Comparison of the data in Table II for polymers A and B with C and D gives some indication of the effect of polymer cationic charge density on dewatering activity. While the percentage of cationic monomer was systematically increased in polymers C and D over that of A and B, the molecular weight was not held constant. Therefore, definition of the relationship between charge density and optimum dosage is not possible. The configuration of a linear polymer in dilute solution should become more extended with increasing cationic character due to repulsion of the charges along the backbone. If the average end-to-end distance is important, the optimum molecular weight should vary with charge density.

The filtration times in Table II generally indicate improvement with increasing charge for all the sludges except paper mill B. For example, although the optimum dosages of polymers B and D were similar, the filtration times for D were 50 to 87% less than those for B. The number of cationic charges per unit length of polymer apparently has an important role in determining its dewatering activity. The charge density may determine the ability of a polymer to destabilize the colloidal particles which could affect floc structure, cake porosity,[9] and cloth blinding. The curves in Figures 7 and 8 suggest that charge density is also involved in determining final cake moisture. Destabilization of the colloidal particles continues as polymer F's dosage was increased beyond the minimum needed for maximum filtration rate (Figure 10). It was during this same dosage range that minimum cake moisture was first observed.

The results of this investigation cannot be extrapolated to other sludges with certainty. Each sludge differs in its composition, particle surface properties and size distribution. The colloidal fraction of a sludge is known to play an important role in determining polymer dosages[10] and the filterability of the untreated sludge.[3] The surface properties of microorganisms can change with the type of organic substances they are metabolizing, sludge age, food to mass ratio, temperature and dissolved oxygen content. The amount of exocellular polymeric materials on the surface of bacteria increases during endogenous growth.[11]

The effect of polymer molecular weight and charge density will vary with each sludge. There are biological sludges which are effectively dewatered by low-charge density polymers such as A and B. Also, there can be more than one optimum polymer structure as evidenced by the fact that some sludges can be effectively dewatered by polymers of widely different structure. Polymer E, for example, which has a different structure from A-D, had greater optimum dosages than polymer D in all systems except the municipality (Table II) and the pet food plant (Figure 4) where it was considerably more effective.

Average specific resistance as defined by the Carman-Kozeny equation[4,12,13] is a means of characterizing the porosity of a sludge filter cake. Porosity depends on many physical properties such as particle size distribution, shape distribution, concentration and compressibility. There are many combinations of the physical properties which can give the same average specific resistance. The equation does not adequately describe the effect of solids concentration[3] and assumes the particles are incompressible. For these reasons caution should be used when comparing the resistances of different sludges. The average specific resistance of a sludge gives no indication as to the surface chemistry of the particles nor their ability to interact with chemical flocculants. This is why the untreated specific resistances of the seven sludges listed in Table II have no value in predicting optimum polymer dosages or filtration times.

Laboratory evaluation of sludge dewatering aids often involves filter leaf, Buchner funnel or capillary suction time tests. The FLT provides the most information, but also is the most difficult to conduct on biological sludges. Before a specific laboratory test is chosen, consideration should be given to the nature of the sludge and the objectives of the testing. Sludges which are grossly heterogeneous in composition or have a very thick consistency are difficult to evaluate with the CST test due to its small sample size. Low solids sludges which settle very rapidly upon flocculation can cause error with the FLT due to nonuniform cake pickup.

For laboratory studies, from which the data are not expected to be applied to a specific type of dewatering apparatus and where cake moisture is not a primary objective, either the BFT or the CST test is convenient. If ultimate application to a vacuum filter is contemplated, one approach would be to determine approximate optimum dosage ranges with the CST test or BFT and then conduct filter leaf tests within those dosage ranges. For specific applications it is best to use the same filter cloth for the FLT that is used on the vacuum filter.

If application to a belt filter is expected, the CST test is convenient and should provide good correlation to plant performance. Complete destabilization of colloidal particles is best for belt filters since there is little mechanical assistance. Consequently, belt filters can sometimes require higher chemical conditioning costs.

There is no convenient laboratory test which will duplicate the action of a centrifuge. One approach for evaluating centrifuge dewatering aids is to conduct BF or CST tests as a function of mixing intensity. Either mix time or speed can be varied so that an indication of the shear strength of the flocs can be obtained. Actually, mixing intensity studies can be useful with any type of dewatering equipment since insufficient or excess shear can be quite detrimental to the performance of a polymer.[7,14,15]

Evaluation of dewatering aids for a filter press is probably best carried out in a small pressure filter which can provide information on the final cake moisture, cycle time and cloth type.[16] Determination of the average specific resistance at the appropriate pressures appears to correlate with cycle time.[6]

Extrapolating results from laboratory tests to performance in a plant should be done cautiously. The order of performance found by the CST test, BFT or FLT will usually hold true in the plant. However, prediction of what level of performance will be obtained in the plant is difficult. When interpreting laboratory results such as those in Figures 1 through 5, it should be realized that for each system (sludge, mixing conditions, flocculant, dewatering equipment) there is a specific filtration rate below which acceptable performance will not occur.[17] For example, for a system such as shown in Figure 5, the indicated order of performance might become academic because none of the polymers are effective enough.

If laboratory tests show a dewatering aid to be ineffective, it is highly probable that it also will not work well in the waste treatment plant's dewatering equipment. However, if the tests indicate good performance, similar results in the plant are not guaranteed. For the purpose of evaluating dewatering aids, laboratory tests should be considered as a means of selecting the best candidate treatments for a plant trial.

ACKNOWLEDGMENT

We thank Mr. G. O. Barraclough for his assistance.

REFERENCES

1. Baskerville, R. C., and R. S. Gale. *J. Inst. Water Poll. Control* 2:3 (1968).

2. Garwood, J. *Effluent Water Treat. J.* 7:380 (1967).
3. Coackley, P., and B. R. S. Jones. *Sew. Ind. Wastes* 28:963 (1956).
4. Swanwick, J. D., and M. F. Davidson. *Water Waste Treatment J.* 8:386 (1961).
5. Baird, R. L., and M. G. Perry. *Filtr. Sep.* 4:471 (1967).
6. Gale, R. S. *Filtr. Sep.* 8:531 (1971).
7. Gale, R. S., and R. C. Baskerville. *Filtr. Sep.* 7:37 (1970).
8. McGrow, G., and L. Pitts. *Poll. Eng.* 7:38 (1975).
9. Dell, C. C., and J. Sinha. *Filtr. Sep.* 2:461 (1965).
10. Roberts, K., and O. Olsson. *Environ. Sci. Technol.* 9:945 (1975).
11. Paroni, J. L., M. W. Tenney and W. F. Eschelberger, Jr. *J. Water Poll. Control Fed.* 44:361 (1972).
12. Kozeny, J. *S. Ber. Weiner Acad. Abt. IIa.* 136:271 (1927).
13. Carman, P. C. *Trans. Inst. Chem. Eng.* 16:168 (1938).
14. Swanwick, J. D., K. J. White and M. F. Davidson. *J. Proc. Inst. Sew. Purif.* Pt. 5:394 (1962).
15. White, J. D., and R. C. Baskerville. *J. Water Poll. Control Fed.* 73:486 (1974).
16. Martin, J. L., and P. L. Hayden. *Proc. Purdue Ind. Waste Conf.* 31 (1976).
17. White, M. J. D., and R. C. Baskerville. *Effluent Water Treatment J.* 14:503 (1974).

CHAPTER 18

MECHANISMS AFFECTING TRANSPORT OF STEROLS IN A FLUVIAL SYSTEM

John P. Hassett* and Marc A. Anderson

Water Chemistry Program
University of Wisconsin
Madison, Wisconsin

INTRODUCTION

Investigations of organic compounds in rivers have dealt largely with measurement of total concentrations and include few attempts to differentiate between particulate and dissolved forms or to identify factors affecting transport of these compounds. This paper identifies and defines major factors influencing the transport of simple, nonionic, lipoidal compounds in a river system. Factors considered include associations with suspended matter, associations with dissolved organic matter, interaction between dissolved and particulate forms and degradation.

The sterols coprostanol and cholesterol (Figure 1) were selected as study compounds. These compounds were chosen because they are not rapidly degraded, are readily analyzed in water and sewage and may simulate behavior of nonpolar compounds such as chlorinated hydrocarbons, phthalate esters and steroid hormones. Both are lipids of low polarity; cholesterol behaves as a slightly more polar compound than coprostanol because of its sterically unhindered equatorial hydroxyl group. Cholesterol is found in most plants and animals and therefore originates from almost any source associated with biological activity, including runoff from vegetated areas, dairy waste, sewage and aquatic organisms. Coprostanol is formed from cholesterol by bacteria in mammalian intestines. The sole reported source of coprostanol to the

*Present address: Environmental Studies Institute, Drexel University, Philadelphia, Pennsylvania.

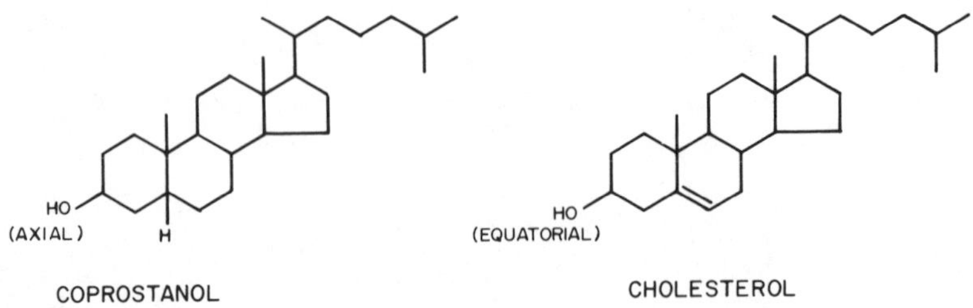

HO
(AXIAL)
H

COPROSTANOL

HO
(EQUATORIAL)

CHOLESTEROL

Figure 1. Coprostanol (5β-cholestan-3β-ol) and cholesterol.

environment is sewage effluent[1-3]; however, it is conceivable that in some situations other sources, such as runoff from livestock feedlots or from land spread with manure, might be significant. Reported values for the aqueous solubility of cholesterol vary. The best estimate appears to be 17 μg/l.[4] No values have been reported in the literature for the solubility of coprostanol.

Several investigators have reported the presence of coprostanol and cholesterol in river water. In Ohio rivers, Murtaugh and Bunch[1] found concentrations of coprostanol ranging from 0.02 to 5.0 μg/l and of cholesterol ranging from 0.5 to 2.5 μg/l. In Canadian rivers discharging to the Great Lakes Dutka *et al.*[3] found concentrations of coprostanol ranging from 0.05 to 2.05 μg/l and cholesterol concentrations ranging from 0.05 to 3.5 μg/l. No distinction was made between the dissolved and particulate phases.

Water samples for analysis were collected from the Menomonee River, located in southeastern Wisconsin near Milwaukee. The specific sampling area was the stretch of Menomonee River extending 8.3 miles upstream from the junction with the Little Menomonee River. There are three sources of sewage along this stretch of river. The first source is the Menomonee Falls-1 sewage treatment plant, which has a trickling filter operating in parallel with activated sludge. Effluents from the two units are combined, chlorinated and discharged to the river. The combined treatment capacity is two million gallons per day (2 mgd). The second sewage source is the Menomonee Falls-2 sewage treatment plant. This is an activated sludge plant followed by a tertiary polishing lagoon and has a treatment capacity of 1 mgd. A significant feature of this plant is the large amount of algae which grows in the lagoon and is subsequently discharged to the river. The third source of sewage to this stretch of river is the Village of Butler sewage bypass. A contract with the City of Milwaukee specifies that 0.4 mgd of Butler sewage will be treated by

Milwaukee sewage treatment facilities. Excess raw sewage is chlorinated and discharged to the river without further treatment.

ANALYTICAL METHODS

Sample Collection and Handling

Samples for organic analysis were collected in hexane-rinsed, 1-gal glass bottles with Teflon®*-lined caps. Samples for chloride and suspended solids analysis were collected in clean 1-liter polyethylene bottles. All samples were stored at 4°C until analyzed or extracted. Suspended solids analyses and extractions were carried out within 36 hr of sample collection.

Suspended solids concentrations were determined by filtering known sample volumes through prewashed, preweighed glass fiber filters (Reeve Angel 934AH). Filters were rinsed with distilled water, dried at 120°C and reweighed to determine the mass of suspended solids present.

Chloride was measured by an argentometric titration procedure.[5] Chloride concentrations upstream, downstream and in each sewage effluent were used to calculate percent contribution of the effluent to river flow by

$$\%\ \text{river flow} = 100 \frac{C_d - C_u}{C_s - C_u} \tag{1}$$

where C_d, C_u and C_s are the chloride concentrations downstream of outfall, upstream of outfall and in sewage effluent, respectively.

Centrifugation and Extraction

Samples for organic analysis were subjected to continuous-flow centrifugation on a Sorvall Superspeed RC2-B refrigerated centrifuge at a force of 31,000 g and a flowrate between 50 and 100 ml/min. Material passing through the centrifuge was operationally defined as dissolved, and material remaining in the centrifuge was defined as particulate.

These samples (3-4 liters) were acidified with 2.0 ml concentrated HCl and extracted in glass separatory funnels with three 100-ml portions of glass-distilled carbon tetrachloride. Particulate matter collected by continuous-flow centrifugation was resuspended in the centrifuge tubes by stirring and transferred into 1-liter separatory funnels. Tubes were rinsed with distilled water followed by carbon tetrachloride and added to the sample. Mixtures were acidified with 1.0 ml concentrated HCl and extracted once with 75 ml and twice with 50 ml of glass-distilled carbon tetrachloride. Solvent from all extracts was evaporated with a gentle air stream and the extracts redissolved in hexane.

*Registered trademark of E. I. du Pont de Nemours and Company, Inc., Wilmington, Delaware.

Extracts were cleaned up by column chromatography on neutral alumina (Brockmann activity grade IV). A slurry of 3.0 g adsorbent in hexane was transferred to a 1.0 cm (i.d.) glass column with Teflon stopcock and glass frit. The sides of the column were gently tapped to settle the adsorbent and 3 g of anhydrous sodium sulfate added. The sample was placed on top of the sodium sulfate in a small volume of hexane and the column eluted according to the scheme in Table I. Cholesteryl acetate (and presumably other steryl esters) eluted in fraction two, coprostanol elute in fraction three and cholesterol eluted in fraction four. Recovery of sterols from this column was greater than 95%.

Table I. Alumina Column Elution Scheme

Fraction	Solvent	Volume (ml)	Compound Eluted
1	hexane	15	-
2	benzene:hexane (1:9)	40	steryl esters
3	benzene:hexane (3:7)	40	coprostanol
4	benzene:hexane (1:1)	40	cholesterol

Gas Chromatography

Fractions from the alumina column were concentrated under a gentle stream of nitrogen, transferred to screw cap vials and evaporated to dryness. Sterols were converted to trimethylsilyl (TMS) ethers by reaction with 50 μl Sil-Prep (Applied Science Laboratories, Inc.) for 1 hr. Excess reagent was evaporated with a stream of nitrogen and the products redissolved in a known volume of carbon tetrachloride.

Samples so prepared were analyzed by gas chromatography using a Perkin Elmer 800 gas chromatograph equipped with a Varian flame ionization detector and Varian solid-state electrometer. A 2-mm (i.d.) glass column packed with 5% OV-101 on Gas Chrom Q (80-100 mesh) was used. The operating temperatures were 300°C for the injector, 275°C for the column and 310°C for the detector. Nitrogen carrier gas flow was 40 ml/min. Hydrogen and air flows to the detector were 20 ml/min and 200 ml/min, respectively. Samples were injected with a microsyringe and compounds identified by comparison of retention times to those of standard coprostanol and cholesterol (Applied Science Laboratories, Inc.). Sterol concentrations were determined by comparing peak heights to a standard curve. The lower limit of detection was 0.02 μg/l. Reproducibility of 10 replicate samples from a station along the river was ±45% (95% confidence limits). For a large sample split into 10 replicate subsamples, 95% confidence limits were ±30%.

Other Procedures

Sterol degradation was studied at 2°C and 23°C. Replicate river samples were taken in 1-gal glass bottles and maintained at the desired temperature. At known intervals, carbon tetrachloride was added to a sample to stop bacterial action. Samples were extracted and analyzed for cholesterol and coprostanol as previously described except that these samples were not centrifuged. Thus, concentrations are expressed as total cholesterol and total coprostanol.

Adsorption-desorption reactions which occur when sewage enters the river were modeled by mixing known amounts of river water and sewage at 20°C for 24 hr. Samples were centrifuged, extracted and analyzed for dissolved and particulate sterols.

Selected water and sewage samples were fractionated by molecular sieve column chromatography. A 1.6-cm (i.d.) glass column was packed to a length of 63 cm with Sephadex G-50 Fine. The carrier solvent (distilled water with 0.02 *M* NaCl and 0.02% NaN_3) was pumped through the column in an ascending mode at 60 ml/hr. Column effluent was monitored for UV absorbance at 254 nm and collected in 10.0 ml fractions by a volumetric fraction collector. The void volume (V_o) of the column was 35 ml for soluble starch and the void plus interstitial volume ($V_o + V_i$) was 95 ml for potassium dichromate.

Samples to be fractionated were centrifuged at 31,000 x g and spiked by placing 4-^{14}C-cholesterol in benzene into a glass vial, evaporating the solvent, adding 10.0 ml of sample, and shaking for 48 hours. A 3.0-ml aliquot was injected onto the Sephadex column and another 3.0-ml aliquot extracted to determine the amount of radioactivity present. This aliquot and fractions eluted from the column were acidified with 2 drops HCl and extracted 3 times with 5.0-ml portions of carbon tetrachloride. ^{14}C activity was determined with a Packard Tri-Carb Liquid Scintillation Spectrometer.

RESULTS AND DISCUSSION

Sewage Outputs

Table II summarizes sewage effluent data for temperature, coprostanol, cholesterol, suspended solids and percent contribution of each sewage source to river flow for sampling dates in January, April, July and November 1976. January cholesterol data have been omitted due to analytical difficulties. In each case, dissolved coprostanol concentrations are higher in winter and late fall than in spring and summer. This may

Table II. Sewage Outputs

Outfall	Date	Coprostanol		Cholesterol		Suspended Solids (mg/l)	River Flow (%)	Temperature (°C)
		Dissolved (μg/l)	Particulate (μg/l)	Dissolved (μg/l)	Particulate (μg/l)			
Menomonee Falls-1	1-13-76	6.9	24.9	-	-	27.1	15.0	12
Menomonee Falls-1	4- 6-76	0.53	11.9	0.53	13.2	13.0	8.0	14
Menomonee Falls-1	7- 7-76	0.43	21.7	0.60	12.8	18.0	33.2	20
Menomonee Falls-1	11-15-76	6.1	50.0	5.6	30.7	23.6	34.2	14
Menomonee Falls-2	1-13-76	1.9	2.9	-	-	9.4	15.0	2
Menomonee Falls-2	4- 6-76	0.096	1.2	0.40	11.9	148	1.6	12
Menomonee Falls-2	7- 7-76	0.034	0.98	0.35	49.7	43.3	16.6	22
Menomonee Falls-2	11-15-76	0.22	0.99	0.61	1.1	17.3	11.7	4
Butler	1-13-76	6.1	256	-	-	104	1.9	4
Butler	4- 6-76	1.2	176	1.8	98.0	73.7	-	13
Butler	7- 7-76	1.2	269	2.0	219	94.5	-	19
Butler	11-15-76	20.9	259	40.2	317	121	1.4	5

be due to increased biological degradation at elevated temperatures, although the same effect is not observed for particulate coprostanol concentrations. Dissolved cholesterol concentrations are also higher in autumn than in spring or summer. Particulate cholesterol concentrations for the Menomonee Falls-2 effluent are higher in spring and summer than in autumn due to cholesterol production by algae growing in the polishing lagoon during the warmer months.

Sterol concentrations reflect the degree of waste treatment (Table II). As expected, the highest concentrations of coprostanol and cholesterol are present in the untreated Butler effluents. The second highest concentrations occur in the secondary treated Menomonee Falls-1 effluents. The principal mechanism of removal appears to be settling of adsorbed sterols since dissolved concentrations are not greatly reduced relative to untreated Butler effluents. Lowest coprostanol concentrations are found in Menomonee Falls-2 effluents. This is due to additional degradation and adsorption in the tertiary polishing lagoon. As already mentioned, cholesterol concentrations in this effluent tend to be elevated during warmer months due to *in situ* synthesis by algae in the lagoon.

Sterols in River Water

Sterols and suspended solids concentrations in the river are presented for January, April, July and November 1976 sampling dates in Figures 2 through 5, respectively. These figures have several features in common. In all cases, the Menomonee Falls-1 sewage effluent (mile 0) is a major source of sterols to the river. The Menomonee Falls-2 effluent (mile 1.9) generally has little effect on sterol concentrations because of lower levels in the effluent and lower contributions to stream flow (Table II). An exception is the large increase in particulate cholesterol downstream of this outfall in July (Figure 4) due to cholesterol-containing algae discharged from the sewage lagoon. The Butler effluent (mile 7.7) is also a significant source of sterols. These sterols tend to be removed rapidly from the water due to settling of dense particulates in the untreated sewage.

Cholesterol and coprostanol concentrations, especially in the dissolved phase, are lower in the spring and summer than in late fall and winter because of lower sterol discharges in the sewage effluents at these times of year. Although sewage is the major source of sterols for this section of river, it is apparently not responsible for the significant increase of particulate cholesterol (Figure 3) and particulate coprostanol (Figure 4) occurring at mile 3.6. These increases in particulate sterol levels are associated with sharp increases in suspended solids concentrations probably attributable to resuspension of sedimented particulate matter.

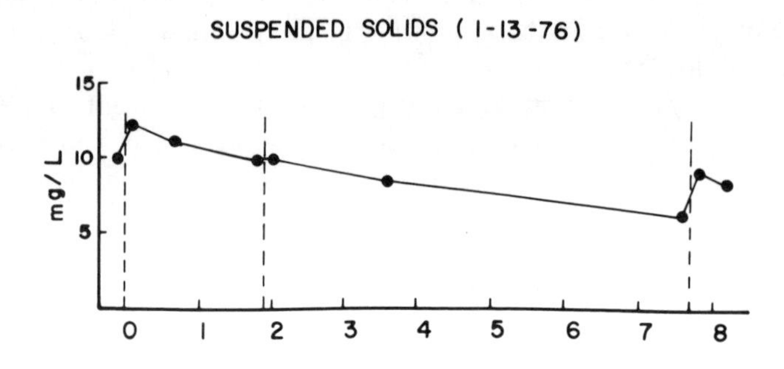

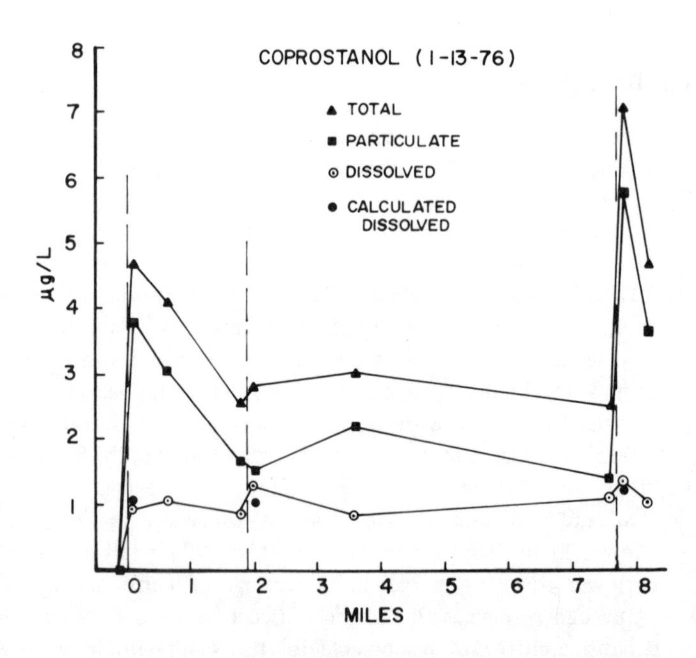

Figure 2. Suspended solids and coprostanol concentrations as a function of distance downstream from the Menomonee Falls-1 sewage outfall for January 13, 1976. Vertical dotted lines indicate location of sewage sources.

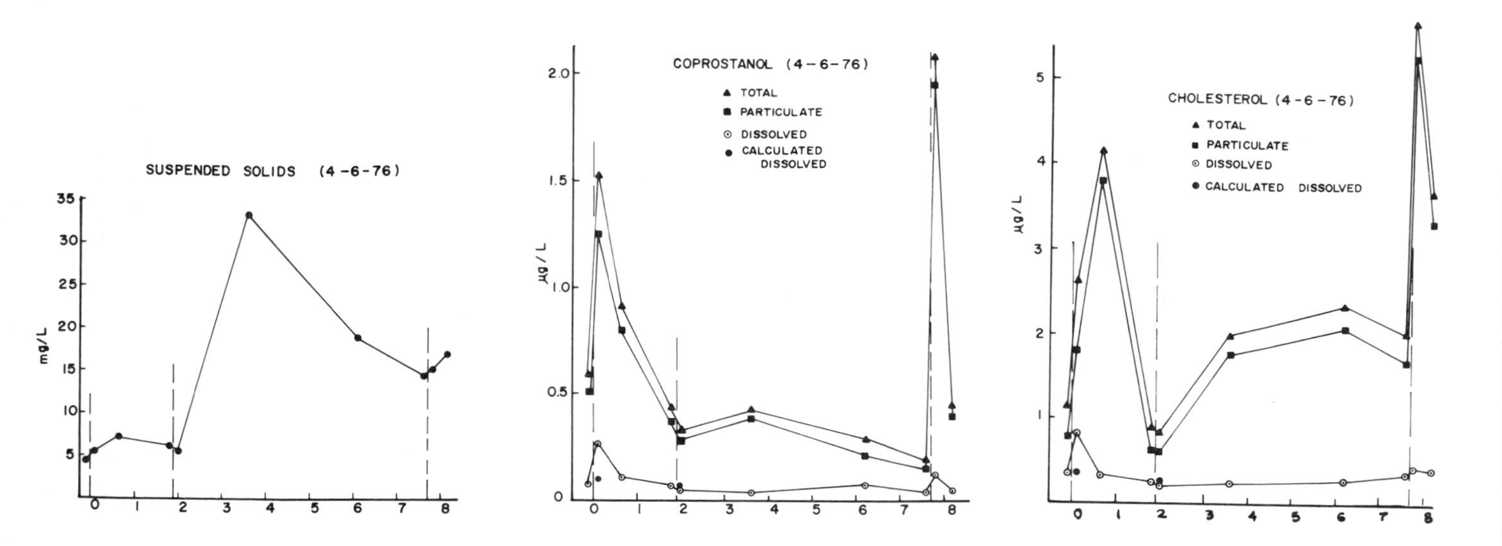

Figure 3. Suspended solids, coprostanol and cholesterol concentrations as a function of distance downstream from the Menomonee Falls-1 sewage outfall for April 4, 1976. Vertical dotted lines indicate location of sewage sources.

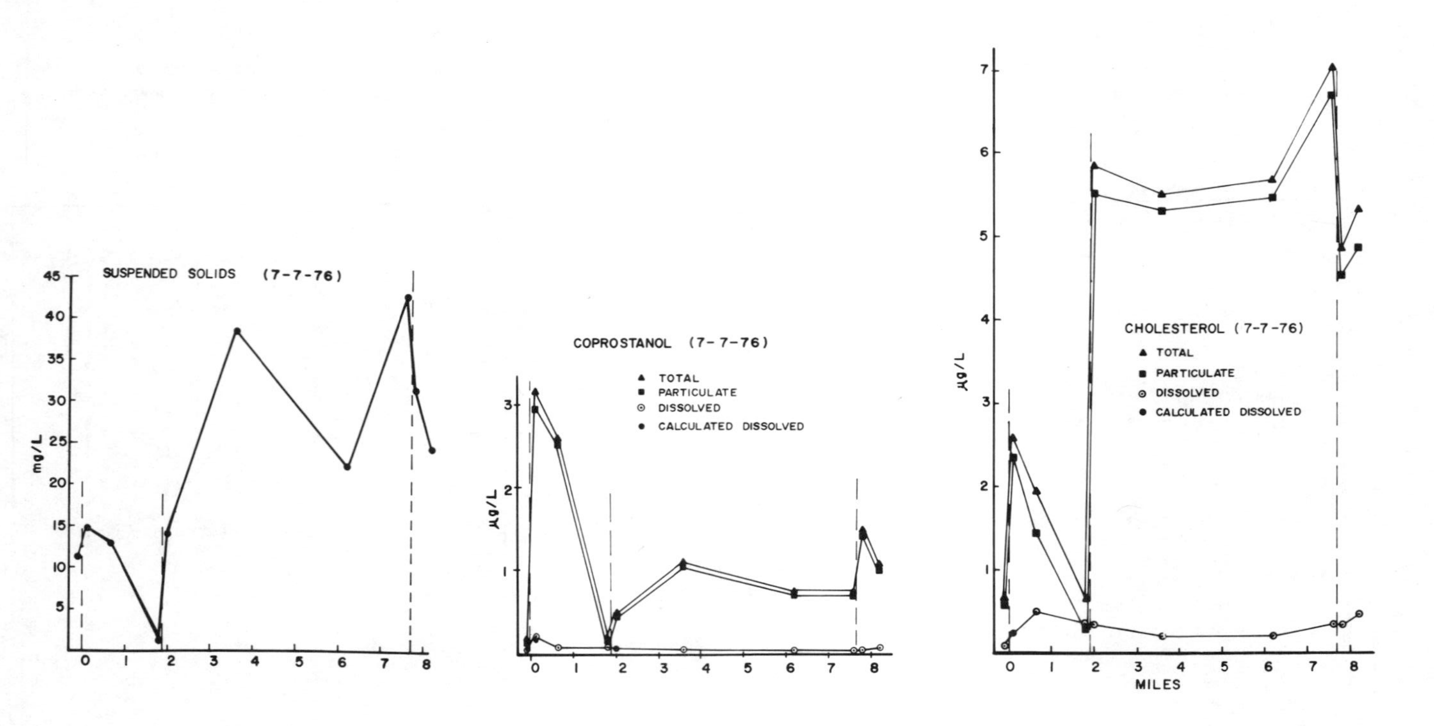

Figure 4. Suspended solids, coprostanol and cholesterol concentrations as a function of distance downstream from the Menomonee Falls-1 sewage outfall for July 7, 1976. Vertical dotted lines indicate location of sewage sources.

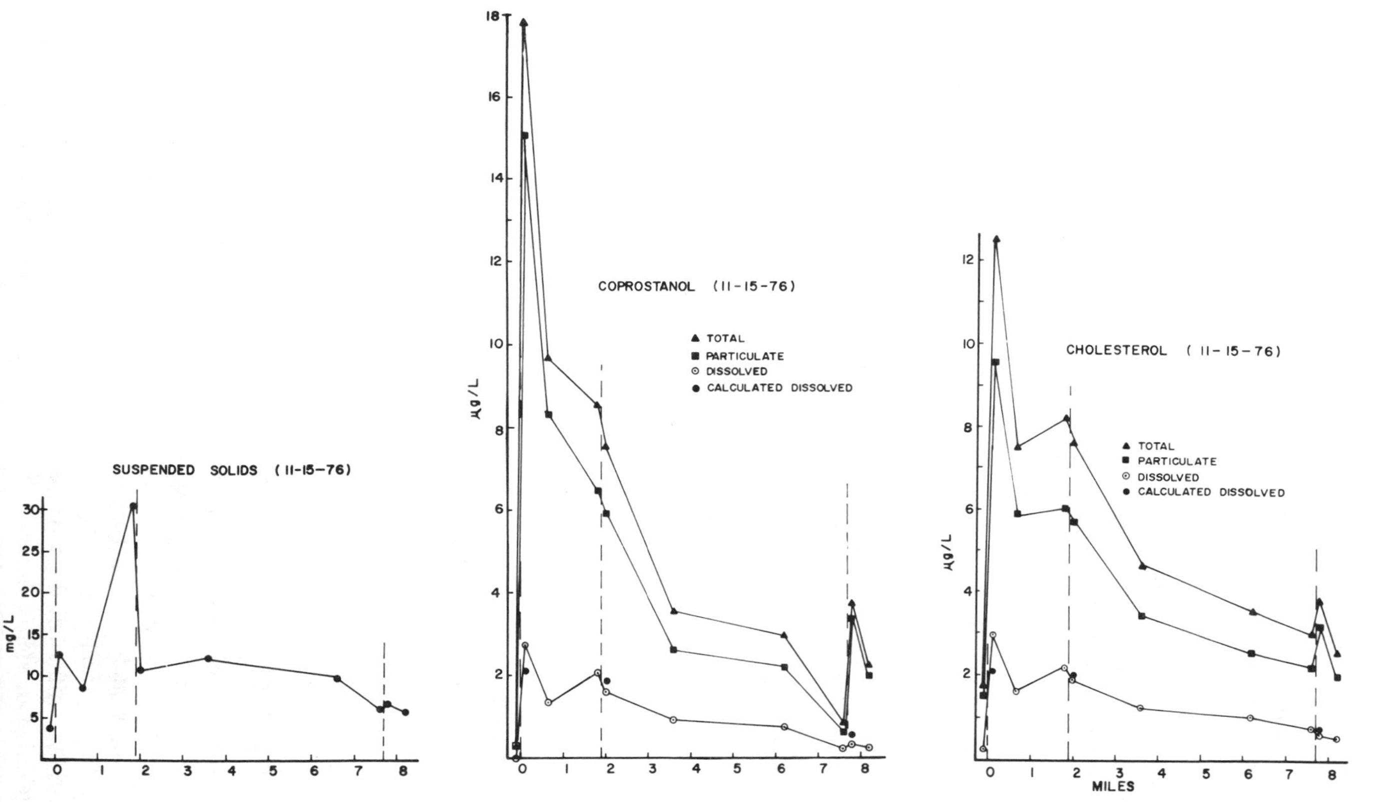

Figure 5. Suspended solids, coprostanol and cholesterol concentrations as a function of distance downstream from the Menomonee Falls-1 sewage outfall for November 15, 1976. Vertical dotted lines indicate location of sewage sources.

Interestingly, while particulate sterol concentrations vary greatly from station to station on a given day, dissolved sterol concentrations remain relatively constant. Data for November (Figure 5), where dissolved concentrations decline gradually downstream, are an exception. Constant dissolved sterol concentrations in the presence of fluctuating particulate levels of these compounds indicate that the system is not at equilibrium with respect to adsorption reactions or that these reactions are not a controlling process in this case. Because many particulate materials, especially those with high organic content such as would be expected in sewage effluent, have high affinities for lipoidal compounds, one might expect to find dissolved species adsorbed onto suspended or sedimentary particulate material. Constant dissolved concentrations of an apolar organic compound in a river have also been observed by Zabik *et al.*[6] They suggest that dissolved DDT concentrations remain constant because they are near the solubility limit of DDT. It would seem probable, however, that such a compound would be more likely to be adsorbed at concentrations approaching its solubility limit. Another possible explanation is an association of lipoidal compounds with soluble organic matter or stable colloids which prevents these compounds from being adsorbed. At present there is insufficient evidence to support or contradict either of these explanations.

Degradation of Sterols

Experiments were performed to determine the importance of sterol degradation in river transport and its effect on long-term laboratory studies. Results are presented in Figure 6. At 2°C there was no significant degradation of total cholesterol or coprostanol after 54 hr. At 23°C there was no significant degradation after 24 hr; however, detectable degradation occurred after 51 hr. These data indicate that degradation rates are not significant with respect to residence times of waterborne sterols in the section of river under study.

Adsorption of Sterols

An attempt was made to interpret river and sewage sterol data according to the Freundlich adsorption model by plotting log (μg/g adsorbed sterol/suspended solids) against log (μg/l dissolved sterol). A straight line would be predicted should the data conform to this model. Results for coprostanol and cholesterol are presented in Figures 7 and 8, respectively. Although these figures indicate a tendency for increased concentration of sterols on particulate material with increased dissolved sterol concentration, any rigorous interpretation of the fit of the model is

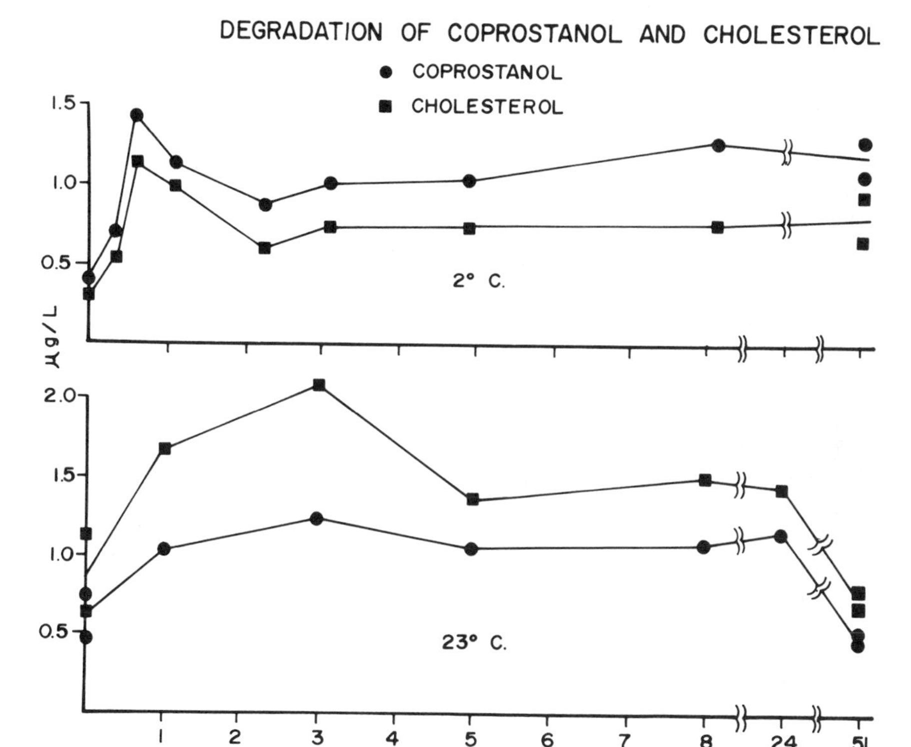

Figure 6. Total coprostanol and total cholesterol concentrations as a function of incubation time at 2°C and at 23°C.

beyond the reliability of the data. Because of many uncontrolled variables among these samples, such as temperature, composition of solids and type of water, it is difficult to interpret these results as evidence for or against adsorption-desorption reactions of sterols in the river. Attempts to interpret these data with other adsorption models, such as the Langmuir equation, were equally unsuccessful.

In an effort to study sterol adsorption behavior under more controlled conditions, river and sewage samples were mixed together in the laboratory. Results for cholesterol are summarized in Figure 9; coprostanol results are not presented because dissolved coprostanol concentrations were below detection limits. The continuous line in this figure connects pure river water and pure sewage values and describes the expected results for river-sewage mixtures if no adsorption or desorption occurs. The actual results

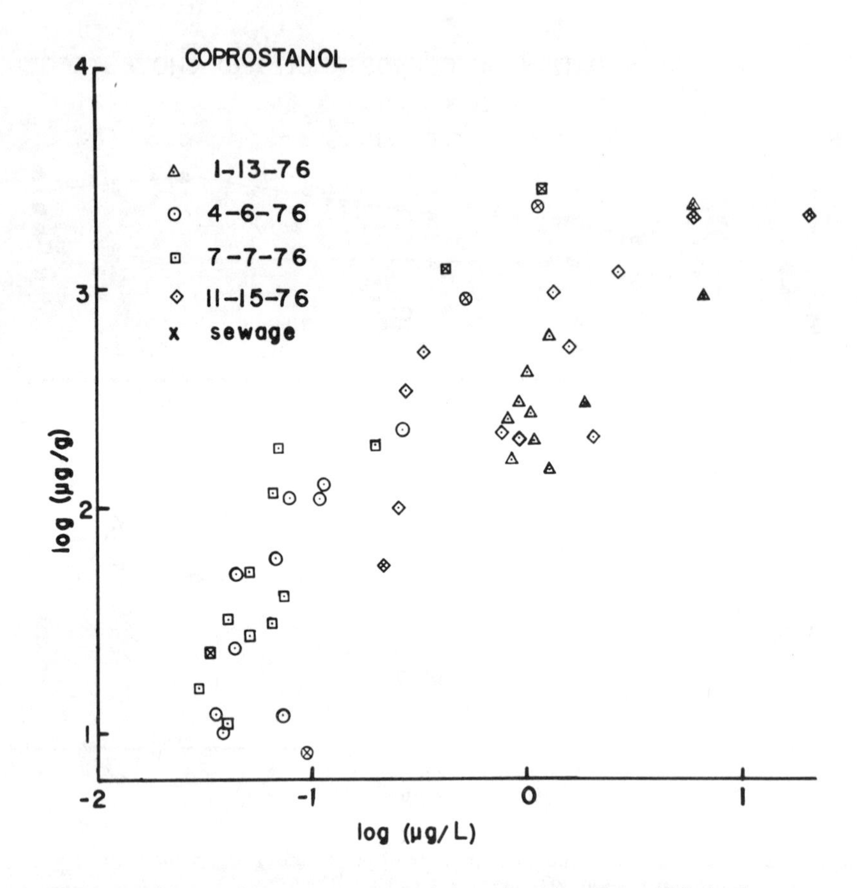

Figure 7. River and sewage coprostanol data plotted as Freundlich isotherms.

tend to fall slightly to the left of this line, indicating adsorption of cholesterol. Results of this experiment do not agree with results for river samples taken immediately downstream of the Menomonee Falls-1 outfall. From chloride concentrations upstream, downstream and in the effluent it was possible to calculate the percentage of river water and sewage in samples taken at this station. It was thus possible to predict dissolved sterol concentrations at this point if no adsorption or desorption took place; *i.e.*, if concentrations were determined solely by mixing. In two cases (Figures 2 and 4) predicted values were equivalent to observed values, suggesting that no adsorption-desorption reactions occurred. In the other two cases (Figures 3 and 5) predicted values were lower than observed values, indicating that desorption had taken place. Because of the conflicting results of these experiments, it is not possible with the

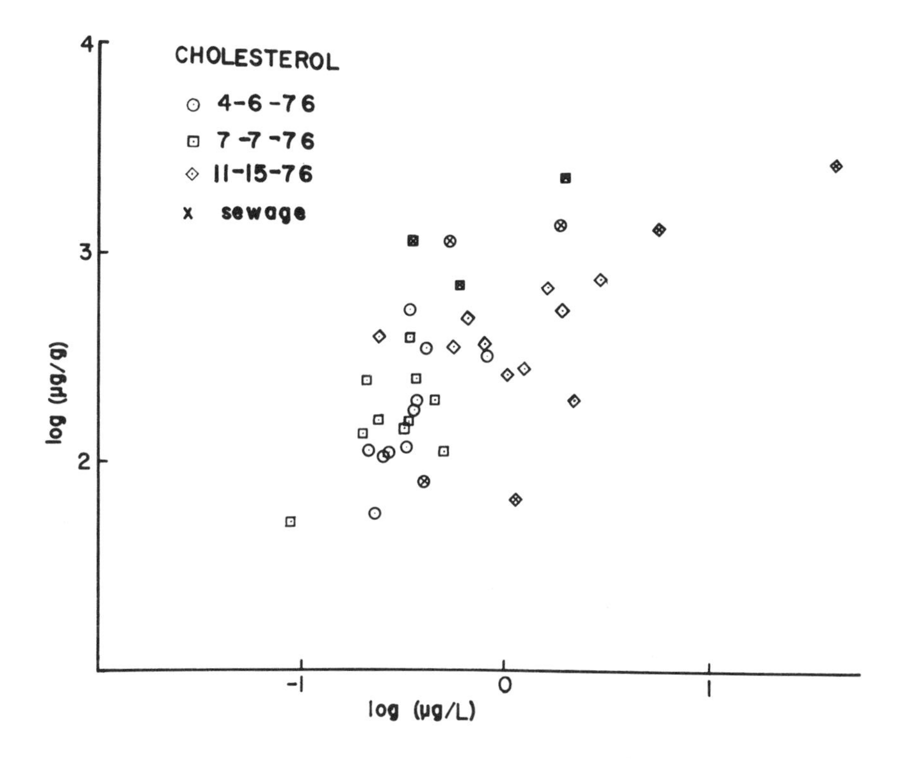

Figure 8. River and sewage cholesterol data plotted as Freundlich isotherms.

available data to determine what role, if any, adsorption-desorption reactions play in the behavior of sterols in this river.

Association With Dissolved Organic Matter

A possible explanation for the apparent lack of interaction between particulate and dissolved sterols may be associations of sterols with dissolved organic matter in the river. To investigate the ability of sterols to form water-soluble associations with dissolved, high-molecular-weight organic matter, selected samples were spiked with ^{14}C-cholesterol and fractionated by molecular sieve chromatography. Results of these experiments are summarized by the chromatograms in Figures 10, 11 and 12. UV absorption indicated the presence of dissolved organic matter, and ^{14}C activity indicates the presence of cholesterol.

Both river water samples (Figure 10) have major ^{14}C peaks in fraction 4, the highest molecular weight fraction (MW 30,000). The upstream

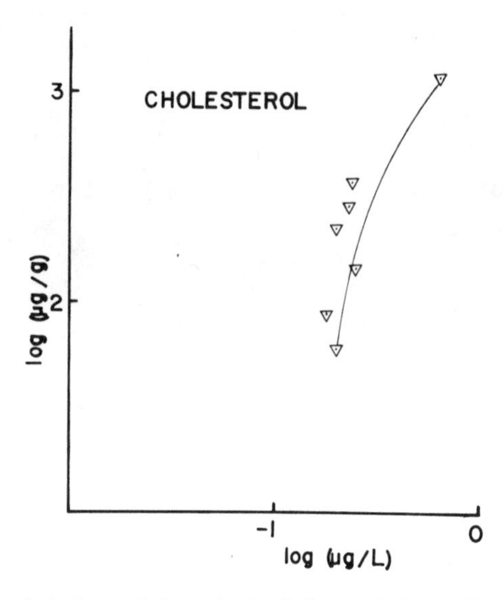

Figure 9. Plot of cholesterol data obtained from mixtures of sewage and river water. Continuous line connects pure river water and pure sewage data and predicts results if no adsorption-desorption occurs.

sample also has a major peak in fraction 11, where material with a molecular weight less than 1500 should elute, while the downstream sample has only a minor peak in fraction 8. There were no significant peaks beyond fraction 11 in either sample. The UV absorbance patterns of these samples are dissimilar in that the downstream sample has a strong peak in fraction 4, but the upstream sample has no peak in this area. Differences between these samples may be due to interactions between material in sewage and river water in the downstream sample.

The sewage sample (Figure 11) has a large ^{14}C peak in fractions 4-5 and a smaller peak in fraction 8. The peak in fraction 8 may be related to the small peak in fraction 8 of the downstream river sample. As in the river water samples, there are no ^{14}C peaks beyond fraction 11.

Figure 12 is a chromatogram of aqueous NaOH soluble material extracted from Menomonee River sediment. This extract has very large ^{14}C peaks in fractions 4 and 10. An interesting aspect of this chromatogram is the minor ^{14}C in fraction 17 and the very large peak in fractions 21-25 (fractions 21-25 were collected as a composite sample; this peak may actually be much higher and narrower). Both peaks elute at the same points as UV-absorbing material, indicating that ^{14}C-cholesterol is associated with this material.

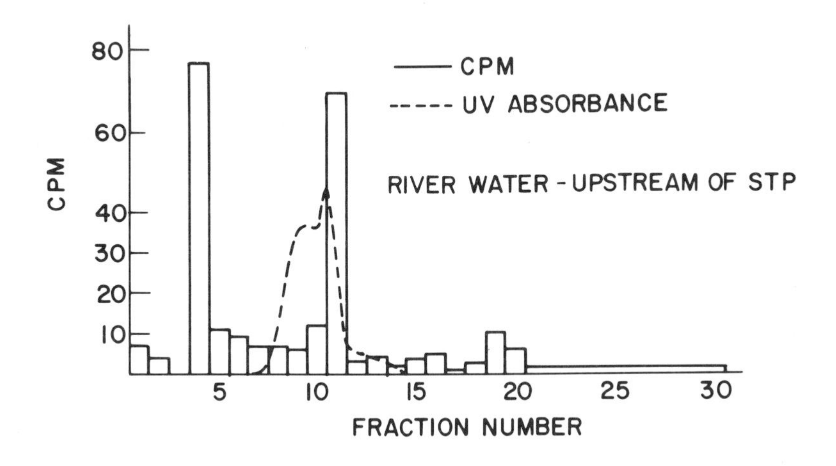

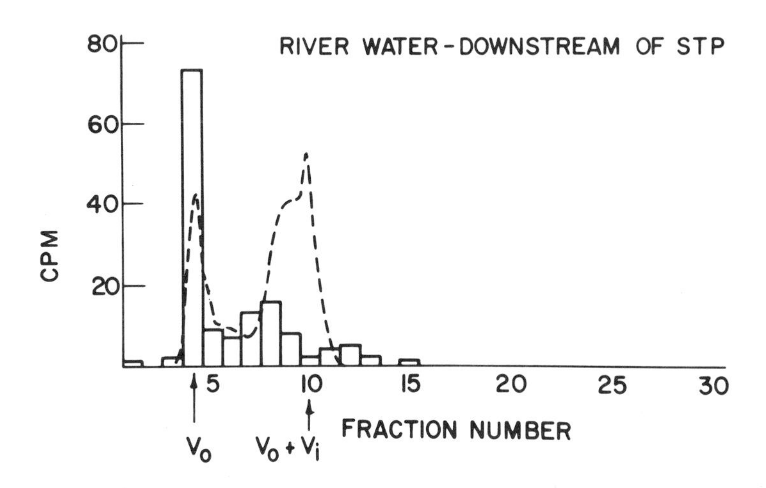

Figure 10. Molecular sieve chromatogram on Sephadex G-50 of 4-^{14}C-cholesterol-tagged river water upstream and downstream of the Menomonee Falls-1 sewage outfall. Each fraction is 10 ml.

All samples discussed thus far have shown a major ^{14}C peak in fraction 4 and all but one have shown a major peak in either fraction 8 or 10. The elution pattern for ^{14}C-cholesterol in distilled water is somewhat different, however (Figure 11). In this case there are small peaks in fractions 4-5 and 8, but the largest peak is in fraction 17. There are also smaller peaks in fractions 13 and 15. It is unclear why a single compound

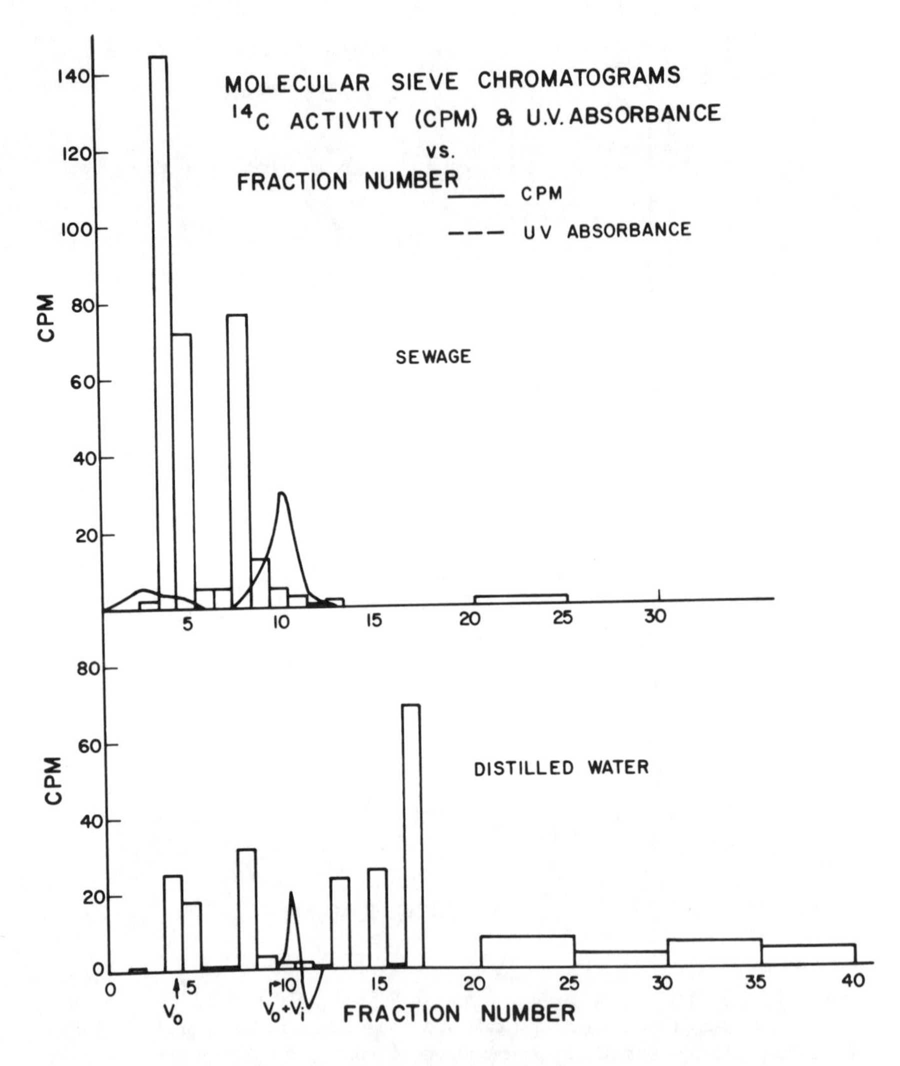

Figure 11. Molecular sieve chromatogram on Sephadex G-50 of 4-^{14}C-cholesterol-tagged Menomonee Falls-1 sewage and distilled water. Each fraction is 10 ml.

in distilled water should elute in several peaks. One possibility is that the distilled water was not entirely pure. A more likely explanation is that some of the cholesterol was present in the water as aggregates rather than as truly dissolved molecules. These aggregates should be fractionated on

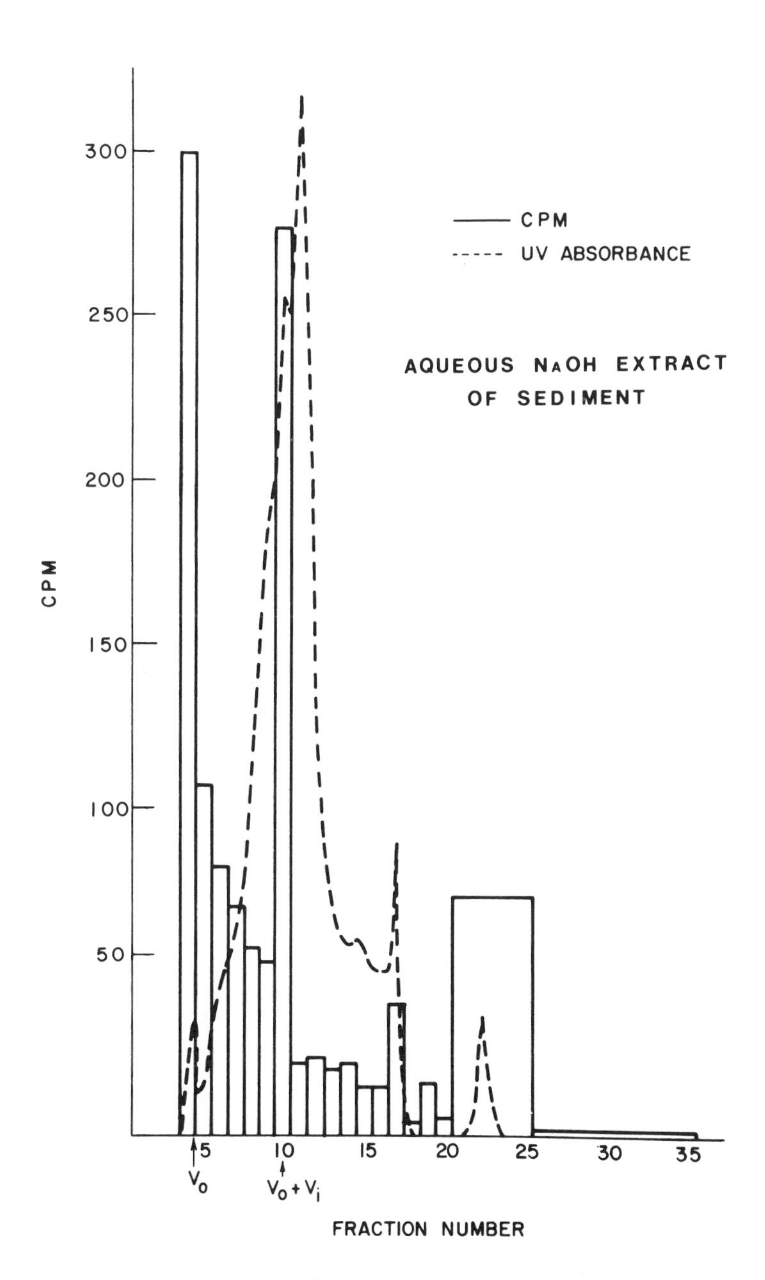

Figure 12. Molecular sieve chromatogram on Sephadex G-50 of an aqueous NaOH extract of Menomonee River sediment tagged with 4-^{14}C-cholesterol. Each fraction is 10 ml.

the column according to size. The fact that some material eluted at volumes greater than the elution volume of low-molecular-weight compounds ($V_o + V_i$) indicates that adsorption-desorption as well as physical size fractionation is occurring on the column. It is not known which, if any, of the peaks represents actual dissolved cholesterol.

Table III summarizes percents of injected ^{14}C material recovered from the molecular sieve column. For sewage and water samples, between 20 and 35% of injected radioactivity was recovered; the remainder was adsorbed on the column and never eluted. Recovery of ^{14}C material in the NaOH sediment extract was much higher. This is probably because concentrated organic matter in the sample, as evidenced by its dark yellow-brown color and high UV absorbance, bound cholesterol and prevented its adsorption by the Sephadex column. Thus, it appears that dissolved natural organic matter is capable of reducing adsorption of a sterol by solids.

These molecular sieve experiments indicate that cholesterol is capable of forming associations with organic material in river water, sewage and sediments, and in distilled water may form aggregates with itself. These aggregates are probably arranged with the hydrophobic hydrocarbon portions of the molecules in the interior of the cluster and the hydroxyl groups on the outside exposed to water. Although these experiments show that cholesterol can associate with other organic material in water, they do not show whether or not this occurs in natural systems. Other types of experiments are required to resolve this.

SUMMARY AND CONCLUSIONS

This study has shown that total sterol concentrations in a fluvial system closely follow fluctuations in particulate sterol concentrations. Sterol concentrations in the dissolved phase are less than concentrations in the particulate phase, but in many cases they form a significant percentage of the total. Dissolved sterol concentrations remain remarkably constant in all samples for a given day, and there appears to be little interaction with the particulate phase. The most important factor determining fluctuations in total sterol concentration among a set of samples for a given day appears to be deposition and resuspension of sterol-bearing particulate matter.

Sterol degradation rates are slow compared to the residence time in the section of river studied; however, degradation would be an important factor in long-term studies, especially during warm weather.

Results of adsorption experiments are ambiguous. Further work is required to determine if adsorption-desorption reactions take place in the river and, if not, why they do not occur.

Table III. Recovery of ^{14}C Activity From a Molecular Sieve Column

Sample	Recovery (%)
River-Upstream of Outfall	33.3
River-Downstream of Outfall	18.2
Sewage	35.8
Distilled Water	20.2
Sediment Extract	113

Molecular sieve experiments indicate that sterols can form associations with dissolved natural organic matter and that these associations can reduce adsorption of sterols, but they do not reveal whether or not such associations are important in natural waters. Unfortunately, detection limits for sterols are not low enough to allow analysis of unlabeled sterols eluted from a molecular sieve column. Other approaches must by used in order to establish the significance of sterol-organic matter interactions.

ACKNOWLEDGMENTS

We thank Aileen Keith and Susan Barta for their excellent technical assistance during this project. This study was performed as part of the efforts of the International Joint Commission Pollution From Land Use Activities Reference Group, established under the Canada-U.S. Great Lakes Water Quality Agreement of 1972. Funding was provided through U.S. EPA grant R00514201. Additional support was supplied by the Water Chemistry Program and the Water Resources Center of the University of Wisconsin.

REFERENCES

1. Murtaugh, J. J., and R. L. Bunch. *J. Water Poll. Control Fed.* 39:404 (1967).
2. Smith, L. L., and R. E. Gouron. *Water Research* 3: 141 (1969).
3. Dutka, B. J., A. S. Y. Chau and J. Coburn. *Water Research* 8:1047 (1974).
4. Madan, D. K., and D. E. Cadwallader. *J. Pharmacol. Sci.* 62:1567 (1973).
5. *Standard Methods for the Examination of Water and Wastewater,* 13th ed. (Washington, DC: American Public Health Association, 1971).
6. Zabik, M. J., B. E. Pape and J. W. Bedford. *Pestic. Monit. J.* 5:301 (1971).

CHAPTER 19

USE OF CLINOPTILOLITE FOR REMOVAL OF TRACE LEVELS OF AMMONIA IN REUSE WATER

Ronald C. Sims and Ervin Hindin

Department of Civil and Environmental Engineering
Washington State University
Pullman, Washington

INTRODUCTION

The presence of low levels of ammonia in water can restrict the utilization of this vital resource. Water-consuming processes and/or carrier water systems sensitive to the presence of ammonia must have even submilligram per liter amounts removed. In one area of aquaculture, the propagation and rearing of fish, only extremely low levels can be tolerated. Nitrogen in the form of molecular NH_3 acts on fish as a true internal poison, entering the fish by way of the gills.[1] Hyperplasia of gill filaments can occur at NH_3 concentrations as low as 0.006 to 0.008 mg/l.[2]

Sources of ammonia-nitrogen in a fish hatchery include: (1) bronchially excreted ammonia, which comprises 60 to 90% of the total nitrogen excreted by fish[3-5] (2) decomposition of excess food[2] and (3) excretion of urine by fish.[3,6] It has been demonstrated that growing rainbow trout excrete 17 mg of NH_3-N per kilogram of body weight per hour.[7] Production of ammonia from excess food results from the decomposition of the organically bound nitrogen by heterotrophic bacteria.[8] The contribution of ammonia from urea is considered to be least significant.[6,8] Although an ammonia buildup from these sources is normally not encountered in unpolluted lakes and streams, such sources can cause lethal concentrations to accumulate in the closed-loop rearing systems of fish hatcheries.

Because ammonia can be very damaging to trout at very low concentrations and because of the trend toward reusing water in the newer fish hatcheries, the need exists to accomplish its rapid and essentially complete removal from the fish-rearing water. The purpose of the research described herein was to determine the feasibility of using the zeolite, clinoptilolite, as a tertiary treatment process for removal of NH_4^+ from fish hatchery reuse water. The feasibility criteria were based on determining the ammonium ion exchange capacity of the clinoptilolite. Another aim was to determine if operational parameters could be adjusted to achieve rapid and essentially complete removal of NH_4^+ from waters of different cationic concentrations likely to be encountered in alpine streams. Those parameters investigated included the source of the clinoptilolite and the effects of the zeolite mesh size, flowrate and water chemistry on the stoichiometry and kinetics of NH_4^+ exchange. The research was directed with an objective of obtaining a water after zeolite treatment of NH_4^+-N concentration of less than 0.5 mg/l (0.036 meq/l). The use of clinoptilolite would be feasible only if the zeolite could be regenerated. The final phase of the research was to devise a regeneration scheme utilizing $NaHCO_3$ to remove NH_4^+ from the zeolite followed by biological nitrification of the regenerant stream.

Clinoptilolite represents a zeolite with unusual selectivity for NH_4^+ in the presence of Ca^{+2}, Mg^{+2} and Na^+ ions. Ames[9] first suggested its possible use as a tertiary treatment process to remove NH_4^+ from wastewaters. Subsequent studies by Mercer *et al.*[10] utilized clinoptilolite for NH_4^+ removal from Richland, Washington, secondary effluent. Two full-scale domestic treatment plants employing clinoptilolite include the Upper Occoquan Regional Plant in Virginia and the Tahoe-Truckee Plant in California.[11] Braico[12] utilized clinoptilolite for NH_4^+ removal from recycled fish hatchery water.

THEORY AND BACKGROUND

Equilibrium Relationship

One approach to a model for describing ion exchange behavior is to consider ion exhange equilibria as a simple chemical reaction. Thus, the law of mass action can be applied. In a binary system

$$bA^{+a} + aBZ_b \rightleftharpoons bAZ_a + aB^{+b} \qquad (1)$$

expresses the reversible equilibrium

where a = valence of ion A
b = valence of ion B
Z = exchange site in the zeolite

Therefore, according to the law of mass action

$$K = \frac{(a)^b_{AZ_a} \cdot (a)^a_B}{(a)^b_A \cdot (a)^a_{BZ_b}} \tag{2}$$

where $(a)^b_A$, $(a)^b_{AZ_a}$ = the activities of ion A in solution and in the resin

$(a)^a_B$, $(a)^a_{BZ_b}$ = the activities of ion B in solution and in the resin.

The selectivity coefficient as given by Koon and Kaufman[13] may be defined by

$$K^A_B \left[\frac{Q}{C_o}\right]^{a-b} = \left[\frac{y}{x}\right]^b_A \left[\frac{x}{y}\right]^a_B \tag{3}$$

where C_o = solution concentration meq/l
Q = total exchange capacity
c = solution phase ionic concentration, meq/l
q = solid phase ionic concentration in meq/g
x = c/C_o
y = q/Q

The separation factor, which describes the interaction of the solid-liquid phases, is also an important parameter. This factor expresses the preference of an ion exchanger for one ion relative to another.[14]

$$Q_{S_B}^{\ A} = \left[\frac{q}{c}\right]_A \left[\frac{c}{q}\right]_B \tag{4}$$

or

$$Q_{S_B}^{\ A} = \left[\frac{y}{x}\right]_A \left[\frac{x}{y}\right]_B \tag{5}$$

where Q^A_B is the separation factor for ion A with respect to ion B. Other symbols are as described previously.

A value of one for Q_S indicates that the zeolite has no preference for ion A over ion B; $Q_S > 1$ indicates that the zeolite prefers ion A over ion B; $Q_S < 1$ indicates that the zeolite prefers ion B over ion A. Thus, the greater the value of Q_S, the more selective the ion exchanger is for ion A. Selectivity quotients readily indicate the applicability of a specific exchanger for the desired removal of one or more ions from water.

The term degree of column utilization refers to the exhaustion cycle; the term efficiency refers to the regeneration cycle.[13] The degree of column utilization is the fraction of the exchange capacity utilized. Efficiency is defined as the ratio of exchange sites theoretically renewable to exchange sites actually renewed per unit volume of regenerant. In column operation, there is a trade-off between degree of column utilization and efficiency. This is because ion exchange reactions are equilibrium reactions, and impracticably large quantities of regenerant are required to drive the reaction (Equation 1) completely to the left. Although increasing regenerant volume per unit weight of zeolite increases the extent of regeneration, the resultant bed capacity is not directly proportional to the increase in regenerant used. Regeneration efficiency is usually sacrificed to achieve a reasonable degree of column utilization.

The effects of temperature on the exhaustion and regeneration kinetics of clinoptilolite are insignificant at normal operating temperatures. Braico[12] concluded that column studies at room temperature (23°C) produce nearly the same results as studies conducted at a fish hatchery water temperature of 12°C.

Regeneration

Regeneration of the column involves removing and replacing the exchanged ion with ions that place the zeolite in the desired form for exhaustion again. Regeneration can be accomplished upflow or downflow. Upflow operation is often used to minimize leakage of NH_4^+ during the initial stages of exhaustion. With upflow regeneration, the bottom of the bed becomes the most thoroughly regenerated portion of the column.[14]

Clinoptilolite has most often been regenerated with a lime solution containing NaCl or $CaCl_2$, or both, to increase regenerant strength.[9-11] Lime contains Ca^{+2}, which can replace NH_4^+ on the zeolite. Using lime also increases the pH, which drives the NH_4^+-NH_3 equilibrium to the NH_3 or gas phase. The addition of NaCl or $CaCl_2$ increases the equivalents of regenerating ions per unit volume of regenerant. However, because of the small hydrated diameter of Na^+ (15.8 A) relative to Ca^{+2} (19.2 A), superior exchange kinetics are obtained with Na^+.[13]

A problem encountered in the process is the tendency of the zeolite to become unstable at high pH and deteriorate. According to Koon and Kaufman,[13] attrition rates are approximately 0.25, 0.35 and 0.55% per operating cycle using regenerants having a pH of 11.5, 12.0 and 12.5, respectively. Thus, zeolite replacement costs constitute a significant part of the total cost of NH_4^+ removal by clinoptilolite if a high pH regenerant is used. A 2% NaCl solution at neutral pH will be used to regenerate clinoptilolite at the Upper Occoquan Regional Plant in Virginia and the Tahoe-Truckee Plant in California.[11] Although 25 to 30 bed volumes are required for regeneration compared with 10 to 30 bed volumes for high pH regeneration, the regenerant will be recovered and reused within a closed-loop system.

Biological regeneration of the zeolite has also been investigated. The employment of nitrification as a means of regenerating the zeolite has been studied by Sims and Little[15] and Semmens.[16]

EXPERIMENTAL METHODS AND MATERIALS

Exchange Capacity and Kinetics

Downflow continuous-column operation was utilized to determine the total NH_4^+ exchange capacity in a system where no competing cations were present. For each test a 1 *M* solution of NH_4Cl was passed through a column of clinoptilolite in the sodium form at a rate of 60 ml/hr until the effluent concentration equaled the influent concentration. The NH_4^+ was eluted from the column with 0.5 *M* NaCl. The NH_4^+ exchange capacity was also determined from breakthrough curves obtained by passing an NH_4Cl solution through a column containing clinoptilolite in the Na form at 15 bed-vol/hr and measuring the effluent concentration as a function of volume of water treated. The exchange capacities of samples of different mesh sizes from different geographical locations were compared.

Multicomponent systems were then used to determine the stoichiometry and kinetics of NH_4^+-exchange with one clinoptilolite sample. Upflow continuous columns were used. Each feedwater was passed through 3 columns in parallel; (1) at 15 bed-vol/hr to determine an exhaustion history, (2) at 30 bed-vol/hr, and (3) at 15 bed-vol/hı to determine the effect of flowrate. Flows through the last two columns were terminated when the effluent NH_4^+ concentration reached 0.035 meq/l. The concentration of NH_4^+ remained constant while the concentrations of competing cations were increased in each of the multicomponent systems. Thus, the effect of competing cations and of flowrate on NH_4^+ exchange could be determined.

Nitrification Tests

Laboratory-scale batch tests utilizing fish hatchery water were used. The two experimental flasks contained polyethylene Koch Ring packing and the two control flasks contained glass beads. The total surface area exposed was the same in the experimental and control reaction flask: 476.6 cm^2/l of water. The surface area per liquid volume at a fish hatchery under study was approximately 526 cm^2/l. The concentrations of NH_4^+ and HCO_3^- were increased by adding regenerant from a column treated with dilute feed waste. The initial concentrations of NH_4^+ and HCO_3^- in all flasks were adjusted to 6 mg/l and 1000 mg/l as $CaCO_3$, respectively.

Nitrifying bacteria cultured in the laboratory were used to inoculate fish hatchery water for the nitrification tests. At the beginning and end of the nitrification tests, the concentration of these microorganisms was determined using an extinction dilution procedure described by Alexander.[17] The rate of nitrification and the population of nitrifying bacteria in the experimental and control flasks were monitored to determine the feasibility of nitrifying the spent regenerant. The order of the reactions in the flasks was also determined.

Analyses

The concentration of calcium and magnesium was determined using Perkin-Elmer model 303 Atmoic Absorption Spectrophotometer equipped with a HGA-2000 graphite furnace and a Perkin-Elmer Model 065 recorder. Potassium and sodium concentrations were determined by flame atomic absorption spectrophotometry. The operational parameters used were those suggested in the Perkin-Elmer manuals.[18,19]

Ammonium and nitrate activities were determined using specific ion electrodes coupled to an Orion model 801 pH/millivolt meter with a model 751 digital printer. The activities were corrected to read in terms of concentrations of nitrogen in mg/l.

Clinoptilolite

Clinoptilolite samples were obtained from the Hector, California, and the Buckhorn, New Mexico, deposits. The mesh sizes studies were minus 4 and 4 x 20 for the California zeolite and 4 x 20 and 20 x 50 for the New Mexico zeolite. These mesh sizes represent commercially available forms.

Clinoptilolite samples were cleaned up by washing with distilled water to remove fines, equilibrating with 1 *M* NaCl and washing with distilled water until the water showed no chloride ion when reacted with $AgNO_3$. The samples were oven dried at 105°C and subsequently placed in column reactors for further testing.

Continuous-Column Tests

Ion exchange columns were constructed from 57 x 1 cm polystyrene and contained from 1 to 20 g of zeolite which was supported on a cotton plug resting on a rubber stopper to help distribute the flow. All columns had bed depth to diameter ratios of 7 to 28. The columns were operated upflow, except for three runs, at 5 to 50 bed volumes per hour (bed-vol/hr) and as a packed bed for exhaustion and regeneration tests, at 5 to 50 and 15 bed-vol/hr, respectively. Flow through the columns was maintained by gravity feed. The influent was free of suspended solids. All exhaustion and regeneration experiments were conducted at room temperature, about 23°C.

Synthetic wastewaters fed to the columns were prepared according to Table I. The water used was dissolved-oxygen-saturated deionized water that had been glass distilled and passed through Dowex 50W-X8 cation exchange resin; the strongly acidic cation exchange resin removed NH_4^+ to a level of 10^{-6} *M*. Chemicals were added to prepare the simulated wastewater in the concentration and composition desired. Column effluent was collected in erlenmeyer flasks that were acid cleaned, rinsed thoroughly with deionized water and oven dried.

Table I. Cation Composition of Synthetic Wastewater Feed Used in Continuous Column Tests

Cation	"Low Strength" (meq/l)	"Medium Strength" (meq/l)	"High Strength" (meq/l)
Na^+	0.070	0.347	0.700
K^+	0.026	0.128	0.256
Ca^{2+}	0.136	0.618	1.360
Mg^{2+}	0.062	0.308	0.617
NH_4^+	0.071	0.071	0.071
Totals	0.365	1.472	3.004

Solutions of 0.05 *M* and 0.10 *M* $NaHCO_3$ were used for regenerating the columns exhausted to 0.035 meq/l of NH_4^+ in the effluent. The regenerant bed volumes required and the efficiency of regeneration as a function of solution concentration were determined. The reasons for using $NaHCO_3$ were twofold. Na^+ has been shown to increase the kinetics of regeneration over Ca^{2+}, as described earlier, and HCO_3^- provides a source of inorganic carbon for bacteria to accomplish biological nitrification. The alkalinity of most alpine streams originating in granitic areas is generally low. This method of regeneration takes advantage of both the cation and anion components of the salt.

RESULTS AND DISCUSSION

Exchange Capacity Studies

Ammonium exchange capacities are summarized in Table II. It can be seen that the exchange capacities for the California and New Mexico zeolites at the three mesh sizes studied were nearly identical. The difference between exchange capacities due to the experimental method was always larger than the difference in exchange capacities between the different sources and sizes of zeolite material used. The average NH_4^+ exchange capacity of clinoptilolite in the sodium form was 1.96 meq/g (dry) zeolite.

New Mexico clinoptilolite of 4 x 20 mesh was chosen for experimentation with the multicomponent systems.

Table II. NH_4^+ Exchange Capacity of Clinoptilolite in the Na^+-Form

	California Clinoptilolite		New Mexico Clinoptilolite	
Test[a]	4 x 20 Mesh (meq/g)	Minus 4 Mesh (meq/g)	4 x 20 Mesh (meq/g)	20 x 50 Mesh (meq/g)
1	1.97	2.02	1.90	1.96
2	1.83	1.88	2.04	1.90

[a]Test 1–Capacity determined by elution of NH_4^+ with NaCl. Test 2–Capacity determined from breakthrough curves for an NH_4Cl solution feedwater.

Breakthrough Curves

Breakthrough curves illustrating the effect of competing cations in the three multicomponent systems studied are presented in Figure 1. Throughputs to complete exhaustion of the clinoptilolite decreased from 3650 to 2350 to 950 bed volumes for the low-, medium- and high-strength feeds, respectively. Throughputs to a breakthrough of 0.035 meq/l NH_4^+ decreased from 3100 to 700 to 200 bed volumes for the low-, medium- and high-strength feeds, respectively. Therefore, increasing the competing cations by 5 and 10 times over the base concentrations reduced throughput volume to breakthrough by about 4 and 15 times, respectively.

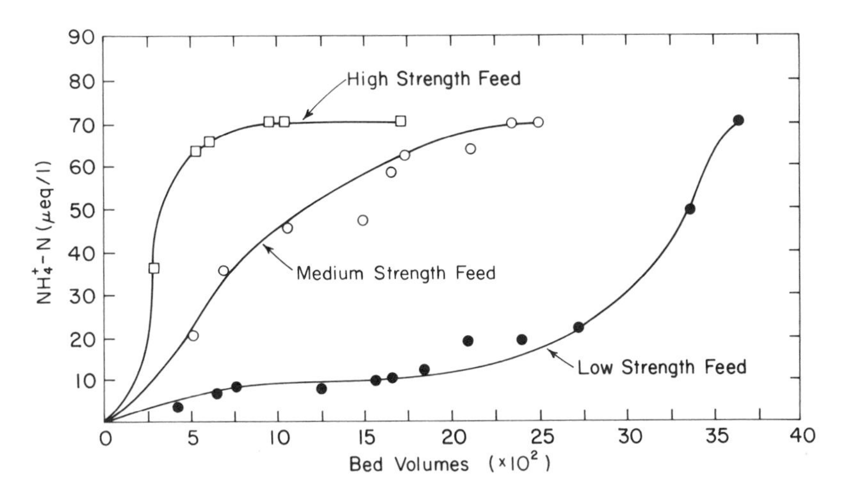

Figure 1. Effect of competing cations on ammonia removal.

As a means of comparing the performance of clinoptilolite among different wastewaters, a compilation of literature values and those obtained in this study are shown in Table III. The volume of low-strength wastewater, which was treated to a breakthrough of 0.035 meq/l NH_4^+, was much greater than all other wastewaters listed. Relatively high throughputs were obtained with the medium-strength synthetic wastewater. The cationic concentration of the high-strength synthetic wastewater was in the same range as that of a domestic sewage. Thus, of the three strengths of synthetic wastewater, the high-strength feed gave the lowest throughput volumes. However, this volume was in the same range as that reported for domestic sewage.

Separation factors and selectivity coefficients were determined from column studies to identify the cations competing most effectively with NH_4^+. The data are summarized in Table IV. Since the zeolite was initially in the Na form, it was not possible to measure the solid-phase Na^+ concentrations. Sodium ion concentrations were calculated by subtracting the sum of the solid-phase concentrations of Ca^{2+}, Mg^{2+}, K^+ and NH_4^+ from the cation exchange capacity of 1.96 meq/g. The disadvantage of this method is that errors in measuring the solid-phase concentration of other cations are accumulated into the value of solid-phase Na^+ concentration.

The separation factors and selectivity coefficients indicate that NH_4^+ removal in the presence of the competing cations studied is favorable except for K^+ and Na^+. In the medium- and high-strength multicomponent

Table III. Performance of Clinoptilolite With Different Wastes

Waste	Cationic Strength (meq/l)	Influent NH_4^+-N (meq/l)	Breakthrough NH_4^+-N (meq/l)	Volume Treatable (bed-vol)	Investigator
Domestic Sewage	3.7	1.34	0.035	350	Koon and Kaufman[13]
Domestic Sewage	5.9	1.07	0.071	150	EPA[11]
Domestic Sewage	8.2	1.35	0.035	250	Koon and Kaufman[13]
Domestic Sewage	8.9	1.35	0.035	150	Koon and Kaufman[13]
Fish Hatchery	1.7	0.18	0.035	1250	Braico[12]
Fish Hatchery	18.6	0.18	0.035	350	Braico[12]
Fish Hatchery	0.365	0.071	0.035	3000	This study
Fish Hatchery	1.54	0.071	0.035	700	This study
Fish Hatchery	3.00	0.071	0.035	200	This study

Table IV. Separation Factors and Selectivity Coefficients for Clinoptilolite

Parameter	Low-Strength Feed	Medium-Strength Feed	High-Strength Feed
Selectivity coefficient			
$K^{NH_4^+}_{Na^+}$	0.22	2.283	1.00
$K^{NH_4^+}_{Ca^{2+}}$	7945	1326	540
$K^{NH_4^+}_{Mg^{2+}}$	8822	2729	1561
$K^{NH_4^+}_{K^+}$	1.08	0.625	0.397
Separation factor			
$Q_S{}^{NH_4^+}_{Na^+}$	0.22	2.28	1.10
$Q_S{}^{NH_4^+}_{Ca^{2+}}$	1.4	1.01	0.92
$Q_S{}^{NH_4^+}_{Mg^{2+}}$	1.61	2.08	2.67
$Q_S{}^{NH_4^+}_{K^+}$	1.08	0.63	0.38

system, K^+ was preferred over NH_4^+. The NH_4^+-Na^+ values were less than 1.0 in the low-strength multicomponent system because of the very low concentration of Na^+ in the influent. Similar results have been found by Braico[12] with fish hatchery water. The separation factors and selectivity coefficients obtained definitely indicate the suitability of clinoptilolite for removing NH_4^+ from the multicomponent systems studied.

The effect of flowrate on NH_4^+ removal from the low-strength synthetic feed was also investigated as shown in Figure 2. At the high flowrate of 25 bed-vol/hr, 2700 bed volumes were treated to a breakthrough of 0.035 meq/l NH_4^+. At the low flowrate of 14 bed-vol/hr, 3000 bed volumes were treated to the same breakthrough.

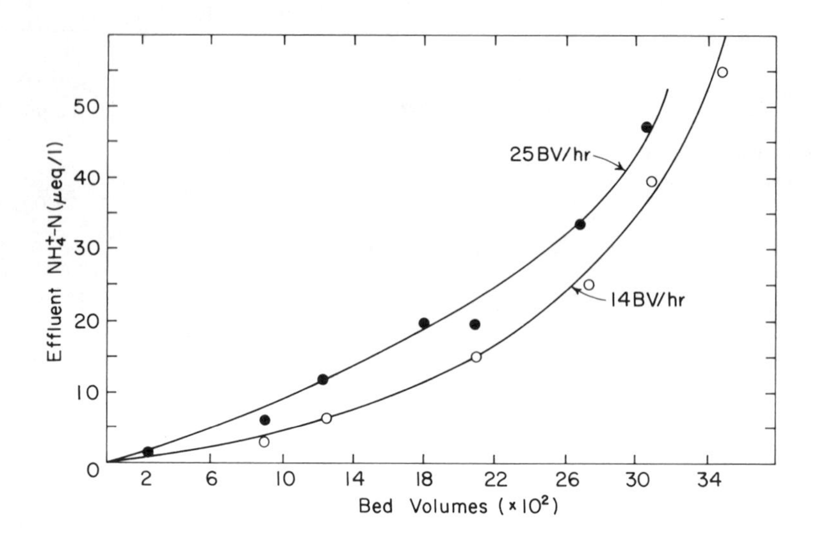

Figure 2. Effect of flowrate on NH_4^+ breakthrough curves with low-strength feed.

Clinoptilolite Requirements

In order to calculate the amount of clinoptilolite required for treatment of water likely to be encountered at a fish hatchery similar to the Dworshak National Fish Hatchery, Figure 3 was developed. This figure, similar to one developed by Koon and Kaufman[13] for domestic waste, illustrates the NH_4^+ exchange capacity for total exhaustion and for a breakthrough of 0.035 meq/l as a function of the cationic strength of the influent wastewater. The NH_4^+ exchange capacities were calculated by integrating the areas above the breakthrough curves. The cationic strength of each waste was determined by means of the equation

$$I_c = \tfrac{1}{2} m_i z_i^2 \qquad (6)$$

where m_i is the concentration of cationic species i and z_i is its charge. Although Figure 3 is empirical and is not applicable to any combination of cations, it demonstrates the effects of possible changes in water quality at a hatchery such as the Dworshak National Fish Hatchery and provides a way to size treatment units for NH_4^+ removal from the reuse water.

Regeneration Studies

Figure 4 presents results for regenerating columns that received the low-strength feed. Fifty bed volumes of 0.10 *M* solution were required to

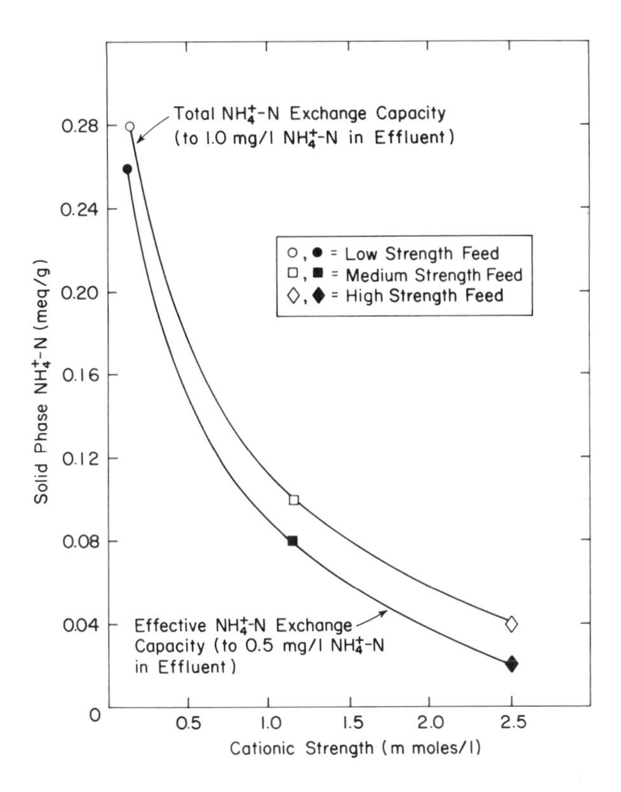

Figure 3. Effect of reuse water strength **on** NH_4^+ exchange capacity.

regenerate the zeolite. This is much larger than the 10 to 30 bed volumes of lime solution that were used to regenerate columns treated with domestic wastes. However, the throughput volume to breakthrough is so much larger than with domestic waste that the percentage of treated water required for regeneration is competitive; that is, about 1%. The amount of 0.05 *M* $NaHCO_3$ required to regenerate the zeolite was 90 bed volumes. This represents about 3% of the volume treated to breakthrough. Using the 0.10 *M* solution resulted in a 44% reduction in the volume of regenerant required.

Figure 5 illustrates the regeneration of those columns receiving the medium-strength synthetic wastewater. The total volumes of 0.10 *M* and 0.05 *M* $NaHCO_3$ required to regenerate the zeolite are similar. Also, the volumes of regenerant required are similar to the requirements for regenerating the low-strength feed treated columns. However, the initial concentrations of the individual regenerant streams are lower for the

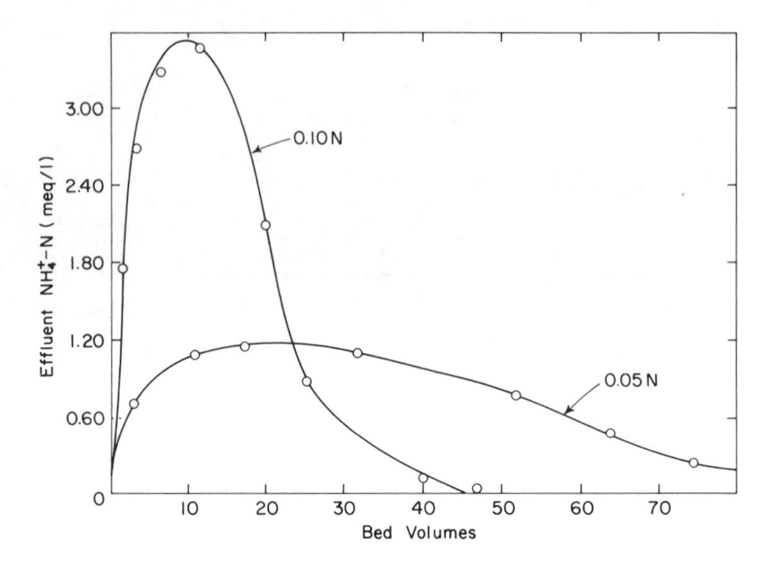

Figure 4. Effect of $NaHCO_3$ concentration on ammonium elution from columns treated with low-strength feed.

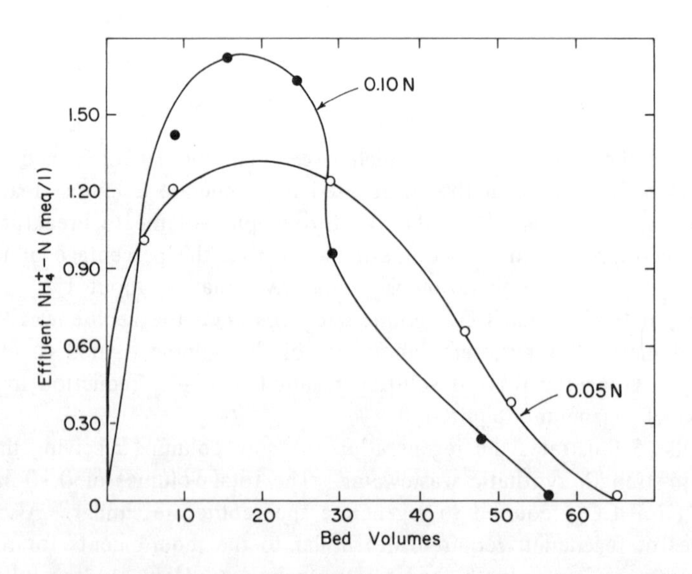

Figure 5. Effect of $NaHCO_3$ concentration on ammonium elution from columns treated with medium-strength feed.

columns treated with medium-strength wastewater. This is to be anticipated, as the NH_4^+ concentration is less in the zeolite treated with medium-strength feed than in the zeolite treated with low-strength feed.

Similar results were obtained for regenerating the columns that received the high-strength feed. The volume of 0.10 *M* solution required for regeneration was 75 bed volumes. This is 21% less than the amount of 0.05 *M* solution required to regenerate the zeolite, which was 95 bed volumes. These volumes for regeneration represent about 24% of the volume of water treated to breakthrough. The required 24% of the volume treated for regeneration could be reduced by greatly increasing the concentration of $NaHCO_3$ in the regenerant solution.

Table V illustrates the relationship between the regeneration efficiency and the degree of column utilization for the regenerant solutions. About 72% of the theoretical exchange capacity of all of the columns had been utilized at breakthrough. The regeneration efficiencies for complete regeneration averaged 25% for the 0.05 *M* $NaHCO_3$ solution and 17% for the 0.10 *M* solution. Efficiencies were higher for the more dilute solution because the total volume requirements of $NaHCO_3$ to regenerate the 72% of the column utilized did not change dramatically with the concentration of regenerant.

From Figures 4 and 5 it is evident that during the initial stages of regeneration there was a definite relationship between regenerant concentration and efficiency. The less dilute the regenerant solution, the higher the NH_4^+ concentration in the effluent regenerant stream. This was expected, because during the initial stages of regeneration, there is a large concentration gradient between NH_4^+ on the zeolite and NH_4^+ in the solution. During this stage of regeneration, the rate of exchange should be a function of the concentration of regenerant solution. The law of mass action also predicts that Equation 2 will shift to the left in proportion to the concentration of Na^+ present. However, because the large selectivity coefficient indicates that the reaction proceeds strongly to the right, infinitely large quantities of regenerant would be required to drive the reaction completely to the left. This is approached during later stages of regeneration. At this stage the concentration gradient in the column treated with high-strength wastewater has been reduced below that in the columns treated with low concentration regenerant. Therefore, the operating characteristics of the exchanger are such that higher regeneration efficiencies were obtained with the more dilute regenerant.

Nitrification of Spent Regenerant

Results of the batch nitrification tests with the experimental reactors containing Koch Ring material are shown in Figure 6. The initial decrease

Table V. Column Efficiency and Utilization as a Function of Feed and Regenerant Concentrations

Strength of Synthetic Wastewater Solution	Regenerant Concentration (*M*)	Column Utilization (%)	Regeneration Efficiency (%)
Low	0.05	72	22
Low	0.10	73	18
Medium	0.05	74	31
Medium	0.10	72	18
High	0.05	70	21
High	0.10	70	15

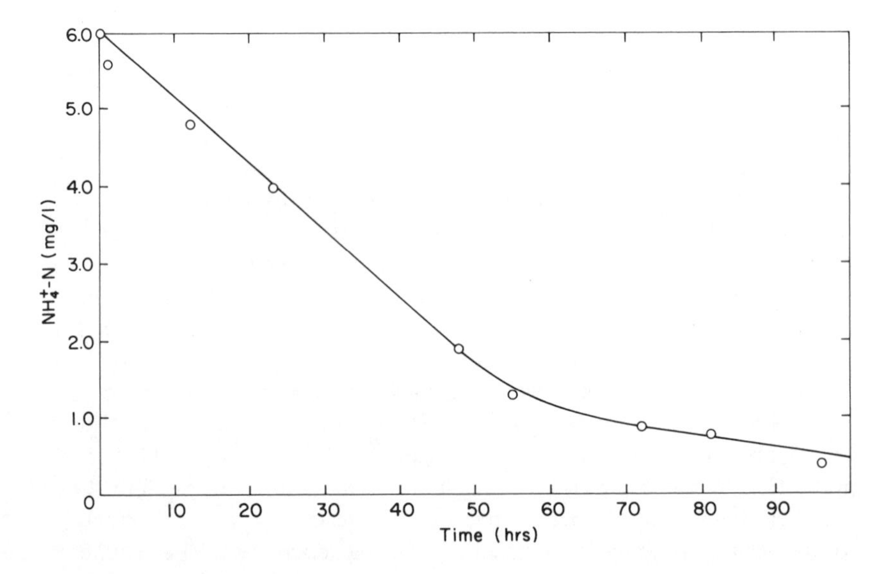

Figure 6. Nitrification with Koch Rings in batch reactors.

in NH_4^+ concentration with time was linear, representing the rate of removal of 0.085 hr^{-1}. The plot indicates that nitrification followed zero order kinetics at NH_4^+-N concentrations above 1.3 mg/l. The reaction, therefore, can be represented as

$$\frac{dN}{dt} = -k_1 \tag{7}$$

where N is the NH_4^+-N concentration and k_1 is the reaction rate constant. Below a concentration of 1.3 mg/l NH_4^+-N, the rate of removal decreased. The data are linear on a semilog plot, indicating that the reaction rate is proportional to the NH_4^+ concentration and is therefore first-order below 1.3 mg/l as N. The reaction can be represented as

$$\frac{dN}{dt} = -k_2 N \tag{8}$$

The rate constant k_2 was 0.01 hr^{-1}. The difference between k_1 and k_2 represents an 8.5-fold decrease in the rate of nitrification. Therefore, a significant reduction in the rate of nitrification occurred when the NH_4^+-N concentration was reduced below 1.3 mg/l.

Results of the nitrification tests with the control reactors are presented in Figure 7. The initial decrease in NH_4^+ concentration with time was also linear. However, the zero order reaction rate constant, k_1, was found to be lower in the control reactors at 0.067 hr^{-1} than in the experimental reactors. Below a concentration of 1.2 mg/l NH_4^+-N, the rate of removal decreased again. As before, a linear plot of time against concentration on semilog paper indicated that the reaction was first order; the relationship can be described by Equation 8. The rate constant, k_2, was 0.012 hr^{-1}. The difference between k_1 and k_2 represents a 5.6-fold decrease in the rate of nitrification when the NH_4^+-N concentration was less than 1.2 mg/l.

Nitrifying bacterial population was also determined for both batch reactors. The initial concentration of nitrifying microorganisms was 10^3/ml and it increased in both systems by a factor of only four during the course of the experimental period. The rate of 0.010 hr^{-1} for k_2 in the experimental reactor is lower than the more typical rate of 0.8 hr^{-1} calculated from data collected at Dworshak National Fish Hatchery under optimum conditions. This difference is believed to be caused by differences between the nitrifying bacterial population when the experiment was performed and those numbers found at Dworshak National Fish Hatchery.

CONCLUSION

This study has demonstrated that the proposed method for removing low levels of NH_4^+ from fish hatchery water is a feasible method to meet standards for water reuse. The proposed scheme utilizes ion exchange of NH_4^+ with clinoptilolite, followed by chemical regeneration with $NaHCO_3$ and subsequent bacterial nitrification of the NH_4^+ in the regenerant stream.

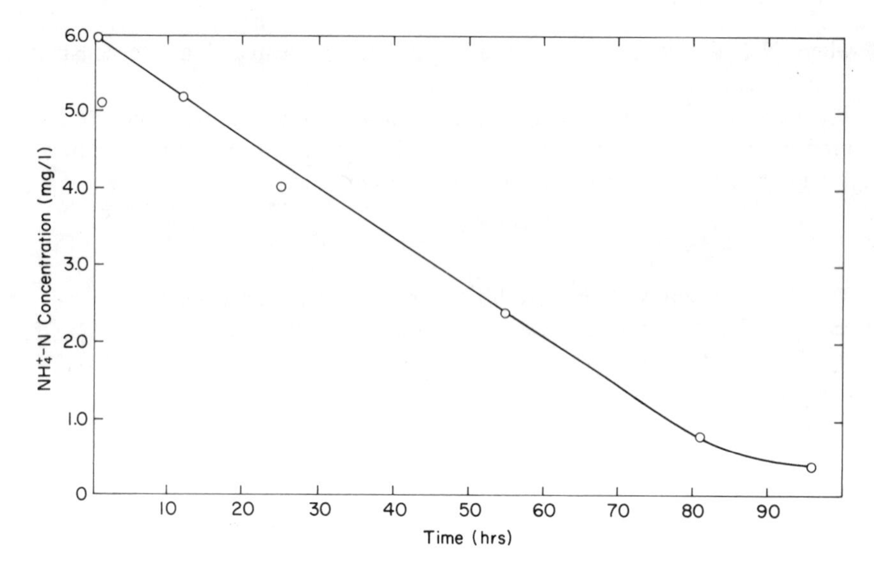

Figure 7. Nitrification with glass beads in batch reactors.

ACKNOWLEDGMENTS

This investigation was supported in part by the U.S. Department of Interior Fish and Wildlife Service and the College of Engineering, Washington State University. Appreciation is extended to the Double Eagle Petroleum and Mining Company for supplying samples of New Mexico clinoptilolite.

REFERENCES

1. Ericksen Jones, J. R. *Fish and River Pollution* (London: Butterworth and Co., 1964).
2. Croker, M. and W. Hess. Paper presented at 2nd Annual Conference of Complete Water Use, Chicago, IL, May 4-8, 1975.
3. Smith, H. W. *J. Biol. Chem.* 81:727 (1929).
4. Wood, J. D. *Can. J. Biochem. Physiol.* 36:1237 (1958).
5. Fromm, R. O. *Comp. Biochem. Physiol.* 10:121 (1963).
6. Delaunay, H. *Compt. Rend. Soc. Biol.* 101:371 (1929).
7. Shirahata, S. *Bull. Fac. Fish.* 17:68 (1964).
8. Spotte, S. *Fish and Invertebrate Culture* (New York: John Wiley and Sons, Inc., 1970).
9. Ames, L. Paper presented at 13th Pacific Northwest Industrial Waste Conference, Washington State University, Pullman, WA, 1967.
10. Mercer, B., L. Ames and C. Touhill. *Am. Chem. Soc. Proc.* 6 (1968).

11. "Process Design Manual for Nitrogen Control," EPA Technology Transfer (1975).
12. Braico, R. Ph.D. Thesis, Montana State University, Bozeman, MT (1972).
13. Koon, J. H., and W. J. Kaufman. "Optimization of Ammonia Removal by Ion Exchange Using Clinoptilolite," SERL Report No. 71-5, Sanitary Engineering Research Laboratory, College of Engineering, and School of Public Health, University of California, Berkeley, CA (1971).
14. *Duolite Ion Exchange Manual* (Redwood City, CA: Diamond Shamrock Co., 1960).
15. Sims, R., and L. Little. *Environ. Lett.* 4:27 (1973).
16. Semmens, M. J. *Environ. Sci. Technol.* 11:255 (1977).
17. Alexander, M. *Methods of Soil Analyses*, Part 2 (Madison, WI: American Society of Agronomy, Inc., 1965).
18. *Analytical Methods for Atomic Absorption Spectroscopy Using the HGA Graphite Furnace* (Norwalk, CT: Perkin-Elmer Corp., 1973).
19. *Analytical Methods for Atomic Absorption Spectrophotometry* (Norwalk, CT: Perkin-Elmer Corp., 1971).

CHAPTER 20

CYANIDE SPECIES AND THIOCYANATE METHODOLOGY IN WATER AND WASTEWATER

N. P. Kelada, C. Lue-Hing and D. T. Lordi

The Metropolitan Sanitary District
of Greater Chicago
Chicago, Illinois

INTRODUCTION

Cyanides are known to be toxic to different forms of life, to man and more so to aquatic life at low concentrations. The chemistry of cyanides in the environment is complex and the uncertainty of the proper identification of toxic cyanide species and their concentration levels is complicated by the fact that the analytical methodology needs improvement. There is a need to obtain valid data to estimate the tolerable cyanide loads contributed by industry and residential communities to municipal sewage treatment plants. In a broader sense there is also a need for more sensitive, accurate and reproducible methods to furnish proper information about the behavior of the different cyanide species in the receiving aquatic environment. This information should be helpful in the assessment by concerned authorities of the cyanide limits for effluents and streams (currently 25 $\mu g/l$ in the state of Illinois).

Cyanides are compounds characterized by having the chemical group $-C \equiv N$. Ideally, cyanides can be classified according to their physicochemical properties as free, simple, complex or organic cyanides. In addition, other terms such as total cyanides, oxidizable cyanides, and cyanides amenable to chlorination are frequently used: in fact, they reflect the corresponding methods of treatment and analysis.

Present Cyanide Methodology

Current cyanide analytical procedures involve three steps: separation, absorption and measurement. Separation of cyanides from the sample is a prerequisite step to eliminate several types of interferences.[1] Separation usually follows acidification and some form of volatilization to break down the complex cyanides to the free form.[2] Several separation techniques are known, such as distillation at atmospheric pressure[3-5] or under reduced pressure[6] and separation by aeration.[3] Automated methods utilize modified techniques such as flash distillation[7] or thin-film distillation.[8] Absorption of the evolved HCN in an alkaline solution such as sodium hydroxide[3,4] or sodium acetate[8] is the next step. Spiral flow gas washers[3] and recirculation[8] techniques are used in manual methods, whereas fractionating columns[7] or absorption columns[8] are used in automated methods.

Measurement of the recovered cyanides can be volumetric, *e.g.*, by titrating the absorbed cyanide with silver nitrate using rhodamine indicator,[9] or colorimetric, using color reagents such as benzidine[10], pyrazolone,[11] or barbituric acid[12] in pyridine solution. Generally, the colorimetric measurement is for low cyanide levels and is applicable to automation. The type of cyanide determined by the different methods is controlled through selective decomposition and distillation.

Standard Methods[3] calls for distillation at atmospheric pressure, absorption of HCN in NaOH using a spiral flow gas washer and pyridine-pyrazolone colorimetric measurement. However, the method is reported[5,13] to have low reproducibility and 100% variation at a cyanide level of 25 μg/l. An automated determination[14] using Technicon flash distillation technique has been reported to be unsatisfactory at cyanide levels below 200 μg/l.[8] The thin-film distillation developed by Goulden *et al.*[8] detects concentrations as low as 0.5 μg/l and achieves complete breakdown of strong metal-cyano-complexes by ultraviolet irradiation. However, the thin-film distillation unit and the absorption column are complicated and the condenser has some serious disadvantages.

The main objective of this study was to develop an automated system to distinguish between different cyanide species at the low levels found in municipal treatment plant effluents and other environmental samples. It was also desirable to develop a reliable method for determination of thiocyanate, CNS^-, in the presence of different cyanide species. This involved the parameter "cyanides amenable to ozonation."

EXPERIMENTAL

Reagents

The acid digestion mixture was prepared from 20% by volume of 85% orthophosphoric acid and 4% of 50% hypophosphorus acid. Sodium hydroxide, 0.02 *M*, was used for HCN absorption and preparation of the cyanide working standards; potassium dihydrogen phosphate buffer solution was prepared at 0.2 *M* (pH 4.2); chloramine-T solution was 0.4% (w/v); and pyridine-barbituric acid solution (P-B) was prepared as described in ASTM[4] and then diluted four times with distilled water. Cyanide standards were prepared every week from KCN alkaline stock solution of 1 g/l CN concentration and working standards of 1 to 100 μg/l CN were prepared daily by dilution of the stock with 0.02 *M* NaOH.

Appartus

A schematic flow diagram of the cyanide automated system is shown in Figure 1. The sampler, pump, colorimeter and recorder were Technicon AutoAnalyzer I modules. Presently an AutoAnalyzer II dual channel with a printer is used: One channel for simple cyanide without UV irradiator, and the second channel for total cyanide. A 50-mm flow cell was used with Colorimeter I and a 15-mm flow cell was used with Colorimeter II. The sampler presently is used with a 20/hr 1:1 cam; however, it was modified to accept an external timer (Industrial Timer Corp. "Dual Trol" recycling timer, equipped with TM-5M 5-min module for the wash cycle and TM-3M 3-min module for sampling) which was used initially to optimize the sampling rate.

Figure 2 shows details of the continuous thin-film distillation unit which is made from borosilicate glass tubing. A is the horizontal thin-film evaporation tube, B is a vertical tube to deliver the gases evolved to the HCN absorption coil (using 0.02 *M* NaOH absorbing solution), and C is a U-shaped liquid waste outlet. C′ is preferred over C, especially in case of samples with high suspended solids in that it does not get clogged and can be easily cleaned. The aluminum bar is heated with a 150-W cartridge heater. Temperature is controlled with a Payne Engineering model 18 D-I-10 solid-state power control equipped with an 18 G proportional temperature controller and thermistor detector. The thermistor must be located close to the cartridge heater to avoid temperature cycling. A Variac may be substituted for the power control provided care is taken to monitor the temperature.

The UV irradiator utilizes a mercury lamp (type A 673-A-36, 550 W Conrad Precision Industries) and a quartz sample coil 3-mm i.d. tubing with 11-cm coil diameter and 30 turns. Details of a comparable UV irradiator are available in the literature.[15]

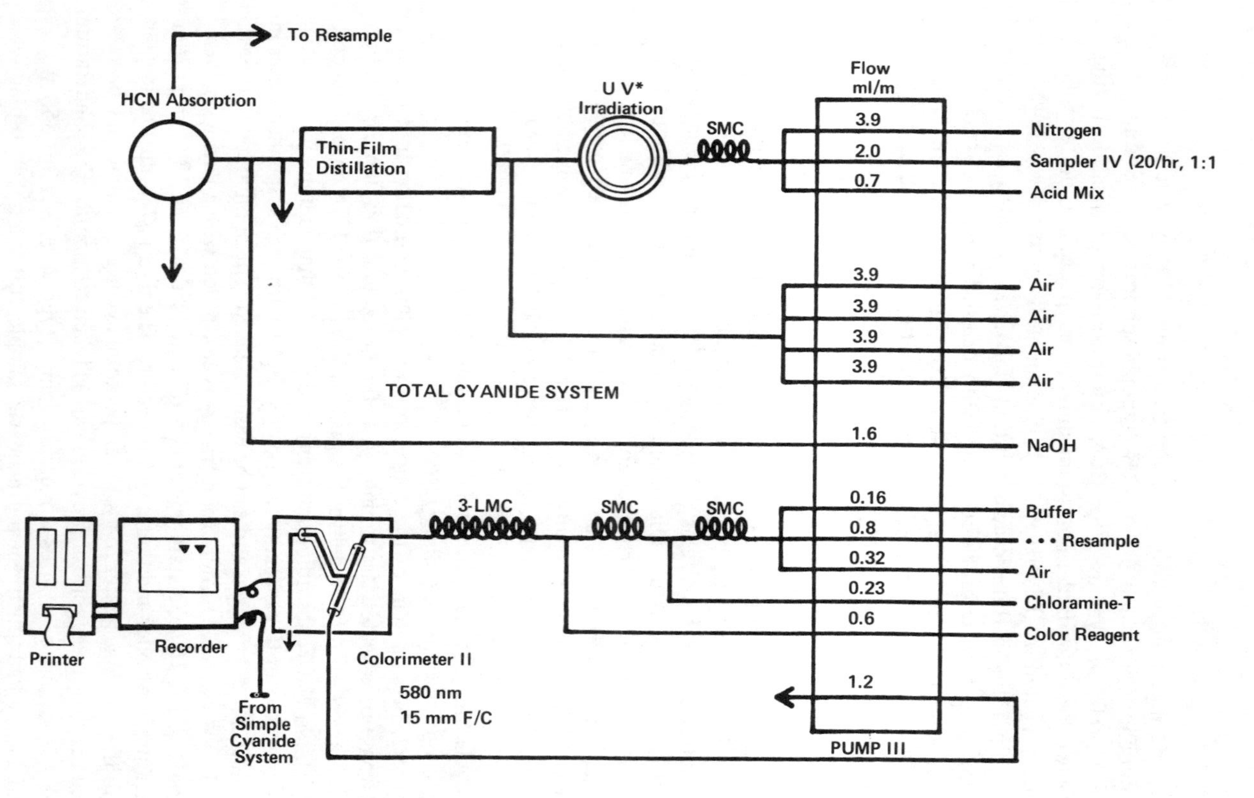

Figure 1. Manifold for automated cyanide system.
*U.V. irradiation on line for total cyanide and bypassed for simple cyanide.

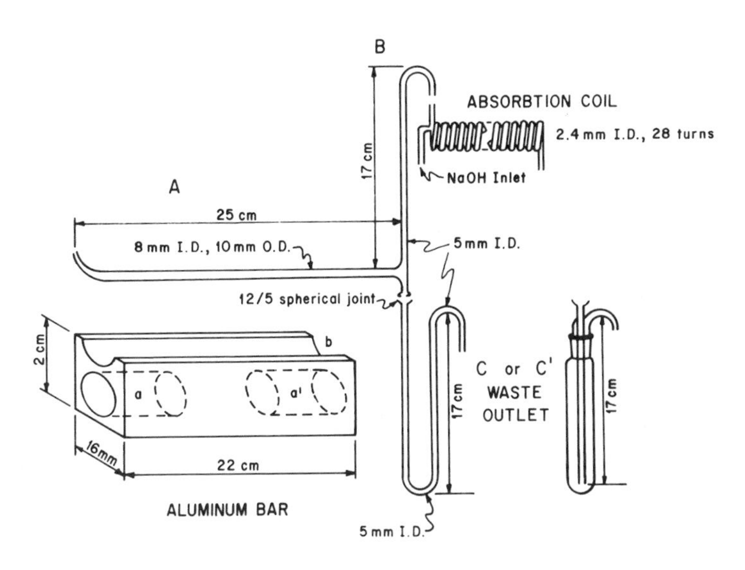

Figure 2. Continuous thin-film distillation unit.

Equipment used for ozonation included a generator (PCI Ozone Corp. model (C2P-6C) and O_2 gas cylinder equipped with a Matheson 3102 high-purity two-stage regulator capable of delivering up to 15 psig. Plastic tubing was used for delivery of O_2 and O_3 with a fritted-glass sparger fitted at the end of the ozone line. A 100-ml glass measuring cylinder or a glass beaker was used for the ozonation of water samples.

Procedure

The system is set into operation and the temperature of the aluminum heating bar of the thin-film distillation unit is adjusted to about 125°C so that approximately 15% of the acidified sample vaporizes. The UV irradiator is required for the total cyanide determination. Nitrogen gas is used instead of air to segment the flow to prevent ozone formation and oxidation of cyanide. The UV coil is bypassed for simple cyanides. For the determination of free cyanide the thin-film distillation unit is also bypassed and the samples or standards (pH $\approx$ 11.5) are fed directly into the colorimeter part of the manifold.

Calibration procedures, using 1 to 100 μg/l CN (or up to 500 μg/l), are made separately for each of the free, simple and total cyanide trains.

Samples can be preserved at pH ⩾ 12 and 5°C, but the samples should be analyzed as soon as possible. Sulfide ion gives positive interference when present in concentrations higher than 20 mg/1. Therefore, samples with higher sulfide content should be diluted with distilled water until the lead acetate paper test just becomes negative, which is about 5 mg/l-S^{2-}.

Sediment samples should be mixed thoroughly and homogenized in a heavy-duty blender, such as a Waring model CB-6. A suitable weight of sample, about 1 to 5 g/l (depending on the cyanide concentration), is transferred quantitatively to a volumetric glass container and adjusted to volume with 0.02 *M* NaOH. The sample is kept in suspension by the use of an ultrasonic disrupter (W 350 Sonifier), then fed into the system.

In this study the ozonation process was always performed in a closed ventilated hood. O_2 gas was passed to the generator and the pressure regulator was adjusted to 5 SCFH. Then the power was turned on and ozone output was controlled at 75%. Under those conditions the ozone generated was approximately 25 mg/min. A 30-to 50-ml homogenized portion of the water sample was put into the 100-ml glass measuring cylinder and ozone was bubbled through the sample by means of the fritted-glass sparger. Generally the samples were ozonated for 10 min.

BASIC SYSTEM DESIGN AND OPERATIONAL ASPECTS

Continuous Thin-Film Distillation

The concept of thin-film distillation was developed by Goulden *et al.*[8] and relies on evaporating HCN and some water of the acidified samples. The water is condensed back and HCN is absorbed through an absorption column. The present investigations have shown that the aforementioned technique has some disadvantages. It was found that considerable amounts of HCN were refluxed with the condensed water and not recovered. Also, it was found that the lower the temperature of the condenser, the lower the cyanide recovery and the less reproducible the performance. The present design is very simple (Figure 2) and the need for condensers is eliminated. The temperature of the aluminum bar is controlled at about 125°C so that 15-17% of the acidified sample is evaporated and does not condense back, but is collected with HCN in the absorber. It should be mentioned also that aeration helps the stripping of HCN gas from solution. The air flow is about 20 ml/min and the acidified sample flowrate is 2.7 ml/min.

HCN Absorption and Recovery

The poor precision of different methods for the determination of cyanide has generally been attributed to imcomplete and irreproducible recovery of HCN, especially at low cyanide concentrations.[8,13] In this work, a very simple and efficient HCN gas absorber was devised, consisting of a 2.4-mm i.d., 28-turns glass coil. The HCN gas is forced with air through the coil simultaneously with 0.02 *M* sodium hydroxide. The HCN recovery was found to be 90% at the present range of interest, 0 to 100 μg/1 CN.

Automation of Pyridine-Barbituric Method

Color reagents such as pyridine-benzidine and pyridine-phenylene-diamene are not presently used because of their potentially harmful effects.[10,16] The pyridine-pyrazolone method metnioned in *Standard Methods*[3] and by other authors[8] requires 30 min for color development. This color reagent is unstable and requires a lengthy daily preparation, whereas the pyridine-barbituric acid reagent is reported[4] to be stable and needs only 8 min for color development. Factors related to the feasibility of the latter method for automation were studied and optimized.

Buffer systems were used to study the effect of pH on color development and a pH of 7 was preferred. The concentration of NaOH solution for HCN absorption was optimized at 0.02 *M*. In reference to the concentration of the pyridine-barbituric acid reagent, the highest sensitivity was found at half the concentration described by the ASTM procedure, yet practically, it is recommended that one-fourth of that concentration be used.

ANALYTICAL BEHAVIOR OF CYANIDES

The behavior of some inorganic metal-cyano-complexes and their analytical response to the developed automated system were evaluated and found to be largely dependent on the pH and their stability constants (Table I). Representative data are shown in Table II, and the metal-cyano-complexes can be classified as very strong, strong, medium and weak complexes.

Very strong metal cyanide complexes such as those of cobalt, platinum and gold are difficult to dissociate and are not oxidized to cyanate by agents such as chlorine or ozone. They do not dissociate in solution or decompose by photoillumination.[17] Cobalt(III) complex does not break down by any of the known distillation procedures and, as expected, there was no cyanide detected by either the free or the simple cyanide

Table I. Metal-Cyano-Complex Ions and Their Stability Constants

Metal	Complex Ion		Stability Constant at 25°C	Reference
Cobalt(III)	hexacyanocobaltate	$[Co(CN)_6]^{3-}$	1×10^{64}	18
Iron(III)	ferricyanide	$[Fe(CN)_6]^{3-}$	1×10^{52}	19
Iron(II)	ferrocyanide	$[Fe(CN)_6]^{4-}$	1×10^{47}	19
Nickel(II)	tetracyanonickelate	$[Ni(CN)_4]^{2-}$	1×10^{22}	17
Cadmium(II)	tetracyanocadmiate	$[Cd(CN)_4]^{2-}$	7.1×10^{16}	17
Manganese(III)	hexacyanomanganate	$[Mn(CN)_6]^{3-}$	5×10^{9}	18

Table II. Analytical Behavior of Cyanide Compounds

Compounds	CN Preparation (μg/1)	CN Measured Free (μg/l)	Simple (μg/l)	Total (μg/l)
Metal-Cyano-Complex				
Hexacyanocobaltate	100	ND[a]	ND	99.0
Ferricyanide	100	3.6	3.4	98.5
Ferrocyanide	100	3.5	3.6	97.0
Tetracyanonickelate	100	13.7	98.8	100.2
Tetracyanocadmiate	100	101.5	98.6	98.5
Hexacyanomanganate	100	101.5	98.5	101.2
Organic Cyanides				
Acetonecyanohydrin	100	98	103.0	101.1
Acetonitrile	1,000	ND	ND	ND
Acrylonitrile	100	ND	ND	15.0
Adiponitrile	1,000	ND	ND	ND
Potential Cyanide-Forming Materials				
Cyanate	1,000	ND	ND	ND
Thiocyanate	100	b	ND	98.0

[a]ND = Not Detected.
[b]Very sensitive to pH variations; 25 and 100 μg/l were obtained at pH 7 and 5, respectively.

determinations. However, the UV irradiator utilized for total cyanides was effective in dissociation of the cobalt complexes, and complete recovery was achieved (Table II).

Strong iron complexes include ferricyanide and ferrocyanide; however, all of the cyanide was recovered from these complexes by the total cyanide system. On the other hand, both free and simple cyanides showed about the same small recovery of 3.6%. This indicates that the thin-film evaporation used for simple cyanides does not break down iron complexes, and what was measured probably was due to some decomposition by aging and photoillumination.[19,20]

Medium-strength complexes such as that with nickel(II) show complete recoveries with both simple and total cyanides, whereas the free cyanide measurement gave significantly lower values. It was also noticed that the percentage of free cyanide measured is inversely proportional to the original concentration: 35% at 10 μg/l CN and 13% at 100 μg/l CN. This could be explained by a higher degree of dissociation or ionization on dilution.

Weaker complexes are formed with cadmium(II) and manganese(III); Table II shows that all the cyanides were recovered with the free, simple and total cyanide determinations. It should be noted that the free cyanide measurement was made at pH 7.

Organic cyanides have one or more cyanide groups attached to organic radicals and include the cyanohydrins and nitriles. Acetonecyanohydrin, $(CH_3)_2C(OH)CN$, dissociates completely in solution yielding its cyanide content as evidenced from the free cyanide measurement. Of course, the simple and total cyanide measurements also showed the release of all of the cyanide content. On the other hand, the nitriles generally do not break down to release cyanides. Acetonitrile, CH_3CN, and adiponitrile, $NC(CH_2)_4CN$, did not show any significant values by free, simple or total cyanide. However, acrylonitrile, $CH_2 = CHCN$ released only about 15% of its cyanide content after UV irradiation. These differences between the cyanohydrins and nitriles should be considered carefully for evaluation of the environmental aspects.

Potential cyanide-forming materials were also examined. The thiocyanate ion is not measured by the simple cyanide determination, but it breaks down completely by UV irridation and gives total cyanide values on an equimolar basis. Cyanate ion was found to be very stable and did not show any positive results. Other compounds, namely urea, glycene, cysteine and nitrobenzene, were also examined and no significant values were obtained.

Analytical Definitions of Cyanide Measurements

It may be beneficial at this point to redefine the types of cyanide determinations and indicate what they actually measure:

1. *Free Cyanides.* Direct color development measures only cyanide ion, CN^-, and hydrocyanic acid, HCN, in neutral aqueous solutions. The

source of the free cyanides may include soluble metal cyanides and dissociated weak metal-cyano-complexes. The free cyanide measurement also includes the organic cyanohydrins.

2. *Simple Cyanides or Acid Dissociable Cyanides.* Utilization of the thin-film distillation technique breaks down medium-strength cyanide complexes. This method includes also the free cyanides.

3. *Total Cyanides.* Utilization of UV irradiation breaks down all strong cyanide complexes. Therefore, the total cyanide measurement will determine all free, simple and complex cyanides.

4. *Complex Cyanides.* The strongly complexed cyanides are calculated by finding the difference between total cyanides and simple cyanides.

The two types of measurements, simple cyanides and total cyanides, offer the most practical considerations for environmental monitoring. On the other hand, the free cyanide measurement requires precautions in interpretation because (1) the cyanide species measured at pH 7 may not be the same as other pH values in the environment, (2) the color developed is largely dependent on and inversely proportional to the ionic strength which is variable in different samples, and (3) it may be subjected to some interferences such as color and turbidity. In contrast to free cyanide, the parameter simple cyanides is not liable to such variations. In addition, the simple cyanide determination yields relatively higher values than the free cyanide determination. Acidification and controlled temperature break down the more readily hydrolyzable cyanide compounds and thus provide an additional increment of environmental protection.

SYSTEM PERFORMANCE CHARACTERISTICS

Recovery From Spiked Samples and Accuracy Evaluation

A variety of water and wastewater samples (*i.e.*, distilled water, tap water, river water, sanitary and ship canal, raw sewage and treated effluents) were chosen with a wide range of cyanide content, up to 35 μg/l as simple cyanide and up to 135 μg/l for total cyanides. The samples were spiked with different levels of simple and complex cyanides and then analyzed again to determine the efficiency of recovery. Some representative data are presented in Table III. Recoveries are defined as

Table III. Recovery of Cyanides from Spiked Samples and Accuracy Evaluation

Sample	CN Added		Cyanide Measured					
			Simple CN		Total CN			
	(μg/l)	(μg/1)	Conc. (μg/1)	Recovery (%)	Conc. (μg/l)	Recovery (%)	$\bar{E}$[a] (μg/l)	RE[b] (%)
Distilled Water	0	0	0		0			
	10	25 Co(III)	9.8	28.0	35.6	101.7	0.6	1.7
	25	25 Co(II)	48.7	97.4	50.0	100.1	0.0	0.0
Des Plaines River	0	0	2.1		11.7			
	10	10 Fe(III)	11.3	46.0	30.7	95.0	-1.0	-5.0
	25	10 Ni(II)	35.8	96.3	45.4	96.3	-1.3	-3.7
Raw Sewage (WSW Influent)	0	0	30.2		119.2			
	25	25 Fe(II)	52.1	43.8	168.2	98.0	-1.0	-2.0
	25	25 Mn(III)	72.7	84.0	166.2	94.0	-3.0	-6.0
Effluent (WSW)	0	0	4.3		25.2			
	25	25 Co(III)	25.6	42.6	76.9	102.4	1.2	2.4
	25	25 Co(II)	40.9	73.2	73.0	94.6	-2.7	-5.4

[a]Mean Error, $\bar{E} = \bar{x} - T.V.$, where $\bar{x}$ is the mean of spike results (all experiments were run in duplicate).
[b]Relative Error, RE = (E/T.V.) x 100.

$$\text{Simple CN} = 100 \times \frac{\text{Simple CN Measured - Sample Simple CN}}{\text{CN Spiked}}$$

$$\%\ \text{Total CN} = 100 \times \frac{\text{Total CN Measured - Sample Total CN}}{\text{CN Spiked}}$$

The simple cyanide determination shows variable recoveries of the total spiked amount, being dependent upon the strength of the metal-cyano-complex and the ratio of the complex in the spike. As expected, recoveries with strong complexes are lower than with weaker complexes. Generally, recoveries with weak complexes should have been 100% which was the case from distilled water, tap water and river water. However, it was not true with raw sewage and plant effluent where the recoveries were lowered by 15 to 30%. It appears that such samples contain sequestering components which complex the cyanides, making them unavailable for the simple cyanide determination. The total cyanide determination shows essentially complete recoveries of the spiked cyanide with all types of samples; the average was 97%.

The total cyanide recoveries were utilized for evaluation of the accuracy of the automated cyanide system. Table III also shows the mean error ($\bar{E}$) and the relative error (RE) of the measured recovered cyanide in relation to the true value (T.V.) of the cyanide spike. The overall errors for 101 experiments using a wide variety of samples and different cyanide types were -0.9 μg/l for the mean error, and -2.4% for the relative error.

Reproducibility and Sensitivity

Reproducibility studies were run for the cyanide range 1 to 100 μg/l using KCN and $[Co(CN)_6]^{3-}$ for simple and total cyanides. Ten to twelve replicates were used for each concentration considered. Some of the peaks obtained are shown in Figure 3. The simple cyanide system gave a standard deviation (σ) $\leqslant$ 0.8 μg/l and a coefficient of variation (COV) $\leqslant$ 2.3%. The total cyanide system gave a $\sigma \leqslant$ 1.8 μg/l and a COV $\leqslant$ 6.8%. Reproducibility studies with field samples, river water, Sanitary and Ship Canal water, raw sewage and treated effluents, gave an overall $\sigma <$ 1.0 μg/l and COV < 5% for both simple and total cyanide determination. There was no apparent relation between reproducibility and the nature of the samples.

The system can analyze up to 20 samples per hour and calibration is linear throughout the ranges 0 to 50 and 0 to 100 (Figure 3) and up to 500 μg/l CN. The lower limit of detection, defined as the signal

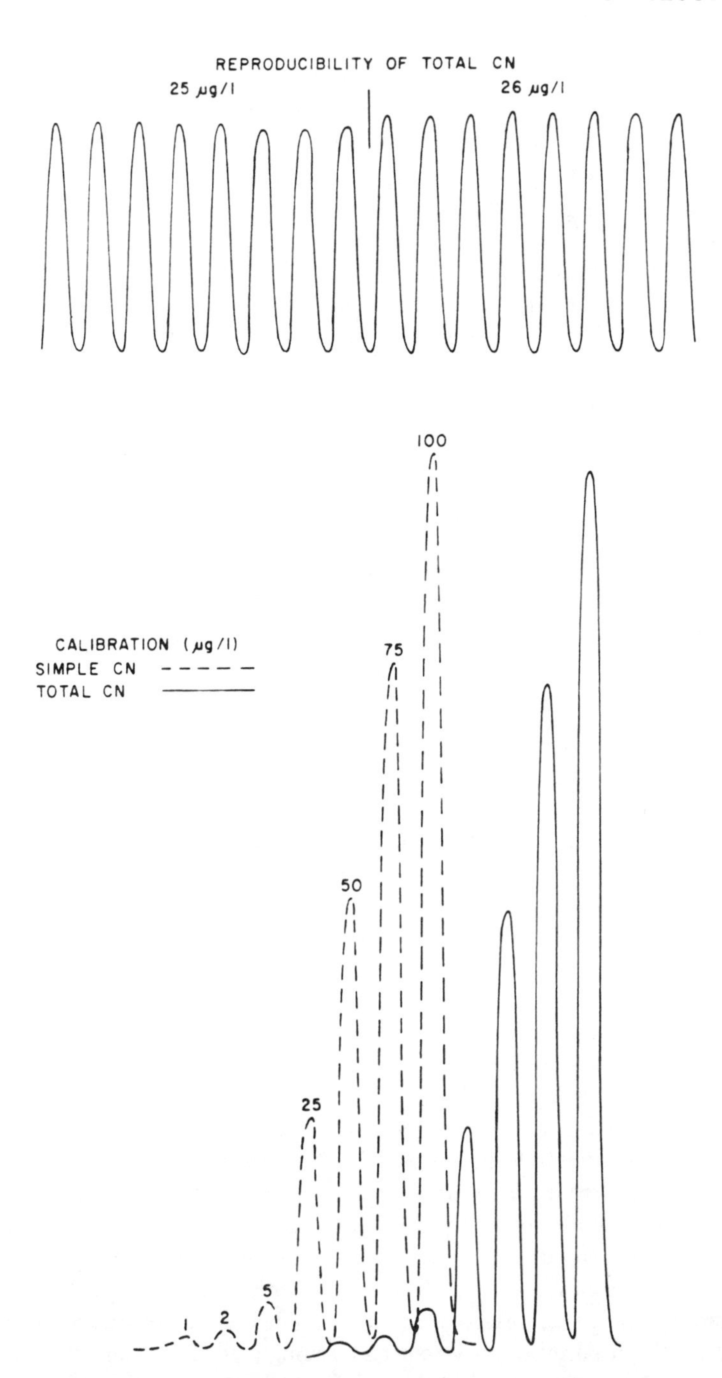

Figure 3. Reproducibility and calibration recorder tracing.

which is double the baseline noise, was found to be 0.25 μg/1 CN. Hence, one analysis requires 3 ml of sample and the absolute detection limit is less than one nanogram, or rather 0.75 ng. The sensitivity, defined as the slope of the calibration curve, was found to be 0.012 A/μg/l (absorption units per μg/l). In other words, one absorption unit is equivalent to 85 μg/1.

The system is capable of distinguishing between small differences in cyanide concentration, such as between 1 and 2 μg/1 and between 25 and 26 μg/1 (Figure 3). Statistical analysis and student t-test of the results of ten replicates of each cyanide concentration indicate that the confidence livel, to distinguish a difference of 1 μg/1 at the levels examined, is higher than 99.9% ($P < 0.001$) for both simple and total cyanides.

Interferences

The effects of substances which have been reported to cause some interferences[3,4,8] and which may be found in natural waters, raw sewage and industrial and treated effluents, were studied with the following results:

1. Turbidity-contributing substances are automatically removed through the thin-film distillation step.

2. Metal cations such as those of iron, zinc, calcium and magnesium did not have any effect at the 1000 mg/1 level. However, mercuric and cuprous chlorides have a significant negative effect and should not be used as catalysts in the distillation step as in some manual procedures.[3-5] The present MSDGC automated method does not utilize any such catalysts, as the breakdown of strong cyanide complexes is by UV irradiation.

3. Anions such as carbonate, sulfate and chloride did not interfere at levels comparable to or higher than commonly found in water and wastes. Sulfide, however, interferes significantly. The reported methods for treatment of sulfide interference[3,4,8] were found to be unsatisfactory. Cyanide losses from 20 to 60% were observed, probably due to absorption of the cyanides on the precipitated materials. The present approach is to eliminate sulfide interference by controlling the concentration of the chloramine-T solution. At a chloramine-T concentration of 0.4% (w/v) no intereference was observed up to a sulfide ion concentration of 20 mg/l on the total cyanide system and 15 mg/l on the simple cyanide system. Thus, when sulfide is present in high concentrations the interference is removed by dilution of samples.

4. Oxidizing materials, such as oxygen, ozone and chlorine, may oxidize both the free and some complex cyanides and result in lower cyanide values. Proposed treatments[3,4,8] with ascorbic acid, sodium sulfite and bisulfite as well as with stannous chloride and hypophosphorus acid were found to be unsatisfactory. They either introduced interferences or did not demonstrate efficient reduction. Oxalic acid was also examined and 2 g/l or more was required to reduce 50 μg/l chlorine. The reaction was shomewhat slow and required low pH. Therefore, oxalic acid, when used, should be added to the sample 20 to 30 min before preservation with sodium hydroxide. Sodium arsenite was found efficient in reduction, only 0.1 g/l being required to reduce 50 mg/l chlorine. In addition, the reaction was rapid and pH-independent. Therefore, the order of addition of sodium arsenite and hydroxide is immaterial. However, in a few cases the treatment of field samples with arsenite resulted in unexpectedly high cyanide values. Accordingly, it is advisable to analyze the samples containing oxidants as soon as possible.

Applicability and Cyanide Species Distribution

The method described above is applicable to a wide variety of samples and has been utilized to monitor cyanide species in different environmental media. In addition to water and wastewater, attention has been given to sediments. Some representative results for simple, complex and total cyanides in several field samples are given in Table IV. It was noticed that sediments from rivers and water channels contained levels of cyanide

Table IV. Cyanide Species Distribution in Water, Wastewater and Sediments

Sample	Total CN (μg/l)	Simple CN (μg/l)	Complex CN (%)
Raw Sewage (SW Influent)	119.2	30.2	75
Treated Effluent (SW)	25.7	4.3	83
Calumet River:			
Surface Water	8.4	2.1	75
Bottom Sediment[a]	0.16[a]	0.02[a]	88
Sanitary & Ship Canal:			
Surface Water	19.6	4.9	75
Bottom Sediment[a]	12.7[a]	0.9[a]	93
Calumet Lagoon Sludge Sediment[a]	160.6[a]	21.6[a]	87

[a]Concentration in ppm on dry weight basis.

significantly higher than those found in the overlying waters. This could be expected as various components are concentrated within the bottom sediments. Lagoon sludge sediments from the Calumet Treatment Plant have also shown high cyanide concentrations, from 80 to 200 mg/l. It is interesting to note that in natural waters, raw sewage, treated effluents and sediments more than 75% of the cyanides present were complexed.

COMPARATIVE STUDIES BETWEEN METHODS

Comparison with *Standard Methods*

The MSDGC automated method was compared for precision with the manual *Standard Methods* procedure which is widely used. A variety of water and wastewater samples were analyzed in duplicate for total cyanides by both methods and the results are given in Table V. Differences between duplicate results were compared statistically and are summarized as follows.

Statistic (μg/l)	MSDGC Method	*Standard Methods*
Range of Differences	0.1-14.1	0.0-51.0
Mean of Differences	2.6	11.4
Standard Deviation	3.9	14.3

The fact that all three statictics were of significantly higher magnitude with the *Standard Methods* procedures indicates that the MSDGC automated method is more precise.

The correlation between the two methods was evaluated by determining the linear correlation coefficient, r, between paired total cyanides for 80 water and waste samples. It was found to be 0.473; this value, though not very high, is statistically significant at the 1% level. Then the data were separated into two sets: the lower 40 cyanide values ($\leqslant$ 35 μg/1) and the higher 40 values (36-165 μg/1). The correlation at the low cyanide level was not significant, $r = 0.245$, but with the higher cyanide values the correlation was significnatly higher, $r = 0.583$.

It was also noticed that the total cyanide values obtained by the MSDGC method were generally higher than with *Standard Methods.* This could be attributed to the complete breakdown of the strong cyanide complexes by UV irradiation as well as efficient cyanide recovery.

Comparison With ASTM

The MSDGC automated method for "simple cyanide" was compared with the ASTM, Designation D2036-B "cyanide amenable to chlorination."

Table V. Comparison Between the MSDGC and *Standard Methods*. Duplicate Analyses in Water and Wastewater Samples

		Total Cyanide Determination			
		MSDGC Automated System		*Standard Methods*	
Sample Type	Sample Number	A(μg/l)	B(μg/l)	A(μg/l)	B(μg/l)
Tap Water, Dechlorinated	1	2.5	1.5	6	7
	2	2.1	1.4	2	5
	3	2.9	2.8	4	0
Des Plaines River	1	12.0	13.0	9	0
	2	8.5	8.8	3	2
Sanitary & Ship Canal	1	20.8	19.5	9	7
	2	26.4	26.6	11	8
Raw Sewage, W.S.W.	1	94.3	98.8	52	39
	2	186.0	185.0	149	130
	3	55.3	56.9	48	55
	4	46.8	47.4	35	37
Lemont Effluent, Nitrified	1	31.8	32.4	12	44
	2	53.7	53.0	18	7
	3	53.8	52.9	10	14
Calumet Effluent, Nitrified	1	116.0	129.8	100	128
	2	158.0	148.8	131	80
	3	89.0	88.2	68	68
	4	148.0	144.4	88	40
Calumet Effluent, Secondary	1	34.5	36.2	40	40
	2	92.8	106.9	92	100
	3	40.8	39.6	45	40
WSW Effluent, Secondary	1	29.0	32.0	32	36
	2	27.6	27.5	31	24
	3	28.2	27.1	17	5

The latter relies on chlorination of a part of the sample under alkaline conditions and is the difference between total cyanide measurements before and after chlorination. Several synthetic cyanide mixtures as well as a variety of wastewater samples were analyzed by the two methods, and some representative results are shown in Table VI. Both methods give comparable results with the synthetic samples, except in the presence of thiocyanate which with the ASTM method gives higher results. Another advantage with the MSDCG method is that simple cyanides amenable to chlorination are estimated as the difference of two total cyanide measurements before and after chlorination.

Wastewater samples, raw sewage, treated effluents and industrial wastes show some erroneous results with the ASTM method (Table VI). Higher

Table VI. Analysis for Cyanides Amenable to Chlorination and Simple Cyanides

Sample	Total CN: Before Chlorination (μg/l)	Total CN: After Chlorination (μg/l)	CN Amenable to Chlorination (μg/l)	Simple CN (μg/l)
Synthetic CN Mixture				
KCN, Mn(III) Cd(II), Ni(II)	100.3	1.6	98.7	101.5
Fe(II), Fe(III) Co(III), AN[a]	76.0	72.7	3.3	2.1
Co(III), Fe(II), SCN^-	99.5	49.4	50.1	1.0
KCN, Co(III) ACH[b], SCN^-	98.6	23.8	74.8	50.4
Wastewater Samples				
Raw Sewage #1	99	89	10	13
#2	185	246	-61	44
Treated Effluent #1	42	82	-40	9
#2	24	24	0	4
Industrial Waste #1	227	990	-763	18
#2	119	10	109	111

[a]AN: acetonitrile.
[b]ACH: acetonecyanohydrin.

total cyanide values were obtained frequently after chlorination, leading to negative values for the cyanides amenable to chlorination. These observations have also been reported by several laboratories and the proper reaction mechanisms have not yet been elucidated. These anomalies were never observed with the MSDGC method for simple cyanide.

DETERMINATION OF OXIDIZABLE CYANIDES

The cyanides amenable to chlorination have been used as an estimate of the easily dissociable cyanides and to evaluate the efficiency of industrial waste treatment process. However, as mentioned above, the method proved to be unsatisfactory with doubts about the chlorination procedure itself and its products which apparently interfere with cyanide measurement. Alternative oxidizing agents such as permanganate, hydrogen peroxide and ozone were evaluated. Ozone was found to be most suitable for the oxidation of cyanides. The products of ozonation do not include undesirable intermediates such as cyanogen chloride, CNCl, a gas produced by chlorination with limited solubility and much more toxicity than cyanide. Nor do the ozonation products appear to interfere with cyanide measurement. Instead, cyanides are oxidized directly to the safer cyanate ion, CNO^-. It should be noted that thiocyanate is eventually oxidized to CNO^- by both chlorination and ozonation. The following equations compare and summarize both oxidation reactions

Chlorination:
$$CN^- + Cl_2 \longrightarrow CNCl + Cl^-$$
$$CNCl + 2OH^- \longrightarrow CNO^- + H_2O + Cl^-$$
$$CNS^- + 3Cl_2 + 2H_2O + 2OH^- \longrightarrow CN^- + SO_4^{2-} + 6HCl$$

Ozonation:
$$CN^- + O_3 \longrightarrow CNO^- + O_2$$
$$CNS^- + 3O_3 + 2OH^- \longrightarrow CN^- + SO_4^{2-} + 3O_2 + H_2O$$

Factors related to the oxidation reaction kinetics with ozone were optimized. At room temperature ozonation was more effective than at higher temperatures (50°C and 80°C). With synthetic samples, 3 min of contact were sufficient. However, with different types of field samples 10 min was required and cyanide complexes such as those of iron and cobalt did not render themselves to oxidation within 30 min.

Oxidation of synthetic solutions of cyanides proceeds to completion at alkaline pH values. However, to introduce effective ozonation of some field wastewater samples it was necessary to raise the pH to 13. Figure 4 shows that KCN and Ni(II) complex as well as thiocyanate were effectively oxidized, whereas the strong metal-cyano-complexes of Co(III) and Fe(III) were not oxidized at any pH.

THIOCYANATE DETERMINATION

From the above discussion it can be concluded that the parameters "simple cyanides" and "cyanides amenable to ozonation" measure the same cyanide species with the exception of the "thiocyanate," which is not measured at all with the simple cyanides, but was included completely in the cyanides amenable to ozonation. In summary, to determine thiocyanate, a given sample is subjected to analyses for total cyanides without ozonation and for total cyanides after ozonation. The difference equals the cyanides amenable to ozonation. Then the third run is for simple cyanides without ozonation. It is expected, within the experimental variations, that the cyanides amenable to ozonation should be equal to or larger than the simple cyanides depending on the presence or absence of thiocyanate in the sample, and the difference between the two equals the thiocyanate.

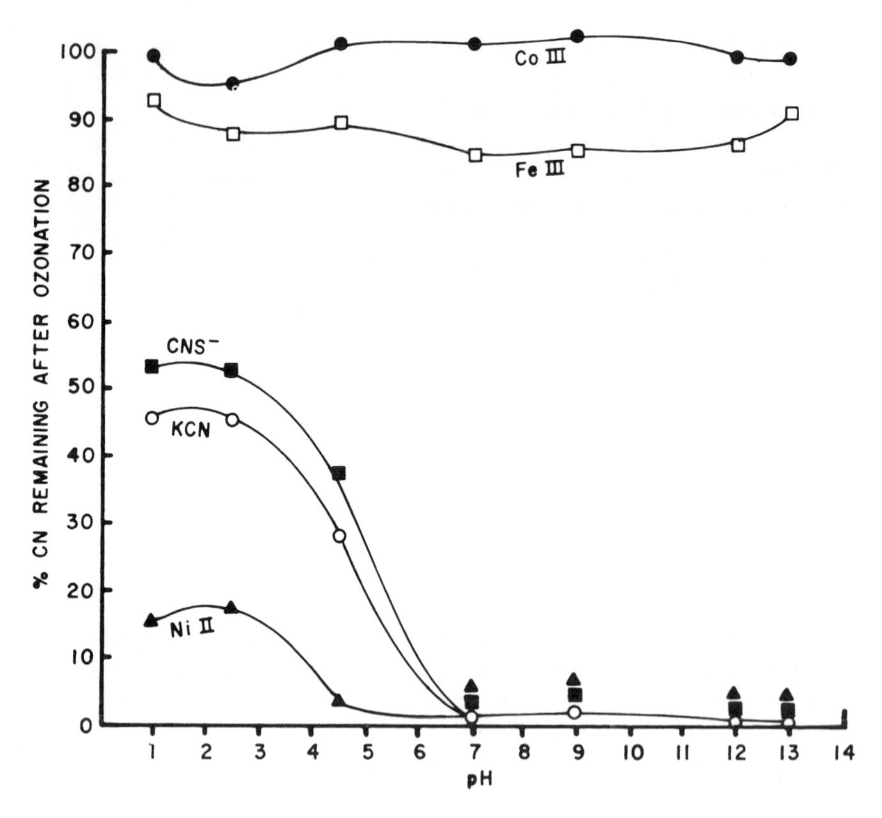

Figure 4. Effect of pH on ozonation of simple and complex cyanides and thiocyanate.

Figure 5 shows peaks obtained from three runs for each of two different samples, North Side Plant treated effluent and Calumet Plant raw sewage. The first sample contained essentially no thiocyanate and the second sample contained thiocyanate of more than 200 μg/1 (measured as CN), which constituted about 58% of its total cyanide content.

The method was applied to synthetic mixtures and the results were excellent as shown in Table VII. The thiocyanate determined was always within about 5% of the expected value. Experience with wastewater from MSDGC treatment plants was satisfactory. Samples of raw sewage and treated effluents from John Egan Water Reclamation Plant, Hanover Park, North Side, West-Southwest and Calumet plants were analyzed for the different cyanide species and thiocyanate and the results are presented in Table VIII. The total cyanide content from areas which are basically residential was significantly lower than from heavily industrialized areas such as Calumet. It could be observed that for residential

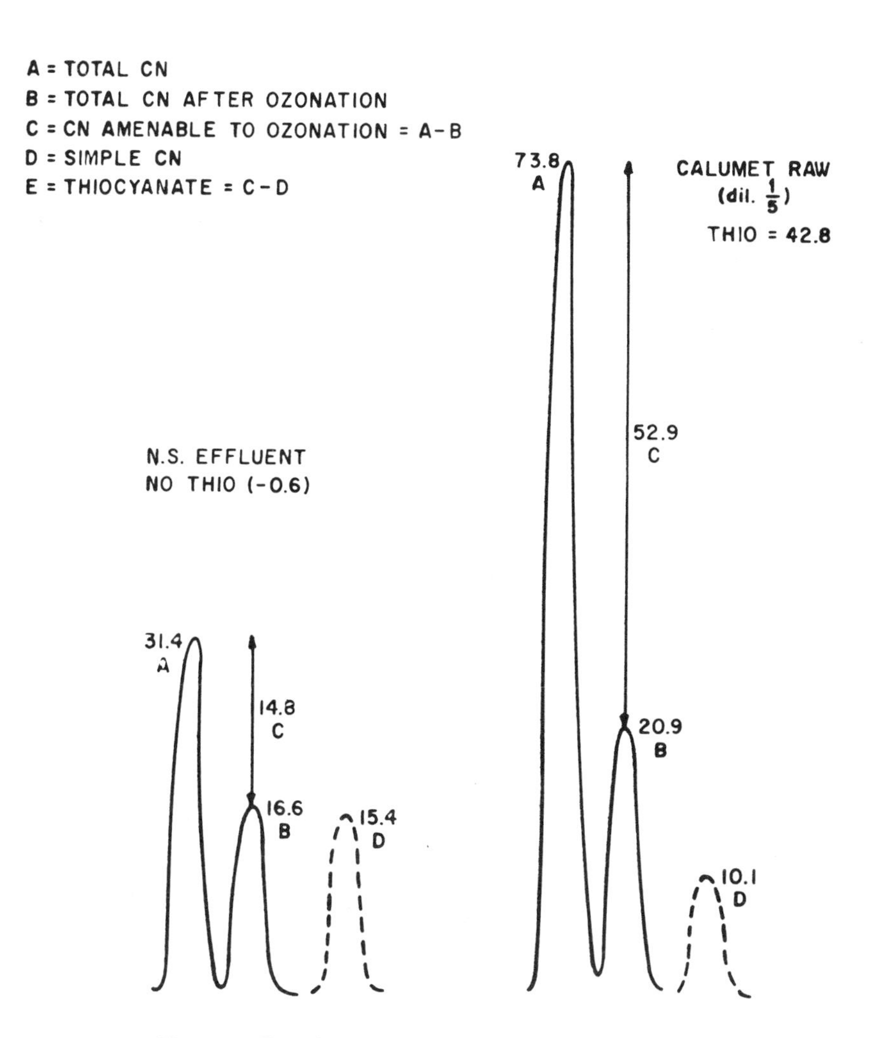

Figure 5. Cyanides amenable to ozonation and thiocyanate.

areas the cyanides amenable to ozonation are not much higher than simple cyanides yielding thiocyanate values of about 15% of the total cyanides, whereas the Calumet plant, which receives discharges from petroleum and steel industries, shows a high level of thiocyanate which amounts to more than 50% of the total cyanides. It is interesting to notice that the thiocyanate content of the Calumet effluent was about 65% before the final treatment with chlorine and then dropped to less than half its concentration (about 30%) after the routine operation of chlorination. Knowing the problem of chlorination products and their undesirable

Table VII. Cyanides Amenable to Ozonation and Thiocyanate in Synthetic Mixtures[a]

Mixture[b]	Total CN	Total After Ozonation	CN Amenable to Ozonation	Simple CN	Thio CNS^-
I.	95.6	47.4	48.2	48.3	-0.1
II.	101.6	48.9	52.7	52.0	0.7
III.	97.6	24.8	72.8	24.6	48.2
IV.	101.6	24.5	77.1	26.8	50.3
V.	102.7	2.3	100.4	50.7	49.7
VI.	96.5	24.0	72.5	23.8	48.7
VII.	102.6	1.8	100.8	102.0	-1.2

[a]Concentration in μg/1 as CN.
[b]Mixture Composition:
I. KCN: 50, Co(III): 49.0
II. KCN: 50, Fe(II): 48.6
III. KCN: 25, Co(III): 24.5, CNS^-: 51.2
IV. Ni(II): 24.7, Fe(III): 24.2, CNS^-: 51.2
V. KCN: 25, Ni(II): 24.7, CNS^-: 51.2
VI. ACH[c]: 24.0, Co(III): 24.5, CNS^-: 51.2
VII. KCN: 25, Cd(II): 26.5, Mn(III): 25.8, Ni(II): 24.7
[c]ACH: Acetonecyanohydrin.

Table VIII. Thiocyanate and Cyanide Species in Raw Sewage and Treated Effluents[a]

Sample	Total CN	Total After Ozonation	CN Amenable to Ozonation	Simple CN	Thio (SCN^-)	Thio Total
Egan Raw	43.5	28.0	15.5	10.5	5.0	11.5%
Egan Effluent	64.7	49.4	15.3	4.1	11.3	17.5%
Hanover Park Raw	69.8	40.6	29.2	18.9	10.3	14.8%
Hanover Park Effluent	29.8	18.1	11.7	6.2	5.5	16.8%
Northside Raw	106.8	39.9	66.9	59.7	7.2	6.7%
Northside Effluent	31.5	16.6	14.9	15.5	-0.6	0.0%
W.S.W. Raw	46.9	26.7	20.2	12.1	8.1	17.3%
W.S.W. Effluent	19.5	13.4	6.1	3.5	2.6	13.3%
Calumet Raw	369.0	104.3	264.7	50.3	213.4	57.8%
Calumet Effluent Before Chlorination	323.8	83.0	240.8	31.5	209.3	64.6%
Calumet Effluent After Chlorination	219.0	86.0	133.0	66.8	66.2	30.2%

[a]Concentration in μg/1 as CN.

effects on cyanide measurements it could be expected that if ozonation were to replace chlorination, both simple cyanides and the cyanides amenable to ozonation would drop significantly. This would decrease also the total cyanide levels to a certain extent depending on the ratio of the strongly complexed cyanides.

Analyses of some industrial wastes for thiocyanate and cyanide species are presented in Table IX. It is clear that in steel and petroleum effluents the thiocyanate content is significantly high, up to 90%, whereas in metal electroplating wastes, as expected, thiocyanate was not found.

Table IX. Thiocyanate and Cyanide Species in Some Industrial Wastes[a]

Industry Type	Total CN	Total After Ozonation	CN Amenable to Ozonation	Simple CN	Thio SCN^-	Thio Total
Steel Mill	17.50[b]	2.59[b]	14.91[b]	3.47[b]	11.44[b]	65.4%
Petroleum Refinery	3.30[b]	0.25[b]	3.05[b]	0.06[b]	2.99[b]	90.6%
Electro Plating	99.2	8.1	91.1	94.0	-2.9	0.0%
Metal Polishing and Plating	165.5	118.0	47.5	20.0	27.5	16.6%
Others	26.2	10.0	16.2	16.1	0.1	0.4%

[a]Concentration in μg/l as CN.
[b]Concentration in mg/1 as CN.

REFERENCES

1. Hewitt, P. J., and H. B. Austin. *Water Poll. Control* 381 (1972).
2. Chamberlin, N. S., and H. B. Snyder, Jr. *Proc. Purdue Ind. Waste Conf.* 10:277 (1955).
3. *Standard Methods for the Examination of Water and Wastewaters* 13th ed., (Washington, D.C.: American Public Health Association, 1971).
4. American Society for Testing and Material. *Annual Book of ASTM Standards,* Part 23, Designation D2036-72 (1972).
5. "Methods for Chemical Analysis of Water and Wastes," USEPA Water Quality Office, Cincinnati, Ohio (1971).
6. Robert, R. F., and B. Jackson. *Analyst* 96:209 (1971).
7. Conetta, A., J. Jansen and J. Salpeter. *Pollution Engineering* (January 1973), p. 36.
8. Goulden, P. D., B. K. Afghan and P. Brooksbank, *Anal. Chem.* 44:1845 (1972).
9. Ryan, A. J., and G. W. Culshaw. *Analyst* 69:370 (1944).
10. Aldridge, W. N. *Analyst* 69:262, (1944).

11. Epstein, J. *Anal. Chem.* 19:272 (1947).
12. Asmus, E., and H. Garschagen. *Z. Anal. Chem.* 138:414 (1953).
13. Lue-Hing. C., D. T. Lordi and E. Knight. "Report on Cyanide Studies," Metro. San. Dist. of Greater Chicago (October 1973).
14. Casapieri, P., R. Scott and E. Simpson. *Anal. Chem. Act.* 49:188 (1970).
15. Afghan, B. K., P. E. Goulden and J. F. Ryan. *Advances in Automated Analysis,* Vol. II, (Miami, FL: Thurman Associates, 1970), p. 241.
16. Bark, L. S., and H. G. Higson. *Talanta* 11:471 (1964).
17. Mirsch, E. *Fortschr-Wasserchem* 1:120 (1964).
18. Sillen, L. G., and A. E. Martell. *Stability Constants of Metal Ion Complexes,* Special Publication 17, Chemical Society (London: Burlington House, 1964), p. 107.
19. Broderius, S. J. Ph.D. Theses, Oregon State University (1973).
20. Henrickson, T. N., and L. G. Daignault. "Treatment of Complex Cyanide Compounds for Reuse or Disposal," Report No. EPA-R2-73-269 (1973).

CHAPTER 21

INVESTIGATION OF CHEMICALLY MODIFIED FORMS OF PEAT AS INEXPENSIVE MEANS OF WASTEWATER TREATMENT

Edward F. Smith* and Harry B. Mark, Jr.
Department of Chemistry
University of Cincinnati
Cincinnati, Ohio

Patrick MacCarthy
Department of Chemistry and Geochemistry
Colorado School of Mines
Golden, Colorado

INTRODUCTION

Methods for removing pollutants from industrial effluents must be efficient but yet as simple and inexpensive as possible. A possible way of achieving this is to utilize naturally occurring materials which are low in cost due to their availability. With this in mind, chemically modified forms of peat have been investigated for their ability to remove impurities from wastewater. In order to maintain the economic advantage of the peat, the modification procedures were kept relatively simple and utilized inexpensive reagents.

Peat may be roughly defined as the product resulting from a slow and partial decay of dead vegetation. It is usually formed in areas where water saturates or completely covers the dead plant material that has accumulated on the ground. The water blocks the action of aerobic bacteria and in turn greatly inhibits the rate of decay of the plant debris. Peat therefore consists of the following components[1]:

*Present Address: Exxon Chemical Company, U.S.A., Baton Rouge, LA.

1. Organic matter in an organized state of preservation and, therefore, identifiable.
2. Organic matter which has undergone considerable breakdown but in which cell structure is still visible.
3. Organic matter which has been degraded below the cellular level, composing what might be termed "humus" and often forming a peat matrix.
4. Inorganic matter derived either from dust or inwash, or from the cells of some plants.

The proportions of each of these components will vary from one peat sample to another depending on the nature of the peat-forming material and the rate of decomposition. Thus, peat is a chemically heterogeneous and very complex material.

Approximately 1.5% of the earth's surface is covered with peat while the largest deposits occur in the northern parts of the northern hemisphere.[1] Canada and the U.S.S.R. account for approximately 80% of the total. In the United States, Alaska has the most extensive peat areas along with large areas in both Minnesota and Michigan.[2] Peat, therefore, is an abundant material which is widely available and relatively easy to obtain.

The principal uses of peat have been as a fuel and agriculturally for soil conditioning. Fogarty *et al*[3] have described other actual and potential uses which have been considered for a variety of environmental and industrial applications. This study involves the use of modified forms of peat as an inexpensive cation exchanger, anion exchanger and oil coalescer which could ultimately be utilized in a flow system.

COMPARISON OF PEAT SAMPLES

As indicated previously, peat is a very complex and heterogeneous mixture whose composition is dependent on both the extent of decay and the nature of the material from which it originated. It is interesting therefore, to compare various peats. In this study, three samples of Irish peat and one of Michigan peat were utilized and their characteristics were compared. Irish blanket-bog peat was obtained at the Irish Agricultural Institute, Peatland Experimental Station, Glenamoy, County Mayo, Ireland. A sample of Michigan peat was obtained from the Anderson Peat Company, Imlay City, Michigan.

The three Irish peat samples were very similar in their appearance and were decomposed to a relatively homogeneous state. These samples, which were stored in sealed containers upon removal from the bog, had approximately the same water content (87%). The Michigan peat, on the other hand, was more heterogeneous in that it contained pieces of stones and wood chips. The water content for this peat was also much lower than for the Irish peat (67%) because it is sun-dried, screened and milled before being bagged for retail distribution.[4]

When the samples were dried, hard conglomerates formed and these had to be ground before sieving. The Michigan peat was much easier to grind than the others. The resulting particles had very irregular shapes and surfaces. No leaching from the dried and sieved Irish peat was observed in acidic solutions. The Michigan peat initially leached extensively in all solutions. The leaching gradually subsided somewhat after extensive rinsing. All of the particles swelled on wetting and at higher pH.

As peat is a natural cation exchanger, the effective sodium and copper capacities are shown in Tables I and II. In comparing the capacities, the samples appear to be very similar as to the number of available exchange sites. The higher capacities for the divalent cation indicate a wide variety in the types of exchange sites. Also, the exchange capacities have been found to increase with the degree of decomposition of the peat.[5]

The infrared spectra as shown in Figure 1 also reflect the basic similarity of the samples. The bands are broad due to the overlapping of absorptions from the wide range of basically similar groups. The bands correlate well with the spectra obtained by Schnitzer and Kahn[6] and

Table I. Comparison of the Effective Sodium Capacities (meq/g) for Sulfuric Acid-Treated Peat[a]

Peat	No Acid Treatment	Dried Peat (18-35 U.S. Mesh)			Wet Peat	
		100°C	150°C	200°C	150°C	200°C
I.P. #1	0.35	0.93	0.77	1.22	1.18	1.37
	0.39	0.95	0.75	1.41	1.18	1.40
	0.42		0.78		1.22	
I.P. #2	0.32	0.82	0.75	1.23	0.41	0.87
	0.34	0.77	0.81	1.20	0.45	0.98
I.P. #3	0.22	0.45	0.87	1.16	0.49	0.58
	0.22	0.64	0.91	1.22	0.54	0.71
	0.21		0.85			
	0.21		1.08			
			1.07			
M.P.	0.39	0.92	0.63	0.90	0.40	0.76
	0.41	0.67	0.71	0.96	0.41	0.81
			0.79			
			0.87			
			0.90			

[a]Treatment of 1.20 g (dry weight) with 4.0 ml of sulfuric acid.

Table II. Comparison of the Effective Copper(II) Capacities (meq/g) for Sulfuric Acid-Treated Peat[a]

	No Acid Treatment		Dried Peat (18-35 U.S. Mesh)						Wet Peat			
			(100°C)		(150°C)		(200°C)		(150°C)		(200°C)	
Peat	Tit.	Pol.	Tit.	Pol.	Tit.	Pol.	Tit.	Pol.	Tit.	Pol.	Tit.	Pol.
I.P. #1	0.57	0.55	1.68	1.63	1.28	1.30	2.02	2.02	2.40	2.40	2.48	2.51
	--	--	1.75	1.75	1.24	1.27	1.66	1.65	2.30	2.41	2.77	2.83
	--	0.79			1.26	1.30			2.38	2.31		
									2.82	2.75		
									2.60	2.60		
I.P. #2	0.48	0.46	1.44	1.33	1.30	1.31	1.94	1.81	0.98	1.01	1.76	1.68
	0.54	0.51			1.43	1.44	1.81	1.71	0.93	0.93	1.98	1.97
I.P. #3	0.43	0.38	0.87	0.81	1.40	1.41	1.92	1.77	0.97	0.97	1.15	1.13
	0.45	1.40			1.47	1.47	2.06	2.02	1.04	1.03	1.40	1.39
	0.36	0.33										
	0.36	0.32										
M.P.	0.80	0.80	1.40	1.27	1.11	1.12	1.81	1.81	0.95	1.00	1.77	1.74
	0.81	0.81			1.27	1.25	1.95	1.88	0.93	0.95	1.79	1.75
					1.66	1.53			1.40	1.40		

[a]Treatment of 1.20 g (dry weight) wigh 4.0 ml of H_2SO_4.

MacCarthy *et al.*[7] for fulvic and humic acids which are extracted from peat. The band at *ca.* 3450 cm^{-1} is due to hydrogen-bonded OH. The two bands around 2900 cm^{-1} indicate aliphatic groups while no aromatic hydrogen is observed. The absorptions at 1630 cm^{-1} and 1725 cm^{-1} are caused principally by carbonyl stretches from ketone and carboxyl groups.

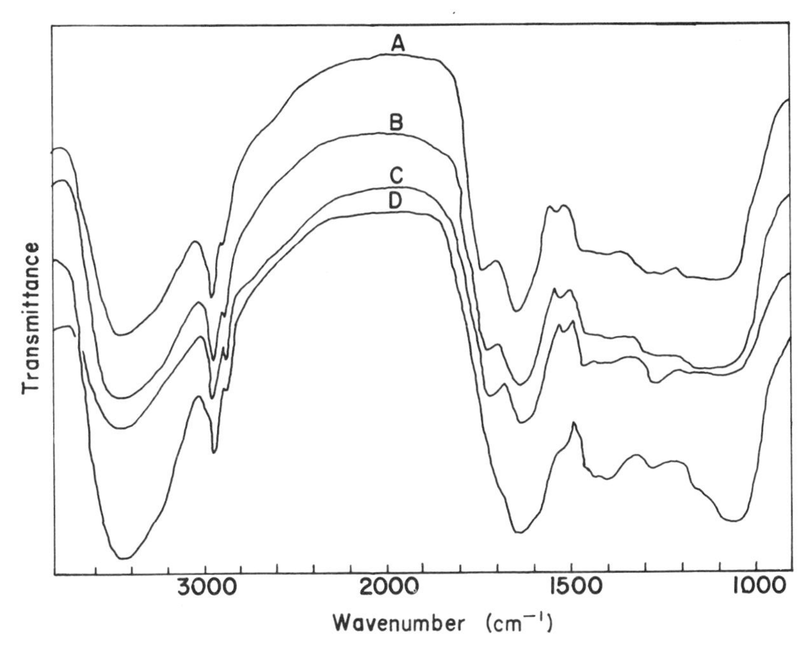

Figure 1. Infrared spectra of dried Irish peat (A–I.P. No. 1; B–I.P. No. 2; C–I.P. No. 3) and Michigan peat (D–M.P.).

The only significant difference between the spectra is the absence of the 1725 cm^{-1} band in the Michigan peat sample. Washing with acid causes this band to appear in the Michigan peat while leaving the Irish peat spectra unchanged. This type of behavior has been reported previously for the reaction of humic acids and peats with metal chlorides.[9] The band was assigned to the carboxyl stretching mode of un-ionized carboxyl groups. Upon neutralization or complexation of metals, the carboxyl groups are ionized and the 1725 cm^{-1} band is not observed. X-ray fluorescence data confirmed this by indicating a much higher calcium and iron content in the Michigan peat. Washing with acid reduced the calcium, zinc and iron content and restored the 1725 cm^{-1} band.

Although the original peat samples appear to be physically different, chemically they are more similar than would be expected. Peat nevertheless constitutes a complex mixture which remains to be resolved. Thus, chemical modification of these peat samples may result in wider variations between them.

CATION EXCHANGER

Introduction

In order to utilize ion exchangers for environmental applications on an industrial scale, low-cost processes for the production of ion exchange or chelating resins must be developed. A possible way for achieving this is to utilize certain inexpensive naturally occurring materials. Various agricultural products and by-products have been shown to possess exchange properties in both modified[10,11] and unmodified[12] forms. Coal treated with acid[13] has also been considered, and a sulfonated form is commercially available.[14]

Peat in its natural form has cation exchange properties.[15] Raw peat, however, is very impervious to flow and can be used in a flow system only after being dried. There have been numerous patents which indicate that peat exhibits enhanced cation exchange capabilities after treatment with acid, especially sulfuric acid.[16-19] The process is both simple and low in cost. Others have described the characterization of sulfuric acid-treated forms of peat from Germany,[20] Rumania,[21] Russia [22,23] and Pakistan.[24] Yet this modified form of peat is not commercially available. This study has investigated the sulfuric acid treatment of peats taken from Ireland and Michigan. In addition, the effects of phosphoric, chromic, formic and nitric acids have been examined. Of these, only chromic acid has been studied previously for the production of a peat cation exchanger.[24] A complete description of the experimental procedures used for this work has been published previously.[25-27]

The approach undertaken has been to optimize the modification processes not only in regard to exchange capacity but also with respect to the resulting physical and chemical characteristics. These are especially important as the product would ultimately be used in a flow system. Of course, the optimization must involve a compromise between the most desirable exchanger characteristics and the cost of production.

Besides the usual physical characteristics, such as swelling and flow, considered in the study of ion exchangers, we have also been concerned with the leaching properties of the modified peat. Fulvic acids, which are considered to be of significant environmental concern, are readily leached from untreated peat with water. Accordingly, if modified peat is to be considered for environmental applications, such leaching, as well as that due to soluble decomposition products from the sulfuric acid treatment, will present a problem by introducing organics into the effluent if proper precautions are not observed.

The Sulfuric Acid Treatment of Peat

Comparison of Effective Capacities

A very simple process was utilized for the sulfuric acid treatment of the peat. Samples in their original form were heated with concentrated acid in open beakers on a hot plate. Samples of peat which were initially dried and sieved to 18 to 35 U.S. mesh (1.0 to 0.5 mm diameter particles) were also used. After rinsing, drying and resieving to 18 to 35 U.S. mesh, the effective capacities of the modified peat were determined. Using the apparatus shown in Figure 2, an excess of a solution of an ion was passed through the H^+ form of the exchanger. Capacities were then determined by titrating the released protons. Alternately, the retained ions were eluted and determined polarographically or by titration. It should be pointed out that the 18 to 35 mesh size was used for these determinations because it is considered to be the lower limit for industrial applications.[28]

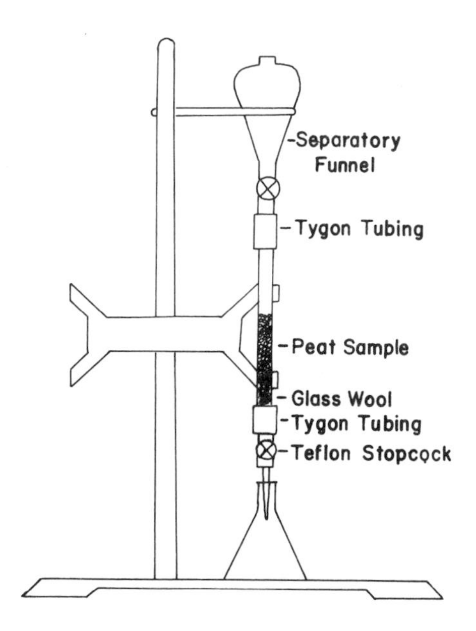

Figure 2. Apparatus for determination of effective capacities.

A comparison of the effective sodium capacities for the treatment of both dried and wet peat is shown in Table I. The data are for the use of 4 ml of H_2SO_4 per 1.20 g of peat (dry weight) for a 2.0-hr reaction period. These conditions were selected because larger amounts of acid and longer reaction times gave only slight increases in the capacities and were more likely to leach. The dried peat is somewhat more consistent and the effective capacity values are higher than the samples prepared from wet peat (except for I.P. No. 1).

The effective capacities for Ba^{2+}, Cu^{2+} and Pb^{2+} were also determined. The capacities increased in the following order:

$$Na^{+} < Ba^{2+} < Cu^{2+} < Pb^{2+}$$

The lead capacities were approximately twice the sodium capacities. Table II contains the effective copper capacities. The number of protons displaced from the columns by the copper and the equivalents of retained copper were found to be approximately equal. These data indicate that the modified peat contains a wide variety of exchange sites.

When determining the Pb^{2+} capacities, nitric acid was used to rinse the columns in place of hydrochloric acid. The strong oxidizing power of nitric acid caused the columns to leach and swell. The Na^{+} capacities were subsequently determined and were found to increase. This can be attributed to the swelling and oxidation of the peat. The oxidative process was minimized by rinsing the columns with water immediately after the acid rinse.

Recovery Efficiency

If modified peat is to be used in environmental applications, its ability to remove trace levels of cations efficiently must be determined. In an attempt to demonstrate the recovery efficiency, 100 ml of a 10 ppm Cu^{2+} solution were passed through the H^{+} form of the columns at a flowrate of 2 ml/min. The recovery efficiency was then calculated from the percentage of copper eluted relative to the total passed through the column. All of the samples shown in Table I had efficiencies of at least 95%. In a similar manner, a large volume (one liter) of a lower concentration of Cu^{2+} (1.33 ppm) was passed through these columns and good recovery efficiencies were again observed ($>$ 90%). The samples used for these determinations ranged from 0.2 to 0.8 g in weight and from 2 to 8 cm in height in the 6-mm i.d. columns.

The effect of the flowrate on the recovery efficiency was determined by varying the flowrate used for passing the 1.33 ppm Cu^{2+} solution though the columns. The samples weighed 0.22 to 0.36 g and had heights of 3.6 to 4.7 cm. Figure 3 is a plot of the results and indicates that relatively fast flowrates and a small amount of exchanger can still give a high recovery.

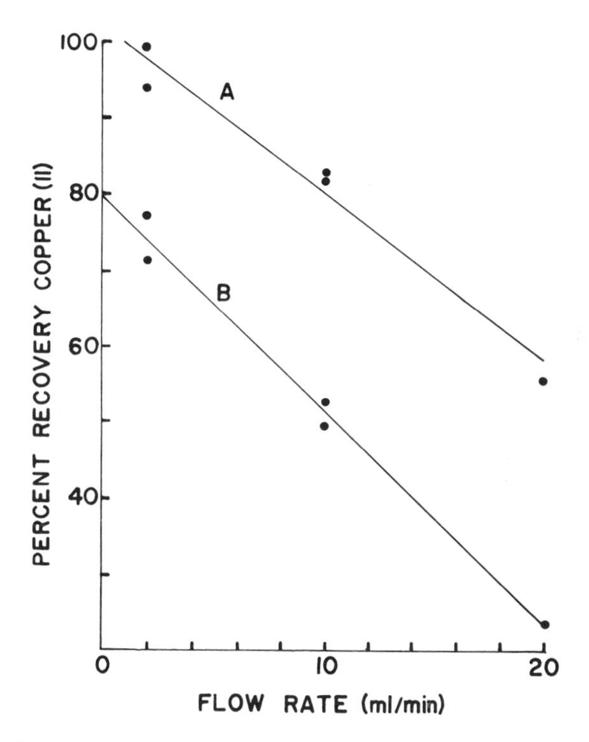

Figure 3. Effect of flowrate on the recovery efficiency of Cu(II) from (A) 1.0 l of 1.33 ppm Cu(II) and (B) 0.1 l of 9.91 ppm Cu(II) and 1.15×10^4 ppm Na(I).*

The selectivity of the recovery efficiency is also shown in Figure 3. The lower plot shows the efficiency for the recovery of Cu^{2+} (9.91 ppm) from 100 ml of a solution which also contained an excess of Na^+ (1.15×10^4 ppm). The presence of the counter ion lowers the efficiency slightly but good recoveries are still achieved.

Physical Characteristics

One of our main concerns was the conditions under which the modified peat would leach. Up to this point in our experimental investigation, only the samples which were dried and reacted with larger amounts of acid were found to leach in distilled water (pH 5-6). Thus, the samples

---

***Reprinted from Reference 26, p. 636, by courtesy of the Water Pollution Control Federation.**

shown in Table I were subjected to 100 ml of four alkaline buffers (pH 7,8,9 and 10) at a flowrate of 2 ml/min. All of the samples began leaching by pH 8.

The effluents that resulted from passing the four buffers through the columns were collected and combined. In order to obtain an indication of the amount of leached material, the absorbance at 260 mm was measured for each sample (Table III). These values give an average of the absorbance of each of the four individual buffer effluents. The data indicate that the milder conditions and the use of dried peat gave the lowest overall leaching. For the Michigan peat (M.P.) and the Irish peat (I.P. No. 3), the leaching properties were greatly improved by the sulfuric acid treatment.

The footnoted values in Table III indicate that those samples appeared to break down at pH 10 and caused a decrease in the flowrate through the column. In some cases the columns became completely blocked. Rinsing the columns with 3 *N* HCl removed most of the leached material and no further leaching was observed in distilled water.

In addition to leaching, the increased pH caused the peat to swell. The percentage change in height that resulted from raising the pH to 10 for each sample is given in parentheses after its absorbance value in Table III. Increased leaching was usually accompanied by increased swelling. Upon rinsing the columns with 3 *N* HCl, the leaching subsided but the samples did not return to their original heights. A significant increase in the effective sodium capacity was also observed due to the swelling. Drying the sample at 70°C returned them to their original sizes and capacity values.

Characterization of the Chemical Modifications

In order to characterize the changes which occur within the peat from the sulfuric acid treatment, infrared spectra were obtained (Figure 4). The carbonyl bands at 1730 cm^{-1} and 1620 cm^{-1} increased dramatically. This, along with the loss of the aliphatic hydrogen bands, suggests that an oxidation of the peat occurs. The band at *ca.* 1220 cm^{-1} is due to sulfur-oxygen stretches. Thus, the spectra suggest that both an oxidation and a sulfonation process are occurring. This conclusion is supported by the elemental analyses of these samples.

The data presented by Diaconescu and Chirac[21] also show that an oxidation of peat from the Dersca-Dorohoi region of Rumania is occurring with sulfuric acid treatment, and Helfferich[13] mentions a similar process occurring with the modification of coal. Thus, it appears that the sulfuric acid may sulfonate some of the available aromatic sites but, at the same

Table III. Leaching and Swelling of Sulfuric Acid-Treated Peat[a]

Peat	No Acid Treatment	Dried Peat (100°C)	Dried Peat (150°C)	Dried Peat (200°C)	Wet Peat (150°C)	Wet Peat (200°C)
I.P. #1	0.24 (22)	0.11 (6)	0.22 (4)	0.50[b] (47)	0.81 (10)	1.06[b] (62)
	0.27 (6)			0.43[b] (30)	0.54 (9)	1.81[b] (62)
					0.29 (8)	
					0.75 (17)	
					0.54 (10)	
I.P. #2	0.48[b] (2)	0.11 (7)	0.23 (5)	0.19 (18)	0.19 (5)	0.48 (6)
				>2.0[b] (26)		1.35[b] (15)
I.P. #3	0.76 (5)	0.29 (13)	0.09 (12)	1.32[b] (32)	1.08 (6)	0.22 (5)
	1.77[b] (7)		1.26 (39)			
M.P.	> 2.0[b] (53)	0.07 (10)	0.32 (6)	0.54 (33)	0.27 (6)	0.63 (9)

[a]Leaching given as absorbance at 260 nm for the combined effluents after passing pH 7, 8, 9 and 10 buffers hrough the columns. Swelling (values in parentheses) is given as percentage change in height.
[b]Column blocked at pH 10.

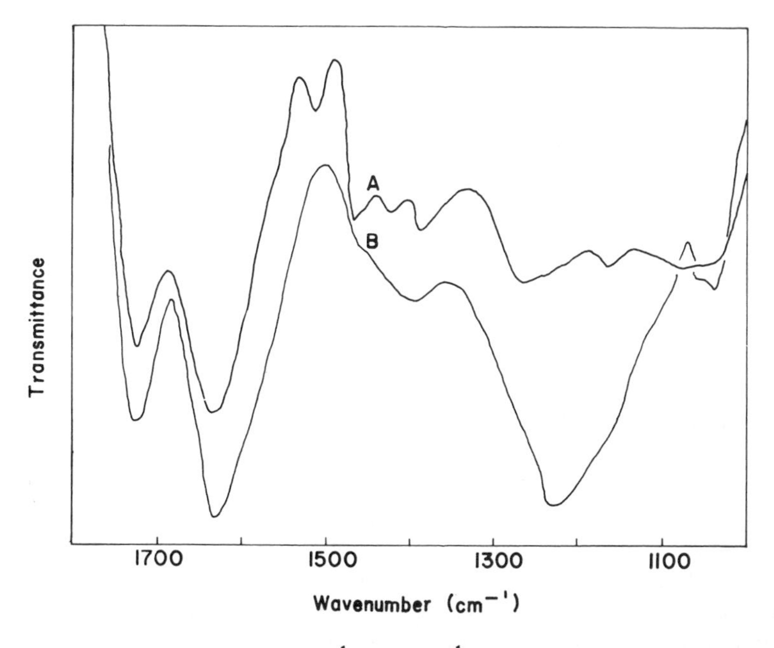

Figure 4. Infrared spectra (1800 cm^{-1} - 100 cm^{-1} region) for dried (A) I.P. No. 3 and (B) sulfuric acid-treated, dried I.P. No. 3 (150°C, 2 hr, 4 ml acid/ 1.20 g peat).

time, it is oxidizing other substituents to carboxyl groups. As the process which is occurring is not just a sulfonation process, we have refrained from referring to the sulfuric acid-treated peat as sulfonated peat; both terms have been used interchangeably in the literature.

Sulfuric Acid Treatment by Refluxing

Diaconescu and Chirac[21] employed refluxing for the sulfuric acid treatment of peat with the view that a constant and excess quantity of acid would be maintained under these conditions. This approach was used with 10-g samples of wet Irish peat (I.P. No. 1). Dried samples were not considered due to the degradation and subsequent leaching which occurs with the use of larger amounts of acid. In comparing these samples with those prepared in open beakers, the refluxed samples gave much lower capacities at 150°C. This could possibly be explained by the relatively large quantity of water present in the samples which is not removed during reflux as it is when open beakers are used. The diluting effect of the water, as well as possible hydrolysis, would be significant under these conditions. The values obtained at 200°C, on the other hand, were comparable for both methods of preparation. Five times more acid was needed to achieve this temperature so that the amount of water would be less significant. The reflux method is an alternative to simply heating wet peat with the acid in open beakers. This procedure would involve recovery and purification steps of the unused acid but would eliminate any waste of acid.

Phosphoric Acid Treatment of Peat

Treatment of the first batch of Irish peat (I.P. No. 1) with phosphoric acid increased the effective sodium capacity by 50%. The most interesting data, though, were the minimal leaching and swelling at pH values as high as 12.[27] The same treatment of the other Irish peat samples and the Michigan peat gave capacities which were the same as or lower than those for the untreated peat. These showed improvement in their leaching and swelling characteristics but were no better than the sulfuric acid-treated forms. Spectral and elemental data were not sufficient to explain the exceptional results for the first Irish peat batch (I.P. No. 1).

In an attempt to improve the physical characteristics of the sulfuric acid-treated peat, samples of peat which had been previously reacted with sulfuric acid, rinsed and dried were then treated with phosphoric acid. This additional acid treatment step increased the effective capacities of the sulfuric acid-treated wet peat by 50%. The phosphoric acid treatment of the peat which was dried before treating it with sulfuric acid showed no effect. As far as physical characteristics, no changes were observed in the amount of leaching or swelling that occurred after this subsequent treatment with phosphoric acid. Thus, this procedure does not appear to offer any distinct advantages.

Other Acid Treatments of Peat

The use of chromic acid in place of sulfuric acid for treating peat in open beakers was examined. There was no significant difference in the effective capacities for the treatment of the wet peat with either of the acids. For dried peat, the use of chromic acid resulted in slightly lower capacities than resulted from the use of sulfuric acid. Thus, the catalytic effect of using chromic rather than sulfuric acid as noted by Khundkar and Iqbal[24] for the treatment of peat from Khula, Pakistan, was not observed for either wet or dry Irish peat.

Dried samples of the different peats were refluxed with formic acid. The reason for using formic acid is that it has been used previously to convert simple alcohols into carboxyl groups.[29] The effective capacities were unaffected by the formic acid treatment. Thus, the expected introduction of carboxyl groups, which would have been indicated by an increase in the copper capacities, was not observed.

Samples of dried peat (1.20 g) were treated with 4 ml of nitric acid. Fuming began immediately and continued until the mixture cooled. Heating the reaction mixture to 150°C resulted in complete breakdown of the granules. Allowing the reaction to take place at room temperature for 2 hr left some of the particles of peat intact. These, however, leached continuously. As noted previously, 3 N HNO_3 causes the sulfuric acid-treated peat to leach. Thus, the oxidizing ability of nitric acid is apparently much too strong for the peat matrix.

Conclusions

The data show that peat has considerable potential for the removal of cationic species from water over a wide range of concentrations. The peat itself can be easily and cheaply obtained. The sulfuric acid-treatment process is both inexpensive in its use of small amounts of acid and simple in the mere heating of the peat with the acid. The resulting mixed cation exchanger has good physical characteristics up to approximately pH 8 for use in a flow system, as well as relatively good capacities. The best effective Na^+ capacities for the modified peat were within 25% of those for the commercially produced Dowex 50W.[30] This sulfonated polystyrene resin shows some leaching under alkaline conditions and swells to a greater extent than the sulfuric acid-treated peat.

As far as the other modification processes, none of the acids were capable of producing the capacities obtained when only sulfuric acid was used. The use of phosphoric acid in combination with sulfuric acid, however, showed some possibility for improving the physical characteristics.

In applying the modified peat to the removal of cations from water systems, it has two advantages which are shared by the commercial organic ion exchangers. One is that it can be easily regenerated by washing with strong acid if recovery of the cations is desired or for reuse of the exchanger. The other is that it could be ashed after use to a very small volume of material and then disposed of easily. With the lower cost of modified peat, these procedures would be more economically feasible than if the synthetic commercial resins now available were used. Thus, the application of modified peat on an industrial scale appears to be highly feasible.

Further investigation of the structure of peat and the changes that are caused by these modifications is needed. In doing this, it may be possible to formulate methods to further improve the characteristics of the resulting modified peat. Methods of extending the usable pH range by curtailing the leaching would be very significant.

ANION EXCHANGER

Introduction

In addition to cations, anionic pollutants must also be removed from water by procedures which are as economical as possible. Fewer anion than cation exchangers are presently available due to their lower chemical and thermal stability.[31] In an attempt to prepare inexpensive anion exchangers, relatively simple chemical modifications of peat have been investigated. Brittin[32] patented a peat-anion exchanger by the reduction of nitric acid-treated peat. An amphoteric ion exchanger has also been produced by treating humic acids with phenylenediamine and then polycondensing with an aldehyde.[33] This study involves the coupling of an amine, ethylenediamine (EDA), to modified peat.[25] The aliphatic amine was chosen because it is not as weakly basic as the aromatic amine. Peat which had been treated with sulfuric acid was utilized because of the additional carboxyl groups introduced by the acid treatment. One disadvantage of this modified form of peat is that it leaches in basic solutions. Thus, it would be desirable to couple the amine by an amide linkage under conditions less severe than refluxing in the basic EDA. For this reason, thionyl chloride was also used to form the acid chloride of the modified peat before treating with EDA. Subsequent formation of the strong-base exchangers by quaternization of the weak-base forms with methyl iodide and dimethyl sulfate was also attempted.

As done previously for the cation exchangers, a consideration of the physical characteristics in relation to the exchange capacity was made. Leaching and swelling with pH changes were again of great concern.

Treatment of Sulfuric Acid-Treated Peat with EDA

By refluxing EDA with peat which had been previously treated with sulfuric acid (150°C, 2 hr, 4 ml H_2SO_4 per 1.20 g dry peat), the cation exchanger was converted into an anion exchanger. Table IV gives the capacity data for the modified samples. The chloride capacity obtained is approximately equal to the sodium capacity of the starting material. The drop in the effective copper capacities also corresponds to the decrease in the sodium capacities. The two methods used for determining the copper capacity do not agree as they did for the cation exchangers. For the anion exchanger, the number of equivalents of copper retained is about four times greater than the equivalents of protons displaced.

As far as the conditions for refluxing with EDA are concerned, it appears that 3 hr and 1 ml EDA per gram of peat are sufficient for the conversion. Longer reflux periods result in lower yields as does the use of peat which was wet rather than dry before treatment with sulfuric acid. Attempts at treating peat which had not been treated previously with sulfuric acid resulted in complete breakdown of the dried particles.

The weak-base anion exchangers that were produced leached initially upon wetting the column packings. No further leaching was observed under nonalkaline conditions. The effects of pH greater than 7.0 were not examined, as weak-base exchangers are ineffective in basic solutions.

The chemical changes that occur from refluxing sulfuric acid-treated peat with EDA can be explained from the infrared spectra (Figure 5) and elemental analysis data. The carbon composition is relatively constant, while the nitrogen and hydrogen content increase considerably. The band at 1730 cm^{-1} disappears from the infrared spectrum, while the 1630 cm^{-1} band broadens from the treatment with base. This indicates the formation of amide linkages with the free carboxyl groups. The decrease in the band at *ca.* 1220 cm^{-1} and a lower sulfur content can be attributed to the loss of sulfonic acid groups through hydrolysis. Not all of the sulfur which was incorporated during the sulfuric acid treatment step is lost, however. The appearance of bands at *ca.* 1180 cm^{-1} and *ca.* 1390 cm^{-1} indicate the possible formation of sulfonamides.

If the increase in the nitrogen content (8.8%) is used to calculate the expected anion exchange capacity, a chloride capacity of 3 meq/g would

Table IV. Effective Chloride Capacities for Sulfuric Acid-Treated Peat After Treatment with Ethylenediamine

	Starting Material				Reaction Conditions			Product			
	Capacity (meq/g)							Capacity (meq/g)			
			Cu^{2}+							Cu^{2}+	
Type of H_2SO_4 Treated Peat	CL^-	Na+	Tit.	Pol.	EDA (ml/g)	Reflux Time (hr)	Yield	Cl^-	Na+	Tit.	Pol.
I.P. #1 (Dried)	0.01	0.83	1.26	1.30	2.0	5.0	50%	0.70	0.05	0.08	0.37
	0.01	0.78	1.17	1.30	2.0	5.0	66%	0.72	0.01	0.09	0.50
	0.01	0.78	1.17	1.30	2.0	3.0	80%	0.60	0.02	0.08	0.46
	0.01	0.78	1.17	1.30	1.0	3.0	85%	0.70	0.01	0.08	0.50
I.P. #3 (Dried)	--	0.85	1.35	1.35	4.0	3.0	82%	0.68	0.01	--	--
M. P. (Dried)	--	0.79	1.35	1.66	4.0	3.0	76%	0.05	0.02	--	--
I.P. #2 (Wet)	0.04	0.45	0.93	0.93	8.0	1.0	45%	0.03	0.03	--	--

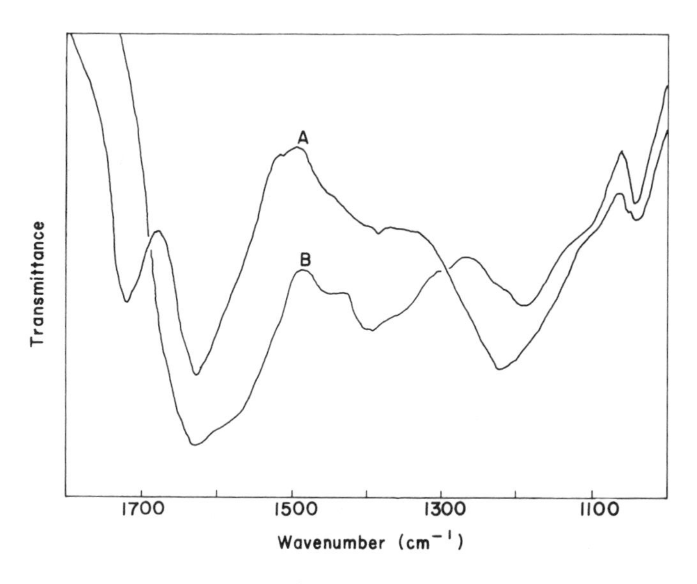

Figure 5. Infrared spectra (1800 cm^{-1} - 100 cm^{-1} region) of sulfuric acid-treated peat (I.P. No. 3) before (A) and after (B) reflux with ethylenediamine.

be predicted. The observed chloride capacity was 0.68 meq/g. Possibly this means that some of the free amine groups are very weak bases. Also polymerization could occur and neither end of the EDA would then be able to ion exchange.

Coupling of EDA Via an Acid Chloride Intermediate

Because of the milder conditions involved in forming an amide via an acid chloride intermediate rather than directly under reflux conditions, the reaction of peat with thionyl chloride ($SOCl_2$) was examined. In order to examine the effects of the thionyl chloride alone, the peat samples were refluxed with $SOCl_2$ and then rinsed with water to return them to their original form. The dried peat samples were unaffected. The Michigan peat continued to leach in all solutions. For sulfuric acid-treated peat samples, one batch of samples (I.P. No. 2) which was treated with sulfuric acid in the wet form showed a 50% increase in capacity. All of the other samples showed only a slight increase. The thionyl chloride is capable of introducing sulfonic groups.[34]

Dimethylformamide (DMF) has been found to catalyze the formation of acid chlorides with $SOCl_2$.[35] At the same time, the possibility of the acid chloride reacting with DMF also exists. When a 20% solution of

$SOCl_2$ in DMF was used, a violent reaction was activated by an initial period of heating. Further heating was resumed when the reaction subsided after a few minutes. Much lower sodium capacities were found when DMF was used. Thus, it appears that the DMF is reacting with the acid chlorides and, in turn, hinders its ability to act as a catalyst. Infrared spectra showed a band at *ca.* 1700 cm^{-1} which is characteristic of the conversion of carboxyls into primary amides. The spectra also indicated that the sulfonic acid groups were removed and/or sulfonamides were formed.

After the peat was treated with $SOCl_2$, it was treated with EDA. A violent reaction occurred that involved the evolution of HCl fumes. The chloride capacities for these samples were very low. All of the sodium capacities were also reduced to minimal amounts. This suggests that the more reactive acid chloride form of peat is polymerizing with the EDA. The only samples which had significant chloride capacities were those which were refluxed with the EDA. These reaction conditions may allow hydrolysis of the polymerized groups so that free amine groups are produced. This procedure, however, defeats the purpose of using the acid chloride intermediate.

Quaternization of Weak-Base Exchangers

Both methyl iodide (CH_3I) and dimethyl sulfate ($(CH_3O)_2SO_2$) were used to quaternize the weak-base exchangers. Dimethyl sulfate is commonly used commercially for producing strong-base anion exchangers.[36] Refluxing the weak-base exchangers with these reagents for several hours produced exchangers with very low strong acid capacities. The weak-base capacities were also greatly reduced (*ca.* 50%). Thus, rather than being methylated, amine groups were apparently removed and/or deactivated. An elemental analysis of one sample showed a 1.01% reduction in the nitrogen composition. This decrease can account for up to 0.36 meq of the chloride capacity. The infrared spectra for the analyzed sample did not show any significant changes. These results indicate that milder methylation procedures are required to produce a strong-base anion exchanger from this modified peat.

Conclusions

A weak-base anion exchanger can be prepared from sulfuric acid-treated peat by refluxing with EDA. Total effective chloride capacities of *ca.* 0.70 meq/g are obtained. These are 25 to 50% of those observed for synthetic anion exchangers with free amine groups.[37] At the same time, this simple modification procedure offers definite cost advantages.

Further work on characterizing this peat-anion exchanger is planned. Other reactions which would produce a more stable product are also going to be examined. This will involve investigation of other derivatization procedures. Of course, cost must remain a primary consideration in order to make the modified form competitive.

OIL COALESCER

Introduction

In order to prevent oil pollution, laws now require ships to limit their discharges of oily waste. Bilgewater must therefore be cleared of an average 0.1% oil (1000 ppm) before being pumped overboard. Shipboard filter cartridges which contain fibrous material are now being tested by the Navy. Suspended oil droplets coalesce in these filters to form larger droplets. When the larger droplets pass through the filter, they can then be effectively separated by gravity. Spielman and Gorew,[38] Davies *et al.*,[39] Langdon *et al.*[40] and others have discussed this technique in further detail.

One problem which is encountered with fibrous bed filters is plugging due to the suspended solids, which have an average concentration of 0.04%. If granular bed filters were used, it would be possible to backwash when plugging occurred. Shore installations have already used large sand and mixed-media filters. More lightweight and efficient systems are required for shipboard use.

Madia *et al.*[41] have suggested the use of polystyrene beads. They demonstrated that Rohm and Haas XAD-2 was an excellent oil coalescer. The XAD-2 was also shown to be more efficient than polypropylene, sand and anthracite coal. This study compared the oil removal capability of dried peat to that of XAD-2 when used in a flow system. Patents now exist for the use of dried and sieved peat for absorbing oil spills from water surfaces.[42,43] Dried Irish peat is suitable for use in a flow system and is stable under nonalkaline conditions. The leaching of dried Michigan peat prevents its use. Sulfuric acid treatment of the Michigan peat and its improved leaching properties allowed its utilization. These forms of peat could provide a granular replacement for the fibrous filter material while being less expensive than the polystyrene beads.

Procedures and Results

The oil removal capability of the peat was determined with the apparatus shown in Figure 6. An oil-in-water emulsion was prepared in distilled water with Exxon 260 diesel fuel (density of 0.84 g/ml at 24°C).

Glass columns (6 mm i.d.) were packed with 1 or 2 g of 0.5 to 1.0-mm diameter particles (18 to 35 U.S. mesh) of the following:

1. XAD-2 (Rohm and Haas)
2. dried Irish peat
3. sulfuric acid-treated Michigan peat (4 ml H_2SO_4 per 1.2 g dried peat, 2 hr, 150°C

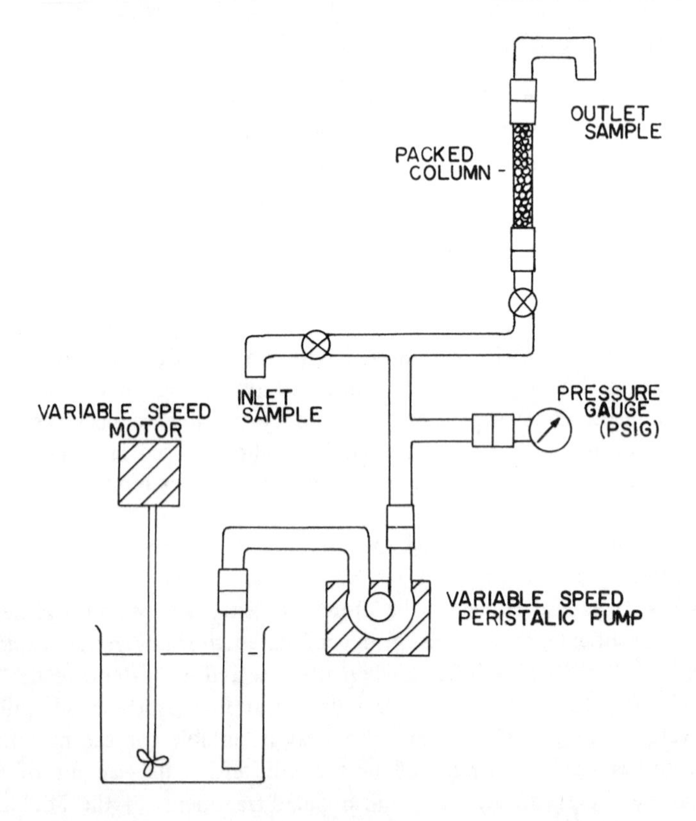

Figure 6. Flow system for the determination of the oil removal capability.*

A flowrate of 25 ml/min was maintained, and alternate inlet and outlet samples were collected. The oil concentration of the samples was determined by the infrared absorbance of CCl_4 extracts. These procedures were based on those developed by Simard *et al.*[44] and Gruenfeld.[45] Complete experimental details have been published previously.[8,25]

As shown in Figure 7, the Irish peat (I.P. No. 3) removed the oil effectively and was comparable to the XAD-2. The average inlet

*Reprinted from Reference 8, p. 730, by courtesy of Marcel Dekker, Inc.

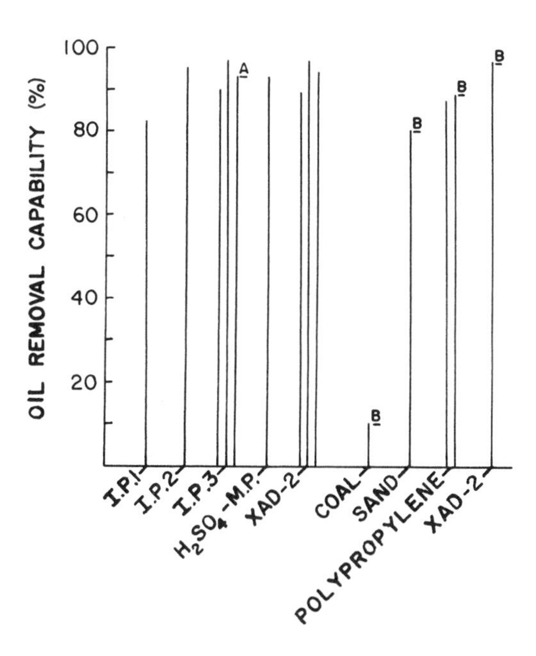

Figure 7. Comparison of the oil removal capability of several granular materials.* (A) Synthetic seawater emulsion. (B) Data of Madia *et al.*[41]

concentration was 250 ppm oil and the samples ranged from 0.5 to 1.5 l. The sulfuric acid-treated peat (H_2SO_4-M.P.) also performed very well. Using a synthetic seawater emulsion (0.5 *M* NaCl and 0.05 *M* $MgSO_4$) gave slightly higher efficiencies for the Irish peat (I.P. No. 3). Longer columns which contained more packing were also found to be more efficient.

The results of Madia *et al.*[41] are also shown in Figure 7 and are in good agreement for the XAD-2. Their data were obtained with a larger-scale system but with similar mesh particles. It is interesting to note the relative ineffectiveness of the coal in comparison to the peat.

Conclusions

The data indicate the feasibility of using both dried and sulfuric acid-treated peat as oil coalescers. Cartridges containing peat would be efficient, lighweight and inexpensive. The only limitation is the leaching of peat under alkaline conditions. Of course, the effect of detergents, which promote emulsification but hinder coalescence,[46] should be examined before these filters are actually implemented.

---

*Reprinted from Reference 8, p. 733, by courtesy of Marcel Dekker, Inc.

SUMMARY

The data presented in this study indicate the feasibility of using modified forms of peat as a cation exchanger, anion exchanger and oil coalescer. The modification processes require relatively inexpensive chemicals in small amounts and only simple procedures are involved. This factor in conjunction with the availability of peat makes these modified systems well suited for large-scale applications.

ACKNOWLEDGMENT

This work was supported by the National Science Foundation (Grant No. CHE76-04321). One of the authors (E. Smith) would like to express his appreciation to the Department of Chemistry and the Graduate Research Council of the University of Cincinnati for fellowships to perform this research.

REFERENCES

1. Moore, P. D., and D. J. Bellamy. *Peatlands* (New York: Springer-Verlag New York, Inc., 1974), pp. 135, 187.
2. Osvald, H. *Vegetation and Stratigraphy of Peatlands in North America,* (Uppsala, 1970), p. 13.
3. Fogarty, W. M., P. J. Griffin and J. A. Ward. *Technol. Ireland* 5:21 (1973).
4. Boffey, P. M. *Science* 190:1066 (1975).
5. Bel'Kevich, P., K. A. Gaiduk and L. R. Chistova. *Rus. J. Phys. Chem.* 46(12):1716 (1972).
6. Schnitzer, M., and S. U. Khan. *Humic Substances in the Environment* (New York: Marcel Dekker, Inc., 1972), p. 72.
7. MacCarthy, P., H. B. Mark, Jr. and P. R. Griffiths. *J. Agric. Food Chem.* 23:600 (1975).
8. Smith, E. F., P. MacCarthy and H. B. Mark, Jr. *J. Environ. Sci. Health,* A11(12):727 Marcel Dekker, Inc., New York (1976).
9. Bel'Kevich, P. I., L. R., Chistova and E. A. Yurkevich. *Vesti Akad. Navuk Balarus. SSR, Ser Khim Navuk* (3):42 (1967); *Chem. Abstr.* 68:38556c (1968).
10. Freeland, G. N., R. M. Hoskinson and R. J. Mayfield. *Environ. Sci. Technol.* 8:943 (1974).
11. Snyder, S. L. and T. L. Vigo. *Environ. Sci. Technol.* 8:944 (1974).
12. Friedman, J., and A. C. Waiss. *Environ. Sci. Technol.* 6:457 (1972).
13. Helfferich, F. *Ion Exchange,* (New York: MacGraw-Hill Book Company, 1962), pp. 17-18.
14. "Zeo-Karb Cation Exchanger," Bulletin 5009-A, The Permutit Co., Paramus, NJ.
15. Szalay, A., and M. Szilagyi. *Acta Phys. Hung. Tom.,* 13:421 (1961).
16. Riley, R. U.S. Patent 2170065 (1939).

17. Liebknecht, O. U.S. Patent 2191060 (1940); U.S. Patent 2206007 (1940).
18. Smit, P. U.S. Patent 2191063 (1940); U.S. Patent 2205635 (1940).
19. Goetz. P. U.S. Patent 2260971 (1941).
20. Schmid, Von E., P. Stipanits and F. Hecht. *Oestr. Chem. Z.* 65(3): 69 (1964).
21. Diaconescu, E., and F. Chirac. *Bul. Inst. Politeh. Iasi* 15 (1-2): 41 (1969); *Chem. Abstr.* 74:128903s (1971).
22. Kotkovskii, A. P., L. U. Kosongova, N. N. Volosovich and S. V. Kashirina. *Pererab. Ispol'z Torfa Sapropelei* 1971:163; *Chem. Abstr.* 78:60579r (1973).
23. Kotkosskii, A. P., L. V. Kosonogova and N. S. Volosovich. *Vesti. Akad. Navuk. Belarus, SSR, Ser. Khim. Nauk.* 1973 (3):86; *Chem. Abstr.* 80:72652p (1974).
24. Khundkar, M. H., and A. F. Iqbal. *Dacca Univ. Studies* 11 (2):41 (1963).
25. Smith. E. F. Ph.D. Thesis, University of Cincinnati, OH (1976).
26. Smith, E. F. P. MacCarthy, T. C. Yu, and H. B. Mark, Jr. *J. Water Poll. Control Fed.* 49:633 (1977).
27. Smith, E. F., P. MacCarthy and H. B. Mark, Jr. *J. Environ. Sci. Health,* A11 (2):179 (1976).
28. Dorfner, K. *Ion Exchangers–Properties and Applications,* A. F. Coers, Ed. (Ann Arbor, MI: Ann Arbor Science Publishers, Inc., 1971), p. 31.
29. Harrison, I. T., and S. Harrison. *Compendium of Organic Synthetic Methods,* (New York: Wiley-Interscience, 1971), pp. 26-27.
30. *Dowex: Ion Exchange* (New York: McGraw-Hill Book Company, 1962), p. 74.
31. Helfferich, F. *Ion Exchange* (New York: McGraw-Hill Book Company, 1962), p. 58.
32. Brittin, W. E. U.S. Patent 2, 260,195 (October 10, 1944); *Chem. Abstr.* 39:3865 (1945).
33. Nazarova, N. I., R. P. Koroleva and N. K. Alybakova. U.S.S.R. Patent 351, 864 (September 21, 1972); *Chem. Abstr.* 78:44473W (1973).
34. March, J. *Advanced Organic Chemistry* (New York: McGraw-Hill Book Company, 1968), p. 403.
35. Bosshard, H. H., R. Mory, M. Schmid and H. Zollinger. *Helv. Chem. Acta* 42:1653 (1959).
36. Michael, M. W. U.S. Patent 2,543,666 (1955).
37. Dorfner, K. In: *Ion Exchangers,* A. F. Coers, Ed. (Ann Arbor, MI: Ann Arbor Science Publishers, Inc., 1972), pp. 56-61.
38. Spielman, L., and S. L. Goren. *Ind. Eng. Chem., Fund.* 11:66 (1972).
39. Davies, G. A., G. V. Jeffreys, F. Ali and M. Afzal. *Chem. Eng.* (London) (266):392 (1972).
40. Langdon, W. M., P. P. Naik and D. T. Wasan. *Env. Sci. Technol.* 6:905 (1972).
41. Madia, J. R., S. M. Fruh, C. Miller and A. Beerbower. *Environ. Sci. Technol.* 10:1044 (1976).

42. Fischer, K. O. P. Canadian Patent 956,921 (October 21, 1974); *Chem. Abstr.* 82:768746 (1975).
43. Schaeffer, J. R., and S. Chuard. Swiss Patent 495,472 (October 15, 1970); *Chem. Abstr.* 74:45397d.
44. Simard, R. G., I. Hasegawa, W. Bandaruk and C. E. Headington. *Anal. Chem.* 23:1384 (1951).
45. Gruenfeld, M. *Env. Sci. Tech.* 7:636 (1973).
46. Kaufman, S. *Env. Sci. Tech.* 10:168 (1976).

CHAPTER 22

WASTEWATER TREATMENT WITH HUMIC ACID-FLY ASH MIXTURES

John B. Green and Stanley E. Manahan
Chemistry Department
University of Missouri
Columbia, Missouri

INTRODUCTION

The inevitable shift from petroleum and natural gas to alternative energy sources, particularly coal and coal-derived synthetic fuels, will result in the production of large quantities of wastewater. Development of technologies for the treatment of this wastewater must give a high priority to processes which are economical and which utilize inexpensive chemicals. This chapter discusses the potential of two such materials—humic acid and fly ash—both of which are obtained from coal. Fly ash is inorganic matter produced by coal combustion, whereas humic acid is an integral part of the organic portion of low-rank coals and can be produced by the partial oxidation of higher-rank coals. Both are complex, poorly characterized materials. Neither have been used for treatment on an industrial scale; however, the effectiveness of both has been demonstrated in the laboratory.

Almost all past and current research on the treatment of water with humic acid has been performed by a small group of Japanese workers using a type referred to as nitrohumic acid.[1,2]

To understand the desirability of using humic acid as opposed to nitrohumic acid for wastewater treatment, it is helpful to consider their preparation. Humic acid is a subcategory of humic matter, which is widely distributed in the environment.[3,4] For example, humic acid may be extracted from soil, natural waters, peat and low-rank coal without any pretreatment.[3-7] Production of humic substances from higher-rank coal, such as bituminous coal, requires an oxidation step prior to extraction.[8-12] Several oxidizing agents have been used, of which nitric acid is the most common.[11-13] The base-extracted product of coal which has been treated with nitric acid is referred to as "nitrohumic acid." This is because some relatively minor nitration of aromatic rings occurs during the oxidation. The properties of nitrohumic acid have not been shown to differ significantly from humic acids from other sources. Thus, there is little justification for the almost exclusive use of nitrohumic acid for wastewater treatment. The large quantities of expensive nitric acid required for the synthesis of nitrohumic acid introduce severe cost constraints. In addition, the emission of copious amounts of NO_2 during synthesis causes air pollution problems. Nevertheless, several patents exist for the production of nitrohumic acid from coal.[14-16]

An alternative route to the production of humic acid from coal is partial air oxidation under alkaline conditions.[8-11] From an economic viewpoint, this is a much more desirable preparation technique. Where applicable, an even better source of coal humic acid is from direct extraction by base of low-rank coals such as lignite or leonardite.

In addition to patents for the use of basic nitrohumic acid solutions for water treatment, several patents have been issued for the preparation and use of nitrohumic acid resins. The most commonly used approach involves mixing basic aqueous solutions of nitrohumic acid and some other polymer, acidification to induce flocculation, heating the filtered floc to form a resin, pelletization, acid washing and conversion to a particular ionic form, such as with calcium. These products are designed to substitute for synthetic cation exchange resins and zeolites for the removal of heavy metals. Manufacture and use of these solid ion exchangers differs considerably from the use of coal humic acid solutions; hence, they are not discussed further.

The application of fly ash to water treatment has been much more extensive than the application of humic acid.[17-28] Although both have an affinity for organics, fly ash generally adsorbs anionic species, including humic acids,[25] whereas humic acid binds cationic species. Therefore, the two materials have complementary roles in wastewater treatment.

PROPERTIES OF HUMIC SUBSTANCES

Briefly, humic substances can be described as natural colloidal chelating agents having molecular weights ranging from 300 to 10^5 and no discrete polymeric units. They have significant aromatic character and phenolic OH and carboxylic acid groups capable of binding metal ions.[3-8]

Humic substances are chemically stable and resistant to microbial degradation. They are thought to be intermediates in the coalification process.[7] Humic substances extracted or prepared from diverse sources (*e.g.*, coal, peat, soil, natural waters) show surprisingly similar properties. This is advantageous from a practical viewpoint because it allows interchangeability of humic substances from different sources in applications such as wastewater treatment.

As shown in Figure 1, humic substances are subdivided according to their solubility characteristics in aqueous and alcoholic media.[3,29] Generally speaking, humic substances which have not been chemically altered by processes such as exhaustive methylation[3,4,6] are insoluble in all non-polar solvents. Some solubility is observed in basic donor-type solvents such as pyridine,[8] but basic aqueous media are notably superior to all others. Humic acid is distinguished from other humic substances by the fact that it is soluble in basic media and insoluble in acidic media. The abrupt precipitation of humic acid at lower pH makes it the most useful of the humic substances for treatment purposes because it can be removed easily.

Elemental and Functional Group Analysis

Table I shows the elemental analyses of selected humic substances from widely different sources. Despite some differences, the C, H and O contents are surprisingly constant in each solubility class. Most of the oxygen present can be accounted for in measured -OH, $-CO_2H$, -C=O and $-OCH_3$ functionality, with the balance believed to be in some type of ether linkage.[3-8] The greater oxygen content of the fulvic acid fraction gives it a higher solubility and exchange capacity.[4] Both aliphatic and aromatic carbon are found in humic substances. The aromatic/aliphatic C ratio is notably higher in humic substances extracted from coal than from other sources and is higher in humic acid as compared to fulvic acid.[30] Humic substances extracted from soil contain appreciable amounts of paraffins and fatty acids.[31] It has not been established whether these are an integral part of the humic substance structure or are simply absorbed.

Both infrared and nuclear magnetic resonance spectral data show that the skeletal hydrogen in humic substances is entirely aliphatic,[4,32] indicating a high degree of substitution on the aromatic rings. The nature of sulfur in humic substances has not been seriously investigated to date, no doubt as a result of its small percentage. Nitrogen in soil-derived humic substances is believed to be largely amine[4,35] as opposed to the nitro nitrogen in nitrohumic acid.[8,11-13] Both nitrogen types would have an effect on cation binding, but the percentage of N is so much lower than the percentage of O that the effect is relatively minor.

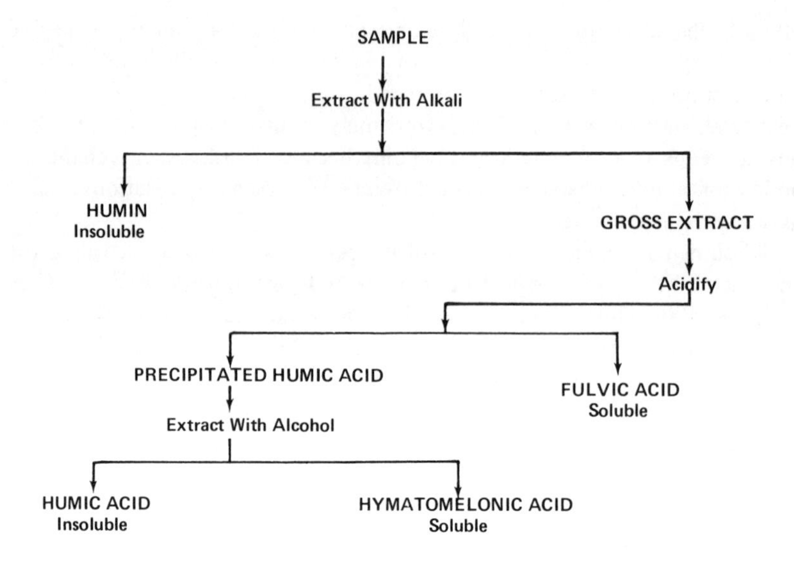

Figure 1. Fractionation of humic substances according to their solubility in aqueous and alcoholic media.[4]

Table I. Elemental Analysis Data for Humic Substances

		Percentage by Weight					
HS[a]	Source	C	H	N	S	O (Diff.)	Reference
FA	Soil	50.9	3.35	0.75	0.25	44.8	6
FA	Soil	42.5	5.9	2.8	1.7	47.1	4
HA	Soil	53.8	5.8	4.1	1.1	32.9	4
HA	Soil	60.4	3.7	1.9	0.4	33.6	4
Humin	Soil	56.3	6.0	5.1	0.8	31.8	4
FA	Water	46.2	5.9	2.6	––	45.3	4
HA	Lake Sediment	53.7	5.8	5.4	––	35.1	4
HA	Coal (HNO_3 ox.)	62.1	3.2	4.8	0.7	29.2	11
HA	Coal (O_2 ox.)	62.1	4.8	1.4	0.8	30.9	11
HA	Coal ($KMnO_4$ ox.)	62.8	3.5	1.9	0.7	31.1	9
HA	Peat	58.7	4.9	2.4	0.4	33.6	33
HA	Coal (no preox.)	63.1	2.8	1.4	0.4	32.3	34

[a]Humic substances: FA is fulvic acid and HA stands for humic acid.

Molecular weight values reported for humic substances vary widely due to several factors including their polydispersity, a tendency to aggregate into larger molecules in certain types of media, and difficulties in removing low-molecular-weight contaminants. Gel filtration studies[4,36] have revealed that molecules having weights varying over several orders of magnitude exist in a given batch of humic material. Therefore, data showing molecular weight distributions are more useful than single average values. Generally, fulvic acids have the smallest average molecular weight and humins the largest. Apparent molecular weight increases at lower pH and higher concentrations of metal ions have been explained by aggregation due to hydrogen and metal ion bridges between humic molecules.[37]

Interaction with Organic Molecules and Metal Ions

As indicated earlier, humic substances have large carbon skeletons with oxygen-containing functional groups located on aliphatic side chains and on aromatic rings on the outside perimeter of the molecules. Thus, organic molecules and other solutes can interact with humic substances by ion exchange or other interaction with functional groups, or by hydrophobic interaction with the carbon skeleton. Since humic substances in soils are believed to bind pesticides and herbicides extensively, considerable research has been devoted to study of humic substance-organic interactions[38-42] Complexation with urea has also been demonstrated.[43] Although both binding mechanisms listed above have been implicated, hydrophobic interactions seem to be the most prevalent.

Metal complexation by humic substances is extremely well documented and has been shown to account for the accumulation of inorganics in coal and peat,[44] to play an important role in metal transport to plants[4] and to explain relatively high concentrations of soluble metals in natural waters.[3,6,45] It also accounts for the effectiveness of nitrohumic acid in removing heavy metals from polluted water. Earlier work in the area was largely qualitative or semiquantitative, but it established that true complexes were formed and also established the relative stabilities of complexes with different metal ions.[4] Generally, the stability of complexes decreases in the following order: $Fe^{3+} > Al^{3+} > Pb^{2+} > Cu^{2+} > Fe^{2+} > Zn^{2+} > Ni^{2+} > Co^{2+} > Mn^{2+} > Ca^{2+} > Mg^{2+}$.[46-48] More recent work has been aimed toward identifying the complexation mechanisms and determining conditional stability constants. One binding mechanism has shown to be chelation by *o*-phenolic and carboxylic groups.[4-6,46,49,50] Chelation in a manner analogous to phthalic acid (*o*-dicarboxylic) is also frequently implicated.[4,49-51] The former structure is believed to be the more stable of the two, and hence the most important where metal concentrations are near trace levels as in natural waters.[6,50] Formation of mixed ligand complexes has been largely overlooked, but stability constants are available for a few systems.[52,53] Studies have shown that the mixed ligand complex is usually more stable than those made up of either ligand alone, even where the auxiliary ligand is NTA. This feature might prove quite advantageous for removing anionic pollutants and explains why humic acid in the presence of Fe(III) effectively removes Cr(VI).[1]

Although considerable progress has been made, substantial variation in metal-humic substance stability constants and assumptions made in calculating them still exists. A major complication in the determination of stability constants is the multitude of acid groupings on humic substance molecules whose acidity is a function not only of nearby functional groups but also of the overall negative charge buildup on a given molecule with increasing ionization.[4,6,50,54] Insufficient data are available to account rigorously for variation in the pK_a of acid groups and its inherent effect on metal ion binding.

In summary, listed below are applications of humic acids for improving the quality of highly polluted water. Many are utilized in patented nitrohumic acid water treatment processes:

1. *Neutralization of Acids.* Calcium, magnesium, sodium and other cations on humates exchange for H^+, causing the resulting humic acid to precipitate and settle.
2. *Removal of Heavy Metals.* Basic, neutral or acidic humic acid chelates heavy metals in solution, and either dissolves or adsorbs onto suspended particles containing metal salts.
3. *Removal of Anions.* Mixed ligand complexation should bring about removal of many anions including phosphate, cyanide and ionized organics.
4. *Removal of Organics.* Humic acid has been shown to sorb many types of organic molecules.
5. *Clarification of Suspended Matter.* Flocculation by synthetic polyelectrolytes is well known; humic acid exhibits similar properties.

The major difficulty which might arise from humic acid treatment is release of soluble organics (fulvic acids). This potential problem is minimized by methods which have been developed for removal of humic substances from water[55-57] and by the fact that because of their microbial resistance[4] humic substances do not greatly increase the BOD of waters, although they add to the COD and TOC. Furthermore, they are naturally present in waters anyway.

PROPERTIES OF FLY ASH

The term "fly ash" usually describes any solid matter which is suspended in the combustion gases produced from burning fuel. Bottom ash is the residue which remains in the combustion chamber. Airborne particulate matter from industrial processes such as cement manufacturing is also referred to as fly ash; hence the term is not limited to combustion. The British use the term "pulverized fuel ash" to describe fly ash originating from fuels.

The chemical composition of fly ash is extremely variable and is dependent on the type of fuel, conditions of combustion and type of ash collection. It typically contains unburned fuel particles (0-5%) and major amounts of oxides of Si, Fe, Al, Ca, Mg, Na, K and S in approximately decreasing order of weight percentage. A variety of trace elements is also present, especially in fly ash originating from coal,[58,59] which accounts for over 90% of the fly ash produced from fuel.[60] A portion of fly ash is magnetic due to the presence of magnetite. The presence of large percentages of Si, Al and Fe oxides indicates that most components of fly ash are water insoluble. Also, incorporation of the normally more soluble constituents, *e.g.*, CaO and Na_2O, into glasses and inside SiO_2, Fe_2O_3 and Al_2O_3 matrices causes the normally soluble matter to be much less soluble than the corresponding pure crystalline materials.

Some components do dissolve in water, however, and a basic pH is usually observed. A slightly acidic ($\sim$ 6) pH is observed with some ashes, especially when the percentage of Fe_2O_3 exceeds the sum of the percentages of alkali and alkaline earth oxides.[61] Hence, fly ashes are often classified as acidic or basic according to the relative abundance of acidic and basic components and the pH of the water suspension. Basicity, elemental and mineral composition, age and particle size distribution are important properties which determine the suitability of a given fly ash for treatment applications.

Fly Ash Particle Size and Shape

Due in part to the presence of vitreous material, fly ash particles are largely spherical in shape.[17,61-63] Some of the glass spheres are hollow and contain gases (largely N_2, CO_2 and H_2O) plus even smaller spheres. Hollow spheres (termed cenospheres) are less dense than water and may be removed by flotation.[63] Besides cenospheres, irregularly shaped microcrystals are observed. Particle size is in the range 0.3 to 100 micrometers.[19,60,64] The size distribution depends, of course, on the particular emission source and means of ash collection. Particle collection efficiency is inversely proportional to size for all collection devices. Many trace elements are volatilized during combustion and condense on the fly ash particle surfaces, as is SO_2.[64-66] Hence, smaller particles have higher trace element and sulfur concentrations due to the increasing surface area-to-volume ratio.

Treatment Applications of Fly Ash

Surface adsorption is the most probable removal mechanism for most pollutants by fly ash. This can be seen in part by the close correspondence of the major components of fly ash to many of the common adsorbents used in classical liquid-solid chromatography. Due to the acidic nature of most of the insoluble fly ash matrix, anionic pollutants have a higher affinity for fly ash than cationic pollutants.[25] However, some cations with higher ionic charges and affinities for hydroxyl groups (Fe^{3+} and Al^{3+}, for example) are also strongly sorbed by fly ash. In those cases, interaction with hydrated Fe_2O_3 and Al_2O_3 surfaces appears likely. Removal of polyelectrolytes by fly ash is analogous to removal of suspended solids by humic acid. In each case, the solid particles are linked by the polyanion, and the aggregate settles out or is filtered. Use of fly ash for water treatment has the additional advantage of improving the filterability of the sludge.[17-19] The fly ash particles cause an open sludge structure which allows water to pass through more freely.

Since humic acid has anionic character and fly ash has cationic character, together they should remove many types of pollutants. However, the interaction between the two prevents their total effect from being simply the sum of their individual effects. For example, the complexing action of humic acid enhances fly ash dissolution, and the affinity of humic acid for fly ash surfaces causes some deactivation of fly ash for adsorbing pollutants. However, the overall interactive effect should be beneficial to water treatment. For example, addition of fly ash should supply Fe^{3+} and Al^{3+} in sufficient amounts to induce humic acid flocculation and easy removal. Hence, addition of alum, $FeCl_3$ or other flocculating agents should be unnecessary. Similarly, the addition of anionic polyelectrolytes in fly ash treatment processes should be unnecessary since humic acid is a polyanion. Since water treatment by separate use of humic acid and fly ash has been researched previously, the authors decided to concentrate their efforts on the interaction between humic acid and fly ash. This and methods for determining the optimum humic acid/fly ash feed ratio are discussed in the remaining portion of this chapter.

REACTION OF HUMIC AND SALICYLIC ACIDS WITH FLY ASH

Experimental

Preparation of Humic Acid

Humic acid was prepared from Illinois No. 6 bituminous coal by oxidation with nitric acid and extraction with NaOH. Details of the procedure have been described elsewhere.[67] One liter of the crude humic acid was precipitated by adjusting the solution to pH 2 with HCl. After settling overnight, the clear liquid was siphoned off and the humic acid plus remaining water was treated 5 times with alternating 80-100 g portions of Dowex 50X8-100 (H^+ form) and Dowex 1X8-100 (OH^- form) ion exchange resins. Separation of the humic acid and resin after each equilibration was effected by vacuum filtration using a coarse fritted glass crucible. Losses of humic acid to the anion resin were minimized by washing the resin with 100 ml of deionized water and adding the wash to the filtrate. A final treatment with Chelex 100 in the H^+ form concluded the purification procedure, and the HA was diluted to two liters. Use of deionized humic acid facilitated the interpretation of the results of later experiments.

Equilibration of Acids and Fly Ash

The lignite fly ash sample used in these experiments was obtained from cyclone dust collectors in the Basin Electric Coop plant, Stanton, North Dakota, on October 17, 1973 (U.S. Bureau of Mines sample No. GF 73-1921). For each acid, 40 to 60 fly ash portions in the range 0.01-0.04 g were weighed into separate 25-ml volumetric flasks. Various amounts of standardized acid (0.01-0.025 *N*) were released into the flasks and diluted with deionized water. The solutions were equilibrated for 1 to 2 weeks with periodic stirring to ensure equilibrium conditions. Then the pH of each solution was measured with a glass electrode, the undissolved fly ash was removed and the concentrations of soluble Na, K, Ca, Mg, Fe and Al were determined.

For the series of solutions containing salicylic acid, simple centrifugation was sufficient for removing the undissolved fly ash. Three aliquots of 1, 2 and 5 ml were taken from each centrifuged sample and diluted to 100, 50 and 10 ml in volumetric flasks containing 1,000 to 2,000 mg/l La, K and Na, respectively. All diluted samples contained 1% (v/v) HNO_3. The following elements were determined: Ca and Mg in the 100-ml flask, Na in the 50-ml flask, and K, Al and Fe in the 10-ml flask. This dilution scheme gave solutions ideal for atomic absorption determination of the respective elements. A linear least-squares computer program was used for data reduction and error analysis.

Humic acid-fly ash mixtures could not be separated by simple centrifugation in many instances because sufficient metal oxides dissolved to induce precipitation of part or all of the humic acid. Thus, a means for separating the two solids was devised which took advantage of their difference in density. A stainless steel centrifuge tube holder was filled with sufficient 2:3 carbon tetrachloride-cyclohexane (density 1.1 g/ml) to fill a graduated plastic 15-ml centrifuge tube (with a 1-to 2-mm hole cut in the bottom) to 3 to 4 ml when the tube was placed in the holder. A 10-ml sample of the humic acid-fly ash mixture was pipetted into the centrifuge tube contained by the holder. The entire assembly was centrifuged for 30 min. During centrifugation the fly ash penetrated both liquid layers and exited through the hole. The solid humic acid settled to the aqueous-organic interface. After centrifuging, a glass tubing-rubber stopper assembly was inserted into the centrifuge tube, and the experimenter's finger was placed over the exposed end of the glass tube. Then the centrifuge tube was lifted out of the holder and the organic layer drained off via careful manipulation of the finger (a separatory funnel can be used by the weak of heart). The aqueous layer was then allowed to drain into a 50-ml digestion flask and the humic acid was digested by adding 3 ml of 2:1 HNO_3-$HClO_4$ to the flask and heating to boiling for approximately 30 min. Some metal humates were invariably lost during transfer to the digestion flask due to a small glob of mixed humic acid-fly ash solid which stuck to the side of the centrifuge tube. The severity of this loss increased with the amount of humic acid added and was estimated to cause a 10% negative error. Since the presence of perchlorates caused erratic absorbance readings during determination of Na and K, the digestion procedure was later modified. Thus, 2 ml of 15 *M* HNO_3 was added and the solution boiled to near dryness. Then an additional 2 ml of HNO_3 was added and the solution again taken to near dryness. Finally, 10 ml of deionized water was added and the digestate was heated to boiling and transferred to a 25-ml volumetric flask while still warm. Aliquots of the diluted digestates were diluted in 1000- to 2000-mg/l La, Na and K solutions and analyzed in a manner analogous to the salicylic acid samples.

Analyses

Weighed portions near 0.05 g of fly ash were digested in a Parr bomb (Cat. # 4745) at 150° for 2 hr using 2 ml aqua regia and 0.1 ml HF. The digestates were transferred to 50-ml volumetric flasks containing 5 ml of 5% boric acid and diluted. The diluted samples were stored in plastic bottles. Complete dissolution of the fly ash was obtained using the digestion method. Known volumes of the stock deionized humic acid were digested as previously described using HNO_3 only. Atomic absorption spectrophotometry using the method of standard additions was used for determination of Na, K, Ca, Mg, Fe and Al.

Solutions of reagent grade salicylic acid were standardized by potentiometric titration with standard sodium hydroxide. The normality was assumed to equal the molarity, meaning that the phenol group (pK_a = 13.4) was not counted as an acidic group for the purpose of defining acid concentration. Standardization of the deionized humic acid was also performed by potentiometric titration with NaOH. The titration results were checked against various ion exchange techniques to be described later.

Results

Table II shows data from analysis of the fly ash and deionized humic acid used for the equilibration experiments. The fly ash was a typical basic lignite with relatively large percentages of Na_2O, K_2O, CaO and

Table II. Elemental Analysis of Fly Ash and Deionized Humic Acid

Fly Ash			Humic Acid		
Constituent	Wt %		Constituent	Crude	Deionized
SiO_2	36.4[a]	34.4 ± 0.4[b]	C (wt %)[c]	--	55.5
Al_2O_3	9.9	11.8 ± 0.5	H (wt %)[c]	--	3.87
Fe_2O_3	10.5	10.6 ± 0.1	N (wt %)[c]	--	5.2
TiO_2	0.5	--	S (wt %)[c]	--	1.07
P_2O_5	0.1	--	O (wt %)[c]	--	34.3
CaO	21.8	22.1 ± 0.5	Na[d]	245	1.58 ± 0.03
MgO	6.7	5.23 ± 0.04	K[d]	1.0	0.10 ± 0.01
Na_2O	7.4	8.6 ± 0.2	Ca[d]	1.0	0.021 ± 0.004
K_2O	1.0	0.98 ± 0.02	Mg[d]	0.20	0.004 ± 0.001
SO_3	3.1	3.07 ± 0.03	Fe[d]	0.80	0.55 ± 0.05
Loss on Ignition	1.6	--	Al[d]	0.16	0.10 ± 0.02
Total	99.0	--			

[a]Bureau of Mines Analysis, Sample No. GF 73-1921.
[b]This work. Uncertainties are standard deviations.
[c]Galbraith Labs, Knoxville, TN.
[d]Mg/l in a 1 g/l HA solution.

MgO. Data from the Bureau of Mines analysis of the fly ash sample are also shown for comparison. Data from analysis of the humic acid both before and after deionization show the high efficiency of ion exchange resins in removing uni- and divalent cations and their relative ineffectiveness in removing Fe^{3+} and Al^{3+}. Since the affinity of the resin for cations increased with the charge on the ion, it is obvious that these metals are tightly held by humic acid.

All raw data from the fly ash-acid equilibrations were normalized based on a 1000-mg/l fly ash suspension (*i.e.,* 0.025 g fly ash/25 ml). Normalization was necessary since exactly 0.0250 g fly ash was not weighed into each 25-ml flask. The dissolved cation concentration data were plotted twice: against solution pH and against meq of acid added. The data obtained are shown in Figure 2. Also indicated on the figure (solid lines) are dissolved cation concentrations obtained when a strong mineral acid (HCl, H_2SO_4 or HNO_3) was used in place of humic or salicylic acid. Data from analysis of the humic acid-fly ash equilibrations were corrected for cations originally present in the deionized humic acid (Table II). The conclusions drawn from the figure are:

1. Close agreement of cation concentrations plotted against pH for humic acid and salicylic acids suggests that carboxylic groups in humic acid ortho to phenol groups have the highest acidity and affinity for metal ions.
2. Discrepancies in the same data plotted against meq H^+/0.025 g ash/25 ml show the presence of additional acid groups on humic acid that are not in *o*-carboxylic-phenol configuration. These have both an affinity for metal ions and an acidity less than that of salicylic acid, as evidenced by the lagging of the humic acid plots for Fe^{3+} and Al^{3+} behind those for salicylic acid.
3. Metal-deficient humic acid has a chelating power superior to that of salicylic acid. This is evident from the humic acid curves' lead over salicylic acid curves for all cations at low meq H^+/ 0.025 g ash/25 ml values.
4. Both chelators effectively solubilize Fe and Al at pH values significantly higher than those necessary for dissolution by mineral acids.
5. Considerable intermixing of the metal oxides in fly ash is apparent from the interrelated dissolution patterns of cations as the acid concentration is increased. For example, nearly uniform K_2O composition in fly ash is indicated by the linear increase in K^+ concentration with meq H^+ added. Also, the definite presence of MgO and CaO intermixed with Al_2O_3 and Fe_2O_3 can be seen by comparing the plots of Mg^{2+}, Ca^{2+}, Al^{3+} and Fe^{3+} against pH. As a rough approximation of this effect, the percentage of each metal oxide in the three most distinct fly ash "fractions" is shown in Table III. More detailed information on intermixing was obtained from mineral acid dissolution experiments, which will be published elsewhere.

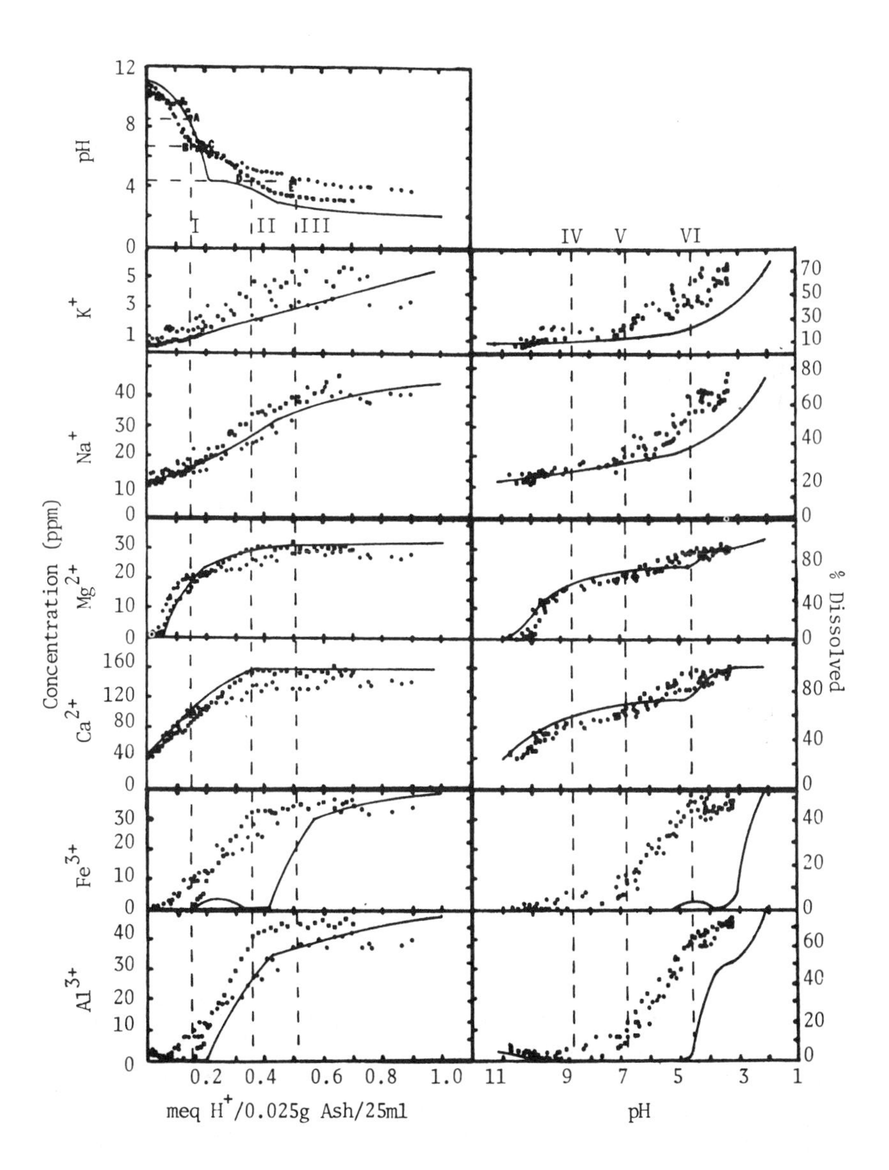

Figure 2. Salicylic and humic acid–fly ash equilibration data (● = humic acid, ■ = salicylic acid).

Table III. Composition of Fly Ash Fractions Sequentially Dissolved by Salicylic and Humic Acids

Fraction[a] No.	Chelating Agent	Na_2O	K_2O	CaO	MgO (mg/g ash)	Fe_2O_3	Al_2O_3	Total[b]
1	HA	27	2.0	139	36	13	15	232
2		22	2.2	55	13	38	58	188
3		8	0.1	7	11	0	4	20
4		28	5.7	23	3	55	40	155
Sum		85	10	224	53	106	117	595
1	SA	22	1.2	124	30	0	0	177
1a		3	0.6	24	7	3	2	40
2		22	3.5	74	16	46	76	237
3		14	0.6	0	0	4	9	28
4		24	4.1	2	1	53	30	114
Sum		85	10	224	54	106	117	596
					(mmol/g ash)			
1	HA	0.43	0.021	2.48	0.89	0.08	0.15	4.05
2		0.35	0.023	0.98	0.32	0.24	0.57	2.48
3		0.13	0.001	0.12	0.02	0	0.04	0.31
4		0.46	0.061	0.41	0.08	0.34	0.39	1.74
Sum		1.37	0.106	3.99	1.31	0.66	1.15	8.59
1	SA	0.35	0.013	2.21	0.74	0	0	3.31
1a		0.05	0.006	0.43	0.17	0.02	0.02	0.70
2		0.35	0.037	1.32	0.40	0.29	0.74	3.14
3		0.23	0.006	0	0	0.03	0.09	0.36
4		0.39	0.043	0.05	0.02	0.32	0.29	1.11
Sum		1.37	0.105	4.01	1.33	0.66	1.14	8.61

[a]1 = alkali; 1a = less soluble (alkali) subfraction; 2 = Al-Fe; 3 = residue; 4 = fly ash not dissolved.

[b]Note that mg/g ash ÷ 10 = % of ash; % SiO_2 34.4.

Discussion

The vertical lines and labeled points A through E on Figure 2 can be used to obtain more detailed information on fly ash, the reaction of acids with fly ash and similarities and differences in dissolution of fly ash by humic and salicylic acids. First, note that lines I, II and III correspond with points A-B, D and E, respectively, on the fly ash titration curves. Also note that lines IV, V and VI, respectively, correspond with points A, B-C and D-E. Table IV gives the coordinates of points A through E.

Table IV. Coordinates of Labeled Points on Figure 2.

Point	pH	meq H^+/0.025 g ash/25 ml	Chelating Agent
A	8.55	0.150	SA
B	6.80	0.150	HA
C	6.80	0.197	SA
D	4.55	0.360	SA
E	4.55	0.510	HA

The metal oxides dissolving at meq H^+/0.025 g ash/25 ml between 0 and 0.15 (line I) are mostly Na_2O, K_2O, CaO and MgO. Hence, the portion of fly ash dissolved within those limits is designated as the "alkali fraction." The composition of the alkali fraction is shown in Table III. Line I intersects the salicylic acid-fly ash titration curve at point A and the humic acid-fly ash curve at point B. The difference in pH at points A and B is due to release of additional hydrogen ion from weakly acidic groups via the following or an analogous reaction:

$$\text{(aromatic ring)}-C(=O)O^- \; (\text{ortho } OH) + M^{n+} \rightleftarrows \text{(aromatic ring)}-C(=O)-O-M-O \; (\text{chelate}) + H^+ \qquad (1)$$

$$M^{n+} = Ca^{2+},\; Mg^{2+},\; (Fe^{3+},\text{ or } Al^{3+}\;?)$$

and also to consumption of base in hydrolysis reactions, for example:

$$(HA)_1-\overset{O}{\overset{\|}{C}}-O-(HA)_2 + OH^- \rightleftarrows (HA)_1-\overset{O}{\overset{\|}{C}}-O^- + HO-(HA)_2 \qquad (2)$$

Equation 2 shows hydrolysis of an ester linkage between two humic acid molecules as an example of a possible hydrolysis reaction. Evidence for hydrolysis will be presented later. In addition, ion exchange evidence will be presented which indicates that hydrolysis results in increased metal ion binding by humic acid.

Equilibrium constants for salicylic acid show that it does not chelate Ca^{2+} or Mg^{2+} (Fe^{3+} and Al^{3+} are not dissolved at point A); hence, only the strongly acidic (pK_a = 2.98) carboxylic groups are neutralized:

$$\text{C}_6\text{H}_4(\text{COOH})(\text{OH}) + M^{n+} \rightleftharpoons \left(\text{C}_6\text{H}_4(\text{COO})(\text{OH})\right)_n \cdot M + nH^+ \qquad (3)$$

The close agreement between the salicylic and the mineral acid data between the origin and line I is further evidence that Ca^{2+} or Mg^{2+} are not chelated by salicylic acid. Thus, the increased soluble cation concentrations for humic acid over salicylic acid at meq H^+ values between 0 and 0.15 indicate significant chelation of divalent and/or trivalent ions by humic acid. Acid groups on humic acid not in a configuration conducive to chelation are assumed to react in a manner analogous to that shown for salicylic acid. Finally, it should be pointed out that the small amounts of Al^{3+} and Fe^{3+} associated with humic acid at point B may not be chelated, but instead exist as adsorbed metal hydroxides.[68]

The portion of fly ash dissolving between lines I and II for salicylic acid and lines I and III for humic acid makes up the "Al-Fe fraction." Line II intersects the salicylic acid-fly ash titration curve at point D, and line III intersects the humic acid-fly ash titration curve at point E. The normally water-soluble metal salts dissolving in this range are intimately mixed with Al_2O_3 and Fe_2O_3 and hence will not dissolve unless the Al and Fe do. If all or most of the acidic groups on humic acid were ortho to phenolic groups, line III would converge on line II. However, since apparently a relatively small fraction of humic acid groups form strong chelates, more humic acid is needed to bring the slightly soluble Al and Fe oxides into solution. Based on the assumption (1) that only Al^{3+} and Fe^{3+} are chelated (via equation 1) at point E and (2) that they are not bound by any other mechanism, calculation of the percentage of acid groups involved in chelation yields a value near 10%. Comparison of the calculated mmol of hydroxide ion released from dissolution of all metal oxides at line III (0.26) with the value for meq H^+ added (0.51) indicates that about half the acidic groups remain protonated. This result is confirmed by the nearly horizontal slope of the humic acid-fly ash titration curve near point E.

Mass and charge balance equations applied to fly ash-salicylic acid data at line II show that all Fe and Al ions are bound as monohydroxy chelates and/or as dichelates. The two structures cannot be distinguished from purely mass-charge considerations; however equilibrium constant data indicate that dichelates are formed. Na^+, K^+, Mg^{2+} and Ca^{2+} ions exist as carboxylic acid salts as before. The calculated mmol salicylic acid (corrected for SO_3 in the fly ash) necessary to achieve the above conditions is 0.363, which agrees well with the meq H^+ value at line II (0.360).

The amount of fly ash dissolved at meq H^+ values to the right of line II for salicylic acid and line III for humic acid is rather small; hence this portion of the fly ash is termed the "residue fraction." Large excesses of both acids are necessary to dissolve any appreciable residual fly ash, thus indicating the slight solubility of the oxides in that fraction. The maximum soluble cation concentrations obtained with salicylic acid are the same as those when 1 *M* HCl is used for ash dissolution. When corrected for metal humate lost during separation of unreacted fly ash (see experimental section), humic acid cation concentrations are also equivalent to 1 *M* HCl values. Boiling 6 *M* HNO_3 is required to dissolve the next major (the magnetic) fraction of fly ash, which is largely Fe. Digestion using HF for dissolution of SiO_2 is required to release all cations present.

Lines IV-VI on the graphs with the pH abscissa merely serve to confirm the previous observations. They do, however, show some things more clearly than the other graphs. As previously mentioned, intermixing of Mg, Ca, Na and K with Al and Fe in the Al-Fe fraction can be seen clearly in the sharp increase in concentration for all ions between lines V and VI. Also, the considerable pH drop in the salicylic acid titration curve between points A and C when correlated with the small increase in cations dissolved between lines IV and V shows that there is a small subfraction of the alkali fraction which is less soluble. Hence, a considerably lower pH is necessary to dissolve it. Inspection of the cation against pH curves between lines IV and V shows that this less soluble subfraction contains basic oxides plus Fe_2O_3 and Al_2O_3.

To summarize briefly the analysis of Figure 2:

1. It is estimated that approximately 10% of the appreciably acidic groups on the humic acid used are carboxylic groups ortho to phenol groups.
2. At large fly ash/humic acid ratios, only the alkali fraction of fly ash is dissolved, and all humic acid groups are exchanged.
3. At the fly ash/humic acid ratio where all of the appreciably soluble Fe_2O_3 and Al_2O_3 is dissolved, only about half of the acid groups on humic acid are exchanged; the other remain protonated.
4. Salicylic acid chelates only Al^{3+} and Fe^{3+}; other ions form carboxylic salts.
5. Once the active Al and Fe oxides are dissolved, the presence of excess salicylic or humic acid results in little additional fly ash dissolution.

CHARACTERIZATION OF FLY ASH AND HUMIC ACID

Experimental

Available Base and Cations in Fly Ash

To determine the available base from alkali metal and alkaline earth hydroxides in fly ash, 0.2 to 0.3 g of precisely weighed fly ash was dissolved in 50 to 60 ml standard (∿ 0.1 *N*) mineral acid (HCl, H_2SO_4 or HNO_3). The ash-acid suspensions were stirred overnight. The amount of excess mineral acid present was determined by potentiometric titration with standard NaOH (∿ 0.2 *N*). A slow rate of back titration was necessitated by formation of Al^{3+} and Fe^{3+} hydroxides, which came to equilibrium rather slowly. All titrations were performed in an N_2 atmosphere. If an extremely long time (2-3 days) was taken per titration, three distinct breaks corresponding to the reactions $Al(OH)_x^{3-x} + OH^- \rightarrow$ $Al(OH)_3$, $OH^- + H^+ \rightarrow H_2O$ and $Mg(OH)_x^{2-x} + OH^- \rightarrow Mg(OH)_2$ were observed. Titrations performed within 1 hr yielded a single inflection: $OH^- + H^+ \rightarrow H_2O$.

Appreciably soluble cations from fly ash were determined by equilibrating ∿ 0.02 g fly ash in 25.00 ml 1 *M* HCl overnight and analyzing for Na^+, K^+, Ca^{2+}, Mg^{2+}, Al^{3+} and Fe^{3+} by atomic absorption. The magnetic and silicate fractions of fly ash were insoluble in 1 *M* HCl.

Acidic Groups in Humic Acid

Direct (NaOH) and back (excess NaOH, HCl) potentiometric titration techniques were used to determine the groups in deionized humic acid having appreciable acidity. Since the two types of titrations gave different results for total acidity, additional intermediate titrations were also carried out. In these intermediate titrations, amounts of NaOH less than that necessary for complete neutralization of acidic groups were added, the solutions stirred overnight and the titrations completed wtih additions of NaOH as in direct titrations. Equilibration after additions of base during direct titrations was quite slow, with 2-3 days usually required to obtain a curve which reflected equilibrium conditions. Back titration equilibration was much faster, with only 2-3 hr required to obtain the portion of the curve needed for determination of the end point. Intermediate titrations usually took about a day to complete.

Ion Exchange Experiments

The procedure for the ion exchange experiments was to add an excess of a 1000 to 3000 mg/l metal chloride or nitrate solution to 15.00 ml of deionized humic acid, adjust the pH to the desired value with known volumes of HCl and/or NaOH and equilibrate for an hour or more with stirring. If the metal ion added was known to form slightly soluble carbonates, the humic acid was purged with N_2 prior to the exchange and kept under N_2 throughout the experiment. Also, the adjusted pH values were never sufficiently high to induce metal hydroxide precipitation. The equilibrated samples were filtered and aliquots were titrated with EDTA using known procedures (Ca^{2+},[69] Cu^{2+},[70] Zn^{2+},[71] Ni^{2+},[72] Fe^{3+},[69] Pb^{2+}[71]). The amounts of metal ions bound in the insoluble humic acid-metal ion flocs were calculated from the titration data and known initial metal ion concentrations. In some experiments, the hydrolyzable groups in the humic acid were hydrolyzed prior to metal equilibration by adding NaOH to pH 13 and stirring overnight. The hydrolyzed samples were first neutralized with HCl and then immediately reacted with metal ions.

Results

The available OH^- in the fly ash based on several types of determinations is shown in Table V. The obvious conclusion is that the larger the percentage of fly ash dissolved, the more OH^- obtained. Dissolution in deionized water yielded the least free base because less ash was dissolved. Dissolution in excess mineral or salicylic acid yielded the highest practical value for free OH^- which could actually be obtained. The available OH^- calculated from total ash dissolution in aqua regia and HF could never be obtained in any industrial application. Close agreement between

Table V. Available Hydroxide Ion as a Function of the Medium Used for Ash Dissolution

Medium[a]	Method Used for Determination	meq OH^-/g ash
DI H_2O	Cation, SO_4^{2-} Analysis[b]	1.9 ± 0.1
DI H_2O	pH	0.9 ± 0.6
SA	Direct Titration, (point A, figure 2)	6.0 ± 0.1
SA	Cation, SO_4^{2-} Analysis (point A, figure 2)	5.9 ± 0.1
SA	Cation, SO_4^{2-} Analysis (point C, figure 2)	7.4 ± 0.1
SA	Cation, SO_4^{2-} Analysis (point D, figure 2)	11.3 ± 0.1
0.1 *N* H_2SO_4	Back Titration	11.6 ± 0.1
0.1 *N* HCl	Back Titration	11.8 ± 0.1
0.1 *N* HNO_3	Back Titration	11.5 ± 0.3
1 *M* HCl	Cation, SO_4^{2-} Analysis	11.9 ± 0.1
Aqua Regia-HF	Cation, SO_4^{2-} Analysis	12.7 ± 0.2

[a]SA = salicylic acid; aqua regia- HF treatment completely dissolves fly ash.
[b]meq OH = mmol(Na^+ + K^+) + 2 mmol (Ca^{2+} + Mg^{2+}) - 2 mmol (SO_4^{2-}).

available base determined from titration and that calculated from soluble cations, assuming that all were in the oxide form in fly ash, shows the validity of that assumption. Similarly, the poor precision and accuracy of the value for available base calculated from the pH of a water-fly ash slurry shows the inapplicability of that method.

Table VI shows the wide range of titration data obtained from initial experiments using 25.00 ml of deionized humic acid. The source of the discrepancy in titration results was revealed by the intermediate titration technique. That is, at pH approximately > 6, the hydrolysis of some group(s) on humic acid occurred. The initiation of base-consuming

Table VI. Initial Data from Potentiometric Titration of Deionized Humic Acid[a]

Total Acidity (meq H^+/ml HA)	Comment	pH at End Point[b]
0.0262	titrated rapidly near the end	8.3
0.0264	point	8.1
0.0272	titrated less rapidly near	8.5
0.0278	end point	8.1
0.0290	titrated slowly near end point	8.7
0.0290		8.8
0.0294[c]		8.2
0.0304	back titration data	8.7
0.0306		8.6
0.0307		9.1

[a] Sample volume = 25.00 ml (202 mg dry weight).
[b] First derivative was used to determine end point.
[c] 25 ml HA + 25 ml 0.1 *F* NaCl were titrated.

hydrolysis can be seen by comparing curves I-III and IV in Figure 3. Equilibration of deionized humic acid and sufficient base (curve IV) to give an initially alkaline solution caused hydrolysis to proceed until an acidic pH (6.6) was obtained and the more reactive groups were cleaved. The net result was a higher value for the total acidity. Equilibration at a quite basic pH (curve V) resulted in maximum hydrolysis, and back titration yielded a substantially higher value for total acidity. Table VII summarizes acidities obtained by potentiometric titration. Table VIII summarizes results of the cation exchange experiments performed on the deionized humic acid. The data are numbered to facilitate comparison with potentiometric data.

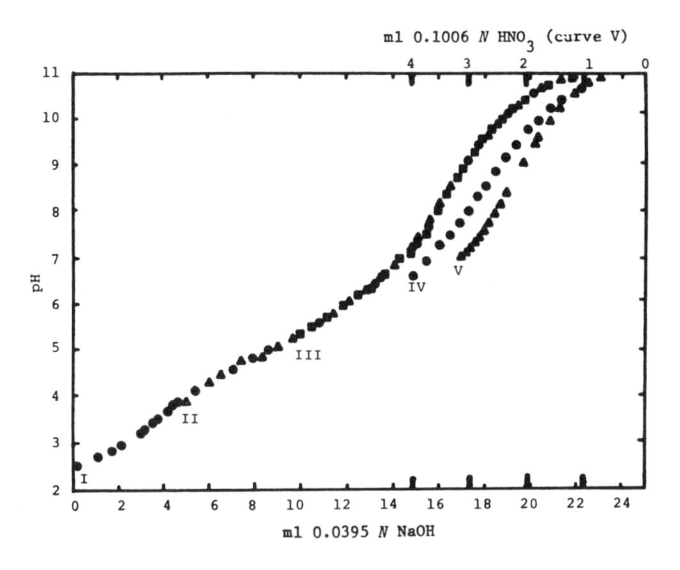

Figure 3. Direct (I), intermediate (II-IV), and back (V) titration curves of 25,00 ml deionized humic acid [Curve (I), (II) and (III) = ●, ▲ and ■ respectively].

Table VII. Quantitative Data from Figure 3

Total Acidity (meq H^+/ml HA)	Curve No.	Interpretation
0.0262[a]	I - III	Total free acid groups (TFAG)
0.0290	IV	TFAG + easily hydrolyzed groups
0.0306	V	TFAG + total hydrolyzable groups

[a]Value for total acidity assumed in fly ash-humic acid equilibration experiments.

Discussion

In most prior work where the available base in fly ash was determined, a sample of fly ash was refluxed with deionized water and an aliquot of the reflux titrated with HCl.[73] Since direct reaction of fly ash with acids was shown to result in a much greater utilization of available base, the error in that procedure should be obvious. This finding has additional significance in the use of fly ash for neutralizing acidic wastewater. For example, the mediocre performance of a water-fly ash slurry in removing SO_2 from stack gases would be greatly improved if the fly ash had been previously dissolved by humic acid. Thus, much more of the basic components of fly ash would have been in solution and the SO_2 neutralized by the basic humates

$$nSO_{2(g)} + nH_2O + M^{n+}_{-}(HA)_{(aq)} \rightleftarrows (HA)\downarrow + M^{n+}_{(aq)} + nHSO^-_{3(aq)} \quad (4)$$

Table VIII. Ion Exchange Data for Deionized Humic Acid

Result No.	Mmol M^{n+} bound/ml HA[a]	Initial $[M^{n+}]$ (*M*)	Final $[M^{n+}]$ (*M*)	Cation	pH	Prior Hydrolysis of Esters?
1	0.0180	0.0213	0.0105	Ca^{2+}	11.4 ± 0.1	yes
2	0.0181	0.0180	0.0102	Cu^{2+}	5.0 ± 0.1	yes
3	0.0154	0.0126	0.0034	Cu^{2+}	5.0 ± 0.1	yes
4	0.0154	0.0180	0.0114	Cu^{2+}	5.0 ± 0.1	no
5	0.0149	0.0152	0.0046	Ca^{2+}	9.5 ± 0.2	yes
6	0.0144	0.0175	0.0113	Zn^{2+}	6.36	yes
7	0.0139	0.0126	0.0043	Cu^{2+}	5.0 ± 0.1	no
8	0.0134	0.0126	0.0046	Cu^{2+}	4.8 ± 0.1	no
9	0.0133	0.0200	0.0144	Ni^{2+}	6.2 ± 0.1	yes
10	0.0129	0.0133	0.0036	Ca^{2+}	9.9 ± 0.1	no
11	0.0125	0.0126	0.0051	Cu^{2+}	4.30	no
12	0.0111	0.0213	0.0146	Ca^{2+}	7.9 ± 0.1	no
13	0.0111	0.0152	0.0073	Ca^{2+}	7.9 ± 0.1	no
14	0.0110	0.0205	0.0156	Fe^{3+}	2.05	yes
15	0.0109	0.0112	0.0037	Ni^{2+}	6.47	no
16	0.0099	0.0126	0.0066	Cu^{2+}	3.71	no
17	0.0097	0.0112	0.0046	Ni^{2+}	6.22	no
18	0.0090	0.0112	0.0050	Ni^{2+}	5.6 ± 0.1	no

[a]The gravimetric concentration of the DI HA was 8.06 ± 0.06 mg/ml. The reproducibility of replicate samples using *identical* conditions was ± 0.5 to 1%. The quantity of DI HA used for each equilibration was 15.00 ml.

Similarly, neutralization of acid in wastewaters by fly ash would be enhanced by prior dissolution of fly ash in humic acid. The proper ratio of fly ash to humic acid would be determined by the available base in the fly ash and the acidity of the humic acid.

Since cation binding by humic acid occurs largely via exchange of hydrogen ion bound by carboxylic and phenolic groups, there should be some relationship between the total acidity of a particular batch of humic acid (as measured by potentiometric titration with NaOH) and its ability or capacity to bind cations. However, this relationship is obscured by several factors including (1) the presence of groups in humic acid which undergo hydrolysis under basic conditions to effectively increase its acidity (Table VII), (2) the presence of several types of groups in humic acid with widely varying tendencies toward ionization and binding of metal ions, (3) the multitude of possible types of humic acid-metal ion interactions due to the structural complexity of humic acid, and (4) the tendency of each metal ion species to interact with humic acid in a manner most compatible with its own chemistry. Although no model could comprehensively account for all the above factors, an attempt is made below to correlate the ion exchange data (Table VIII) and the measured deionized humic acid acidities (Table VII) using a relatively simple (and hence approximate) model. The first point dealt with in the model is the apparent discrepancy in the maximum values for humic acid acidity obtained with titration vs ion exchange techniques.

The highest value for exchange (Nos. 1 and 2, Table VIII) with Ca^{2+} or Cu^{2+} multiplied by two equals 0.0360 meq/ml, which greatly exceeds the highest titration acidity (0.0306 meq/ml, Table VII). Although numerous explanations exist for this discrepancy, the two most probable are:

$$\text{Ar(-COO}^-\text{)(-OH)} + M^{2+} \leftrightarrows \text{Ar(-C(=O)-O-M-O-)} + H^+ \quad (5)$$

$$(HA)_1-\overset{\overset{\large O}{||}}{C}-O-(HA)_2 + OH^- \rightarrow (HA)_1-\overset{\overset{\large O}{||}}{C}-O^- + HO-(HA)_2 \quad (6a)$$

$$(HA)_1-\overset{\overset{\large O}{||}}{C}-O^- + HO-(HA)_2 + X^- + M^{2+} \rightleftarrows (HA)_1-\overset{\overset{\large O}{||}}{C}-O-M----\overset{H}{\underset{\underset{X}{\vdots}}{O}}-(HA)_2 \quad (6b)$$

$$(HA)_1-\overset{\overset{\large O}{||}}{C}-O^- + HO-(HA)_2 + M^{2+} \rightleftarrows (HA)_1-\overset{\overset{\large O}{||}}{C}-O-M-O-(HA)_2 + H^+ \quad (6c)$$

In reaction 5 or an analogous one, a single group (carboxylic) having a titratable proton binds a divalent metal ion with the aid of a nearby group of weak acidity (pK_{a2} of salicylic acid = 13.4). If equation 5 were the only binding mechanism, one would observe 1:1 equivalence between the titration acidity and the mol of M^{2+} bound. This has in fact been observed for some humic acids.[37,46]

Equations 6 describe two possible forms of cation binding following hydrolysis of an intra- ($HA_1=HA_2$) or intermolecular ($HA_1 \neq HA_2$) ester linkage. Similar mechanisms could be formulated by other bonds cleaved by OH^-, but in light of the structural evidence available, ester (and lactone) linkages appear the most probable in accounting for the observed humic acid hydrolysis. In either 6b or 6c, hydrolysis consuming a single hydroxide ion leads to the binding of a divalent ion. Hence, the relationship between titration acidity and mol of M^{2+} bound is the same for reaction 5. Based on relative hydroxyl acidities, 6b is more probable for esters formed from aliphatic alcohols, and 6c is more probable for those from phenols.

Thus, reactions analogous to 5 and 6 must occur to account for the high divalent exchange values (0.0180 mmol/ml, 0.0360 meq/ml) obtained. Assuming that the total hydrolyzable groups in humic acid is the difference between the high and low values in Table VII [0.0306-0.0262 = 0.0044 (meq/ml)], and that the maximum divalent cation capacity is 0.0181 mmol/ml; the difference between them (0.0137 mmol) should be the sum of the divalent metal ions bound via reactions 5 and 7.

$$(HA)_1{-}COOH + (HA)_2{-}COOH + M^{2+} \rightleftarrows (HA)_1{-}\overset{\overset{\displaystyle O}{\|}}{C}{-}O{-}M{-}O{-}\overset{\overset{\displaystyle O}{\|}}{C}{-}(HA)_2 \qquad (7)$$

Comparison of data in Table VI and VII with data in Table VIII leads to the conclusion that nearly complete exchange of acidic groups in humic acid occurs with addition of excess metal ions. Thus, in all probability, binding via the mechanism shown in Equation 7 often occurs between two different humic acid molecules (*i.e.*, $HA_1 \neq HA_2$). That is, since current structural evidence indicates the presence of isolated acidic groups in humic acid which are sterically unable to chelate divalent and trivalent metal ions with the aid of nearby groups on the same molecule,[33,74,75] formation of intermolecular metal ion bridges with similarly isolated groups on other molecules seems to be the only logical explanation. Formation of intermolecular bridges via metal ion bonding is further supported by the observed formation of humic acid-metal ion precipitates at pH as high as 11.5 (Table VIII). Considering the high degree of ionization and hydrolysis of groups in humic acid under basic conditions, precipitation under such conditions could occur only in the presence of strong intermolecular bonding via interactions with metal ions. The role of intermolecular metal ion and hydrogen bonding in humic acid precipitation is discussed further in a recent paper.[37] In cases where $HA_1 = HA_2$, appreciable binding analogous to phthalic acid is probable since the presence of significant proportions of phthalate groups in humic acid has been demonstrated.[4,74,76] Since only phenol (pK_a phenol = 10.0) groups near electron withdrawing groups (*e.g.*, the nitro group; pK_a 2,4-dinitrophenol = 4.08) have sufficient acidity to contribute to total acidity as determined by titration with NaOH, it is assumed that most groups reacting as in Equation 7 are carboxylic.

Assuming, as calculated earlier, that 10% of the acid groups react according to Equation 5, Equation 7 accounts for the bulk of the ion exchange by the particular batch of humic acid used in this work. The estimated contribution from the types of binding shown in Equations 5-7 is shown in Table IX.

Some correlation of this discussion with data in Table VIII can be seen. For example, metal uptake by samples of humic acid previously hydrolyzed is always greater than for samples without prior hydrolysis (compare, for example, results 5 and 10, 2 and 4, 9 and 15). However, the increase in metal ion uptake in hydrolyzed samples varies with experimental conditions. This variation can be accounted for by assuming partial reformation of linkages previously hydrolyzed. Linkage reformation would naturally be favored at low pH and low metal ion concentrations. For example, esterification is catalyzed by hydrogen ion[77]

$$H_2O + (HA)_1{-}\overset{\overset{\displaystyle O}{\|}}{C}{-}O{-}(HA)_2 \overset{H^+}{\leftrightarrows} (HA)_1{-}\overset{\overset{\displaystyle O}{\|}}{C}{-}OH + HO{-}(HA)_2 \qquad (8)$$

and hence the equilibrium is shifted to the right by the presence of hydroxide or metal ions

$$(HA)_1{-}\overset{\overset{\displaystyle O}{\|}}{C}{-}OH + OH^- \rightarrow (HA)_1{-}\overset{\overset{\displaystyle O}{\|}}{C}{-}O^- \qquad (9)$$

$$(HA)_1{-}\overset{\overset{\displaystyle O}{\|}}{C}{-}OH + (HA)_2{-}OH + M^{2+} \rightarrow 2H^+ + (HA)_1{-}\overset{\overset{\displaystyle O}{\|}}{C}{-}O{-}M{-}O{-}(HA)_2 \qquad (10)$$

Table IX. Estimated Contributions from Different Binding Mechanisms To the Deionized Humic Acid Exchange Capacity

Mmol M^{2+} Exchanged/ml HA	Mechanism	Equation # (see text)	% of Total CEC[a]
0.0123	Exchange with strongly acidic groups	7	68
0.0014	*o*-carboxylic-phenolic chelation	5	8
0.044	Exchange after hydrolysis	6 a-c	24
0.0181[b]	Total		100

[a]CEC = cation exchange capacity.
[b]2.24 mmol M^{2+}/g HA.

Thus, the increased humic acid acidity due to hydrolysis may be partially or totally lost at low pH and/or concentration of free metal ion *(cf.* result 14, Table VIII). For example, comparison of results 2 and 3 indicates that reversal of hydrolysis has occurred to a larger extent in result 3. Also, comparison of results 4 and 7 suggests that hydrolysis in the absence of NaOH pretreatment may be induced at higher metal ion concentrations by shifting Equation 8 to the right via Equation 10.

The occurrence of some data in Table VIII which correlate well with the numerical estimates of the different binding modes (Table IX) may only be coincidence, but it is worth pointing out. For example, several results are obtained which are close to 0.0137 mmol/ml, the number estimated for the contribution via Equation 5 and 7 to the total exchange capacity. Also, the several data near 0.0110 mmol/ml day may indicate saturation of a particular type of group (*e.g.*, all aromatic carboxyls) via the mechanism shown in Equation 7. Data between values in Table IX are explained by combinations of the various types of reactions. As a single example of this, result 11 could be explained as the sum of the total participation of chelating groups (0.0014 mmol Cu^{2+}/ml) plus the participation Equation 7 type groups to the extent of 0.0111 mmol Cu^{2+}/ml.

Finally, it should be apparent that the selection of a value for the acidity of the deionized humic acid is somewhat arbitrary. For the purpose of plotting the humic acid-fly ash data (Figure 2), the total acidity was taken to be 0.0262 meq H^+/ml HA (Table VII) because most of the data occurred at low pH and low metal ion concentrations.

Summary and Recommendations

Obviously, the effective acidity of a humic acid solution and its relation to ion exchange capacity is complicated by several factors. Thus, the reader is cautioned to view published methods[4,6,33,34,46,54,78-81] which claim specificity for carboxylate groups, for example, with some skepticism. Other researchers have also expressed doubts concerning specificity of methods.[75,82,83] Similarly, data from nonaqueous titrations[80,81] may not accurately reflect the effective humic acid acidity in water. As an estimate of the effective total acidity of a given humic acid, reaction with excess Cu^{2+} at pH 5.0 ± 0.1 is recommended. The unbound Cu^{2+} can be determined as in the experimental section or by any convenient method. Hydrolysis with NaOH may be used in conjunction with Ca^{2+} (pH 11.4) or Cu^{2+} (pH 5.0) exchange to check for presence of hydrolyzable groups. Other exchange conditions may be used which more realistically reflect the conditions under which the humic acid-metal ion reactions occur in a given application.

Exchange, filtration of metal humate, and determination of unbound metal ions are steps which can be rapidly and conveniently carried out. Special sample preparation is unnecessary, although prewash with acid may be advisable to remove metal ions already in the humic acid. Washing with acid is a commonly used, simple and effective procedure for removing most metals bound to humic acid.[84] Extraction with dithizone in chloroform or CCl_4 is also effective.[85,86]

In contrast to other published methods, the presence of free acids or bases does not interfere with the type of ion exchange procedure suggested here. For example, potentiometric titration techniques require that the sample be free from all acids or bases. Hence, extensive treatment with ion exchange resins or some similar procedure is a required pretreatment. Titration in nonaqueous solvents imposes the additional requirement that the humic acid be in dry solid form. Ion exchange techniques in which the hydrogen ion released by the exchange is measured also require humic acid samples free from other acids or bases.

Methods for measuring acid functionality based on formation of organic derivatives, hydrolysis of the derivative and, finally, measurement of the nonhumic acid product are impractical due to the large amount of time and effort they require. Titration techniques are also tedious both in time spent obtaining the titration curve and in removing interfering acids or bases. Acetate salt exchange techniques require a lengthy distillation step for separation of the acetic acid product. Thus, in spite of the sometimes pronounced effect of experimental conditions on the result obtained, the exchange procedure suggested here is the only available one which is practical for routine measurement of humic acid acidities.

REFERENCES

1. Matsuda, Y., *et al. Kogyo Yosui* 204:16 (1975); *Chem. Abstr.* 84: 140368k.
2. Adhikari, M., *et al. Res Ind.* 20:70 (1976).
3. Schnitzer, M. In *Organic Compounds in Aquatic Environments,* S. J. Faust and J. V. Hunter, Eds. (New York: Marcel Dekker, Inc. 1971).
4. Schnitzer, M., and S. U. Khan. *Humic Substances in the Environment,* (New York: Marcel Dekker, Inc., 1972).
5. Manahan, S. E. *Environmental Chemistry,* 2nd ed. (Boston: W. Grant, 1976), pp. 97-100.
6. Gamble, D. S., and M. Schnitzer. In *Trace Metals and Metal Organic Interactions in Natural Waters,* P. C. Singer, Ed. (Ann Arbor, MI: Ann Arbor Science Publishers, Inc., 1973).
7. Manskaya, S. M., and T. V. Drozdova. *Geochemistry of Organic Substances* (New York: Pergamon Press, Inc., 1968).

8. Howard, H.C. In *Chemistry of Coal Utilization,* H. H. Lowry, Ed. (New York: John Wiley and Sons, Inc., 1945).
9. Bailey, A. E. W., *et al. Fuel* 33:209 (1954).
10. Friedman, L. D., and C. R. Kinney. *Ind. Eng. Chem.* 42:2525 (1950).
11. Thomson, G. H. *J. Appl. Chem.* 2:603 (1952).
12. Lilly, V. G., and C. E. Garland. *Fuel* 11:392 (1932).
13. Mazumdar, B. K., *et al. Fuel* 46:379 (1967).
14. Higuchi, K., *et al.* Japanese Patent 70,23,321 (1970); *Chem. Abstr.* 73:122248q.
15. Shibuya, H., *et al.* Japanese Patent 71,06,609 (1971); *Chem. Abstr.* 75:57265w.
16. Mizmoto, K. Japanese Patent 75,05,301 (1975); *Chem. Abstr.* 84: 138891g.
17. Eye, J. D., and T. K. Basu. *J. Water Poll. Control Fed.* 42:R125 (1970).
18. Ballance, R. C., *et al.* U.S. Bureau of Mines Report Invest. No. 6869, Washington, D. C., (1966).
19. Tenny, M. W., and T. G. Cole. *J. Water Poll. Control Fed.* 40: R281 (1968).
20. Adhikari, M., *et al. J. Inst. Chem.* 47:165 (1975).
21. Adhikari, M., *et al. J. Indian Chem. Soc.* 52:127 (1975).
22. Randall, C. W., *et al. Proc. Water Resources Poll. Control Conf.,* 20:208 (1971).
23. Mancy, K. H., *et al. Proc. Purdue Ind. Waste Conf.,* 146 (1964).
24. Kolar, L. *Chem. Prum.* 17:378 (1967); *Chem. Abstr.* 67:101717c.
25. Kolar, L. *Rostl. Vyroba* 12:1055 (1966); *Chem. Abstr.* 66:49533c.
26. Katoh, S., and Y. Kimura. *Kami Pa Gihyoshi* 25:168 (1971); *Chem. Abstr.* 74:143503e.
27. Kawamura, Y., *et al.* Japanese Patent 73,34,367 (1973); *Chem. Abstr.* 81:14322r.
28. Otto, R., and P. Hecht. *Isotopenpraxis* 7:412 (1971); *Chem. Abstr.* 76:37164b.
29. Kononova, M. M. *Soil Organic Matter,* 2nd ed. (New York: Pergamon Press, Inc., 1966).
30. De Borger, R. *C. R. Acad. Sci., Ser. D* 269:1564 (1969); *Chem. Abstr.* 72:54117y.
31. Khan, S. U., and M. Schnitzer. *Soil Sci.* 112:231 (1971).
32. Schnitzer, M. In *Soil Biochemistry,* A. D. McLaren and J. J. Skujins, Eds. (New York: Marcel Dekker, Inc., 1971).
33. Bailey, N. T., and G. J. Lawson. *Fuel* 44:63 (1965).
34. Brigs, G. C., and G. J. Lawson. *Fuel* 49:39 (1970).
35. MacCarthy, P., and S. O'Cinneide. *J. Soil Sci.* 25:420 (1974).
36. Schnitzer, M., and S. I. M. Skinner. Proc. Symp. Use Isotopes and Rad. in Soil Org. Mat. Studies, International Atomic Energy Agency, Vienna, 1968.
37. Green J. B., and S. E. Manahan. *J. Can. Chem.* 55:3248(1977).
38. Khan, S. U. *Residue Rev.* 52:1 (1974).
39. Gaillardon, P. *Weed Res.* 15:393 (1975).

40. Khan, S. U. *J. Environ. Qual.* 3:202 (1974).
41. Khan, S. U. *Can. J. Soil Sci.* 53:199 (1973).
42. Khan, S. U., and R. Mazurkewich. *Soil Sci.* 118:339 (1974).
43. Banerjee, S. K., *et al. J. Indian Chem. Soc.* 53:186 (1976).
44. Szalay, A., and M. Szilagyi. In *Advances in Organic Geochemistry*, Proc. 4th Int. Meeting Org. Geochem., P. A. Schenk, and I. Havenarr, Eds. (New York: Pergamon Press, Inc., 1968).
45. Benes, P., *et al. Water Res.* 10:711 (1976).
46. Green, J. B. M.A. Thesis, University of Missouri, Columbia, MO (1975).
47. Khan, S. U. *Soil Sci. Soc. Am. Proc.* 33:851 (1969).
48. Schnitzer, M. *Soil Sci. Soc. Am. Proc.* 33:75 (1969).
49. Buffle, J. A. E. *T.S.M.–L'EAU* 72:3 (1977).
50. Gamble, S. S., *et al. Can. J. Chem.* 48:3197 (1970).
51. Stevenson, F. J. *Soil Sci.* 123:10 (1977).
52. Manning, P. G., and S. Ramamoorthy. *J. Inorg. Nucl. Chem.* 35: 1577 (1973).
53. Ramamoorthy, S., and P. G. Manning. *J. Inorg. Nucl. Chem.* 36: 695 (1974).
54. Gamble, D. S. *Can. J. Chem.* 48:2662 (1970).
55. Mangravite, F. J., *et al. J. Am. Water Works Assoc.* 67:88 (1975).
56. Cassell, E. A., *et al. Water Res.* 9:1017 (1975).
57. Mantoura, R. F. C., and J. P. Riley. *Anal. Chim. Acta* 76:97 (1975).
58. Babu, S. P., Ed. "Trace Elements in Fuel," *Adv. Chem. Ser.* 141, (Washington, D.C.: Am. Chem. Soc., 1975).
59. Ondov, J. M., *et al. Anal. Chem.* 47:1102 (1975).
60. Perkins, H. C. *Air Pollution* (New York: McGraw-Hill Book Company, 1974).
61. Manz, O. E. In U.S. Bureau of Mines Information Circular 8650, G. H. Gronhovd, and W. R. Kube, compilers (1974), pp. 204-219.
62. Fisher, G. L., *et al. Science* 192:553 (1976).
63. Pedlow, J. W. In U.S. Bureau to Mines Information Circular 8640, J. H. Faber, *et al.*, compilers, (1974), pp. 33-43.
64. Davison, R. L., *et al. Environ. Sci. Technol.* 8:1107 (1974).
65. Linton, R. W., *et al. Science* 191:852 (1976).
66. Natusch, D. F. S., *et al. Science* 183:202 (1974).
67. Green, J. B., and S. E. Manahan. *J. Inorg. Nucl. Chem.* 39:1023 (1977).
68. Ghassemi, M., and R. F. Christman. *Limnol. Oceanog.* 13:583 (1968).
69. Cheng, K. L., *et al. Anal. Chem.* 24:1640 (1952).
70. Flaschka, H. *Mikrochemie Ver. Mikrochim. Acta.* 39:38 (1952); 40:42 (1952).
71. Korbl, J., *et al. Collection Czechoslov. Chem. Commun.* 22:961 (1957).
72. Harris, W. F., and T. R. Sweet. *Anal. Chem.* 24:1062 (1952).
73. Martens, D. C. *Compost Sci.* 12(6):15 (1971).
74. Wagner, G. H., and F. J. Stevenson. *Soil Sci. Soc. Am. Proc.* 29: 43 (1965).
75. Van Dijk, H. In *Use of Isotopes in Soil Organic Matter Studies*, Suppl. to *J. Appl. Rad. Isotopes*, 1st ed. (New York: Pergamon Press, Inc., 1966), pp. 129-141.

76. Neyroud, J. A., and M. Schnitzer. *Can. J. Chem.* 52:4123 (1974).
77. Morrison, R. T., and R. N. Boyd. *Organic Chemistry,* 2nd ed. (Boston: Allyn and Bacon, Inc., 1966), pp. 675-679.
78. Schaffer, H. N. S. *Fuel* 49:197 (1970).
79. Borggaard, O. K. *Acta Chem. Scand.* 28:121 (1974).
80. Van Dijk, H. *Sci. Proc. Royal Dublin Soc., Ser. A,* 1:163 (1960).
81. Maher, T. P., and H. N. S. Schaffer. *Fuel* 55:138 (1976).
82. Fuchs, W., and A. G. Sandhoff. *Fuel* 19:45 (1940); 19:69 (1940).
83. Stevenson, F. J., and K. M. Goh. *Soil Sci.* 113:334 (1972).
84. Basu, A. N., *et al.* *J. Indian Soil Sci.* 12:311 (1964).
85. Hodgson, J. F., *et al.* *Soil Sci. Soc. Am. Proc.* 29:665 (1965).
86. Kloecking, R., and D. Muecke. *Z. Chem.* 9:453 (1969).

CHAPTER 23

CHEMISTRY OF HEAVY METALS IN ANAEROBIC DIGESTION

Thomas L. Theis and Thomas D. Hayes
Department of Civil Engineering
University of Notre Dame
Notre Dame, Indiana

INTRODUCTION

Anaerobic digestion is a very old and efficient means of handling concentrated organic wastes. Traditionally, it has been reserved for the treatment of municipal sludges and high-strength industrial wastes; however, because of the production of a usable and convenient energy by-product (methane), it is currently being examined more critically for application to the degradation of other types of organic wastes. Several excellent reviews on the biochemistry of the anaerobic digestion process exist[1-3]; therefore, this aspect will not be considered further here.

One difficulty associated with anaerobic digestion stems from the concentrated nature of the wastes which are treated. In most wastewater treatment schemes, a variety of substances, many of them toxic, are concentrated within the sludge. For example, several researchers have noted the magnification of chromium, copper, nickel, zinc, mercury, cadmium, lead, manganese and iron in wastewater sludges[4-9]. In some cases the metals were concentrated by several hundred times on a volumetric basis[10-12]. This is often the root cause of digester failures the reasons for which, while sometimes evident, often remain unclarified. This has brought about an undertone of distrust of the process because of its seemingly capricious and inexplicable nature.

In spite of the widespread usage of anaerobic digestion and the length of time for which it has been practiced there have been surprisingly few studies relating to a basic understanding of the chemical interactions which occur among sludge components. This is especially evident in those studies on heavy metals. In general, most researchers have been concerned with documenting the outward effects on the process.

Probably the most complete effort of this type was reported by Barth[13] on a series of studies for Cr, Cu, Ni and Zn. Several studies have demonstrated the ability of aerobic bacteria to sorb heavy metals from solution[14-17]; however, little has been reported previously on metal uptake by anaerobic bacteria. Gould and Genetelli[18] were able to fractionate anaerobic sludges according to size ranges, and they performed analyses for metals on the various fractions. They, along with other researchers, found most of the heavy metals to be removed within the sludge. Previous work by the authors has verified this finding.

In this study the chemistry of Cd, Cr, Cu, Pb, Ni and Zn in anaerobic digestion was examined with particular attention focused on the distribution of the metals in the digester sludge. Based upon work already performed,[19] the following conclusions appear warranted:

1. For the metals studied, their distribution is almost exclusively between the controlling inorganic precipitate and the sludge biomass fraction.
2. Equilibria involving the distribution of heavy metals are established rapidly.
3. Evidence strongly suggests the active transport of metals to the interior of the bacterial cell.

In this chapter, the information obtained previously will be treated in a quantitative manner in order to establish a basic understanding of the chemistry of these heavy metals in the anaerobic digester environment.

CHEMICAL EQUILIBRIA INVOLVING HEAVY METALS

The chemistry of heavy metals in natural waters is generally quite complex. However, the specialized environment within the anaerobic digester simplifies considerably the equilibria which need to be considered. Table I summarizes relevant operating parameters which affect heavy metal interactions. It is evident that, among inorganic precipitates, the hydroxides, carbonates, phosphates and sulfides are of greatest importance. The low oxidation reduction potential implies that only reduced species would be formed, while the relatively narrow pH operating values will limit the need for considering reactions outside of this range. Furthermore, the high concentrations of coordinating ligands favor formation of an inorganic

Table I. Chemical Environment Maintained in Conventional Anaerobic Digestion (As Reported in References 20 and 21)

Variable	Range
pH	6.5 to 7.5
E_H (mv)	-300 to -530
Alkalinity (mg/l as $CaCO_3$)	1,500 to 5,000
Total Phosphorus (mg/l as P)	350 to 750
Sulfide	100 to 400

solid phase. Chemical equilibria of interest for heavy metals are given in Table II.

Table II. Chemical Equilibria for Heavy Metals

Reaction	log K	Reference
Cadmium		
$Cd^{+2} + 2e^- = Cd_{(s)}^\circ$	-13.66	23
$Cd(OH)_{2(s)} = Cd^{+2} + 2OH^-$	-13.75	23
$CdCO_{3(s)} = Cd^{+2} + CO_3^{-2}$	-11.32	23
$Cd_3(PO_4)_{2(s)} = 3Cd^{+2} + 2PO_4^{-3}$	-32.6	24
$Cd\ S_{(s)} = Cd^{+2} + S^{-2}$	-27.19	23
Chromium		
$Cr^{+2} + 2e^- = Cr_{(s)}^\circ$	-30.9	24
$Cr^{+3} + e^- - = Cr^{+2}$	-6.9	24
$HCrO_4^- + 7H^+ + 3e^- = Cr^{+3} + 4H_2O$	60.6	24
$HCrO_4^- = H^+ + CrO_4^{-2}$	-6.5	24
$Cr(OH)_{2(s)} = Cr^{+2} + 2OH^-$	-17	24
$Cr(OH)_{3(s)} = Cr^{+3} + 3OH^-$	-30.31	23
$Cr(OH)_{3(s)} + OH^- = Cr(OH)_4^-$	-0.4	24
$Cr\ PO_{4(s)} = Cr^{+3} + PO_4^{-3}$	-21.62	22
Copper		
$Cu^+ + e^- = Cu_{(s)}^\circ$	8.80	23
$Cu^{+2} + 2e^- = Cu_{(s)}^\circ$	11.4	23
$Cu\ O + 2H^+ = Cu^{+2} + H_2O$	7.9	23
$2Cu\ O + 2H^+ + 2e^- = Cu_2O + H_2O$	22.7	23
$Cu_2O_{(s)} + 2H^+ = 2Cu^+ + H_2O$	-1.68	23

Table II. Continued

Reaction	log K	Reference
$Cu_2(OH)_2\ CO_{3(s)} = 2Cu^{+2} + 2OH^- + CO_3^{-2}$	-33.9	26
$Cu_3(PO_4)_{2(s)} = Cu\ O_{(s)} + H_2O = Cu\ O_2^{-2} + 2H^+$	-32.2	23
$3Cu^{+2} + 2PO_4^{-3}$	-36.9	24
$Cu\ S_{(s)} = Cu^{+2} + S^{-2}$	-36.17	23
$Cu_2\ S_{(s)} = 2Cu^+ + S^{-2}$	-48.94	23
Lead		
$Pb^{+2} + 2e^- = Pb_{(s)}^{\circ}$	-4.27	23
$Pb\ O_2 + 4H^+ + 2e^- = Pb^{+2} + 2H_2O$	49.15	23
$Pb^{+2} + 2H_2O = H\ PbO_2^- + 3H^+$	-28.08	25
$Pb\ CO_{3(s)} = Pb^{+2} + CO_3^{-2}$	-13.0	23
$Pb_3(PO_4)_{2(s)} = 3Pb^{+2} + 2PO_4^{-3}$	-42.1	24
$Pb\ S_{(s)} = Pb^{+2} + S^{-2}$	-28.16	23
Nickel		
$Ni^{+2} + 2e^- = Ni_{(s)}^{\circ}$	-8.47	23
$Ni(OH)_{2(s)} = Ni^{+2} + 2OH^-$	-15.88	23
$Ni^{+2} + 2H_2O = H\ NiO_2^-$	-30.48	26
$Ni\ CO_{3(s)} = Ni^{+2} + CO_3^{-2}$	-6.80	25
$Ni_3(PO_4)_{2(s)} = 3Ni^{+2} + 2PO_4^{-3}$	-30.3	24
$\alpha Ni\ S_{(s)} = Ni^{+2} + S^{-2}$	-20.68	23
Zinc		
$Zn^{+2} + 2e^- = 2n_{(s)}^{\circ}$	-25.9	23
$Zn(OH)_{2(s)} = Zn^{+2} + 2OH^-$	-15.86	26
$Zn^{+2} + 2H_2O = ZnO_2^{-2} + 4H^+$	-40.83	25
$Zn\ CO_{3(s)} = Zn^{+2} + CO_3^{-2}$	-9.85	23
$Zn_3(PO_4)_{2(s)} = 3Zn^{+2} + 2PO_4^{-3}$	-32.04	24
$\alpha Zn\ S_{(s)} = Zn^{+2} + S^{-2}$	-25.13	23

A convenient way of summarizing chemical information is through graphical representation of equilibria. Figure 1 (A) through (F) shows E_H (or $p\epsilon$) vs pH diagrams for each of the metals in this study. These diagrams were constructed using the data given in Table II plus acid-base equilibria for the carbonate, phosphate and sulfide systems. Ligand concentrations are approximately those which exist in typical anaerobic digester liquor ($C_T = P_T = S_T = 10^{-2}$ *M*). From these plots it can be seen that, with the exception of Cr, the sulfides of the metals exert primary inorganic solubility controls. Chromium, which is controlled by chrome(III) hydroxide, is not known to form a stable sulfide. For copper, equilibrium considerations indicate that cuprous sulfide, Cu_2S, controls solubility. All other metals would be expected to form simple sulfides.

EXPERIMENTAL METHODS

The experimental portion of the research was carried out in several bench-scale anaerobic digesters. A schematic of the setup is shown in Figure 2. Digesters were operated on a 10-day detention time at a temperature of 35°C. Feed consisted of anaerobic sludge from the South Bend Wastewater Treatment Plant. After a period of stabilization as indicated by baseline parameters (pH, alkalinity, volatile dry weight, TOC, volatile acids, gas production rate and composition), heavy metal addition commenced. Dosing consisted of adding the desired amount of a single metal to a feed aliquot, which was allowed to equilibrate for 1 hr, and then adding the aliquot to the digester. Several dosage levels were employed.

Mass balances were performed on each metal experiment. Metal fractions examined were the soluble, inorganic insoluble and biomass insoluble. This last fraction was further divided between intracellular and exocellular metals. The fractionation of the sludge was achieved through a scheme consisting of sieving, elutriation, filtration and washings in nitric acid and EDTA. Through this methodology a good quantitative separation of cell biomass from the sludge was achieved. All metal analyses were performed by atomic adsorption spectrophotometry. A more detailed explanation of the fractionation scheme can be found in Hayes and Theis.[19]

Some conclusions from this study were summarized previously. Metals were found to be distributed almost exclusively between the inorganic precipitates and cell biomass. EDTA washing of cells yielded few metals; however, cell lysis followed by analysis gave quantitative recovery of metals as indicated by total mass balances.

ANALYSIS OF RESULTS

The fact that an insignificant fraction of heavy metals is truly soluble in an anaerobic digester creates some interesting implications regarding their chemical behavior. If one considers the two major repositories of metals to be the inorganic precipitate and the organic cell complex, then for the former

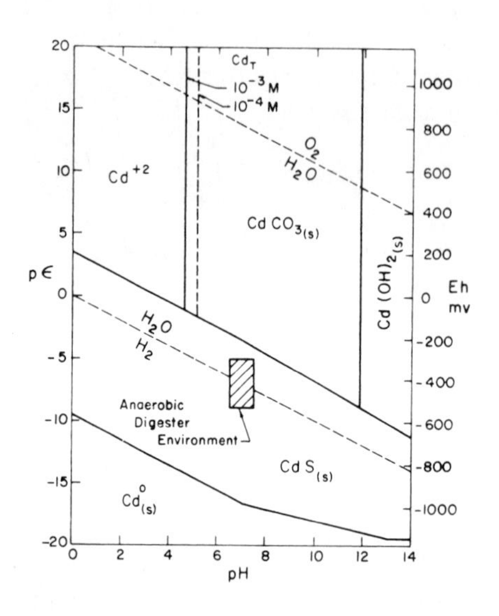

Cd_T
10^{-3} M
10^{-4} M
Cd^{+2}
O_2
H_2O
$Cd\,CO_{3(s)}$
$Cd\,(OH)_{2(s)}$
p∈
Eh
mv
H_2O
H_2
Anaerobic
Digester
Environment
$Cd\,S_{(s)}$
$Cd^{o}_{(s)}$
pH

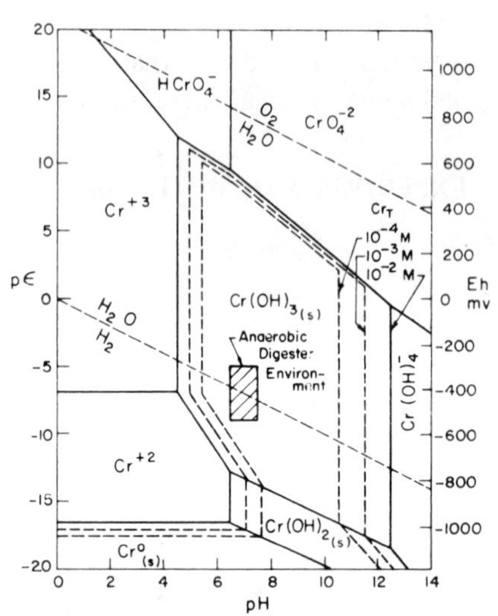

$HCrO_4^-$
O_2
H_2O
CrO_4^{-2}
Cr^{+3}
Cr_T
10^{-4} M
10^{-3} M
10^{-2} M
p∈
Eh
mv
H_2O
H_2
$Cr(OH)_{3(s)}$
Anaerobic
Digester
Environment
$Cr(OH)_4^-$
Cr^{+2}
$Cr(OH)_{2(s)}$
$Cr^{o}_{(s)}$
pH

A **B**

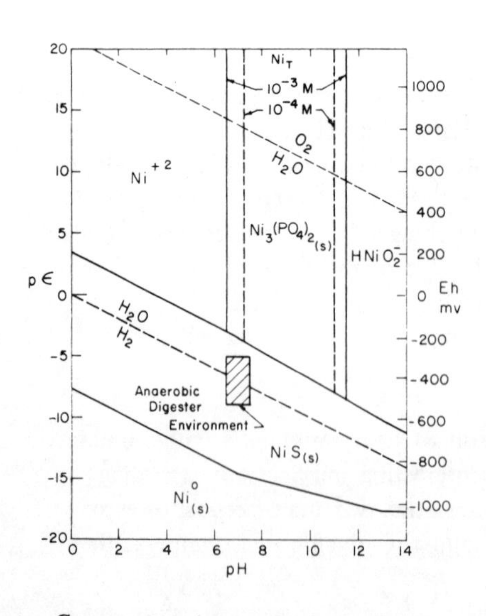

Ni_T
10^{-3} M
10^{-4} M
O_2
H_2O
Ni^{+2}
$Ni_3(PO_4)_{2(s)}$
$HNiO_2^-$
p∈
Eh
mv
H_2O
H_2
Anaerobic
Digester
Environment
$NiS_{(s)}$
$Ni^{o}_{(s)}$
pH

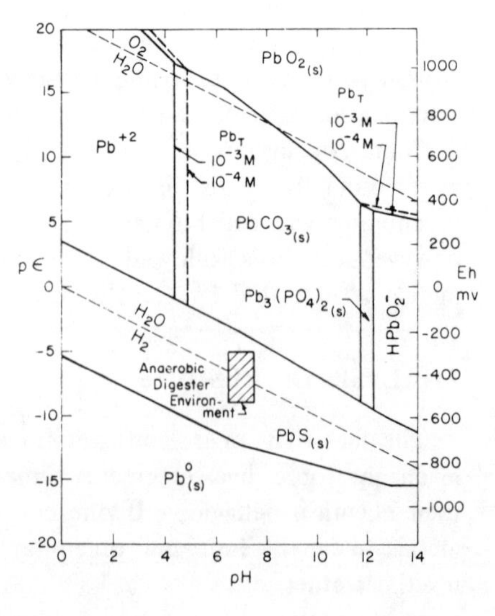

O_2
H_2O
$PbO_{2(s)}$
Pb_T
10^{-3} M
10^{-4} M
Pb^{+2}
Pb_T
10^{-3} M
10^{-4} M
$PbCO_{3(s)}$
p∈
Eh
mv
$Pb_3(PO_4)_{2(s)}$
$HPbO_2^-$
H_2O
H_2
Anaerobic
Digester
Environment
$PbS_{(s)}$
$Pb^{o}_{(s)}$
pH

C **D**

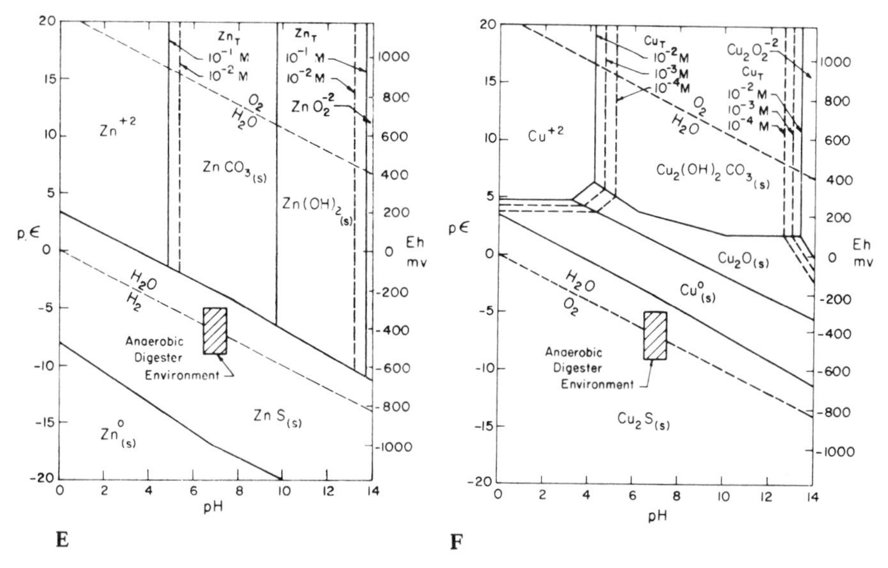

Figure 1. Predominance area diagrams in terms of pE and pH for heavy metals. Ligand concentrations are $C_T = P_T = S_T = 10^{-2}M$. (A) cadmium, (B) chromium, (C) nickel, (D) lead, (E) zinc and (F) copper.

$$nMe^{m+} + mL^{n-} = Me_nL_{m(s)}, \; K = \frac{1}{K_{so}} \tag{1}$$

and for the latter,

$$Me^{m+} + CELL = Me{-}CELL \tag{2}$$

If it is further accepted that the metals are actively transported to the cell interior, then

$$Me{-}CELL + A = Me{-}B + CELL + A \tag{3}$$

where A is an active transport complex of some form and Me–B represents metals bound within the bacterial cell membrane. Combining reactions 2 and 3

$$Me^{m+} = Me{-}B, \; K_B = \frac{[MeB]}{[Me^{m+}]} \tag{4}$$

Insofar as the metal is incorporated within a membrane it is not likely to be easily removed, at least within the range of the chemical environment, in an operating anaerobic system. Thus reaction 4 should probably be considered as essentially irreversible except in instances of extraordinary disruptive forces.

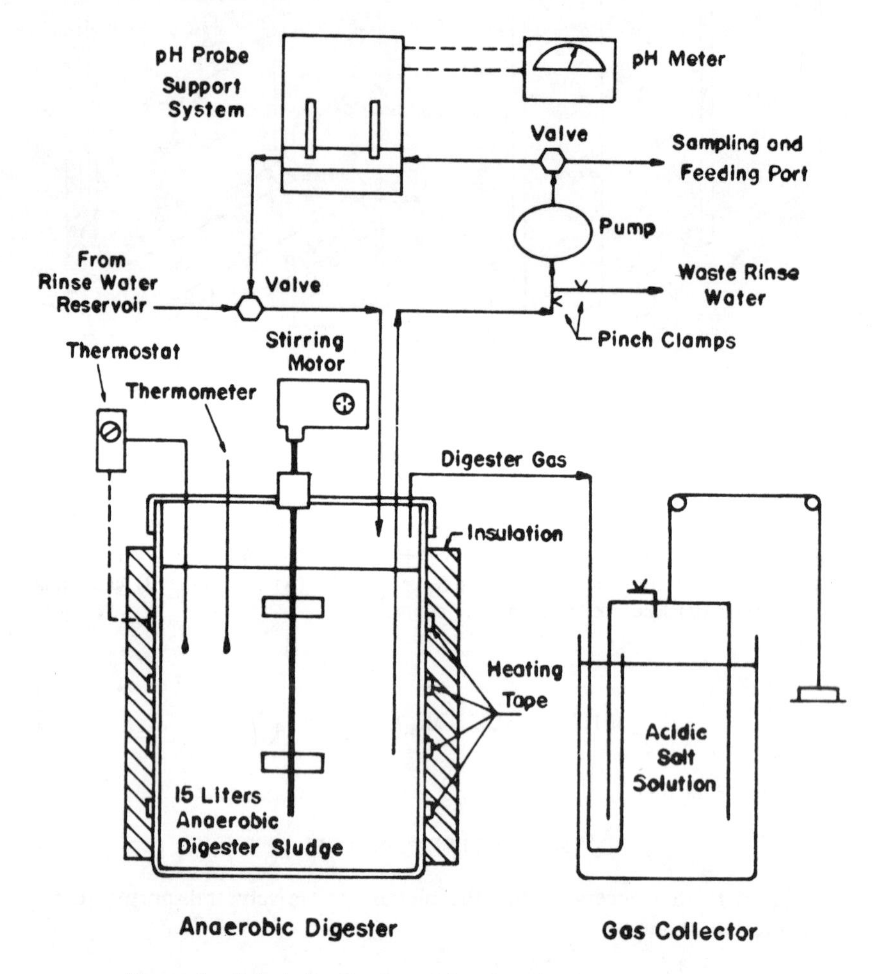

Figure 2. Schematic diagram of bench-scale anaerobic digester.

A good indication of the relative magnitude of K_B for each metal can be found by noting the type of inorganic precipitate, as given previously, and its solubility product. Writing a mass balance on total metals (ignoring soluble species)

$$[Me]_T = \left[Me_nL_{m(s)} \right] + [Me{-}B] \tag{5}$$

it is possible to solve for K_B. Combining mass action expressions for reactions 1 and 4 and the mass balance (Equation 5), one obtains an expression for K_B

$$K_B = \frac{[Me-B]\,[L^{n-}]^{m/n}}{[K_{so}\,([Me]_T - [Me-B])]^{1/n}} \tag{6}$$

The precise form for this expression depends upon the specific solid phase which exists. For the simple sulfides, ligand concentration (S^{2-}) is a function of pH. Therefore, for CdS, NiS, PbS and ZnS, Equation 6 reduces to

$$K_B = \frac{\delta_2\,[S(\text{-II})]_T}{K_{so}\,(\eta\text{-}1)}, \quad \eta = \frac{[Me]_T}{[Me-B]} \tag{7}$$

where δ_2 is the second acid-base distribution coefficient for the sulfide system. For $Cr(OH)_3$ one obtains

$$K_B = \frac{[OH^-]^3}{K_{so}\,(\eta\text{-}1)} \tag{8}$$

For cuprous sulfide the expression is slightly different because of the quadratic term. Equation 6 becomes

$$K_B = \left[\frac{\delta_2\,[S(\text{-II})]_T}{K_{so}}\,\eta'\right]^{1/2}, \quad \eta' = \frac{[Cu-B]^2}{[Cu_2S]} \tag{9}$$

In the experiments performed, each of the variables in Equations 7, 8 and 9 were measured. Average pH for all runs was 7.01 while total sulfide was approximately 5 x 10^{-3} *M*. The ratio η (η' for copper) can be determined from the metal fractionation scheme given previously. Figure 3 (A) through (E) shows bacteria-bound metals as a function of total metals. A least-square fit of the data indicated a high positive correlation. The slope of these lines, then, is η. Figure 3 (F) gives a similar plot for copper. The data, again, correlate well; the slope of the line is η'.

Using this information, cell-metal affinity constants can be computed. These are presented in Table III for each metal. It is clear from the magnitude of these constants that the bacteria within an anaerobic digester present surfaces which are highly active in a chemical sense.

DISCUSSION

The concentration of heavy metals within actively metabolizing anaerobes is of direct consequence to their tolerance limits. As has been shown, for most operations sulfide controls the availability of metals to bacteria. A somewhat more quantitative view of the effects of digester sulfide on metal accumulations in bacteria is given in Figure 4. In this plot, the relative degree of metal enrichment in the organic phase is expressed as a

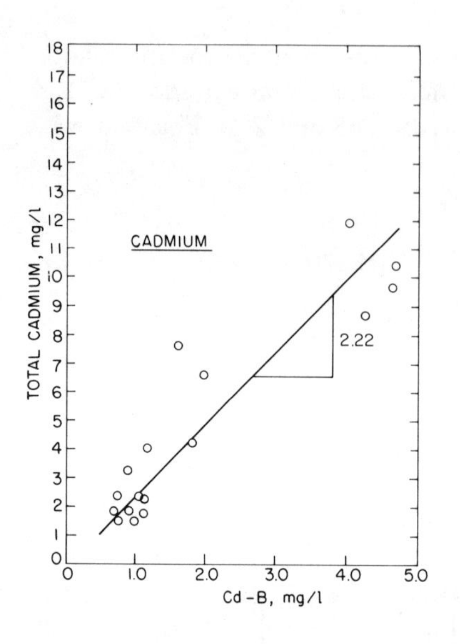
CADMIUM
TOTAL CADMIUM, mg/l
Cd-B, mg/l
2.22
0 1 2 3 4 5 6 7 8 9 10 11 12 13 14 15 16 17 18
0 1.0 2.0 3.0 4.0 5.0

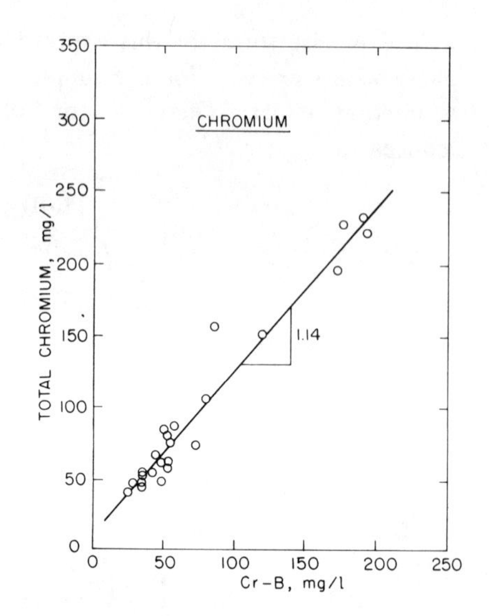
CHROMIUM
TOTAL CHROMIUM, mg/l
Cr-B, mg/l
1.14
0 50 100 150 200 250 300 350
0 50 100 150 200 250

A

B

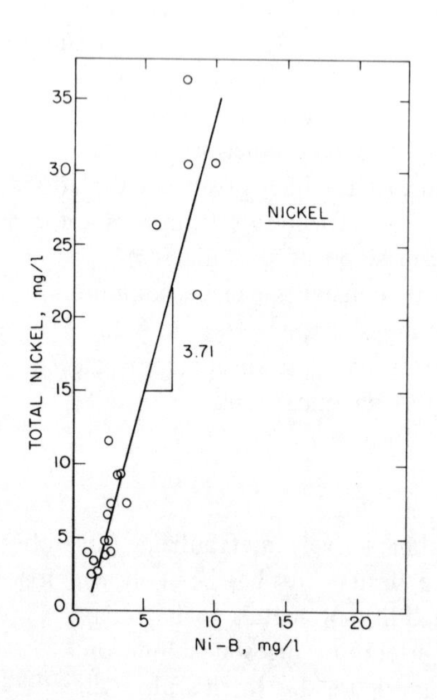
NICKEL
TOTAL NICKEL, mg/l
Ni-B, mg/l
3.71
0 5 10 15 20 25 30 35
0 5 10 15 20

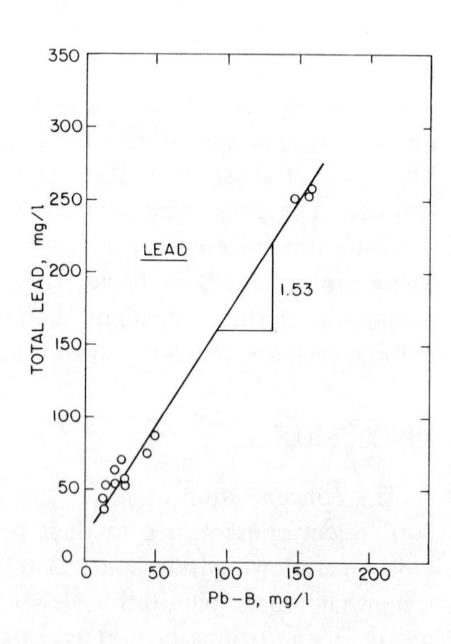
LEAD
TOTAL LEAD, mg/l
Pb-B, mg/l
1.53
0 50 100 150 200 250 300 350
0 50 100 150 200

C

D

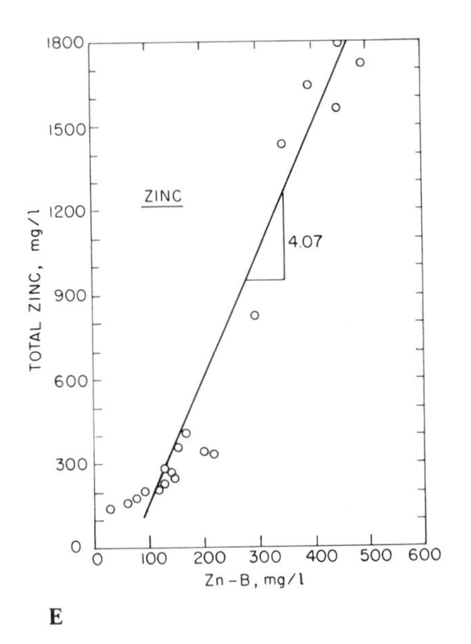

E

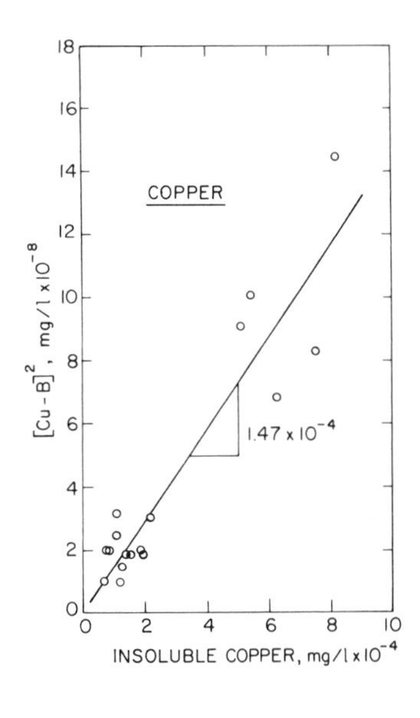

F

Figure 3. (A) Cadmium taken up by biomass in digester. Slope is η of Equation 7 in text; (B) Chromium taken up by biomass in digester. Slope is η of Equation 8 in text; (C), (D) and (E) Nickel, lead and zinc taken up by biomass in digester. Slopes are η of Equation 7 in text; (F) Copper uptake by biomass in digester. Slope of this plot is η' in Equation 9 in text.

Table III. Cell-Metal Affinity Constants, K_B

Metal	K_B
Cadmium	3.60×10^{17}
Chromium	1.46×10^{10}
Copper	6.06×10^{17}
Lead	7.89×10^{18}
Nickel	5.10×10^{10}
Zinc	1.26×10^{15}

ratio of metals in bacteria at some arbitrary sulfide concentration to bacterial metals at a reference sulfide level, in this case 5×10^{-3} *M*. This ratio is plotted as a function of total sulfide. Thus, all lines pass through 1.0 at -2.3 log units. It has been proposed that one way for controlling metal toxicity in digesters is to add large amounts of sulfide in order to inactivate metals in an inorganic precipitate.[27] Figure 4 gives some measure of the effectiveness of such a strategy for each metal. Roughly speaking, doubling the sulfide levels will bring about a 25% (for copper and

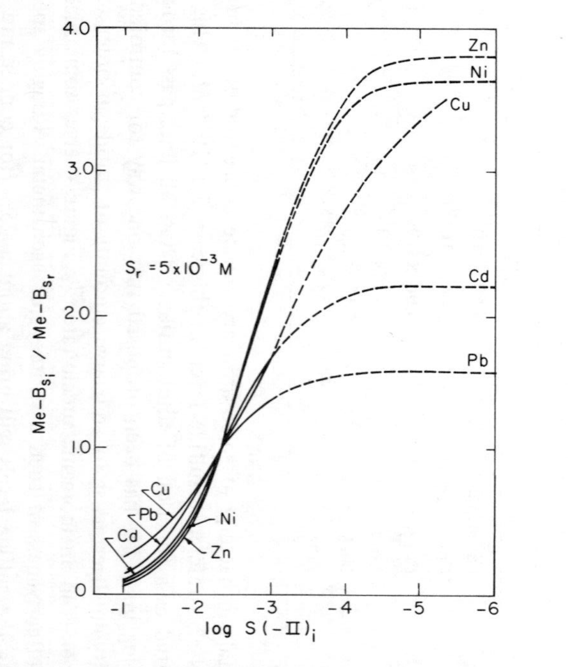

Figure 4. Curves for metal accumulations in digester biomass as a function of sulfide variation compared to a reference sulfide level of 5×10^{-3} *M*. pH is 7.0. Dotted lines suggest a ligand other than sulfide probably controls solubility in this range.

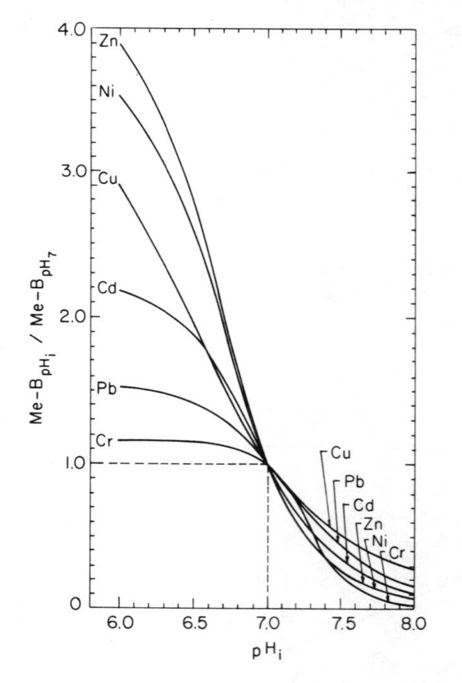

Figure 5. Metal accumulations in biomass referenced to pH 7.0 as a function of pH variation. Sulfide ≠ 5×10^{-3} *M*.

lead) to a nearly 50% reduction (for zinc) in metals associated with bacteria. Greater reductions at still higher concentrations are possible. It should be noted, however, that sulfide itself is toxic to bacteria when high amounts are present.[28] Of course this method of controlling metals has no effect on chromium. The approach outlined in Figure 4 can be used in conjunction with threshold toxicity values for a given sludge to optimize sulfide addition.

As sulfide levels decrease the metals in bacteria increase sharply. The dotted line portions below approximately 10^{-3} M sulfide are meant to suggest the likelihood that sulfide is no longer capable of precipitating all metals present. At still lower levels other ligands control metal solubility.

One of the more important and time-honored parameters in anaerobic digestion is pH. It is interesting to observe the relationship which exists between hydrogen ion concentration and metal distribution. In Figure 5 an enrichment ratio similar to that used in Figure 4 but based on pH variations is plotted against pH. Typical digester pH values would vary between 6.5 and 7.5 with 7.0 as the standard operating value. Increasing the pH decreases the solubility of the metals in solution; thus, fewer metals become associated with the biomass. A rise of only 0.3 pH units is capable of bringing about a 30% reduction in biomass copper and a 56% reduction in nickel and zinc. It is generally cheaper to adjust pH rather than other ligand concentrations in a wastewater treatment process, especially if expensive and generally unpleasant sulfide salts are used. In most instances, it is important to generate sufficient sulfide to exert major solubility controls; however, decreasing toxic effects through further sulfide addition may not be necessary. In addition, pH adjustment will affect chromium distribution in the digester. The standard operating procedure of dosing failing (or "sour") digesters with lime is based on sound chemical as well as aesthetic reasoning.

Figure 5 suggests a mode of failure which touches on a matter considered earlier, that of seemingly irrational digester behavior. Even a normally operating system, well acclimated to the metal feedings which it receives, could experience difficulty from a slight depression of pH, as for instance a small organic shock loading. For a system operating at pH 7.0, a decrease to 6.8 will bring about a 60% increase in nickel and zinc associated with the biomass perhaps inducing additional complications. It is not necessary that the system receive a sudden increase in total metals to show signs of distress; rather it is the distribution of metals which must be considered.

Figures 4 and 5 suggest further implications for metal availability in the disposal environment or in metal removal schemes. Although the literature is unclear on the matter, some sources claim organically bound metals to be generally less available than other forms.[29,30] If this is true, and if metal toxicity or accumulation is a concern in the disposal environment, then digester operation at a slightly acidic pH, if possible, would be desired. Others have proposed sludge acidification as a means of solubilizing and eventually recovering metals.[31] In such cases the higher the operating pH, the greater the inorganic fraction and hence ease of recovery at low pH. Those metals which are organically bound, however, would require a more prolonged exposure under more extreme conditions to bring about significant removals. Published recoveries of metals are generally poor and supportive of the work presented here.

Although the effects of temperature were not studied in the research, an analysis based upon the solubility characteristics of the solid phases can be made. All of the inorganic dissolution reactions are strongly endothermic; that is, as heat is supplied they become more soluble. Thus, at higher temperatures more of the metals become available for uptake by the biomass fraction, the implication being that high-rate systems are more susceptible to metal toxicity. Many operators report difficulties occurring at the onset of warm weather conditions.

The relative order of the metal distributions in Figures 4 and 5 is not obvious. Solubility, of course, is an important factor. For example, nickel, which is the most soluble of the metal sulfides at neutral pH and the most toxic, displays a rapid accumulation in cells as pH is lowered. Lead, which is considerably less soluble, shows less accumulation under the same conditions. However, the affinity between metal and cell is also of importance. Copper ion forms a highly insoluble precipitate and yet is rapidly taken up as pH is lowered. This is due in part to the freeing of 2 moles of copper for each mole of sulfide and is reflected in the steeper slope in both Figures 4 and 5. Copper also has an extraordinarily high affinity for the biomass. In contrast, chromium, which is comparatively soluble, shows little variation because of its lower tendency to complex with the bacterial fraction.

Table IV summarizes in a qualitative way the chemical interactions of heavy metals in anaerobic digestion. The first column, biomass fraction at pH 7, gives the ranking of each metal within this fraction under normal operating conditions. Column 2 gives the relative solubilities for the metals under typical digester conditions. The toxicity ranking in column three is derived from previous work by the authors[19] and is in agreement with other reports. The biomass affinity in column 4 is simply the ranking according to the K_B values of Table III. Column 5, variability, is essentially the proclivity of each metal toward incorporation with the biomass or inorganic fraction within a range of digester operating parameters. The ranking follows directly from Figures 4 and 5.

Table IV. Summary of Heavy Metal Interaction in Anaerobic Digestion

	Biomass Fraction, pH 7	Solubility	Relative Toxicity	Biomass Affinity	Variability
Greatest	Cr	Cr	Ni	Pb	Zn
	Pb	Ni	Cu	Cu	Ni
	Cu	Zn	Pb	Cd	Cu
	Cd	Cd	Cd	Zn	Cd
	Ni	Pb	Cr	Ni	Pb
Least	Zn	Cu	Zn	Cr	Cr

From Table IV it is clear that the simple static distribution between biomass and precipitate is not sufficient to account for the toxic effects of the metals in the digester environment. With the exception of zinc, the variability of metal distribution between biomass and precipitate is a better indication. This is consistent with the foregoing discussion. Zinc is somewhat unusual in that its reported toxic limits are more than an order of magnitude above the others.[13,19] In this study it is possible that sulfide was no longer controlling zinc solubility at very high zinc concentrations ($>$ 2000 mg/ℓ) and so the assumptions made in deriving Figures 4 and 5 may not be valid for this metal at or near toxic levels.

SUMMARY

In this study, heavy metal interactions in anaerobic digestion are envisioned as occurring primarily between two insoluble fractions: the controlling inorganic precipitate and the digester biomass. Both toxicity and metal availability to further sludge-handling operations are interpreted via this model. Affinity constants between metal and cells have been computed and indicate the relative degree of attraction which exists for each metal. These constants are only approximations and cannot be considered as true equilibrium constants. They are, perhaps, more closely related to linear adsorption coefficients between metal and cell surface, although it appears that active transport to the cell interior does occur. In this respect, various short-lived soluble complexes with metals which aid in the transport step are undoubtedly formed. Delineation of more exact mechanistic pathways for metals in sludge would require a considerably greater research effort.

It would be interesting to note the effects of temperature on toxic susceptibility as noted previously. Other operational parameters of interest could be average cell residence time and mode of digester operation. Other means of metal control could include the addition of chemical complexing agents or a metal-sorbing solid phase (*e.g.*, metal oxides). It is hoped that further work will suggest the need for modifications in the quantitative groundwork which has been considered in this chapter.

ACKNOWLEDGMENT

A portion of this research was supported by the Whirlpool Corporation.

REFERENCES

1. Hobson, P. N., S. Bousfield and R. Summers. *Critical Rev. Environ. Control* 4:131 (1974).
2. McCarty, P. L. *Public Works* 95:107 (1964).
3. Ghosh, S. *J. Water Poll. Control Fed.* 44:948 (1972).
4. Chen, K. Y., G. S. Young, T. K. Jan and N. Rohatgi. *J. Water Poll. Control Fed.* 46:2663 (1974).
5. Ghosh, M., and P. Zugger. *J. Water Poll. Control Fed.* 45:424 (1973).
6. Mosey, F. E. *Water Poll. Control* 70:584 (1971).
7. Brown, H. G., C. P. Hensley, G. L. McKinney and J. L. Robinson. *Environ. Lett.* 5:103 (1973).
8. McDermott, G. N., M. A. Post, B. N. Jackson and M. B. Ettinger. *J. Water Poll. Control Fed.* 37:163 (1965).
9. Adams, C. E. Jr., W. W. Eckenfelder Jr. and B. L. Goodman. Paper presented at the Heavy Metals in the Aquatic Environment Conference, Vanderbilt University, Nashville, TN, December 4-7, 1973.
10. Davis, J. A., III, and J. Jacknow. *J. Water Poll. Control Fed.* 47:2292 (1975).
11. Klein, L. A., M. Lang, N. Nash and S. L. Kirschner. *J. Water Poll. Control Fed.* 46:2653 (1974).
12. Mytelka, A. I., J. S. Czachor, W. B. Guggino and H. Golub. *J. Water Poll. Control Fed.* 45:1859 (1973).
13. Barth, E. F., M. B. Ettinger, B. V. Salotto and G. N. McDermott. *J. Water Poll. Control Fed.* 37:86 (1965).
14. Neufeld, R. D., and E. R. Hermann. *J. Water Poll. Control Fed.* 47:310 (1975).
15. Parsons, A., and P. Dugan. *Appl. Microbiol.* 21:657 (1971).
16. Cheng, M. H., J. W. Patterson and R. A. Minear. *J. Water Poll. Control Fed.* 47:362 (1975).
18. Gould, M. S., and E. J. Genetelli. *Proc. 30th Industrial Waste Conference,* Purdue University (Ann Arbor, MI: Ann Arbor Science Publishers, Inc., 1976).
19. Hayes, T. D., and T. L. Theis, *J. Water Poll. Control Fed.* 50:61 (1978).
20. Pohland, F. G., and K. H. Mancy. *Biotechnol. Bioeng.* 11:683 (1969).

21. Metcalf and Eddy, Inc. *Wastewater Engineering* (New York: McGraw-Hill Book Company, 1972).
22. Udy, M. J. *Chromium: Chemistry of Chromium and Its Compounds* Vol. I (New York: Reinhold Publishing Corp., 1956).
23. Rossini, F. D., *et al.* National Bureau of Standards Circulation 500, U. S. Dept. of Commerce, Washington, DC (1952).
24. Sillen, L. G., and A. E. Martell. *Stability Constants of Metal Ion Complexes* Special Publication No. 17 (London: The Chemical Society, 1964).
25. Latimer, W. M. *Oxidation Potentials* 2nd ed. (New York: Prentice-Hall, Inc., 1952).
26. Garrels, R. M., and C. L. Christ. *Solutions, Minerals, and Equilibria* (San Francisco: W. H. Freeman and Company Publishers, 1965).
27. Lawrence, A. W., and P. L. McCarty. *J. Water Poll. Control Fed.* 37:392 (1965).
28. Mosey, F. E., J. D. Stanwick and D. A. Hughes. *J. Inst. Water Poll. Control* 6:2 (1971).
29. Leeper, G. W. Special report prepared for the U. S. Army Corps of Engineers, Washington, DC (1972).
30. Chaney, R. L. Proc. of Recycling Municipal Sludges and Effluents on Land, National Association of State University and Land Grant Colleges, Washington, DC (1973).
31. Scott, D. S., and H. Horlings. *Environ. Sci. Technol.* 9:849 (1975).

CHAPTER 24

STOICHIOMETRY OF AUTOTROPHIC DENITRIFICATION USING ELEMENTAL SULFUR

Bill Batchelor

Environmental Engineering Division
Civil Engineering Department
Texas A & M University
College Station, Texas

Alonzo Wm. Lawrence

Koppers Company, Inc.
Pittsburgh, Pennsylvania

INTRODUCTION

Fixed forms of nitrogen in wastewater discharges can pollute receiving waters by their ability to stimulate eutrophication[1-3] and by their toxicity to humans[4,5] and aquatic life.[6,7] Wastewater treatment processes have been developed which use physical-chemical mechanisms to remove nitrogen compounds. However, problems of high costs and treatment of inorganic residuals have limited the application of these processes.[8,9] Biological nitrogen removal systems have also been developed which have been shown to be generally reliable and cost-effective.[8,9] An organic chemical such as methanol must be added to the denitrification stage of these processes if they are to achieve high levels of nitrogen removal. The rising cost of such chemicals has lessened the attractiveness of biological nitrogen removal processes.

In hopes of retaining the advantages of biological nitrogen removal systems without incurring the increased cost of organic chemicals, the alternative of using reduced sulfur compounds in an autotrophic denitrification process was studied. *Thiobacillus denitrificans* is the microorganism responsible for autotrophic denitrification.[10] This bacterium can oxidize a wide variety of reduced sulfur compounds (S^{2-}, S, $S_2O_3^{2-}$, $S_4O_6^{2-}$, SO_3^{2-}) while reducing nitrate to elemental nitrogen gas.[11] The process is autotrophic because *T. denitrificans* uses inorganic carbon to form cellular material.

The literature contains little quantitative information on the growth characteristics of *T. denitrificans* that could be used to evaluate the engineering feasibility of autotrophic denitrification. To develop such information, a program of theoretical and experimental investigations of the autotrophic denitrification process was planned and completed.[12] This program evaluated the stoichiometric, kinetic and physical characteristics of the process. This paper will focus on the results of that project which are concerned with the stoichiometry of autotrophic denitrification using elemental sulfur. This substance was chosen because its low cost and ease of handling indicate that it is the most promising one for general use. Other sulfur substrates such as sulfide and thiosulfate have also been studied in the same laboratory.[13]

BACKGROUND

The transformations of compounds involved in microbial growth can be represented quantitatively by a balanced stoichiometric equation. The coefficients in such an equation can be used to determine the relationships among rates of reaction for all products and reactants

$$\nu_a A + \nu_b B \longrightarrow \nu_c C + \nu_d D \tag{1}$$

$$\frac{1}{\nu_a} R_a = \frac{1}{\nu_b} R_b = \frac{1}{\nu_c} R_c = \frac{1}{\nu_d} R_d \tag{2}$$

where ν_a, ν_b, ν_c and ν_d are the stoichiometric coefficients for components A, B, C and D, respectively.

and R_a, R_b, R_c and R_d are the rates of removal of components A and B and rates of production of C and D, respectively, $[M/L^3T]$.

The stoichiometry of microbial reactions in wastewater treatment has often been described in terms of observed yield coefficients, rather than the stoichiometric coefficients shown in Equation 1. An observed yield coefficient relates the production or removal of one component to the production or removal of another component. The component used as a reference is usually the pollutant of primary concern in the treatment process. For example, the primary goal in the treatment of sewage is the removal of oxygen-demanding organics measured as chemical (COD) or biochemical (BOD) oxygen demand. A major problem of these systems is handling the excess biomass produced. The stoichiometric relationship between these two concerns is usually expressed as an observed biomass yield coefficient that relates the amount of biomass produced per mass of COD or BOD removed. In general, the observed yield coefficient for some component C using component A as reference can be defined as follows

$$Y_{obs}^{c} = \frac{\nu_c \text{ (molecular weight of C)}}{\nu_a \text{ (molecular weight of A)}} \tag{3}$$

where Y_{obs}^{c} is the observed yield coefficient for component C using component A as reference, [M/M].

The equation representing the observed stoichiometry of microbial growth can be considered to be a linear combination of two subordinate equations.[14] These subordinate equations represent the two basic processes involved in microbial growth: energy conversion and cell synthesis. Simple elemental and charge balances are used to construct the subreactions. In all but photosynthetic growth both subreactions are oxidation-reduction reactions. Therefore, it is convenient to express them on a basis of a one-electron transfer to facilitate construction of the overall reaction

$$\text{Observed Reaction} = f_e\text{(Energy Reaction)} + f_s\text{(Synthesis Reaction)} \tag{4}$$

where f_e and f_s are fraction of electron equivalents in observed reaction allocated to energy and synthesis subreactions, respectively.

$$f_e + f_s = 1.0 \tag{5}$$

Microbial growth does not normally display a constant stoichiometry. This is due to variations in the composition of cell material and differences in the efficiency with which the microbes couple energy transformation with cell synthesis. The manner in which these processes are coupled depends on environmental variables and is expressed in the values of f_e and f_s. A high efficiency (high values of f_s) occurs at maximum growth rates when growth conditions are optimal. As growth rate declines, a smaller fraction of the energy made available in the energy reaction is effectively used to produce biomass.

The growth rate of microorganisms in a biological waste treatment system is the primary variable related to process performance, so it is easy to relate changes in stoichiometry to process operations.[15] The growth rate in these systems is usually expressed indirectly through the operational variable called the solids residence time (θ_c). This variable is equal to the reciprocal of the growth rate and is calculated by dividing the total amount of biomass in the system by the amount removed per unit time.[15]

Several reports on the effect of growth rate on microbial stoichiometry are available.[16-20] In most instances the effect of θ_c on one stoichiometric parameter is reported. Since the observed stoichiometric equation is a linear combination of two other balanced equations, specifying any one stoichiometric parameter determines all others. The parameter most often used to express stoichiometry in biological wastewater treatment is the observed biomass yield (Y_{obs}), which relates the amount of biomass produced per substrate utilized. In most instances Y_{obs} decreases with increasing θ_c.

Microbial stoichiometry can also be used to calculate oxygen uptake rates[19] and efficiencies of nitrogen and phosphorus removal in wastewater treatment systems.[21] In microbial processes such as nitrification and denitrification, where hydrogen ions are produced or destroyed, the stoichiometric equation can be used to estimate process performance from alkalinity measurements.

THEORETICAL PREDICTION OF MICROBIAL STOICHIOMETRY

A method has been developed by McCarty to predict microbial stoichiometry by estimating f_e and f_s using theoretical arguments.[22] The basis of the method is a balance on the primary energy storage component of the cell, ATP. Most reactions which release energy produce ATP. Most reactions which require energy consume ATP

$$\text{ATP-energy produced by cell} = \text{ATP-energy used by cell} \tag{6}$$

Estimates of the ATP-energy produced by the cell are made from thermodynamic analysis of the energy available from the energy reaction. The change in Gibbs Free Energy (ΔG) for a reaction is the best measure of the available energy released or required by that reaction. The value of ΔG for any given reaction will vary with changing environmental conditions such as temperature, pH, and relative amounts of reactants and products. For the range of these parameters usually encountered in microbial systems, the variation in ΔG is small. Therefore, a value of ΔG measured under standard conditions (ΔG°) is used in these calculations.

Microorganisms are not completely efficient in producing or in utilizing ATP. Therefore, a factor representing the efficiency of energy conversion must be used to determine ATP-energy from thermodynamic energy. This efficiency factor could vary with changing environmental factors and could be different for the energy production and utilization processes. However, in this analysis it will be considered constant. A value for k of 0.6 has been found to be the best estimate for the energy efficiency of a variety of microorganisms.[22] The ATP-energy balance incorporating k is

$$f_e k \begin{pmatrix} \text{change in Gibbs Free} \\ \text{Energy for one} \\ \text{electron equivalent} \\ \text{of energy reaction} \end{pmatrix} = f_s \begin{pmatrix} \text{ATP-energy required} \\ \text{for synthesis of one} \\ \text{electron equivalent} \\ \text{of cells} \end{pmatrix} \quad (7)$$

where k is the efficiency factor for microbial energy conversion.

The energy requirement for synthesis is estimated from experimental data. Several different microorganisms have been found to require approximately 7.5 kilocalories of ATP-energy to produce one electron equivalent of cells from appropriate biosynthetic intermediates. Pyruvate was chosen as the synthesis intermediate for this ATP-energy balance because it appears in both biosynthetic and catabolic pathways in several microorganisms.[22] The ATP-energy required for synthesis consists of the sum of the ATP-energy necessary to convert the carbon source to pyruvate plus that required to convert pyruvate to cell material. Conversion of the carbon source to pyruvate will sometimes release ATP-energy. In this case it should be subtracted from the ATP-energy necessary to convert pyruvate to cell material to obtain the ATP-energy required for synthesis. If a nitrogen source is used by the microorganism which is not at the oxidation level of ammonia, the energy required to transform it to that level must also be included in calculating the energy required for synthesis

$$f_e k(-\Delta G_e^\circ) = f_s\left(\frac{\Delta G_p^\circ}{k^m}\right) + \Delta G_c^\circ \quad , \; m = \begin{cases} +1 & \Delta G_p^\circ > 0 \\ -1 & \Delta G_p^\circ < 0 \end{cases} \quad (8)$$

where ΔG_e° = the standard free energy change for oxidation of one electron equivalent by energy reaction, (kcal/electron equivalent);

ΔG_p° = the standard free energy change for conversion of one electron equivalent of carbon source to pyruvate, (kcal/electron equivalent);

ΔG_c° = the ATP-energy required to convert one electron equivalent of pyruvate to cell material, and has a value of 7.5 kcal/electron equivalent;

m = the coefficient indicating whether energy is released (-1) or used (+1) by reaction which converts carbon source to pyruvate.

Solving Equation 8 for the ratio f_e/f_s gives

$$f_e/f_s = \frac{\frac{\Delta G_p^\circ}{k^m} + \Delta G_c^\circ}{k(-\Delta G_e^\circ)} \quad (9)$$

Individual values for f_e and f_s can be calculated by noting that they are fractions of a whole

$$f_e/f_s = \frac{1}{f_s} - 1 \quad (10)$$

THEORETICAL PREDICTION OF DENITRIFICATION STOICHIOMETRY

A slight modification of McCarty's method is necessary to apply it to autotrophic denitrification. *Thiobacillus denitrificans* reduces carbon dioxide to cell mass in the same manner as photosynthetic cells. It has been shown that, from an energetics viewpoint, it is best to assume that water is the electron donor for this reduction even in nonphotosynthetic cells.[22] Photosynthetic cells excrete oxygen as a by-product of this reaction, but this cannot occur in autotrophic denitrification because it is an anaerobic process. Thus, oxygen produced in the synthesis reaction during autotrophic denitrification must be reduced by electrons from the energy reaction. Figure 1 shows a schematic

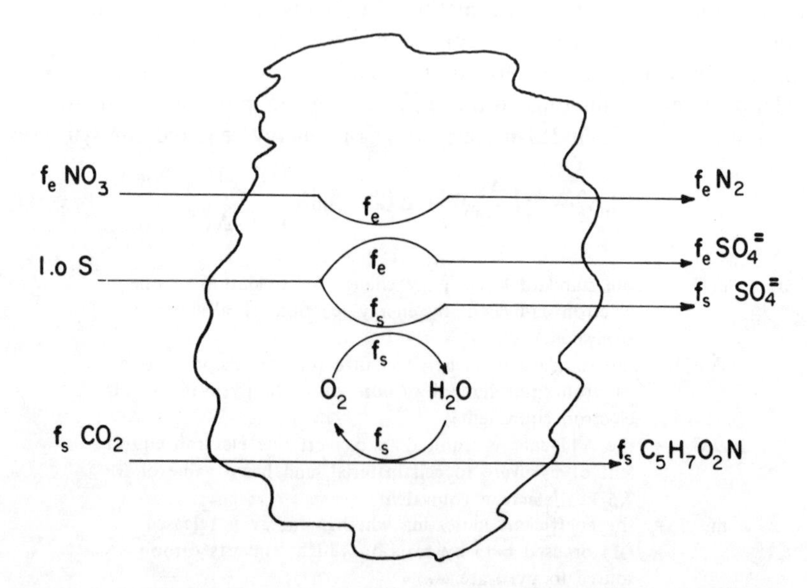

Figure 1. Proposed electron flow for autotrophic denitrification under anaerobic conditions with water the electron donor for synthesis.

of the proposed electron flow in autotrophic denitrification. In the overall reaction stoichiometry, sulfur is the apparent electron donor for synthesis rather than water, since internal recycle excludes oxygen from the overall stoichiometry.

The balance on ATP-energy for autotrophic denitrification using the above assumptions is

$$\text{ATP-energy produced by cell} = \text{ATP-energy used by cell} \quad (6)$$

$$k[f_e(-\Delta G^{\circ}_{sn}) + f_s(-\Delta G^{\circ}_{so})] = f_s\left[\frac{\Delta G^{\circ}_p}{k} + \Delta G^{\circ}_c\right] \quad (11)$$

$$f_e/f_s = \frac{\Delta G^{\circ}_p/k + \Delta G^{\circ}_c + k\Delta G^{\circ}_{so}}{k(-\Delta G^{\circ}_{sn})} \quad (12)$$

where ΔG°_{sn} = the standard free energy change for oxidation of one electron equivalent of sulfur by nitrate, and has a value of 21.78 kcal/electron equivalent

ΔG°_{so} = the standard free energy change for oxidation of one electron equivalent of sulfur by oxygen; and is equal to 23.33 kcal/electron equivalent

The energy and synthesis reactions for autotrophic denitrification using elemental sulfur can be expressed on a one electron equivalent basis

Energy Reaction

$$1/5\ NO_3^- + 1/6\ S + 1/15\ H_2O \rightarrow 1/10\ N_2 + 1/6\ SO_4^{-2} + 2/15\ H^+ \quad (13)$$

Synthesis Reaction

$$1/4\ CO_2 + 1/6\ S + 1/20\ NH_4^+ + 4/15\ H_2O \rightarrow 1/20\ C_5H_7O_2N + 1/6\ SO_4^{-2} + 23/60\ H^+ \quad (14)$$

These reactions and Equation 4 can be used to calculate Y_{obs} from the ratio f_s/f_e. This stoichiometric yield coefficient relates the amount of biomass, measured as organic nitrogen, produced per mass of nitrate-nitrogen removed

$$Y_{obs} = \frac{f_s\ (1/20)\ (14)}{f_e\ (1/5)\ (14)} = \frac{f_s}{f_e}\frac{1}{4} \quad (15)$$

An observed yield coefficient of 0.084 mg organic-N/mg NO_3-N can be calculated using Equations 12 and 15 and a value for k of 0.6.

EXPERIMENTAL MEASUREMENT OF DENITRIFICATION STOICHIOMETRY

A complete discussion of the experimental methods and materials used in this study is presented elsewhere.[12] The experimental plan followed in this study included batch, semicontinuous, and continuous reactor experiments. These reactors used enrichment cultures of *Thiobacillus denitrificans* which were developed from inocula obtained from natural environments and a wastewater treatment system. The cultures were fed a medium consisting of dechlorinated tapwater enriched with potassium nitrate, powdered elemental sulfur, sodium bicarbonate and various inorganic nutrient salts. Figure 2 is a schematic of the continuous culture reactor system used during these experiments.

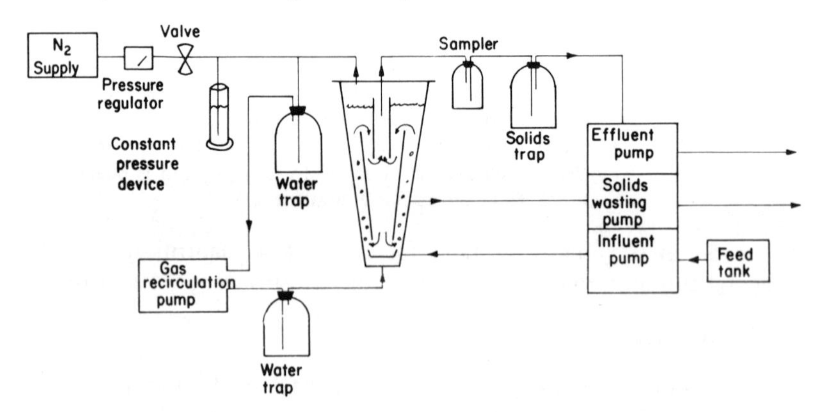

Figure 2. Schematic of continuous culture reactor system.

One of the first series of experiments was conducted to determine batch reaction stoichiometry. The results of these experiments are shown in Table I. The results are expressed in terms of observed yield coefficients for biomass, sulfate-sulfur and alkalinity, which use nitrate-nitrogen as a reference. Organic nitrogen was used as a measure of biomass in these and later experiments because solid elemental sulfur interfered with determination of total biomass by gravimetric methods. The yield coefficients are subscripted with "obs" to emphasize that they are the values observed under a given set of conditions. Batch stoichiometric yield coefficients represent average values calculated from initial and final conditions and cannot represent instantaneous values.

The kinetics and stoichiometry of the autotrophic denitrification process were determined in continuous culture experiments. The effects of solids retention time (θ_c), ratio of sulfur concentration to biomass

Table I. Batch Yield Coefficients

Experiment Number	Y_{obs} $\left(\frac{\text{mg organic-N}}{\text{mg } NO_3\text{-N}}\right)$	$Y_{obs}^{SO_4\text{-}S}$ $\left(\frac{\text{mg } SO_4\text{-S}}{\text{mg } NO_3\text{-N}}\right)$	Y_{obs}^{alk} $\left(\frac{\text{meq}}{\text{mg } NO_3\text{-N}}\right)$
1	0.096	2.27	0.088
2	0.075	2.29	0.120
3	0.095	2.49	0.132
Average	0.089	2.35	0.113

concentration (S/X), and temperature were evaluated in three series of experiments. Results from these studies are shown in Tables II, III and IV.

Contrary to previous experiences with microbial growth, the observed yield was not affected by θ_c (Figure 3). More variation in Y_{obs} was found in the other experiments (Tables III and IV), but no definite effect of temperature or S/X on Y_{obs} is apparent. The stoichiometry of autotrophic denitrification, therefore, can be considered constant over a range of solids retention times (7.6-30 days), values of S/X (45-194 mg S/mg N) and temperatures (12-30°C).

Table V presents the average values for the stoichiometric yield coefficients measured in the batch and continuous culture experiments. These values agree reasonably well, considering the widely different growth conditions under which they were measured.

The theoretical prediction for the observed yield coefficient agrees very well with the average values measured in batch and continuous cultures. However, yields calculated from energetic arguments should be used with some discretion. The method assumes a constant efficiency of both energy production and utilization, regardless of environmental factors such as temperature and pH. Application of the method to autotrophic denitrification also requires the assumption of an energetically unfavorable means of carbon dioxide reduction. However, the excellent agreement of predictions with experimental results indicates the usefulness of the procedure.

APPLICATIONS OF AUTOTROPHIC DENITRIFICATION STOICHIOMETRY

It is convenient to represent the stoichiometry of a microbial reaction in terms of a balanced stoichiometric equation rather than a table of

Table II. Summary of Continuous Culture Results at Various θ_c [a]

Operating and Kinetic Parameters			Mixed Liquor Characteristics		Effluent Characteristics		Stoichiometric Parameters		
θ_c (days)	θ (days)	S/X $\left(\frac{mg\ S}{mg\ N^b}\right)$	X $\left(\frac{mg\ N^b}{1}\right)$	S $\left(\frac{mg\ S}{1}\right)$	NO_3-N $\left(\frac{mg\ N}{1}\right)$	NO_2-N $\left(\frac{mg\ N}{1}\right)$	Y_{obs} $\left(\frac{mg\ N^b}{mg\ N}\right)$	ΔSO_4-S $\left(\frac{mg\ S}{1}\right)$	ΔAlk $\left(\frac{meq}{1}\right)$
10	0.254	142	83 (10)	11,800 (1,700)	0.14 (0.13)	0.14 (0.11)	0.071	94 (2)	3.64 (0.02)
15	0.253	149	133 (14)	19,700 (6,000)	0.07 (0.04)	0 (n = 7)	0.075	94 (13)	3.67 (0.16)
20	0.250	145	171 (32)	24,800 (5,600)	0.01 (0.01)	0 (n = 5)	0.071	103 (1)	3.94 (0)
25	0.238	139	231 (10)	32,100 (1,200)	0.06 (0.01)	0 (n = 8)	0.073	103 (11)	3.81 (0.06)
30	0.242	150	234 (19)	35,100 (1,200)	0 (n = 8)	0 (n = 8)	0.063	108 (8)	4.34 (0.38)

[a]Numbers in parentheses are standard deviations of measurements. If measured value is zero, the number of replicates is given.
[b]Organic-N.

Table III. Summary of Continuous Culture Results at Various S/X[a]

Kinetic Parameters		Operating Parameters		Mixed Liquor Characteristics		Effluent Characteristics		Stoichiometric Parameters		
S/X	$U_{m,a}$	θ_c	θ	S	X	NO_3-N	NO_2-N	Y_{obs}	ΔSO_4-S	ΔAlk
$\left(\frac{\text{mg S}}{\text{mg N}^b}\right)$	$\left(\frac{\text{mg N}}{\text{mg N}^b\text{-day}}\right)$	(days)	(days)	$\left(\frac{\text{mg S}}{1}\right)$	$\left(\frac{\text{mg N}^b}{1}\right)$	$\left(\frac{\text{mg N}}{1}\right)$	$\left(\frac{\text{mg N}}{1}\right)$	$\left(\frac{\text{mg N}^b}{\text{mg N}}\right)$	$\left(\frac{\text{mg S}}{1}\right)$	$\left(\frac{\text{meq}}{1}\right)$
194	2.13	8	0.255	17,330 (750)	90 (8)	0 (n = 5)	0 (n = 5)	0.095	96 (8)	3.80 (0.14)
142	1.62	10	0.254	11,800 (1,700)	83 (10)	0.14 (0.13)	0.14 (0.11)	0.071	94 (2)	3.64 (0.02)
100	1.22	13	0.241	15,200 (1,960)	152 (12)	0.02 (0.02)	0.003 (0.003)	0.094	93 (6)	3.54 (0.26)
56	0.74	20	0.247	11,300 (500)	203 (7)	0.01 (0.004)	0 (n = 5)	0.084	100 (9)	3.81 (0.12)
45	0.64	30	0.244	12,270 (240)	275 (14)	0.04 (0.02)	0.001 (0.0004)	0.075	93 (12)	3.33 (0.12)

[a]Numbers in parentheses are standard deviations of measurements. If the measured value is zero, the number of replicates is given.
[b]Organic-N.

Table IV. Summary of Continuous Culture Results at Various Temperatures[a]

	Kinetic Parameters		Operating Parameters		Mixed Liquor Characteristics		Effluent Characteristics		Stoichiometric Parameters		
T	$U_{m,a}$	S/X	θ_c	θ	S	X	NO_3-N	NO_2-N	Y_{obs}	ΔSO_4-S	ΔAlk
(°C)	$\left(\frac{mg\ N}{mg\ N^b\text{-days}}\right)$	$\left(\frac{mg\ S}{mg\ N^b}\right)$	(days)	(days)	$\left(\frac{mg\ S}{1}\right)$	$\left(\frac{mg\ N^b}{1}\right)$	$\left(\frac{mg\ N}{1}\right)$	$\left(\frac{mg\ N}{1}\right)$	$\left(\frac{mg\ N^b}{mg\ N}\right)$	$\left(\frac{mg\ S}{1}\right)$	$\left(\frac{meq}{1}\right)$
12	0.97	144	15	0.258	25,140 (1,150)	175 (10)	0.01 (0.01)	0 (n = 6)	0.100	92 (2)	3.87 (0.11)
21	1.62	142	10	0.254	11,800 (1,700)	83 (10)	0.14 (0.13)	0.14 (0.11)	0.071	94 (2)	3.64 (0.02)
30	3.92	141	7.6	0.242	10,570 (530)	75 (8)	0 (n = 6)	0 (n = 6)	0.080	88 (6)	3.16 (0.21)

[a]Numbers in parentheses are the standard deviations of the measurements. For values equal to zero, the number of measurements is given.
[b]Organic-N.

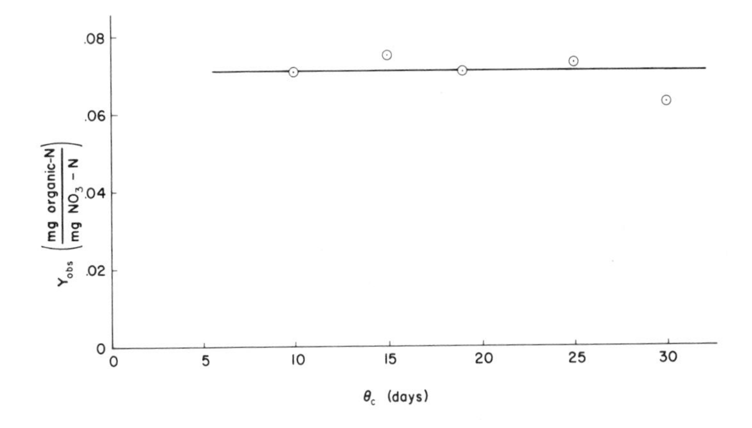

Figure 3. Observed biomass yield versus solids retention time.

Table V. Average Yield Coefficients

	Y_{obs} (mg organic-N / mg NO_3-N)	$Y_{obs}^{SO_4\text{-}S}$ (mg SO_4-S / mg NO_3-N)	Y_{obs}^{alk} (meq / mg NO_3-N)
Continuous	0.080	3.21	0.124
Batch	0.089	2.35	0.113
Theoretical	0.084	2.54	0.093

yield coefficients. This requires that an empirical formula be used to represent the elemental composition of microbial cell mass. The formula $C_5H_7O_2N$ was developed from experimental measurements[23] and has often been used to construct stoichiometric equations.[14,19,22,24] Use of an approximate cell mass formula might introduce some error into stoichiometric calculations, but it also tends to smooth experimental errors and offers a concise method of expressing overall stoichiometry.

A balanced stoichiometric equation for autotrophic denitrification can be constructed with the cell formula $C_5H_7O_2N$ and the average observed biomass yield measured in continuous culture

$$1.0\ NO_3^- + 1.10\ S + 0.40\ CO_2 + 0.76\ H_2O + 0.080\ NH_4^+ \rightarrow 0.080\ C_5H_7O_2N + 0.50\ N_2 + 1.10\ SO_4^{-2} + 1.28\ H^+ \quad (16)$$

The behavior of one reactant or product must be known before Equation 16 can be used to calculate the change in concentration of any other component. Nitrate is the logical choice as a reference component since denitrification systems are primarily concerned with its removal and available kinetic data can be used to calculate the change in its concentration. Microorganisms responsible for autotrophic denitrification can also use oxygen and nitrite to oxidize sulfur. Since nitrate, nitrite and oxygen could all be present in the influent to a denitrification system, they should all be considered in stoichiometric calculations. A convenient method for expressing the combined effect of these compounds is to define the equivalent nitrate-nitrogen concentration (CEN). This variable represents the concentration of nitrate-nitrogen that would exist if all the microbially available electron acceptors were converted to nitrate without changing the concentration of total electron equivalents

$$\mathrm{CEN} = \mathrm{N} + 0.6\ \mathrm{NI} + 0.35\ \mathrm{DO} \tag{17}$$

where CEN = the concentration of equivalent nitrate-nitrogen, $[M/L^3]$;
N = the concentration of nitrate-nitrogen, $[M/L^3]$;
NI = the concentration of nitrite-nitrogen, $[M/L^3]$;
DO = the concentration of dissolved oxygen, $[M/L^3]$.

Calculation of the equivalent nitrate concentration is based on the assumption that the stoichiometry is the same for any biologically available oxidant, when expressed on an electron equivalent basis. This is probably not strictly correct because the free energy available from different oxidations is not the same. However, this assumption will not cause major errors, since in most wastewaters the concentrations of oxygen and nitrite will be small when compared to nitrate on an electron equivalent basis.

Reaction kinetics and stoichiometry can be applied to various reactor configurations through material balances. A schematic of a slurry reactor system using elemental sulfur is shown in Figure 4. This system includes an anaerobic basin for denitrification and a gravity clarifier to separate the sulfur-biomass solids from the effluent and recycle them to the basin. Table VI is a summary of the pertinent material balance equations for this system.[12] These equations are expressed in terms of nitrate-nitrogen, but if other biologically available electron acceptors are present, equivalent nitrate-nitrogen can be used instead.

One of the most important uses of the reaction stoichiometry expressed in Equation 16 is in calculating the required sulfur feed ratio for a slurry reactor. The sulfur feed ratio controls the rate of denitrification because it sets the value of the sulfur to biomass ratio (S/X) in the reactor.

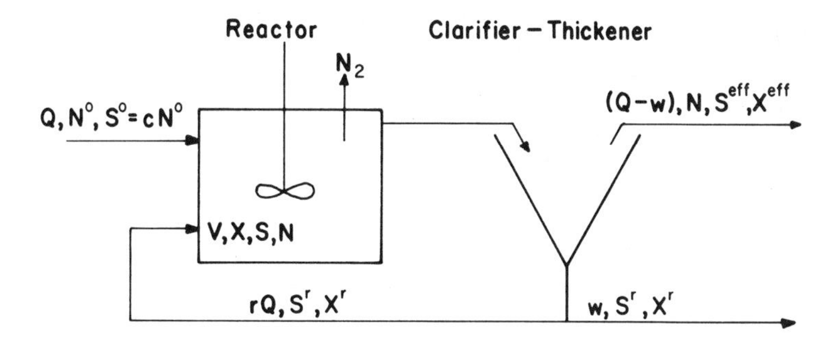

N^o, N = concentration of nitrate-nitrogen in influent and reactor respectively, (M/L^3);

S^o, S, S^r, S^{eff} = concentration of elemental sulfur in influent, reactor, recycle line, and effluent respectively, (M/L^3);

X, X^r, X^{eff} = biomass concentration in reactor, recycle line, and effluent respectively, (M/L^3);

Q, rQ, w = volumetric flow for influent, recyle and wastage flows respectively, (L^3/T).

Figure 4. Schematic of slurry reactor system.

This ratio has been shown to be the primary kinetic variable in autotrophic denitrification systems using elemental sulfur and its value determines whether effective nitrate removal can occur.[12] The material balance equation for S/X using equivalent nitrate-nitrogen as a reference is

$$S/X = \frac{CEN^o\,(c - Y^S_{obs}) + Y^S_{obs}\,CEN}{Y_{obs}\,(CEN^o - CEN)} \tag{18}$$

where CEN^o and CEN are the concentrations of equivalent nitrate-nitrogen in influent and effluent, respectively.

Since nearly complete denitrification was observed in all experiments (Table II-IV), Equation 18 can be simplified and rearranged to facilitate calculation of c

$$c = Y_{obs}\,(S/X) + Y^S_{obs} \tag{19}$$

Values for the observed yield coefficient calculated from the balanced stoichiometric equation (Equation 16) can be used in Equation 19 to

Table VI. Summary of Steady-State Material Balances on Slurry Reactor System

Constituent	Material Balance	Reactor Concentration
Nitrate	$R_n = \frac{N^o - N}{\theta}$	$N = N^o - U\theta X$
Biomass	$R_x = \frac{X}{\theta_c}$	$X = Y_{obs} (N^o - N) \frac{\theta_c}{\theta}$
Sulfur	$R_s = \frac{cN^o}{\theta} - \frac{S}{\theta_c}$	$S = [N^o(c - Y^s_{obs}) + Y^s_{obs}N] \frac{\theta_c}{\theta}$
Sulfur to Biomass Ratio	---	$S/X = \frac{N^o(c - Y^s_{obs}) + Y^s_{obs}N}{Y_{obs} (N^o - N)}$

R_x, R_n, R_s = rate of production of biomass and removal of nitrate-nitrogen and sulfur, respectively, $[M/L^3 - T]$

θ = hydraulic retention time, equal to volume of reactor divided by volumetric flowrate, [T]

U = unit rate of nitrate-nitrogen removal, equal to rate of nitrate-nitrogen removal divided by biomass concentration, [M/M - T]

c = sulfur feed ratio, equal to sulfur feed ratio divided by nitrate-nitrogen feed ratio, [M/M]

Y^s_{obs} = observed yield coefficient for elemental sulfur using nitrate-nitrogen as reference, [M/M]

calculate the sulfur feed ratio required to maintain any value of S/X. The value of S/X desired in the reactor is set by the characteristics of the wastewater, reaction kinetics and effluent limitations.

Knowledge of the extent of alkalinity destruction in autotrophic denitrification is an important aspect of process control. Alkalinity is destroyed because sulfuric acid is the product of microbial sulfur oxidation. Equation 16 can be used to calculate the observed yield coefficient for alkalinity, and this factor can be combined with an estimate of the change in equivalent nitrate-nitrogen to predict the amount of alkalinity destroyed

$$\Delta\, Alk = Y^{alk}_{obs} (CEN^o - CEN) \tag{20}$$

where Δ Alk = the change in concentration of alkalinity across reactor (meq/1)

Y^{alk}_{obs} = observed yield coefficient for alkalinity using nitrate-nitrogen as reference, (meq/mg NO_3- N).

Equation 20 can be used in the design and operation of a system to maintain reactor pH in the range suitable for microbial growth. Measurements of alkalinity destruction can be combined with reaction stoichiometry to estimate the efficiency of nitrate removal. Since the measurement of alkalinity is easier than that of nitrate, this offers an easy method for process monitoring. This technique can also be applied to nitrification and heterotrophic denitrification processes.

Inorganic carbon is used by the microorganisms in autotrophic processes to produce cellular material. The balanced stoichiometric equation enables design engineers to estimate the carbon requirements for such systems

$$\Delta\ CO_2\text{-}C = Y_{obs}^{CO_2\text{-}C}\ (CEN^\circ\text{-}CEN) \tag{21}$$

where $\Delta\ CO_2\text{-}C$ = the change in concentration of inorganic carbon across reactor, $[M/L^3]$;
$Y_{obs}^{CO_2\text{-}C}$ = the observed yield coefficient for inorganic carbon using nitrate-nitrogen as reference, [M/M].

Analysis of Equation 16 indicates that if sufficient alkalinity is added to offset acid production and the wastewater is allowed to reach equilibrium with carbon dioxide in air, a sufficient supply of inorganic carbon will be made available to the microorganisms. This can be done by adding base to the wastewater before it is aerated in the nitrification stage.

Equation 16 can also be used to estimate the amount of sulfate produced

$$\Delta\ SO_4\text{-}S = Y_{obs}^{SO_4\text{-}S}\ (CEN^\circ\text{-}CEN) \tag{22}$$

where $\Delta\ SO_4\text{-}S$ = change in concentration of sulfate-sulfur across reactor, $[M/L^3]$.

Sulfate enrichment of receiving waters is a problem due to possible stimulation of microbial sulfide production and deterioration of drinking water quality. Equation 16 predicts an increase of 75 mg/l SO_4-S in a completely denitrified wastewater originally containing 30 mg /l NO_3-N. An average domestic waste contains about 25 mg/l SO_4-S,[19,25] so most of the effluents would not meet the old drinking water standard of 83 mg/l SO_4-S.[26] This old standard does not seem to be based on taste or on any other physiological basis, except for a laxative effect on new users.[6] Present U.S. drinking water standards do not contain a sulfate regulation.[27] Therefore, a sulfate concentration of about 100 mg/l SO_4-S should not cause serious deterioration of drinking water quality, especially when dilution effects are considered.

Certain microorganisms in natural water systems can reduce sulfate to sulfide under anaerobic conditions if sufficient organic carbon and sulfate are available. Production of sulfides causes an odor problem and precipitation of heavy metals. Insolubilization of some highly toxic metals such as mercury would be beneficial. Formation of metallic sulfides could theoretically cause release of phosphorus previously immobilized by the metal. However, experimentation has failed to show that this occurs.[28]

Nitrogen removal is especially important in coastal and estuarine waters. There, the effects of sulfate enrichment would be negligible due to interactions with seawater containing high concentrations of sulfate (900 mg/l SO_4-S).[29] Although sulfate enrichment does not improve water quality, there does not seem to be sufficient evidence of harmful effects to generally restrict the application of autotrophic denitrification processes.

CONCLUSIONS

Based on the results of this study on the stoichiometry of autotrophic denitrification using elemental sulfur it can be concluded that:

1. Stoichiometry for autotrophic denitrification using elemental sulfur is relatively constant over a range of solids retention time (7.6-30 days), values of S/X (45-194 mg S/mg N) and temperatures (12-30°C).

2. The balanced stoichiometric equation for the process is:

$$1.0\ NO_3^- + 1.10\ S + 0.40\ CO_2 + 0.76\ H_2O + 0.080\ NH_4^+ \rightarrow$$
$$0.080\ C_5H_7O_2N + 0.50\ N_2 + 1.10\ SO_4^{-2} + 1.28\ H^+$$

3. The balanced stoichiometric equation can be used to calculate the amount of excess biomass and sulfate produced and the amount of elemental sulfur, supplemental alkalinity and inorganic carbon required by the process.

4. A theoretical method for estimating reaction stoichiometry can be applied to autotrophic denitrification and the predicted value of the biomass yield coefficient (0.084 mg N/mg N) agrees very well with the values measured in batch (0.089 mg N/mg N) and continuous (0.080 mg N/mg N) cultures.

REFERENCES

1. Vollenweider, R. A. *Scientific Fundamentals of the Eutrophication of Lakes and Flowing Waters, with Particular Reference to Nitrogen and Phosphorus as Factors in Eutrophication* (Paris: 1970) Organization for Economic Cooperation and Development.

2. Ryther, J. H., and W. M. Dunstan. *Science* 171:1008 (1971).
3. Goldman, J. C. *Water Res.* 10:97 (1976).
4. Winton, E. F., R. G. Tardiff, and L. J. McCabe. *J. Am. Water Words Assoc.* 63:95 (1971).
5. Hill, M. J., G. Hawksworth, and G. Tattersall. *Br. J. Cancer* 28:562 (1973); *Biol. Abstr.* 58(4):20689 (1974).
6. McKee, J. E., and H. W. Wolf. *Water Quality Criteria,* 2nd ed, Resources Agency of California, State Water Quality Control Board, Publication No. 3-A (1963).
7. "Proposed Criteria for Water Quality," Vol. 1 U.S. Environmental Protection Agency, Washington, DC (1973).
8. "Process Design Manual for Nitrogen Control" U.S. Environmental Protection Agency, Technology Transfer Division, Washington, DC (1975).
9. Ehreth, D. J., and J. V. Basilico. *Water–1974, AIChE Symp. Ser. No. 145,* 71:57 (1974).
10. Buchanan, R. E., and N. E. Gibbons, Eds. *Bergey's Mannual of Determinative Bacteriology,* 8th ed. (Baltimore: Williams and Wilkins, 1974).
11. Baalsrud, K., and K. S. Baalsrud. *Arch. Microbiol.* 20:34 (1954).
12. Batchelor, B. Ph.D. Thesis, Department of Civil and Environmental Engineering, Cornell University, Ithaca, New York (1976).
13. Driscoll, C. T. M.S. Thesis, Department of Civil and Environmental Engineering, Cornell University, Ithaca, New York (1976).
14. McCarty, P. L. In: Proc. International Conference Toward a Unified Concept of Biological Waste Treatment, Atlanta, Georgia (1972).
15. Lawrence, A. W., and P. L. McCarty. *J. Sanit. Eng. Div.,* ASCE 96(SA3):757 (1970).
16. Heukelekian, H., H. E. Orford and R. Manganelli. *Sew. Ind. Wastes* 23(6):945 (1951).
17. Weston, R. F., and W. W. Eckenfelder. *Sew. Ind. Wastes* 27(7): 802 (1955).
18. van Ulden, N. *Ann. Rev. Microbiol.* 23:473 (1969).
19. Metcalf & Eddy, Inc., *Wastewater Engineering,* (New York: McGraw-Hill Book Company, 1972).
20. Sherrard, J. H., and E. D. Schroeder. *J. Water Poll. Control Fed.* 45(9):1889 (1973).
21. Sherrard, J. H., and L. D. Benefield. *J. Water Poll. Control Fed.* 48(8):562 (1976).
22. McCarty, P. L. In: *Organic Compounds in Aquatic Environments,* S. D. Faust and J. V. Hunter, Eds. (New York: Marcel Dekker, Inc. 1971).
23. Porges, N., L. Jasewicz, and S. R. Hoover. In: *Biological Treatment of Sewage and Industrial Wastes, Vol 1: Aerobic Oxidation,* J. McCabe and W. W. Eckenfelder, Eds. (New York: Reinhold, 1956).
24. Busch, A. W., and N. Myrick. *Proc. Purdue Ind. Waste Conf.,* 15:19 (1960).
25. Schroeder, H. A. *J. Am. Med. Assoc.* 172(17):1902 (1960).
26. U.S. Public Health Service. "Drinking Water Standards," *Federal Register* 27(44):2152 (1962).

27. U.S. Environmental Protection Agency. "National Interim Primary Drinking Water Regulations," *Federal Register* 40(248):59566 (1975).
28. Olson, D. M. "The Effect of Sulfate and Manganese Dioxide on the Release of Phosphorus from Lake Mendota Sediments," Water Resources Center, Wisconsin University, Madison, WI, abstracted in *Government Reports Announcements and Index* 75(23):66 (1965).
29. Riley, J. P., and G. Skirrow. *Chemical Oceanography,* Vol. 1, (New York: Academic Press, 1965).

CHAPTER 25

PREDICTION OF pH STABILITY IN BIOLOGICAL TREATMENT SYSTEMS

Frederick G. Pohland and Makram T. Suidan
School of Civil Engineering
Georgia Institute of Technology
Atlanta, Georgia

INTRODUCTION

General Perspective

The transformation of organic and inorganic pollutants in lakes, rivers and other aquatic ecosystems can cause significant alteration in the quality of the water resource. These changes in quality are often measured by routine physical and chemical analyses of which pH is an important and frequently reported parameter. Deviations from a normal or optimum pH may upset the symbiotic balance requisite for the survival of a particular trophic level within the ecosystem and thereby promote a general deterioration in its quality and availability for use.

Biological treatment systems, as simulations of their natural environment counterparts, also encounter similar pH disturbances with the inevitable retardation or destruction of a desirable dominant microbial population. Such disturbances, unless occasioned by the influx of strongly acidic or alkaline wastes, are often mediated by indigenous biochemical interactions. Accordingly, the redox process associated with nitrogen and sulfur transformations as well as photosynthesis, respiration and methane fermentation all contribute to the pH response through formation of the acidic and basic products of microbial metabolism and their simultaneous reaction within the aqueous environment. Furthermore, these acids and bases form specific buffer regions, the strongest and most concentrated serving to regulate the pH. Therefore, the magnitude of change in pH is dependent upon the types of acids and bases present and their concomitant ability to form or displace the dominating buffer system.

Initial Considerations

Anaerobic biological waste stabilization processes present a convenient focus because of the generally accepted sequential pattern of substrate conversion through acid and methane fermentation, the recognized interactions between the acidic and basic products of the system and the significance of buffers and pH control. Moreover, the buffer systems established within these processes can be identified and are therefore amenable to evaluation by acid-base equilibrium analysis.

Drawing from some initial investigations on the anaerobic stabilization of waste sludges,[1] the central role of acid and alkaline products of substrate conversion in mediating pH response and treatment efficiency, particularly during retarded conditions, can be dramatically illustrated. Utilizing pH, volatile acids and alkalinity as selected indicator parameters, Figure 1 and 2 demonstrate the interrelationship between the major acid-base equilibria and pH during long-term anaerobic sludge stabilization at both mesophilic (36°C) and thermophilic (52°C) conditions. The important facet of these data is that as volatile acids were internally generated and accumulated when the system experienced a loading stress, the volatile acids tended to displace the normal buffer system with its own thereby lowering the pH by a process analogous to an internal acid-base titration. The presence of sufficient buffer, as indicated by the alkalinity, prevented a radical change in pH as was the case throughout the 35°C studies. However, when excess volatile acids were generated, as indicated at 52°C for the period when free acids accumulated, a dramatic pH change was observed which coincided with a point at which the pH lowered below the pK value ($\approx$6.35) for the bicarbonate buffer system. The predicted pH in Figures 1 and 2 have been calculated from the acid-base equilibrium expression developed subsequently in this presentation.

In either case, the existence of a period during which free acids accumulated was determined as presented in the original work (and discussed subsequently herein) in terms of volatile acids alkalinity calculated by converting the total volatile acids to equivalent bicarbonate alkalinity and corrected for the effect of the pH end point utilized in the total alkalinity determination. Therefore, as long as the bicarbonate alkalinity was greater than the volatile acids alkalinity expressed similarly, the anaerobic stabilization process as affected by a displacement of the normal buffer system should not be retarded.

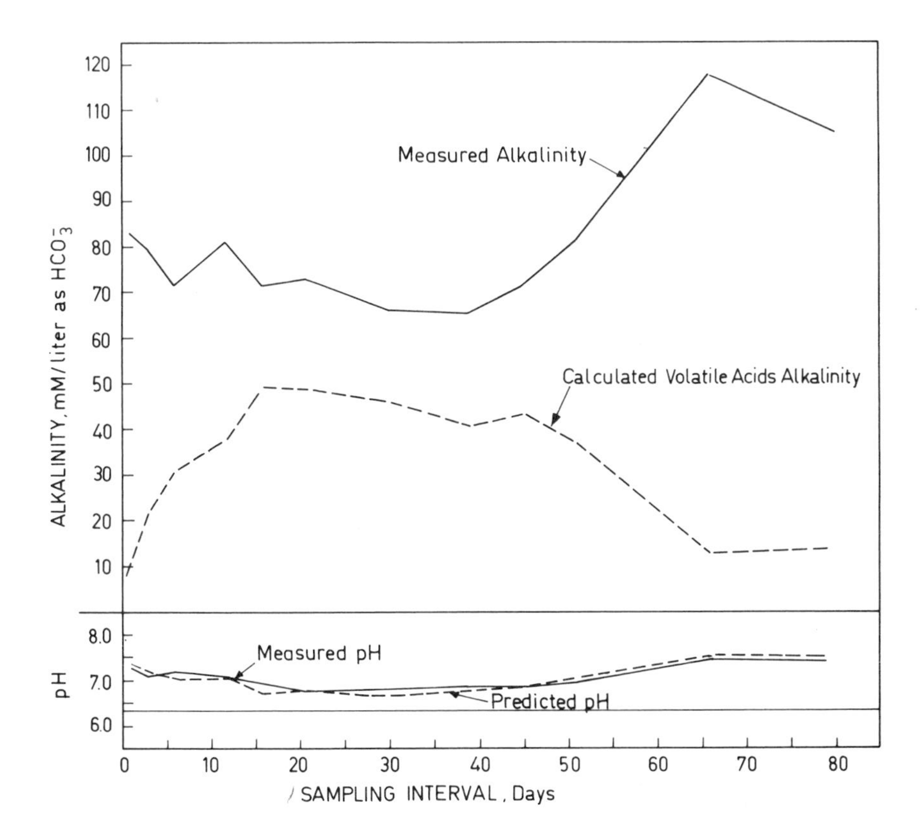

Figure 1. Alkalinity and pH relationships during anaerobic sludge stabilization at 36°C.

NATURE AND ORIGIN OF BUFFER SYSTEMS

Based upon the preceding initial considerations, the identification of the vital role of the major acidic and basic products of substrate conversion during anaerobic stabilization has emerged. These concepts have been discussed in more detail elsewhere[2,3] but can be simply reviewed by considering the microbially mediated conversion and stabilization of waste organic substrates containing the major elements of carbon, hydrogen, oxygen, nitrogen and sulfur as indicated in Figure 3. From such a conceptual illustration, it is not difficult to recognize that depending on the character of the initial substrate, its availability for microbial utilization and the relative rates of conversion of its components and/or intermediates, certain products will either momentarily appear or persist; their nature and acid-base characteristics would then determine the pH and its effect on overall

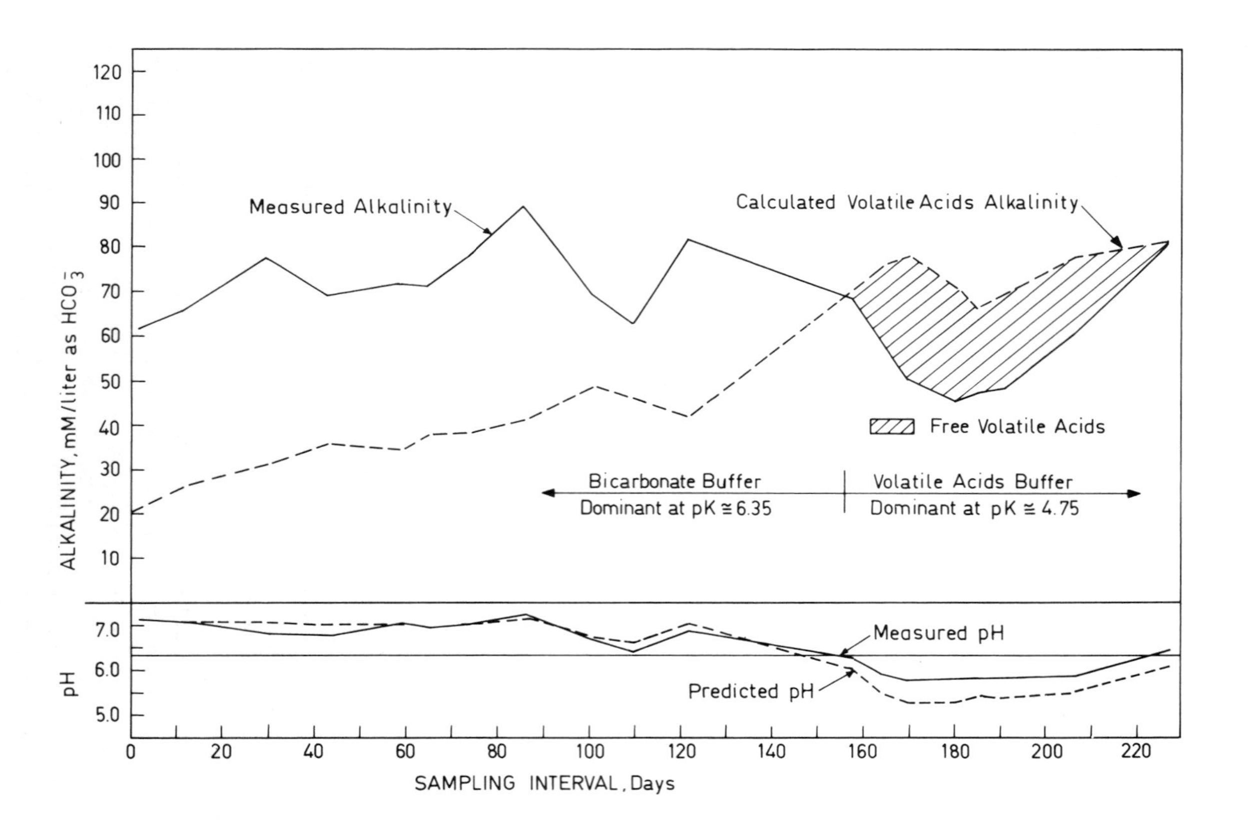

Figure 2. Alkalinity and pH relationships during anaerobic sludge stabilization at 52°C.

<u>Equilibria</u>:

$$CHONS + H_2O \left\{ \begin{array}{l} CH_4 \uparrow \\ + \\ CO_2 \uparrow \rightleftharpoons H_2CO_3 \rightleftharpoons H^+ + HCO_3^- \rightleftharpoons H^+ + CO_3^= \\ + \\ NH_3 \rightleftharpoons NH_4^+ + OH^- \\ + \\ RCOOH \rightleftharpoons H^+ + RCOO^- \\ + \\ H_2S \rightleftharpoons H^+ + HS^- \rightleftharpoons H^+ + S^= \end{array} \right.$$

<u>Alkalinity-Volatile Acids Relationships</u>:

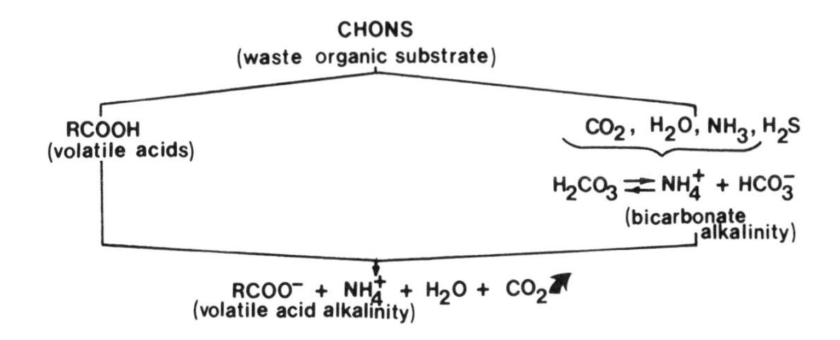

Figure 3. Major acid-base equilibria and their influence during anaerobic biological conversion of waste organic substrates.

treatment efficiency. Accordingly, as illustrated previously in Figures 1 and 2, high volatile acids (RCOOH) concentrations accumulating during retarded conditions would tend to decrease the alkalinity as well as the related bicarbonate buffer capacity, and the pH would be depressed toward the buffer region established by the volatile acids ($pK \approx 4.75$).

A similar result could be presumed to occur in those cases where high ammonia concentrations would tend to increase the pH by displacement of the bicarbonate system in favor of that equilibrium condition ($pK \approx 9.25$). As dramatic an effect would probably not be recorded by hydrogen sulfide or the phosphates and other possible species if present due to their relatively low concentrations, but these would be integral contributors to the overall pH response.

It is also important to recognize that although methane is a desirable conversion product, it is also accompanied by the release of carbon dioxide in varying amounts with more CO_2 being produced during retarded conditions at low pH due to an inhibition of the CH_4 formers and release of "stored" CO_2 as the pH is decreased. This latter aspect has been conceptually illustrated in Figure 3 by the accumulation of volatile acids and their internal titration with the substrate-generated bicarbonate alkalinity. The CO_2 generated originally and again released due to acid production would undergo gas-liquid interchange and affect the aqueous pH based upon partial pressure concepts. Accordingly, the magnitude and state of this CO_2 pool would be a major contributor to the pH response both under normal conditions, when bicarbonates predominate and other contributor concentrations are negligible, and when it is released back into the gas phase due to accumulation of volatile acids.

BUFFER CAPACITY EVALUATION

Model Development

A rigorous study of buffer capacity in aqueous environments was not undertaken until relatively recently with the original formulation by Van Slyke[4] and subsequent review and application particularly by Sillen,[5] Weber and Stumm,[6] Kleijn,[7] Stumm and Morgan[8] and Pohland and Bolton.[9] By considering buffer capacity as indicative of the change in pH with incremental additions of acid or base to a closed system at equilibrium, certain general approaches have been developed to describe both homogeneous and heterogeneous regimes.

The buffer capacity so defined is an expression of the inverse slope of an acid-base titration curve. Therefore, this definition may be applied to the analysis of the aqueous environment of the anaerobic stabilization process, recognizing that a similar change in pH will occur therein upon internal generation or external addition of acids or bases. Utilizing the product acids and bases previously identified together with the indispensable nutrient, phosphorus, the important buffer systems can be represented by the respective availability and dissociation of CO_2, CH_2COOH (HAc), H_2S, NH_3 and H_3PO_4, as indicated in Table I. In this analysis, acetic acid was chosen to represent all the lower volatile fatty acids since the pK values for those acids usually present (acetic, butyric and propionic) in measurable concentration are essentially the same (*i.e.*, 4.75 to 4.87). Moreover, as presented, the difference between the concentrations and activity of the various species was considered negligible, and the pH was not influenced by other indigenous reactions, *i.e.*, complexation, ion exchange and/or sorption.

Table I. Pertinent Acid-Base Equilibria Generated During Anaerobic Microbial Stabilization

Equilibria	Analytical sums
$[H^+]\ [OH^-] = K_w$	
$[H_2CO_3^*] = K_H\, p_{CO_2}$	
$[H^+]\ [HCO_3^-] = K_{c_1}\ [H_2CO_3^*]$	$C_c = [H_2CO_3^*] + [HCO_3^-] + [CO_3^=]$
$[H^+]\ [CO_3^=] = K_{c_2}\ [HCO_3^-]$	
$[H^+]\ [Ac^-] = K_a\ [HAc]$	$C_a = [HAc] + [Ac^-]$
$[H^+]\ [HS^-] = K_{s_1}\ [H_2S]$	
$[H^+]\ [S^=] = K_{s_2}\ [HS^-]$	$C_s = [H_2S] + [HS^-] + [S^=]$
$[NH_4^+]\ [OH^-] = K_n\ [NH_3]$	$C_n = [NH_3] + [NH_4^+]$
$[H^+]\ [H_2PO_4^-] = K_{p_1}\ [H_3PO_4]$	
$[H^+]\ [HPO_4^=] = K_{p_2}\ [H_2PO_4^-]$	$C_p = [H_3PO_4] + [H_2PO_4^-] + [HPO_4^=] + [PO_4^\equiv]$
$[H^+]\ [PO_4^\equiv] = K_{p_3}\ [HPO_4^=]$	

$[H_2CO_3^*]$ = analytical sum of $[CO_2(aq)]$ and $[H_2CO_3]$

Temperature adjustments:

For $K_w, K_H, K_{c_1}, K_{c_2}, K_a, K_{p_1}, K_{p_2}$; $pK = A + \frac{B}{T} + \frac{C}{T^2}$

For K_{s_1}, K_{s_2}; $\ln\left[\frac{K_{T_2}}{K_{T_1}}\right] = -\frac{\Delta H}{R}\left[\frac{1}{T_2} - \frac{1}{T_1}\right]$ using ΔH and K_{T_1} at 25°C

For K_{p_3}; K_{p_3} at 25°C assumed constant

Since the buffer capacity reflects the relative predominance of the contributing species at a particular pH, a buffer index, β, can be formulated which simulates the titration analogy and which in an operational sense can be expressed in terms of alkalinity, or

$$\beta = \frac{d[\text{Base}]}{dpH} = -\frac{d[\text{Acid}]}{dpH} \equiv \frac{d[\text{Alk}]}{dpH} \qquad (1)$$

Accordingly, with the equilibria of Table I,

$$\beta = 2.303 \Big\{ [H^+] + \frac{K_W}{[H^+]} + \frac{C_A K_A [H^+]}{([H^+] + K_A)^2} + \frac{C_P K_{P_1} [H^+]}{(K_{P_1} + [H^+])^2} + \frac{C_P K_{P_2} [H^+]}{(K_{P_2} + [H^+]^2)} + \frac{C_P K_{P_3} [H^+]}{(K_{P_3} + [H^+])^2} + \frac{C_S K_{S_1} [H^+]}{(K_{S_1} + [H^+])^2} + \frac{C_S K_{S_2} [H^+]}{(K_{S_2} + [H^+])^2} + \frac{C_N K_N K_W [H^+]}{(K_N [H^+] + K_W)^2} + \frac{C_C K_{C_1} [H^+]}{(K_{C_1} + [H^+])^2} + \frac{C_C K_{C_2} [H^+]}{(K_{C_2} + [H^+])^2} \Big\} \quad (2)$$

Similarly, with alkalinity as a usually measured input variable which here must be defined in terms of the end point pH selected during titration or $pH_f = -\log[H^+]_f$, then

$$[Alk] = \{[HCO_3^-] - [HCO_3^-]_f\} + 2\{[CO_3^=] - [CO_3^=]_f\} + \{[A_c^-] - [A_c^-]_f\} + \{[HS^-] - [HS^-]_f\} + 2\{[S^=] - [S^=]_f\} + \{[H_2PO_4^-] - [H_2PO_4^-]_f\} + 2\{[HPO_4^=] - [HPO_4^=]_f\} + 3\{[PO_4^{\equiv}] - [PO_4^{\equiv}]_f\} + \{[NH_3] - [NH_3]_f\} + \{[OH^-] - [OH^-]_f\} - \{[H^+] - [H^+]_f\} \quad (3)$$

The buffer system may be either closed or open and in equilibrium with the carbon dioxide partial pressure in the gaseous phase. If the aqueous phase is closed, then the carbonate concentration, C_C, remains equal to the concentration, C_{C_o}, in equilibrium with the CO_2 partial pressure at the initial system pH, or with $pH_o = -\log[H^+]_o$,

$$C_{C_o} = K_H \, P_{CO_2} \left\{ 1 + \frac{K_{C_1}}{[H^+]_o} + \frac{K_{C_1} K_{C_2}}{[H^+]_o^2} \right\} \quad (4)$$

If the carbonate species concentration is assumed to be in continuous equilibrium with the CO_2 partial pressure during titration, then the total carbonate species concentration, C_C, may be determined by

$$C_C = C_{C_0} + 2.303 \int_{pH_0}^{pH} \left\{ \frac{K_{C_1} K_H P_{CO_2}}{[H^+]} + \frac{2K_{C_1} K_{C_2} K_H P_{CO_2}}{[H^+]^2} \right\} dpH \qquad (5)$$

The concentration of the different chemical species as a function of the total concentration of the different species, pH, the partial pressure of CO_2 in the gas phase and the temperature may be computed from the chemical equilibria presented in Table I. These same equilibria when substituted into the expression for alkalinity, [Alk], result in a relationship between the alkalinity and hydrogen ion concentration, or

$$[Alk] = \left[\frac{K_H K_{C_1} P_{CO_2}}{[H^+]} \left\{1 + \frac{2K_{C_1}}{[H^+]}\right\} + \frac{C_A}{\left\{\frac{[H^+]}{K_a} + 1\right\}} + \frac{C_S}{\left\{1 + \frac{[H^+]}{K_{S_1}} + \frac{K_{S_2}}{[H^+]}\right\}} \left\{1 + \frac{2K_{S_2}}{[H^+]}\right\} + \frac{C_N}{1 + \frac{K_N[H^+]}{K_W}} + \frac{C_P}{\left\{\frac{[H^+]}{K_{P_2}} + \frac{K_{P_2}}{[H^+]} + \frac{[H^+]^2}{K_{P_1} K_{P_2}} + 1\right\}} \left\{\frac{[H^+]}{K_{P_2}} + 2 + \frac{3K_{P_3}}{[H^+]}\right\} + \frac{K_W}{[H^+]} - [H^+] \right]_{[H^+]_f}^{[H^+]} \qquad (6)$$

As indicated in Table I, the effect of temperature on the dissociation constants was also considered. The constants for the hydrogen sulfide system were adjusted for temperature using enthalpy data and the Van't Hoff equation; the third dissociation constant for the phosphate system was assumed to be independent of temperature because of lack of enthalpy data; and the temperature effects on the remaining constants were determined by a mathematical fit of available data to the indicated empirical formula.

The equilibrium model for alkalinity may be solved for hydrogen ion concentration or pH since all other parameters would be known input variables. Accordingly, the data required by the model includes: carbon dioxide partial pressure in atm, volatile acids, sulfide, phosphate, and ammonia concentrations, all in mol/l; alkalinity in mol/l of hydrogen ions, the end point pH of the alkalinity titration, and temperature, °C. A computer solution for $[H^+]$ may then be accomplished by first using a rough grid search of one-pH-unit intervals and then the Newton-Raphson iterative search technique until the actual root has been identified. Combining these two approaches results in a rapid solution, thereby reducing computer time and effort.

Once the pH is known, the amount of acid or base required to modify the system pH to any desired level can be determined by numerical integration of the buffer capacity expression using a Runge-Katta integration procedure. Such an evaluation may be made for both the closed system and the system in continuous equilibrium with the carbon dioxide partial pressure.

Model Confirmation and Discussion

To determine the validity of the equilibrium-based model, reported data, specifically selected as representative of anaerobic stabilization of various types of waste organic substrates under alkaline, acid and neutral pH conditions were assembled as indicated in Tables II through V. In those instances where sulfide and phosphate concentrations were not measured, respective concentrations of 0.1 and 2.0 m*M*/l were assumed and used with the other input variables. In all cases, the end-point pH for alkalinity titrations was taken into account in applying the model for pH prediction. The comparison between the predicted and measured pH of these data sets is illustrated in Figure 1 and 2 separately for the previously presented studies of Tables II and III and in Figure 4 collectively for all the data sets.

Inspection of these comparisons indicates a generally acceptable correlation between predicted and measured pH, particularly in the neutral pH range. Deviations during acid and basic conditions suggest either problems with the sufficiency of choice of equilibria to be included for pH prediction with the model, or adequacy and/or reliability of analytical results. Although certain refinements to the model were considered, the additional complexity thereby imposed was not believed warranted at this point, and interpretation of the indicated deviations between predicted and observed pH was focused on the latter issue.

The sophistication of the predictive model was considered to far exceed the normal precision and accuracy associated with routine sampling and analytical procedures used to assess changes in indicator parameters. Part of this problem relates to the time of measurement after the sample was obtained and, more important, whether the conditions at the time of analysis truly duplicated the conditions *in situ* and when and how the pH was measured. Normally, pH is one of the first analyses after a sample is taken, whereas other procedures may be delayed and then subject to conditions of storage which may be different from those at sampling or within the aqueous environment of the process. This is certainly true of the effects of partial pressure of CO_2 in the gas phase above the reaction mixture vs that to which the sample is exposed during storage and analysis;

Table II. Summary of Indicator Parameters During Anaerobic Sludge Stabilization at 36°C[1]

Sample Interval (days)	Gas Production (% CO_2)	Total Volatile Acids (m*M*/l)	Ammonia Nitrogen (m*M*/l)	Total Alkalinity (m*M*/l)[a]	Measured pH	Predicted pH[b]	ΔpH
1	24.2	12.0	72.00	82.40	7.3	7.36	+0.06
3	28.8	30.67	72.93	79.40	7.1	7.18	+0.08
6	27.2	43.67	72.93	71.40	7.2	7.05	-0.15
12	31.7	54.58	72.93	81.00	7.05	7.01	-0.04
16	32.0	70.67	78.71	71.60	6.9	6.73	-0.17
21	35.0	69.83	80.36	73.00	6.75	6.73	-0.02
30	32.0	66.17	77.21	66.00	6.8	6.69	-0.11
39	31.5	58.00	76.43	65.40	6.85	6.78	-0.07
45	32.0	61.92	73.00	71.20	6.85	6.82	-0.03
51	32.0	52.83	75.00	81.00	6.95	7.02	+0.07
66	24.0	18.50	103.00	117.60	7.45	7.51	+0.06
79	23.5	19.67	102.93	105.80	7.4	7.47	+0.07

[a] Alkalinity titration to pH 4.4.

[b] Using equilibrium model assuming constant sulfide and phosphate concentrations of 0.1 and 2.0 m*M*/l, respectively.

Table III. Summary of Indicator Parameters During Anaerobic Sludge Stabilization at 52°C[1]

Sample Interval (days)	Gas Production (% CO_2)	Total Volatile Acids (m*M*/l)	Ammonia Nitrogen (m*M*/l)	Total Alkalinity (m*M*/l)[a]	Measured pH	Predicted pH[b]	ΔpH
1	32.0	29.50	44.71	62.00	7.1	7.10	0.0
12	34.2	37.08	45.71	65.60	7.05	7.06	+0.1
30	41.5	43.33	51.86	77.80	6.8	7.06	+0.26
44	36.4	50.17	48.43	68.80	6.75	6.96	+0.21
59	40.3	48.50	57.20	71.60	7.0	6.96	-0.04
65	40.0	53.17	54.36	71.00	6.9	6.92	+0.02
74	39.6	54.17	62.29	77.20	6.95	6.99	+0.04
86	33.8	57.83	67.86	88.80	7.2	7.14	-0.06
101	40.6	68.67	52.50	69.00	6.65	6.71	+0.06
110	40.5	65.33	50.00	62.40	6.4	6.61	+0.21
122	41.0	59.42	73.93	80.60	6.85	6.97	+0.12
158	40.0	99.17	62.86	68.60	6.25	5.98	-0.27
164	40.6	107.67	72.14	59.60	5.9	5.47	-0.43
170	41.5	110.17	62.29	50.60	5.75	5.24	-0.51
181	42.0	99.00	78.57	45.60	5.8	5.24	-0.56
185	43.5	93.50	73.57	47.20	5.8	5.34	-0.46
191	42.0	98.23	64.29	48.40	5.8	5.31	-0.49
206	42.0	108.83	64.98	60.60	5.85	5.48	-0.37
227	42.0	114.33	73.07	80.60	6.4	6.06	-0.34

[a] Using equilibrium model assuming constant sulfide and phosphate concentrations of 0.1 and 2.0 m*M*/l, respectively.

[b] Alkalinity titration to pH 4.4 end point.

Table IV. Summary of Indicator Parameters During Anaerobic Stabilization at 35°C of Sludge Pretreated by Wet Oxidation

Sample Interval (days)	Gas Production (% CO_2)	Total Volatile Acids (mM/l)	Ammonia Nitrogen (mM/l)	Total Alkalinity (mM/l)	Total Carbonate (mM/l)	Total Phosphate (mM/l)	Total Sulfide (mM/l)	Measured pH	Predicted pH a	Predicted pH b	ΔpH a	ΔpH b
1	27.06	2.0	46.4	122.5	78.0	33.2	15.9	7.4	7.39	7.47	-0.01	+0.07
2	24.50	2.0	55.0	102.5	73.2	19.0	8.1	7.6	7.40	7.63	-0.20	+0.03
4	21.00	3.0	64.3	117.5	73.9	7.42	2.8	7.5	7.58	7.68	+0.08	+0.18
5	21.00	6.0	102.2	100.0	75.5	7.42	2.5	7.7	7.49	7.72	-0.21	+0.02
6	20.25	6.0	72.9	105.0	85.8	7.1	1.9	7.9	7.54	7.85	-0.36	-0.05
12	19.97	22.2	100.0	132.5	84.4	7.42	4.4	7.85	7.59	7.86	-0.26	+0.01
17	20.22	34.4	107.9	135.0	79.5	5.81	3.1	7.75	7.56	7.78	-0.19	+0.03
24	30.80	27.0	58.6	130.0	95.7	37.09	18.8	7.6	7.34	7.55	-0.26	-0.05
25	31.50	27.0	59.3	125.0	91.2	33.2	14.4	7.35	7.34	7.38	-0.01	+0.03
31	32.60	4.0	58.6	107.5	82.5	23.5	11.9	7.45	7.28	7.46	-0.17	+0.01
40	30.76	2.0	65.8	90.0	81.9	6.77	4.1	7.5	7.30	7.51	-0.20	+0.01
54	31.54	2.2	80.8	107.2	93.4	8.39	5.9	7.55	7.36	7.56	-0.19	+0.01
61	33.00	10.2	80.1	101.5	85.8	4.84	3.8	7.5	7.30	7.52	-0.20	+0.02
68	37.02	17.7	60.5	83.0	56.8	4.84	3.4	7.35	7.12	7.41	-0.23	+0.06
75	34.44	13.2	50.0	67.8	49.8	0.64	5.9	7.25	7.07	7.30	-0.18	+0.05
110	31.38	12.0	50.0	66.8	56.2	5.16	2.2	7.6	7.18	7.62	-0.42	+0.02
117	33.70	2.0	53.5	78.5	65.5	3.23	3.4	7.7	7.22	7.71	-0.48	+0.01
124	31.90	1.5	60.7	87.8	62.3	6.45	5.6	7.55	7.26	7.64	-0.29	+0.09
131	31.86	2.5	61.4	85.8	73.9	3.87	2.8	7.75	7.28	7.73	-0.47	-0.02

[a] Using equilibrium model, measured CO_2 partial pressure and alkalinity titration end point of pH 4.0.

[b] Using equilibrium model, computed CO_2 partial pressure from measured carbonate concentration and alkalinity titration end point of pH 4.0.

Table V. Summary of Indicator Parameters During Anaerobic Refuse Stabilization at 35° and 60°C

Reference Source	Gas Production (% CO_2)	Total Volatile Acids (mM/l)	Ammonia Nitrogen (mM/l)	Total Alkalinity (mM/l)[a]	Temperature (°C)	Measured pH	Predicted pH[b]	ΔpH
Pfeffer[10]	30.3	7.0	21.5	28.0	35	6.77	6.77	0.00
	41.4	5.4	21.5	36.8	35	6.75	6.79	+0.04
	42.8	4.63	21.5	38.0	35	6.75	6.80	+0.05
	46.2	4.7	21.5	46.0	35	6.77	6.85	+0.07
	46.2	5.5	21.5	60.0	35	6.87	6.97	+0.10
Khan[11]	45.14	3.0	21.5	24.0	60	6.93	6.76	-0.17
	45.48	3.17	21.5	32.0	60	6.95	6.89	-0.06
	45.32	2.33	21.5	31.4	60	6.95	6.90	-0.05
	47.63	5.67	21.5	21.6	60	6.83	6.65	-0.18
	44.40	8.17	21.5	36.3	60	7.0	6.92	-0.08
	45.34	13.0	21.5	38.8	60	7.1	6.90	-0.20
	42.8	7.67	21.5	30.8	60	6.81	6.85	-0.04
	45.2	9.17	21.5	40.0	60	7.0	6.95	-0.05
	46.3	11.5	21.5	52.2	60	7.16	7.06	-0.10

[a] Alkalinity titration to pH 4.5 end point.

[b] Using equilibrium model, assuming constant sulfide and phosphate concentrations of 0.1 and 2.0 mM/l, respectively; ammonia nitrogen is average of range reported.

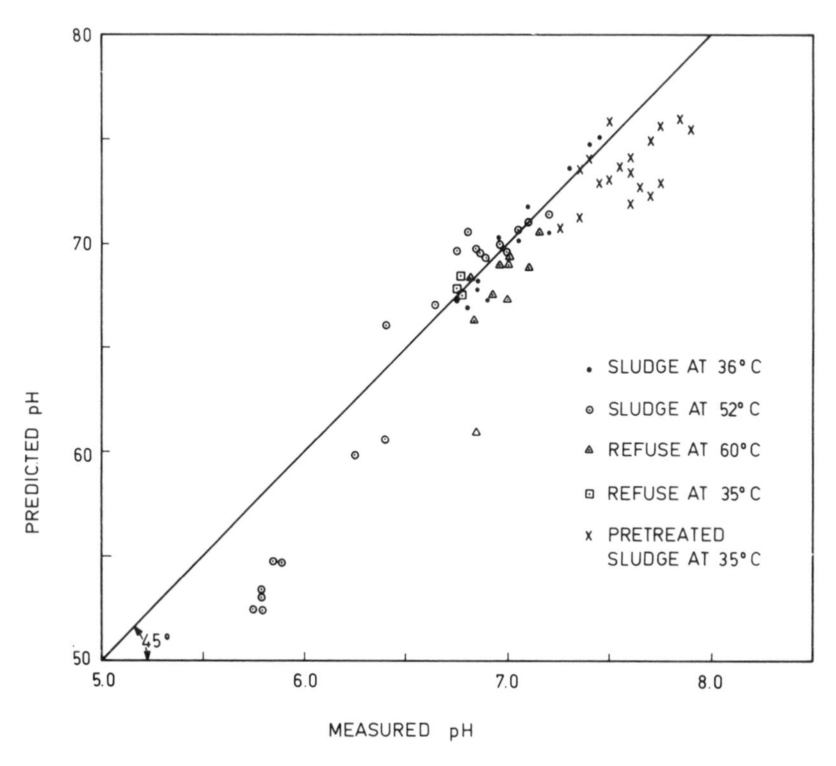

Figure 4. Comparison of predicted and measured pH during anaerobic stabilization before adjustments.

since this effect on pH is an integral part of the model solution, some adjustment is necessary when such procedures are used.

Unfortunately, most published analytical data have not been obtained in such a manner that all parametric analyses are conducted under the same conditions of sampling and storage. Since *in situ* sample analyses other than temperature and gas composition are generally not performed on the process, it can be presumed that conditions did vary and that the comparison between predicted and measured pH as determined herein must also vary. To illustrate how these conditions would influence results, the data with the pretreated sludge (Table IV) were scrutinized in terms of the effects of CO_2 on pH utilizing CO_2 pressures based on *in situ* gas-phase analysis and on carbonate analysis on samples removed from the reacting mixture and on which pH and other parametric analyses were also performed. As indicated in Figure 5, most of the deviation between measured and predicted pH could be attributed to the apparent differences

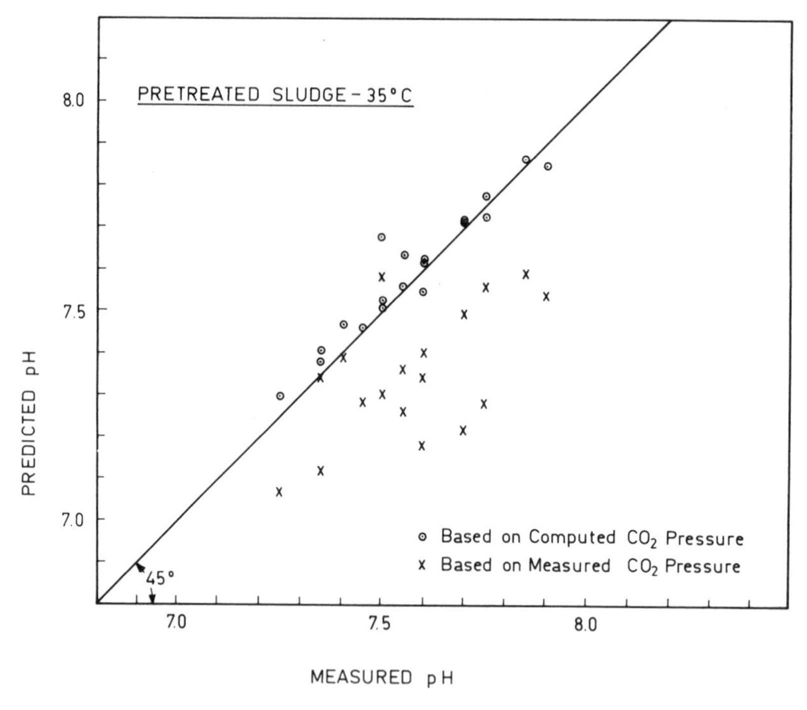

Figure 5. Comparison of predicted and measured pH based upon computed and measured CO_2 partial pressures.

encountered in conditions at the time of analysis, and an average adjustment of 0.25 pH units resulted.

It is likely that similar conditions influenced the reported results of the other studies which would be most pronounced with the greatest departure of pH from that of the normal bicarbonate buffer system around neutrality. Accordingly, the data for retarded conditions for sludge stabilization at 52°C when low pH values were recorded would require a similar adjustment before the observed and measured pH could be fairly compared. Unfortunately, total carbonate analyses were not available from these studies, but if a similar correction were applied to the measured pH values, again a much better agreement results, as indicated in Figure 6.

Since the predictive model includes both CO_2 partial pressures and alkalinity as two of the input variables, it is also most important to recognize the effect of the end-point pH used during alkalinity determinations on the magnitude of the total alkalinity recorded. This is dramatically illustrated in Figure 7, where a specified sample simulating the supernatant from an anaerobic stabilization process was exposed to varying CO_2 partial pressures

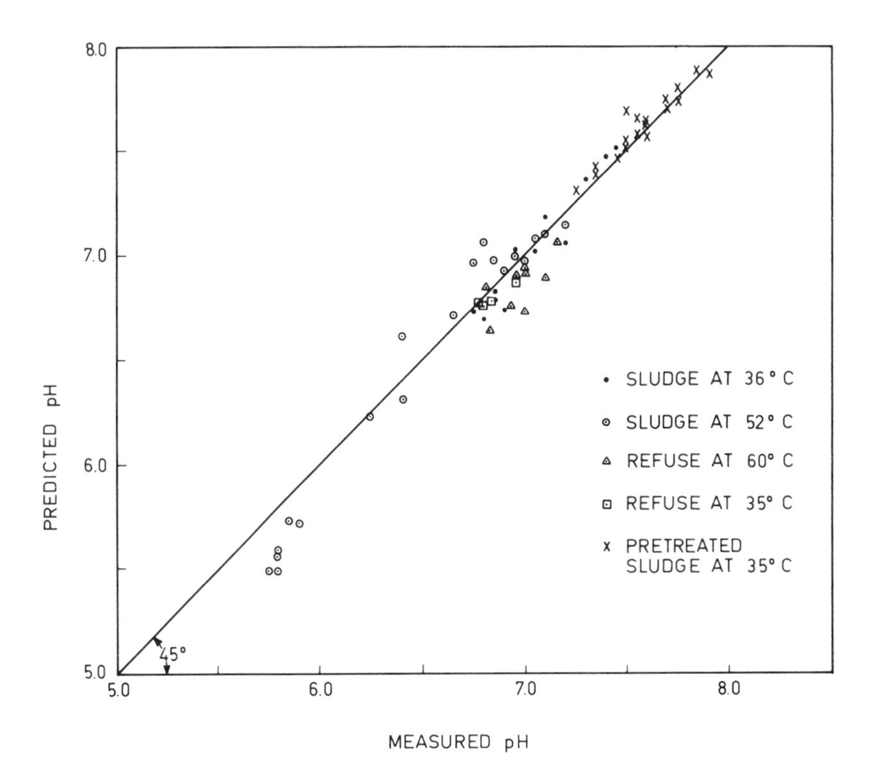

Figure 6. Comparison of predicted and measured pH during anaerobic stabilization after adjustments.

and alkalinity determinations made with end-point pH varying between 4.0 and 4.7. Accordingly, with a reference end point of 4.3, a sample subjected to CO_2 partial pressures from 0.1 to 0.4 atmospheres would vary in alkalinity concentration from 1048 to 2976 mg/l as $CaCO_3$. Moreover, these same samples if titrated to pH values other than the indicated reference would undergo corresponding increases or decreases in reported values of the magnitude indicated in Figure 7, depending on whether the new end point was above or below the reference. The absolute amount of increase or decrease would depend upon the concentrations of acids or bases being measured as parts of the total alkalinity response.

As indicated previously, a considerable fraction of the alkalinity could be contributed by volatile acids accumulations which are not completely accounted for by titration to the indicated reference pH of 4.3. In fact, at this pH and 35°C, only 74.4% of the volatile acids alkalinity has been measured; this percentage would decrease at higher pH and increase at

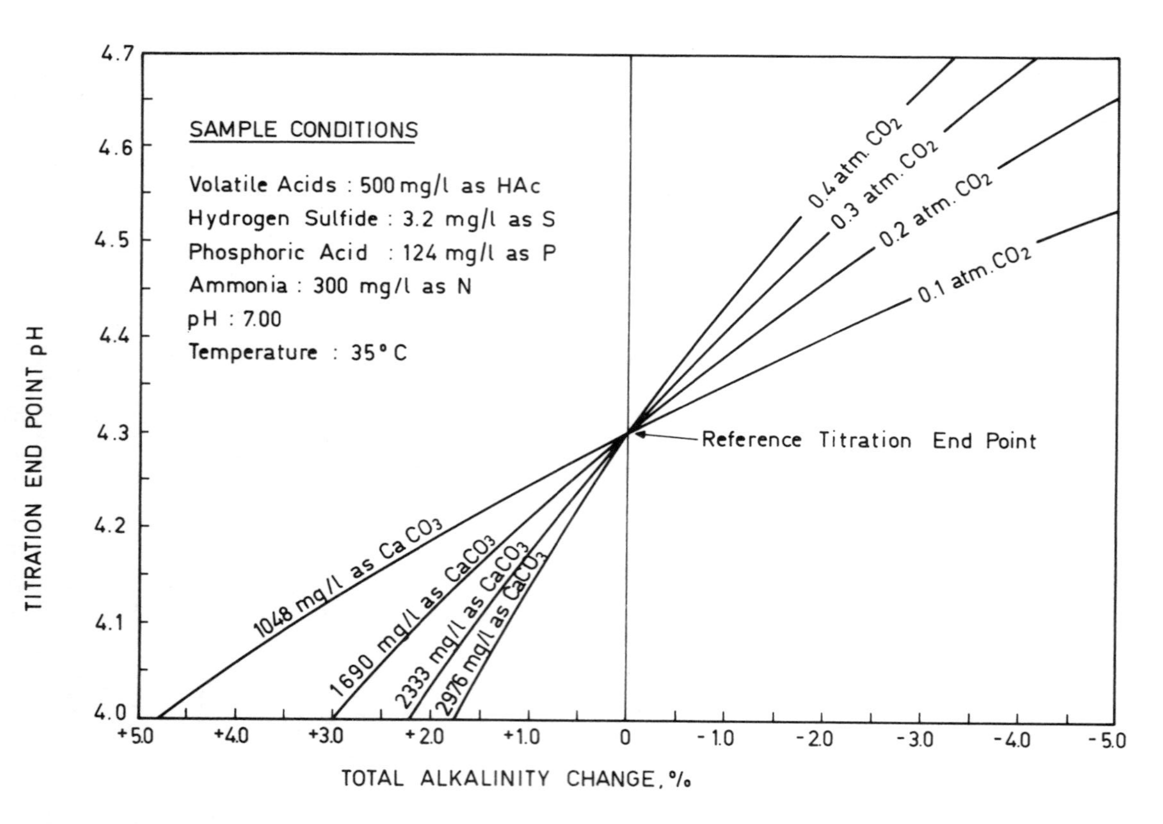

Figure 7. Effect of CO_2 partial pressure and titration end-point pH on total alkalinity determination for specified supernatant sample.

lower pH end points. Furthermore, the often quoted 85% for a pH 4.0 end point is strictly valid only at 35°C, which is not the temperature condition of most alkalinity determinations conducted external to the process environment. Similarly, although a pK value represented by that for acetic acid was assumed for all volatile acids present, there is probably enough difference to alter the percentage of each respective volatile acid included in the alkalinity determination, and therefore influence the pH predicted by the model. These effects are emphasized, assuming a constant pK value for all the volatile acids but at different titration temperatures and pH end points, in Table VI. Such considerations must be included in any attempts to adjust the pH of an anaerobic stabilization process with buffer additions based upon alkalinity determinations. Here the original relationship devised initially by Pohland and Bloodgood[1] would need to be modified as suggested by McCarty[12] but for the proper titration end-point pH and also temperature prevailing in the process. Moreover, the model does not include an adjustment for ionic strength, which may exert a significant influence on results.

Table VI. Fraction of Volatile Acids Accounted for at End Point of Alkalinity Titration

Titration End Point	Temperature (°C)								
	20	25	30	35	40	45	50	55	60
4.0	0.850	0.851	0.851	0.853	0.855	0.857	0.859	0.862	0.866
4.1	0.819	0.819	0.819	0.822	0.824	0.826	0.829	0.833	0.837
4.2	0.782	0.782	0.782	0.785	0.788	0.791	0.794	0.798	0.803
4.3	0.740	0.741	0.741	0.744	0.746	0.750	0.754	0.758	0.764
4.4	0.694	0.694	0.694	0.698	0.700	0.704	0.709	0.714	0.719
4.5	0.643	0.643	0.643	0.647	0.650	0.654	0.659	0.665	0.671
4.6	0.588	0.589	0.589	0.593	0.596	0.601	0.605	0.611	0.618
4.7	0.532	0.532	0.532	0.536	0.540	0.544	0.549	0.556	0.562

OPERATIONAL APPLICATION

In spite of the limitations on acquiring reliable data, the proposed model for explicating the pH response during anaerobic biological stabilization of waste organic substrates was considered sufficiently valid for extension to the area of process control through pH adjustment. Recognizing the uncertainties in alkalinity determinations as influenced by sampling and analytical techniques, rather than providing an oversimplified formulation, it was considered more appropriate to develop a series of control curves based upon constituent acids and bases contributing to the overall pH response in the system and which could be measured independently by existing analytical methods. Consequently, the amount of acid or base to be added for pH adjustment to some desired environmental optimum could be obtained directly for each component species and the total quantity required determined by summation. Example control curves for the acids and bases considered in the model development are indicated in Figures 8 through 12 for a temperature of 35°C, since most anaerobic stabilization processes are operated in the mesophilic temperature range. Accordingly, if the pH, partial pressure of CO_2 in the gas and the volatile acids, ammonia, hydrogen sulfide and phosphoric acid concentrations are known, the neutralizing requirement can be determined directly and related to an appropriate chemical additive for pH adjustment. For example, if the pH of the aqueous environment within an anaerobic stabilization process is 5.5, the CO_2 partial pressure is 0.425 atmospheres, and the volatile acids, ammonia, hydrogen sulfide and phosphoric acid concentrations are 50 m*M*/l, 42 m*M*/l, 0.1 m*M*/l and 2.5 m*M*/l respectively, 27 m*M*/l base would need to be added to adjust the pH to 7.0. The corresponding amount determined by solution of the equilibrium model for buffer capacity was computed to be 27.11 m*M*/l.*

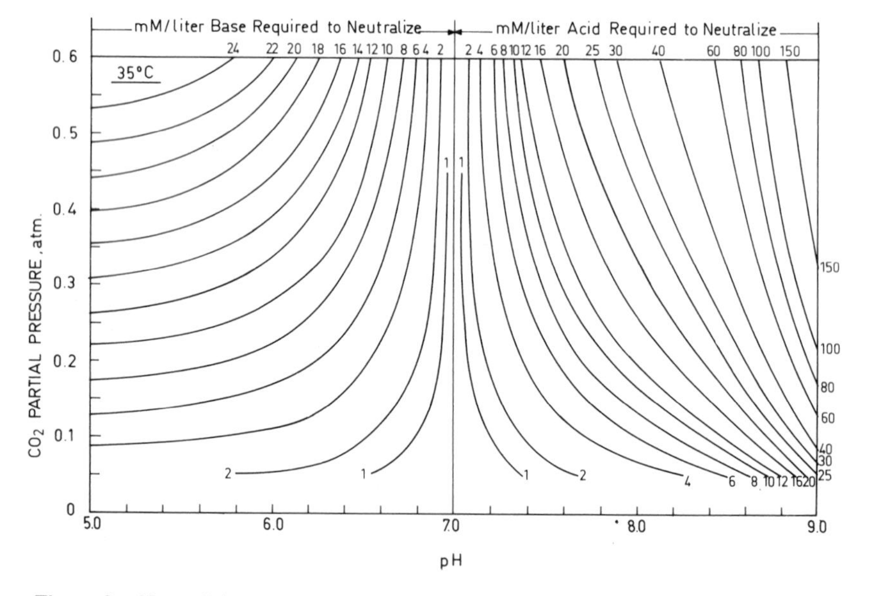

Figure 8. Neutralizing requirement for aqueous carbon dioxide system including buffer capacity of water.

*A copy of the program and the computer printout for this calculation are available from the authors.

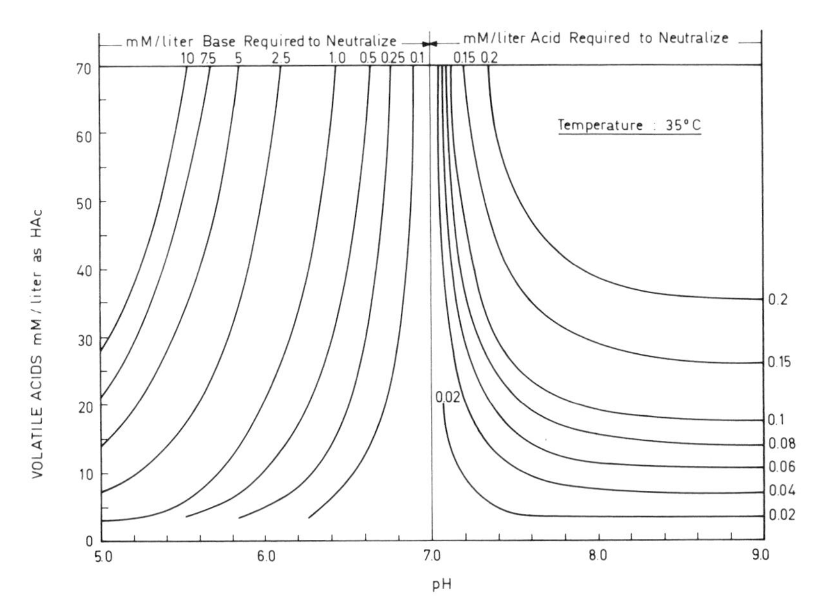

Figure 9. Neutralizing requirement for aqueous acetic acid system excluding buffer capacity of water.

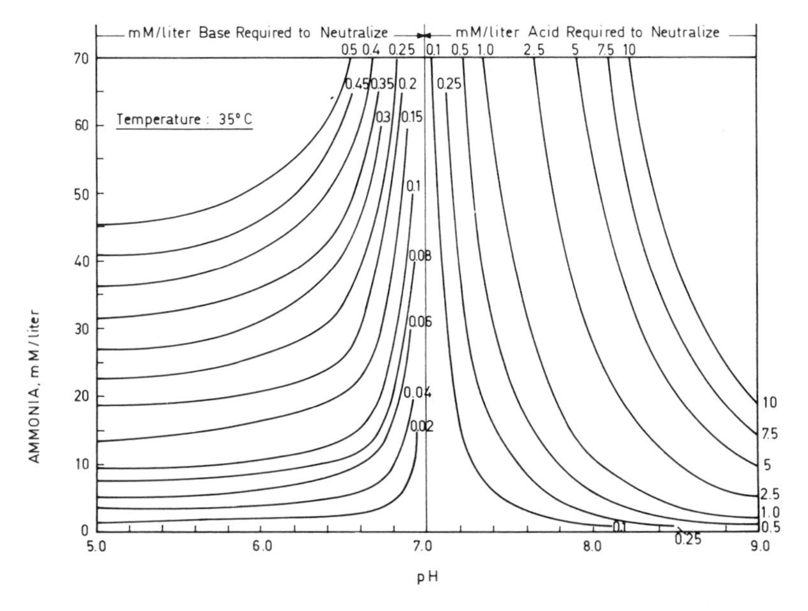

Figure 10. Neutralizing requirement for aqueous ammonia system excluding buffer capacity of water.

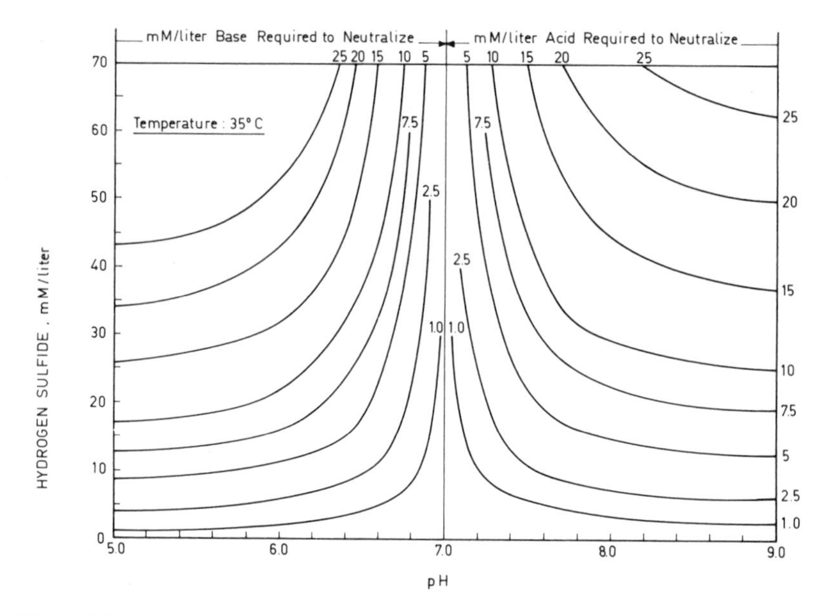

Figure 11. Neutralizing requirement for aqueous hydrogen sulfide system excluding buffer capacity of water.

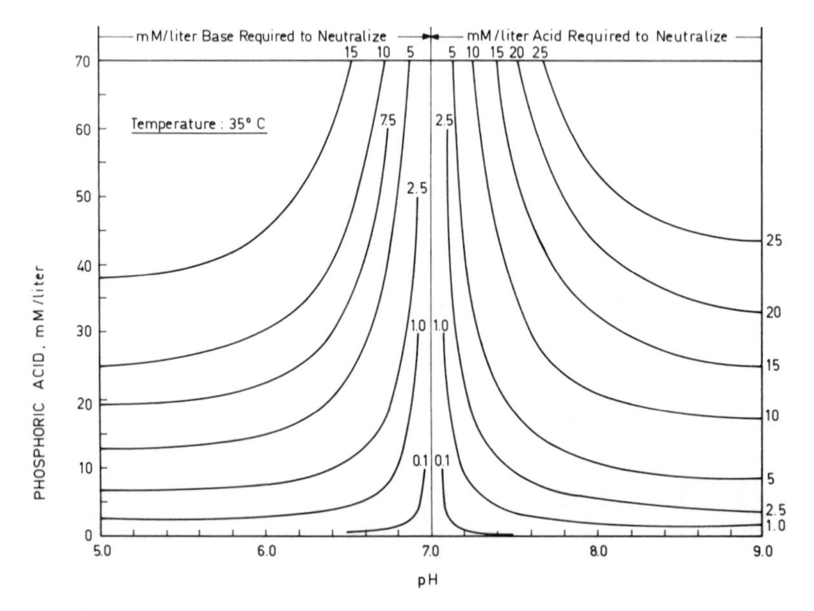

Figure 12. Neutralizing requirement for aqueous phosphoric acid system excluding buffer capacity of water.

CONCLUSION

When compared to the chemical complexity of most biologically mediated processes, a relatively simple acid-base equilibrium model was conceived both to define the origin of and to predict the pH response to major acidic and basic constituents of the anaerobic stabilization process. The model also permitted an evaluation of buffer capacity and the acid or base required to adjust the existing pH to any desired level.

Although comprehensive analytical data to fit the model were relatively scarce, selected data provided a reasonable correlation between predicted and observed pH levels in anaerobic processes treating a variety of waste organic substrates. Deviations were attributed to sampling and analytical deficiencies and differences between internal and external measurements. Additional effort is being directed toward an improvement of the accuracy and precision of data acquisition and an extension of the model to include the effects of ionic strength and other analytical parameters which will better define the aqueous and gaseous phases of the process environment. This effort will eventually culminate in a procedure for process diagnosis and for simple and accurate determination of the relative acceptability and effectiveness of techniques for chemical adjustment of the aqueous environment of biological waste treatment systems to an optimum pH level.

REFERENCES

1. Pohland, F. G., and D. E. Bloodgood. *J. Water Poll. Control Fed.* 35:11 (1963).
2. Pohland, F. G., and K. H. Mancy. *Biotechnol. Bioeng.* 11:683 (1969).
3. Pohland, F. G., and S. Ghosh. *Biotechnology and Bioengineering Symp.*, Vol. 2 (New York: John Wiley & Sons, Inc., 1971), p. 85.
4. Van Slyke, D. D. *J. Biol. Chem.* 52:525 (1922).
5. Sillen, L. G. *Proc. International Oceanographic Congress,* N67 (Washington, DC: American Academy for the Advancement of Science, 1961).
6. Weber, W. J., Jr., and W. Stumm. *J. Am. Water Works Assoc.* 55:1553.
7. Kleijn, H. F. W. *Internat. J. Air Water Poll.* 9:401 (1965).
8. Stumm, W., and J. J. Morgan. *Aquatic Chemistry* (New York: Wiley-Interscience, 1970).
9. Pohland, F. G., and W. R. Bolton. "Buffer Capacity in Aquatic Ecosystems," Environ. Research Center, Georgia Institute of Technology, ERC 0474 (March 1974).
10. Pfeffer, J. T. "Reclamation of Energy from Organic Refuse," EPA-670/2-74-016 (March 1974).
11. Khan, K. M.S. Thesis, University of Illinois (1976).
12. McCarty, P. L. *Public Works* (September 1964), p. 107.

CHAPTER 26

APPLICATION OF CARBONATE EQUILIBRIA TO HIGH-PURITY OXYGEN AND ANAEROBIC FILTER SYSTEMS

James A. Mueller and Jack Famularo
Manhattan College
Bronx, New York

Thomas J. Mulligan
Hydroscience, Inc.
Westwood, New Jersey

INTRODUCTION

The design and analysis of biological waste treatment systems are normally based on evaluation of biological rate parameters, stoichiometric coefficients for electron acceptors and organism growth and physical properties of the biomass. An understanding of the chemical changes associated with the biological changes occurring in the reactor provides additional insight into system performance. For anaerobic systems such as the anaerobic filter, this insight is required to ensure that the pH is controlled within a specific range to support growth of the relatively fragile methane formers and prevent the reactor from going sour. In some aerobic systems, such as those obtaining nitrification, pH control is also required to allow optimal biological growth. With the advent of the covered-tank high-purity oxygen system in the late 1960s, an understanding of the chemical interactions was useful not only for pH control but also for optimum design of the gas transfer equipment.

This chapter summarizes the practical application of the chemical interactions occurring in the aerobic high-purity oxygen system and the anaerobic filter system. The basic approach used to describe the chemical interactions occurring in these systems is to incorporate into a mathematical model the mass balance equations for all parameters affecting both alkalinity and total inorganic carbon (TIC) with pH determination using the carbonate equilibria system.

MASS BALANCE EQUATIONS

Although the aerobic and anaerobic systems differ in their use of electron acceptor in the biological reaction, the anaerobic filter and high-purity oxygen systems are similar in that they both contain a gas phase flowing concurrently with the liquid phase as shown in Figure 1. Two mass balance equations are therefore constructed for each parameter in the system, for the liquid phase and for the gas phase, respectively.

$$\frac{dS_{i,n}}{dt} = \frac{Q}{V_n^1}(S_{i,n-1} - S_{i,n}) + \Sigma R_{i,n}^1 \tag{1}$$

$$\frac{dp_{i,n}}{dt} = \frac{G_{n-1}p_{i,n-1} - G_n p_{i,n}}{V_n^g} + \frac{V_n^1 \bar{R}T}{V_n^g M_i} \Sigma R_{i,n}^g \tag{2}$$

where

S, p	=	parameter concentration in the liquid and partial pressure in the gas
V^1, V^g	=	liquid and gas volumes
Q, G	=	liquid and gas flow rates
$\bar{R}, T, M$	=	gas constant, absolute temperature and molecular weight
i, n	=	species identification and reactor stage location
R^1, R^g	=	sources and sinks in the liquid and gas phases

For each parameter of concern, both the liquid phase concentration and gas phase partial pressure are unknown with an equation (1 and 2) available for each. However, an additional unknown always exists in the gas phase equation: the gas flowrate, G. The final equation defining a constant total pressure in the gas phase allows simultaneous solution of the equation set

$$\Sigma p_{i,n} = p_T - p_V \tag{3}$$

where p_T and p_V are the total and vapor pressures of the gas phase. Because of the nonlinear nature of the gas phase equation, the equation set is normally numerically integrated to steady state which has proven to be a relatively simple, convenient and reliable technique. Development of a steady-state solution which requires less computer time has been

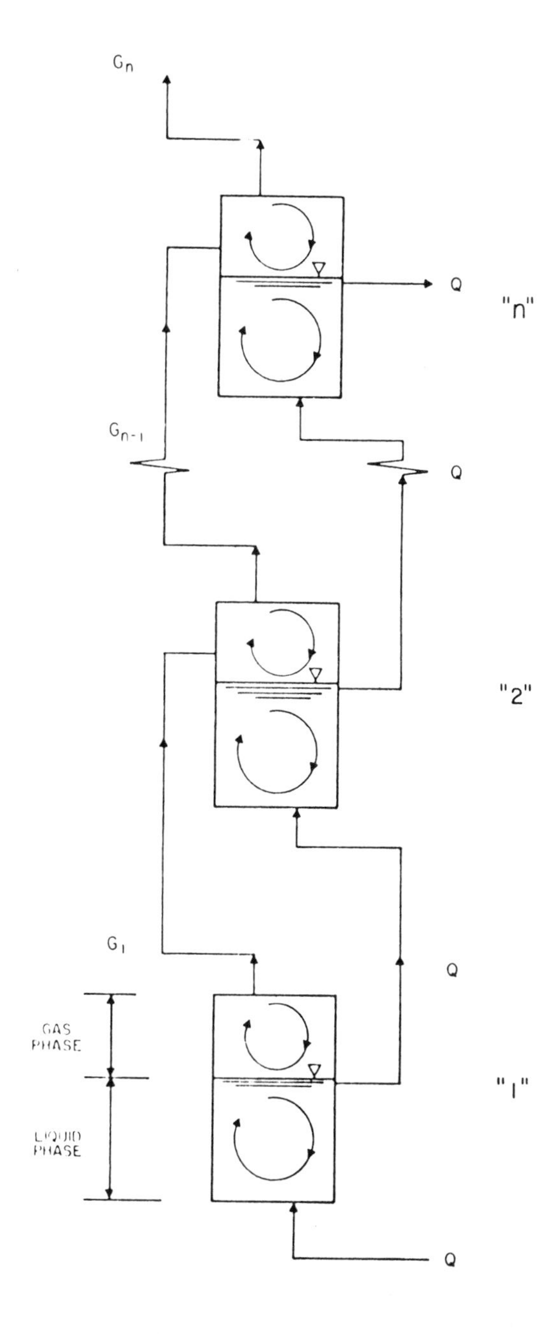

Figure 1. Completely mixed series reactor schematic for concurrent flow of liquid and gas phases.

accomplished for the high-purity oxygen aeration system. This effort required a significant amount of development and computer program debugging time before a viable tool useful for system analysis and design was obtained.

The source and sink terms in Equation 1 incorporate both bacterial growth and decay, organic nitrogen hydrolysis, and interphase gas transfer. In Equation 2, no reaction occurs in the gas phase, the only source and sink terms being interphase gas transfer. Schematics of the aerobic and anaerobic reactions for both systems are shown in Figures 2 and 3. Details of the specific source and sink terms for the high-purity oxygen system can be found in Mueller, Mulligan and DiToro[1] and for the anaerobic filter in Mueller and Mancini.[2] Subsequent to the above publications, the high-purity oxygen system model has been modified to incorporate heterotrophic growth, ammonia utilization and organic nitrogen hydrolysis with their effects on alkalinity.

BIOLOGICAL REACTION KINETICS

In the high-purity oxygen system, first-order removal kinetics have been used for BOD oxidation while zero-order kinetics have been used for sulfite oxidation and nitrification. From experience with pilot studies, there are indications that the major portion of soluble BOD removal occurs in the first stage of the series reactors with the following stages used for oxidation of absorbed BOD and endogenous respiration similar to the kinetic model recently proposed by Busby and Andrews.[3] However, the relatively simple first-order model has adequately simulated observed gas phase data.[1] Before adopting the more complicated function, adequate verification should be obtained.

With respect to the anaerobic filter model, two types of reaction kinetics have been analyzed: Michaelis kinetics based on the individual biological solids fractions and first-order kinetics based on total measured volatile solids as given in Table I.

The Michaelis growth equation for the methane formers, X_V, is based on the nonionized fraction of volatile acids and contains an inhibition function, I_V, after Andrews.[4,5] This feature is desirable to incorporate the effect of pH change on reaction rate; however, no experimental verification of I_V and K_V values exists. Figure 4 presents a comparison of the pH inhibition factors calculated for two volatile acids concentrations with the coefficients used by Andrews to those measured by Clark and Speece.[6] The shape of the curves follows the data except at the peak, where no inhibition of gas production was measured over a relatively wide range of pH, from 6.0 to 8.0. This limits the usefulness of this inhibition function for the anaerobic filter.

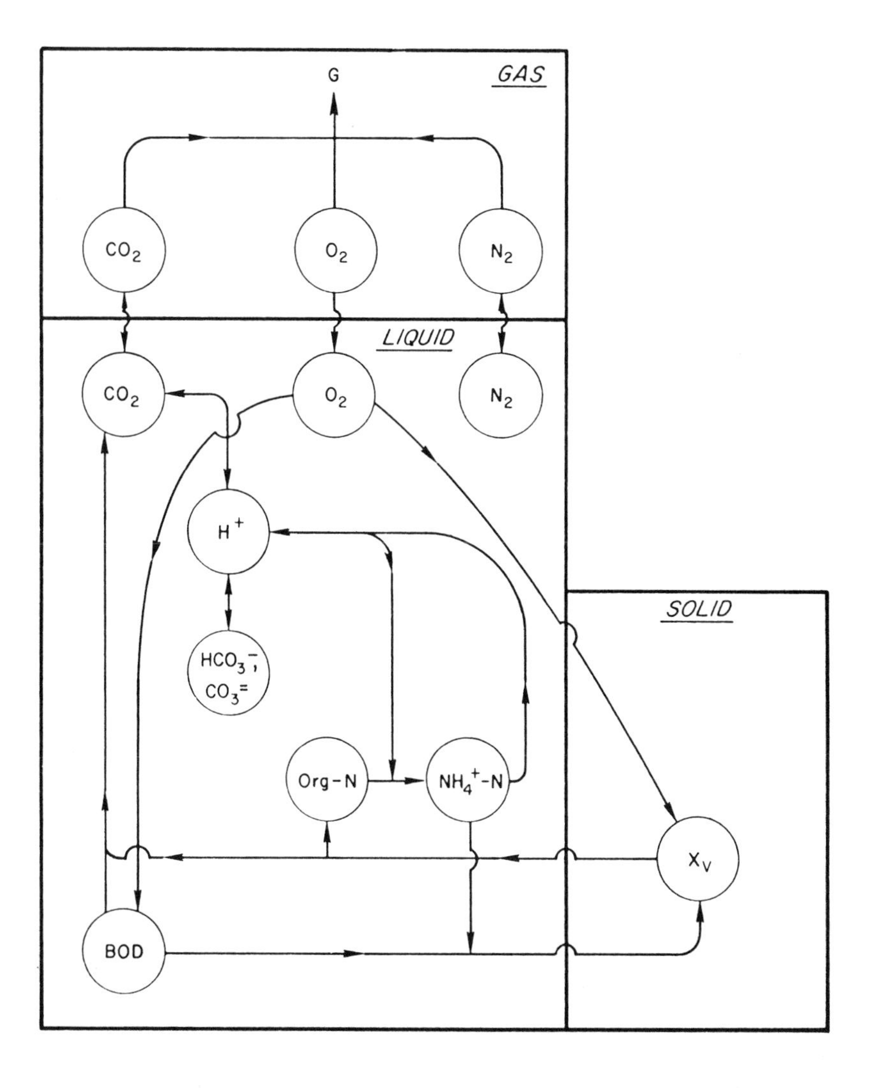

Figure 2. Aerobic reaction schematic.

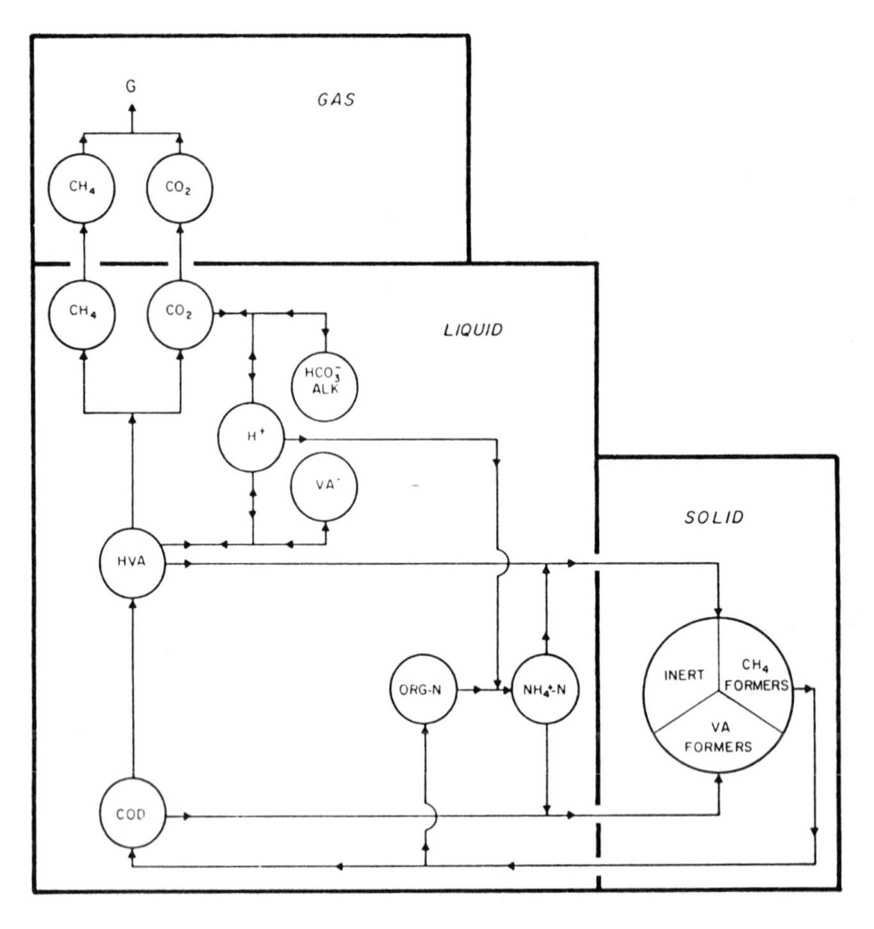

Figure 3. Anaerobic reaction schematic.

As seen from Figure 5, adequate data fit was obtained with the Michaelis kinetics using the data of Young and McCarty.[7,8] However, due to the two degrees of freedom in the model, sufficient data fit was obtained with an order of magnitude change in the Michaelis coefficient and and over 100% change in the unit substrate removal rate for the complex COD. Figures 6, 7 and 8 show that the first-order model adequately simulated the COD and organic nitrogen data. In Figures 6 and 7, the volatile acid fit was poorer than the other constituents, probably due to lack of incorporation of the solids fractions in the model. When only volatile acids were fed to the reactor, the first-order model was adequate (Figure 8).

The form of the kinetic function to be used in simulating the above system has generally been found not to affect the results seriously. In

Table I. Reaction Kinetics Analyzed in Anaerobic Filter Model

Reaction	Michaelis Kinetics	First-Order Kinetics
R_c, growth rate of X_c	$\dfrac{\hat{\mu}_c S_1 X_c}{K_c + S_1}$	$k_c S_1 X_t$
R_v, growth rate of X_v	$\dfrac{\hat{\mu}_v X_v}{1 + \dfrac{K_v}{S_2} + \dfrac{S_2}{I_v}}$	$k_v S_{32} X_t$
R_N, organic N reaction rate	$\dfrac{\hat{k}_N S_5 X_c}{K_N + S_5}$	$k_N S_5 X_t$

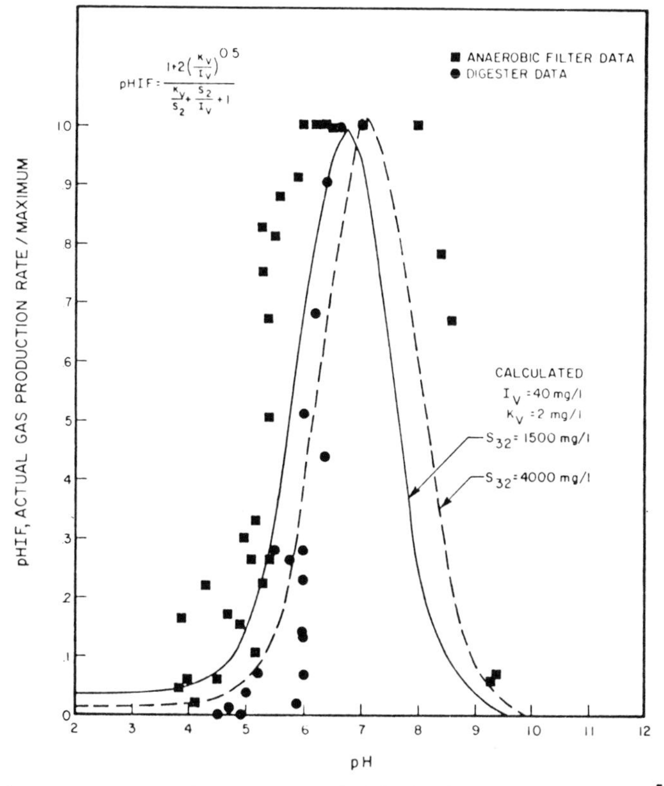

Figure 4. Comparison of calculated pH inhibition factors using Andrews[5] model to observed data of Clark and Speece.[6]

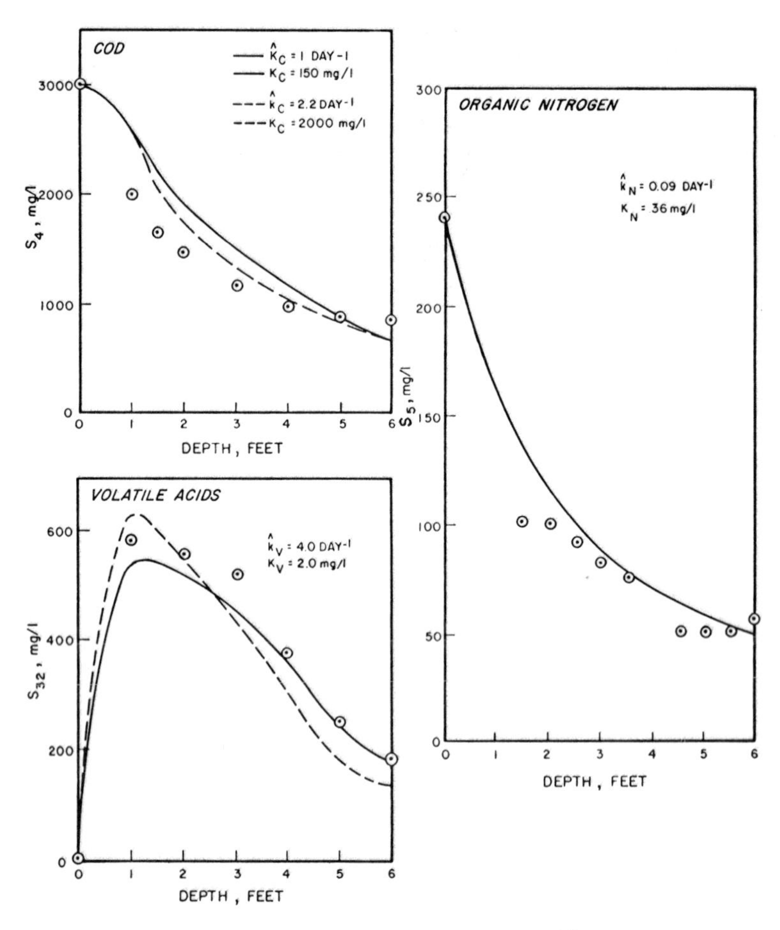

Figure 5. Michaelis model–Young and McCarty[7,8] pc 212 data.

view of the fact that mass transfer resistances through biofilm are neglected in the anaerobic filter model, the simpler kinetic function, first-order kinetics based on observed biomass concentration, has been used in further evaluation of the anaerobic filter system.

CHEMICAL INTERACTIONS

With the exception of the volatile acids production in the anaerobic filter model, the chemical interactions for the two systems are similar. Since the ionic reactions occurring in these systems are orders of magnitude

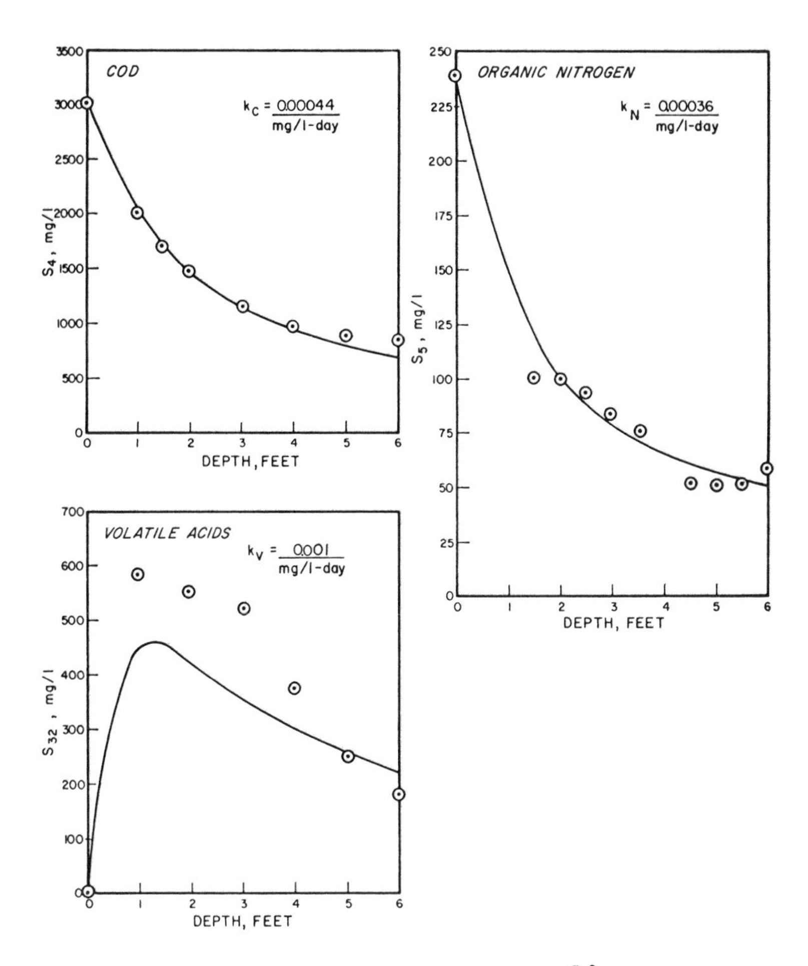

Figure 6. First-order model–Young and McCarty[7,8] pc 212 data.

faster than the biological reactions, they will be considered to be at equilibrium. The chemical interactions are discussed in detail below, first considering the factors affecting alkalinity followed by the carbonate equilibria equations used to predict dissolved CO_2 and pH.

Alkalinity

As the biochemical reactions proceed, changes occur in the concentrations of components which have the ability to remove hydrogen ions from solution. In the high-purity oxygen and anaerobic filter systems, the

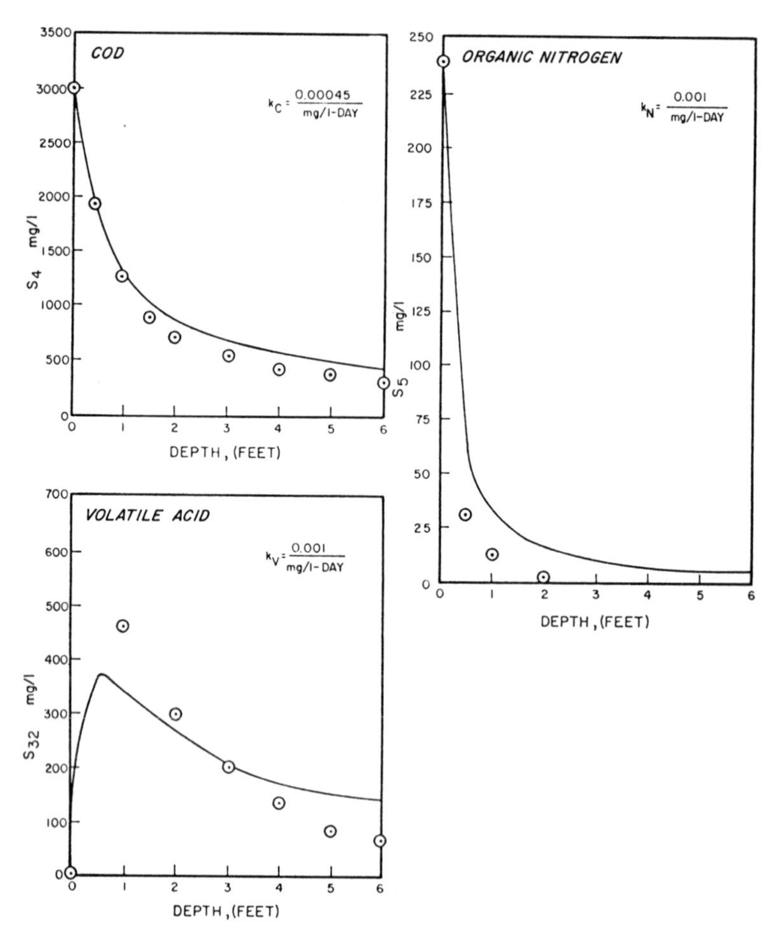

Figure 7. First-order model–Young and McCarty[7,8] pc 53 data.

components of interest are: soluble organic nitrogen (SON), ammonia nitrogen (NH_3-N) and volatile acids (VA).

Changes in the concentration of soluble organic nitrogen may have significant effects on alkalinity depending on the initial charge of the compound, effects which are often neglected in evaluation of chemical changes in biological systems. The organic nitrogen in solution is considered as amino acids having a Zwitteronic nature which ionizes in two steps

$$R(NH_2)CHCOO^- + H^+ \rightleftarrows R(NH_3^+)CHCOO^-, \; pK_{2N} \quad (4)$$

$$R(NH_3^+)CHCOO^- + H^+ \rightleftarrows R(NH_3^+)CHCOOH, \; pK_{1N} \quad (5)$$

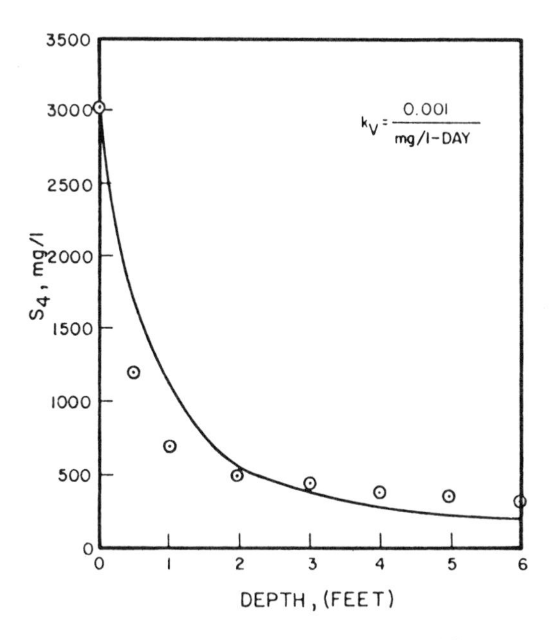

Figure 8. First-order model–Young and McCarty[7,8] va 53 data.

In Equations 4 and 5, the two ionic species containing the negatively charged carboxyl group are capable of removing hydrogen ion from solution and therefore constitute forms of amino acid alkalinity. The remaining species containing the un-ionized carboxyl group and the charged amino group do not contribute to amino acid alkalinity. To simplify the development, the following substitutions will be made

$$X = R(NH_2)CHCOO^-$$

$$Y = R(\overset{+}{N}H_3)CHCOO^-$$

$$Z = R(NH_2)CHCOOH$$

Since there are 14,000 mg of nitrogen for each gram mole (g mol) of the above ionic species, the total SON concentration in mg/l, designated S_5, is given by the equation

$$\frac{S_5}{14000} = [X] + [Y] + [Z] \tag{6}$$

For very dilute solutions, the dissociation equilibria can be expressed in terms of molarities using the equations

$$[H^+][Y] = K_{1N}[Z] \tag{7}$$

$$[H^+][X] = K_{2N}[Y] \tag{8}$$

Since alkalinity is measured by titrating with H_2SO_4 to an end point of approximately pH 4.5-4.25 depending on TIC concentration, some X and Y ions will remain in solution at the end of the titration. The amount remaining will depend on the assumed values of K_{1N} and K_{2N} as given by the equations

$$[Y]_e = \frac{K_{1N}}{[H^+]} [Z]_e \tag{9}$$

$$[X]_e = \frac{K_{1N}K_{2N}}{[H^+]_e^2} [Z]_e \tag{10}$$

Alkalinity is normally expressed in units of mg/l as $CaCO_3$ with an equivalent weight of 50,000 meq/1 for $CaCO_3$. To calculate the alkalinity associated with SON, we need only to relate the molarity of the X and Y species to gram-equivalents of hydrogen ion. The resultant equation for SON alkalinity is

$$\frac{A_{SON}}{50{,}000} = 2\ ([X] - [X]_e) + ([Y] - [Y]_e) \tag{11}$$

where it should be recognized that each mol of X is capable of removing two mol of hydrogen ion and each mol of Y can remove one hydrogen ion. The development is completed by expressing each of the molarities on the rhs of Equation 11 in terms of SON concentration, pH, pK_{1N} and pK_{2N}. The resultant equations for [X] and [Y] are

$$[X] = \frac{S_5}{14{,}000} \left\{ 1 + \frac{[H^+]}{K_{2N}} + \frac{[H^+]^2}{K_{1N}K_{2N}} \right\}^{-1} \tag{12}$$

$$[Y] = \frac{S_5}{14{,}000} \left\{ 1 + \frac{K_{2N}}{[H^+]} + \frac{[H^+]}{K_{1N}} \right\}^{-1} \tag{13}$$

The equations for $[X]_e$ and $[Y]_e$ are the same as Equations 12 and 13 with the exception that $[H^+]$ is replaced by $[H^+]_e$.

The calculation of changes in SON alkalinity resulting from an increase in X and Y must also include a consideration of resultant changes in hydrogen ion concentration; thus

$$\frac{\Delta A_{SON}}{50{,}000} = 2\Delta([X] - [X_e]) + \Delta([Y] - [Y]_e) - \Delta[H^+] \qquad (14)$$

Changes in $[H^+]$ occur due to Reactions 4 and 5 and therefore can be related to changes in [X] and [Z]

$$\Delta[H^+]_1 = \Delta[X]$$

$$\Delta[H^+]_2 = -\Delta[Z]$$

$$\Delta[H^+] = \Delta[H^+]_1 + \Delta[H^+]_2 = \Delta[X] - \Delta[Z] \qquad (15)$$

After substituting Equation 15 into 14, the alkalinity change takes the form

$$\frac{\Delta A_{SON}}{50{,}000} = \Delta([X] + [Y] + [Z]) - 2\Delta[X]_e - \Delta[y]_e \qquad (16)$$

Equations 6, 12 and 13 can now be employed to relate changes in molarity to changes in SON concentration. The resultant equation for ΔA_{SON} is

$$\Delta A_{SON} = 3.57\ f_5\ \Delta S_5 \qquad (17)$$

where $f_5 = 1 - (2/E_x) - (1/E_y)$, and E_x and E_y are end-point correction factors defined by

$$E_x = 1 + \frac{[H_e^+]}{K_{2N}} + \frac{2[H^+]_e}{K_{1N}K_{2N}}$$

$$E_y = \frac{K_2}{[H^+]_e} E_x$$

The constant 3.57 is merely the quotient 50,000/14,000 and has units of mg $CaCO_3$ per mg N.

The above approach can be utilized with as many pK values as required depending on the specific amino acids to be evaluated. Table II summarizes the f_5 values for various amino acid groupings using an end-point pH_e of 4.25 typical of the TIC concentration in the anaerobic filter system. From the results in Table II, it is seen that only the dicarboxylic acids, aspartic and glutamic, significantly increase alkalinity when removed from solution due to the preponderance of the negative charge on these compounds. When the amino acids are combined into protein, the effect on alkalinity will depend on the overall net charge on the compound.

Table II. Summary of f_5 Values for Various Amino Acids

Amino Acid	mol N per mol Amino Acid	Iso electric Point pI	Acidity Constant pK_1	pK_2	pK_3	pK_4	f_5
Aspartic & Glutamic	1	3.0	2.1[a]	3.96[a]	9.64[a]		-0.66
Diiodo-Tyrosine	1	4.29	2.12	6.48	7.82		+0.0015
Cystine	2	5.02	1.04	2.05	8.00	10.25	+0.003
Tyrosine	1	5.63	2.20	9.11	10.07		+0.0088
11 Monocarboxylic	1	5.7-6.1	2.27[a]	9.65[a]			+0.010
Arginine	4	10.76	2.01	9.04	12.48		+0.25
Histidine	3	7.6	1.77	6.1	9.18		+0.323
Lysine	2	9.47	2.18	8.95	10.53		+0.50

[a]Average value.

Having completed the development of equations for SON alkalinity (Equations 8 and 13), we can now consider the alkalinity associated with ammonia-nitrogen. In biological systems, NH_3 is utilized as a source of nitrogen for microorganism growth and as an energy source in nitrifying systems; it is produced from organic nitrogen hydrolysis. Ammonia contributes to total system alkalinity through its ability to remove a hydrogen ion from solution in accordance with the following ionic reaction

$$NH_3 + H^+ \rightleftarrows NH_4^+, \; pK_A \tag{18}$$

The total ammonia concentration, S_6, expressed in units of mg/l-N is related to the molarity of the undissociated and dissociated species in Equation 18

$$\frac{S_6}{14{,}000} = [NH_3] + [NH_4^+] \tag{19}$$

As in the discussion of SON alkalinity, ideal solution is assumed and equilibria are expressed in terms of molarities rather than activities, yielding

$$[H^+][NH_3] = K_A[NH_4^+] \tag{20}$$

Combination of Equations 19 and 20 results in

$$\frac{S_6}{14{,}000} = [NH_3](1 + \frac{[H^+]}{K_A}) \tag{21}$$

and since each mol of NH_3 is capable of removing one equivalent of H^+, it follows that ammonia alkalinity is

$$A_{NH_3} = 3.57\ S_6(1 + \frac{[H^+]}{K_A})^{-1} \tag{22}$$

The change in ammonia alkalinity associated with total NH_3-N concentration includes the change in hydrogen ion concentration and is given by

$$\Delta A_{NH_3} = 50{,}000\ (\Delta [NH_3] - \Delta [H^+]) \tag{23}$$

However, from Equation 18 it is seen that a loss of one hydrogen ion results in the production of an ammonium ion: $-\Delta [H^+] = \Delta [NH_4^+]$. Therefore, the change in alkalinity is

$$\Delta A_{NH_3} = 50{,}000\ \Delta([NH_3] + [NH_4^+])$$

$$= 3.57\ \Delta S_6 \tag{24}$$

In the derivation of the ammonia alkalinity equations, the end-point pH correction was neglected because pK_A is equal to 9.22, resulting in negligible NH_3 concentration at the alkalinity end point.

The last form of alkalinity required in the anaerobic filter model is that due to volatile acids. Since volatile acids contain an average pK value of 4.75, they must be accounted for in the alkalinity titration similar to the amino acids.

$$VA \rightleftarrows VA^- + H^+,\ pK_V \tag{25}$$

letting $x = [VA]$ and $y = [VA^-]$, then

$$\frac{S_{10}}{60{,}000} = x + y \tag{26}$$

where S_{10} is the measured volatile acid concentration expressed as acetic acid.

The volatile acid alkalinity will then be

$$\frac{A_{VA}}{50{,}000} = y - y_e \tag{27}$$

Since

$$y[H^+] = K_V x, \tag{28}$$

Equations 26, 27 and 28 can be combined to yield

$$\frac{A_{VA}}{50{,}000} = \frac{S_{10}}{60{,}000}\left[\left\{1 + \frac{[H^+]}{K_V}\right\}^{-1} - \left\{1 + \frac{[H^+]_e}{K_V}\right\}^{-1}\right] \quad (29)$$

For normal pH conditions above 6.5, substantially all volatile acid will be ionized with Equation 29 simplified to

$$A_{VA} = 0.833\ S_{10}\left\{1 + \frac{K_V}{[H^+]_e}\right\}^{-1} \quad (30)$$

The change in alkalinity is equal to

$$\frac{\Delta A_{VA}}{50{,}000} = \Delta\ (y - ye) - \Delta\,[H]^+ \quad (31)$$

Since $\Delta\ [H^+] = \Delta\, y$

$$\frac{\Delta A_{VA}}{50{,}000} = -\Delta\, ye$$

$$= -0.833\ \Delta S_{10}\left\{1 + \frac{[H^+]}{K_V}\right\}^{-1} \quad (32)$$

At the pH_e for the anaerobic filter of 4.25 and with $pK_v = 4.75$, then $A_{VA} = 0.63\ S_{10}$ and $\Delta\ A_{VA} = -0.20\ \Delta\ S_{10}$.

Carbonate Equilibria and pH

Inorganic carbon is present in biological treatment systems in the form of dissolved carbon dioxide, bicarbonate ion and carbonate ion. It is produced biochemically by the oxidation of soluble carbonaceous BOD and volatile suspended solids. Although produced in the form of CO_2, partial dissociation to HCO_3^- and $CO_3^=$ occurs instantaneously to establish ionic equilibrium. The dissociation reactions and pK values at 25°C are as follows

$$H_2CO_3 \rightleftarrows H^+ + HCO_3^-,\ pK_1 = 6.35 \quad (33)$$

$$HCO_3^- \rightleftarrows H^+ + CO_3^=\ ,\ pK_2 = 10.33 \quad (34)$$

The TIC concentration in mg/l-C is defined by

$$\frac{S_2}{12{,}000} = [CO_2] + [HCO_3^-] + [CO_3^=] \quad (35)$$

Since the system is in a state of ionic equilibrium, the molarities of the components in the above equation are related by

$$[H^+][HCO_3^-] = K_1 [CO_2] \tag{36}$$

$$[H^+][CO_3^=] = K_2 [HCO_3^-] \tag{37}$$

It follows that

$$S_{CO_2} = 44{,}000 [CO_2] = \frac{S_2}{0.273} \left\{ 1 + \frac{K_1}{[H^+]} + \frac{K_1 K_2}{[H^+]^2} \right\}^{-1} \tag{38}$$

Since each bicarbonate ion is capable of removing one hydrogen ion from solution, the bicarbonate alkalinity, A_{HCO_3}, is given by

$$A_{HCO_3} = 50{,}000 [HCO_3^-]$$

$$= \frac{S_2}{0.24} \left\{ 1 + \frac{[H^+]}{K_1} + \frac{K_2}{[H^+]} \right\}^{-1} \tag{39}$$

Each carbonate ion is equivalent to two parts of alkalinity, hence

$$A_{CO_3} = 50{,}000 \times 2 [CO_3^=]$$

$$= \frac{S_2}{0.12} \left\{ 1 + \frac{[H^+]}{K_2} + \frac{[H^+]^2}{K_1 K_2} \right\}^{-1} \tag{40}$$

The carbonaceous alkalinity, A_C, is the sum

$$A_C = A_{HCO_3} + A_{CO_3} \tag{41}$$

It is determined by subtracting all other alkalinities, A_{SON}, A_{NH_3} and A_{VA} from the total alkalinity which is evaluated by considering changes associated with biochemical reaction. Knowledge of A_C and S_2 completely determines the carbonate-bicarbonate equilibrium system and permits the simultaneous solution of Equations 38 to 41 for pH, S_{CO_2}, A_{HCO_3} and A_{CO_3}

CALIBRATION

Prior to model application, calibration of the kinetic parameters in the model must be obtained. Calibration of the high-purity oxygen model have been obtained by Mueller *et al.*[1] for both domestic and industrial wastewaters. In these calibrations a respiratory quotient (RQ) of 1.0 was used for domestic wastewater while 0.85 and 0.60 were required to fit higher F/M ratios used for the industrial wastewater. Additional calibration of the chemistry interactions was more recently obtained on a pulp and paper mill wastewater as shown in Figures 9 and 10. An RQ of 0.90 was used to fit the data. Although the ammonia nitrogen data was scarce, a lower requirement for growth was indicated by the data than is typically expected for biological systems. Due to secrecy agreements, only the relative gas phase data are presented.

Calibration of the chemistry effects in the anaerobic filter model[2] is shown in Figures 11 and 12 for two COD loading rates of 220 and 1680 lb COD/1000 ft^3-day. The f_5 values required to fit the alkalinity

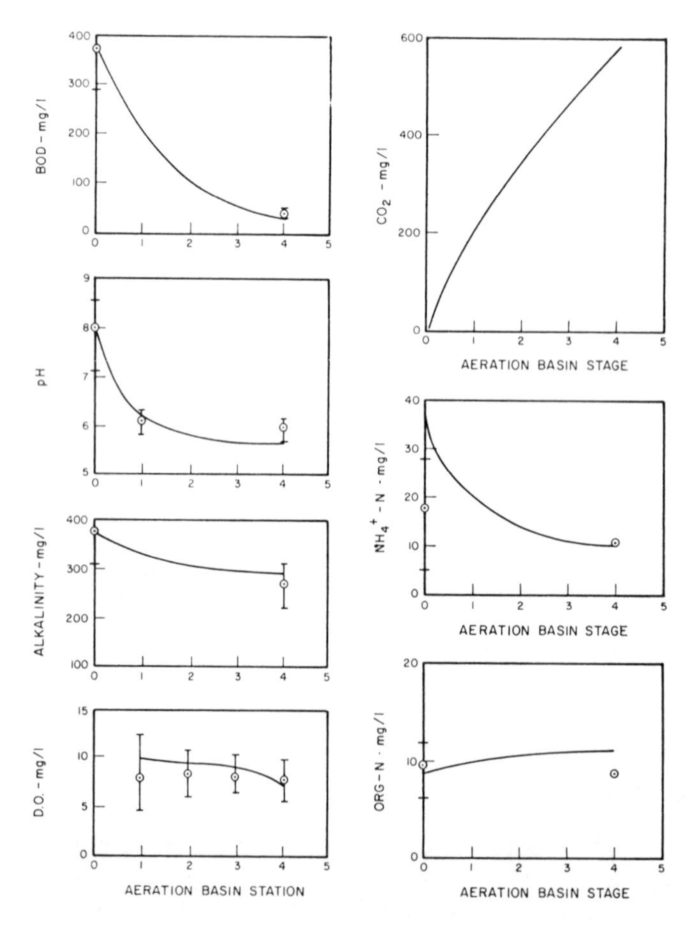

Figure 9. Pulp and paper pilot study liquid-phase data high-purity oxygen model verification.

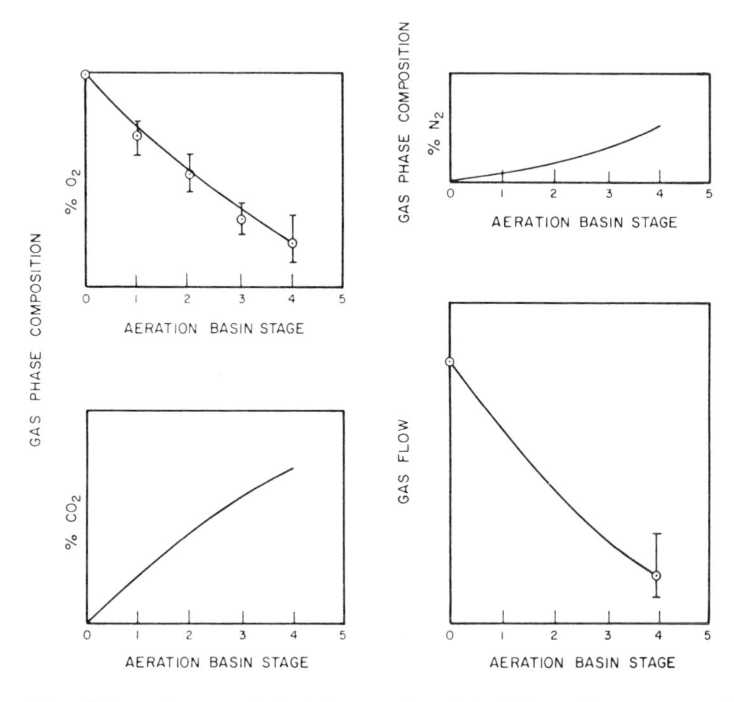

Figure 10. Pulp and paper pilot study gas-phase data high-purity oxygen model verification.

data were -0.72 and -0.92, indicating a significant additional source of alkalinity due to the organic nitrogen in the nutrient broth feed.

APPLICATIONS

The mass balance, kinetic and chemistry equations of the previous sections have been incorporated into a model of the covered-tank high-purity oxygen system. The physical system involves recycle of mixed liquor from the secondary clarifer to the first stage of the reactor. Changes observed in the secondary clarifer on the dissolved components are incorporated in the model. Aerator housepower requirements to obtain a specified level of dissolved oxygen in each stage are calculated in the model for a specified aerator rated capacity, N_O, with the computer program capable of being used as a design or simulation package.

To illustrate application of the model it will be applied to treatment of a wastewater from a chemical plant with the following conditions:

Q = 15.4 MGD
BOD_5 = 1144 mg/1 at 24-hr peak load

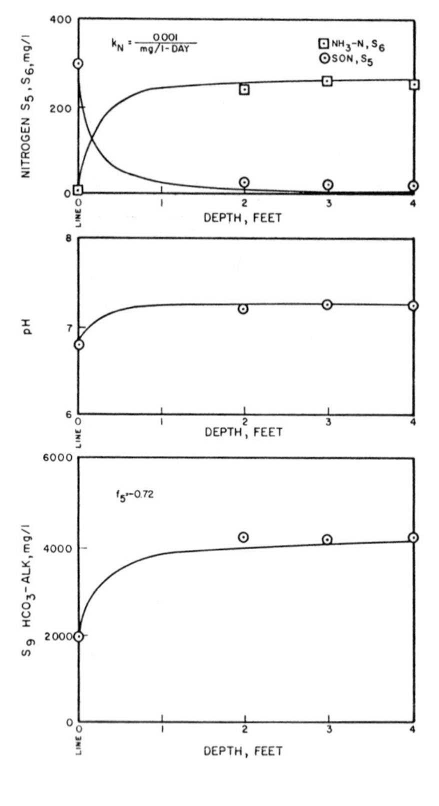

Figure 11. Anaerobic filter first-order model PC 220 nitrogen, alkalinity and pH data.[2]

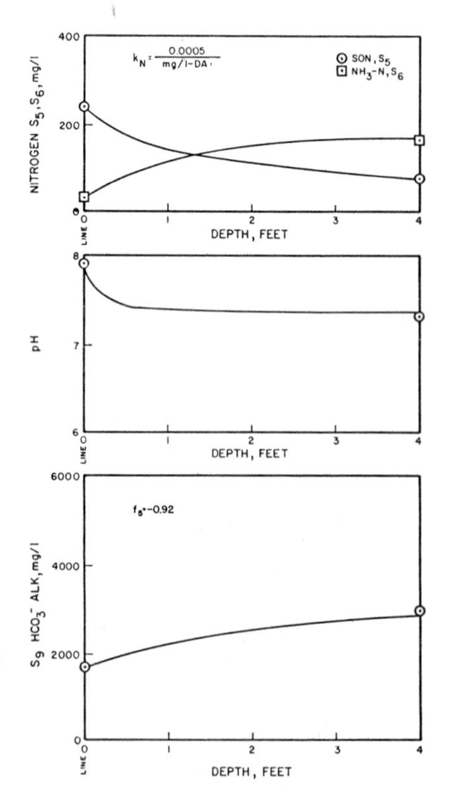

Figure 12. Anaerobic filter first-order model PC 1680 nitrogen, alkalinity and pH data.[2]

BOD_5 = 608 mg/l at average daily load
ALK = 500 mg/l
pH = 10.3

Using a 2-1-1 tank configuration and an N_O of 3.0 lb O_2/hp hr with 80% BOD_5 removal, the design option of the program was utilized to compare the chemistry effects and horsepower requirements of air vs oxygen systems. Figure 13, using a respiratory quotient of 0.63, illustrates that the major chemistry difference in the two systems is the high CO_2 concentration carried in the liquid phase of the oxygen system due to the covered-tank configuration not allowing CO_2 to strip off into the atmosphere as it does in the air system. The pH is therefore decreased to about 6.0, while it is above 7.0 in the air system. For the maximum load condition the oxygen system horsepower requirements, including a generation hp for the oxygen of 1.25 hp/scfm as well as aerator hp, are shown in Figure 14 to be less than those for the air system. This is due to the high-oxygen partial pressures of greater than 60% existing in all stages of the oxygen system. The larger aerator hp requirements in the air system would also require larger volume reactors using a completely mixed configuration. At higher aerator transfer capabilities in wastewater ($\propto N_O$), differences between air and oxygen horsepower requirements become smaller.

In the design of pure oxygen systems, a trade-off can be made between oxygen gas flowrate and aeration hp in achieving optimum operation at a desired DO level. The optimum aeration hp design should result in a minimization of the total treatment hp consisting of the sum of aeration power plus power to produce influent oxygen. Modern oxygen generation plants require between 0.88 and 1.29 hp per scfm of oxygen.* Curve (a) of Figure 15 is a design curve for the peak load of 1144 mg/l BOD described in the previous section. Portions of the curve at low aeration hp correspond to high-influent oxygen flowrate. A power charge of 1.25 hp/scfm of oxygen was employed to develop curve (a). As aeration hp is increased, less oxygen is required to maintain a DO of 2 mg/l, and the percentage of oxygen utilization represented in curve (b) increases. It is apparent that optimum operation corresponds to use of as little oxygen as possible, a direct consequence of the power charge of 1.25 hp/scfm. The plots previously shown in Figures 13 and 14 were taken from the plant simulation at an aeration hp of 450. Total hp requirements for average conditions using the same mixer hp required for the maximum conditions are shown as curve (c) for 90% oxygen utilization.

In Figure 16, a design curve corresponding to maintenance of 4 mg/l of DO is shown for this average loading. The aeration hp of 250, adequate to maintain a DO of 4 mg/l under average loading conditions, is

*UNOX brochure.

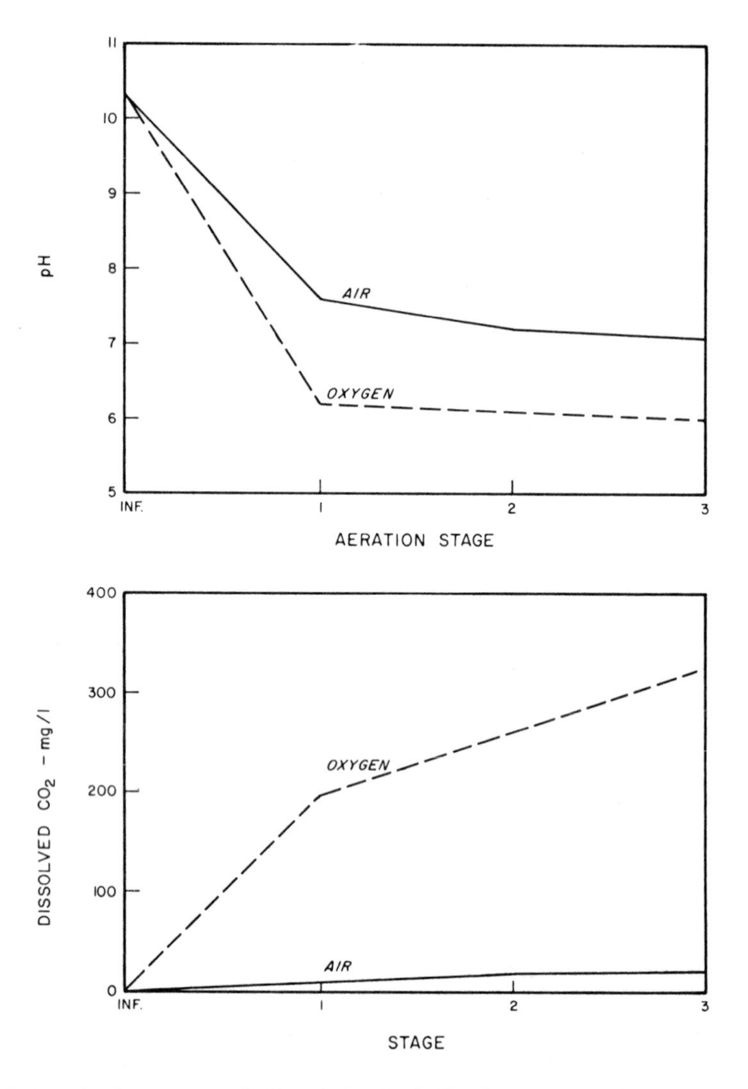

Figure 13. Comparison of pH and dissolved CO_2 for oxygen and air systems.

unable to achieve a DO of 2 mg/l during periods of peak loading. An examination of Figure 15 reveals that at least 390 aeration hp is required to accomplish this.

For the plant under study, if aerators are employed with nonvariable power input, the permissible design must be in the range of 390 to 458 aeration hp. During periods of average influent BOD, gas flow can be reduced. However, due to the excessive aeration hp, high DO levels between 12-17 mg/l result, equivalent to an oxygen utilization of 90%.

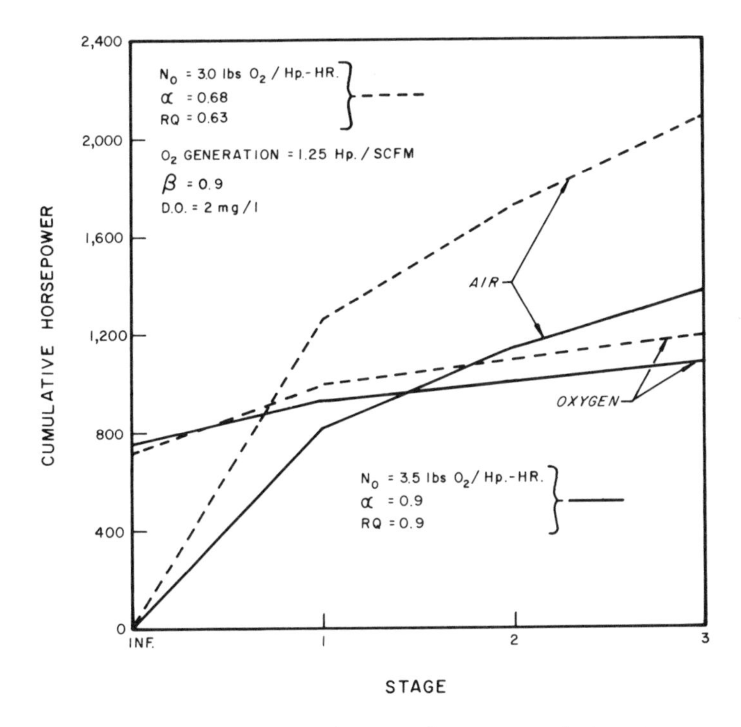

Figure 14. Horsepower requirements for oxygen and air systems.

Total treatment power lies in the range of 820 to 900 hp, as shown in curve (c) of Figure 15. Clearly, the peak demands of this system suggest the need for a variable-hp aerator, such as a turbine aerator, which can be operated at 450 hp during peak-load conditions and reduced to 250 hp at average loading.

The anaerobic filter model is applied to system evaluation considering the chemisty effects for various wastewater alkalinities. Figure 17 indicates that for a loading condition of 220 lb COD/1000 ft^3-day, the initial alkalinity in the wastewater has a negligible effect on system pH due to alkalinity production during organic nitrogen conversion to ammonia. Supernatant liquors generated from low-pressure oxidation of sludge should fall into this category, requiring little-to-no supplemental alkalinity additions to the feed contrary to a recent study of Haug.[9] This fact should make anaerobic filter treatment of these recycle streams more economically attractive from both an operating and a capital cost standpoint.

When the organic nitrogen source of alkalinity is not present in the waste, an alkalinity requirement of 0.25 to 0.5 the influent COD is

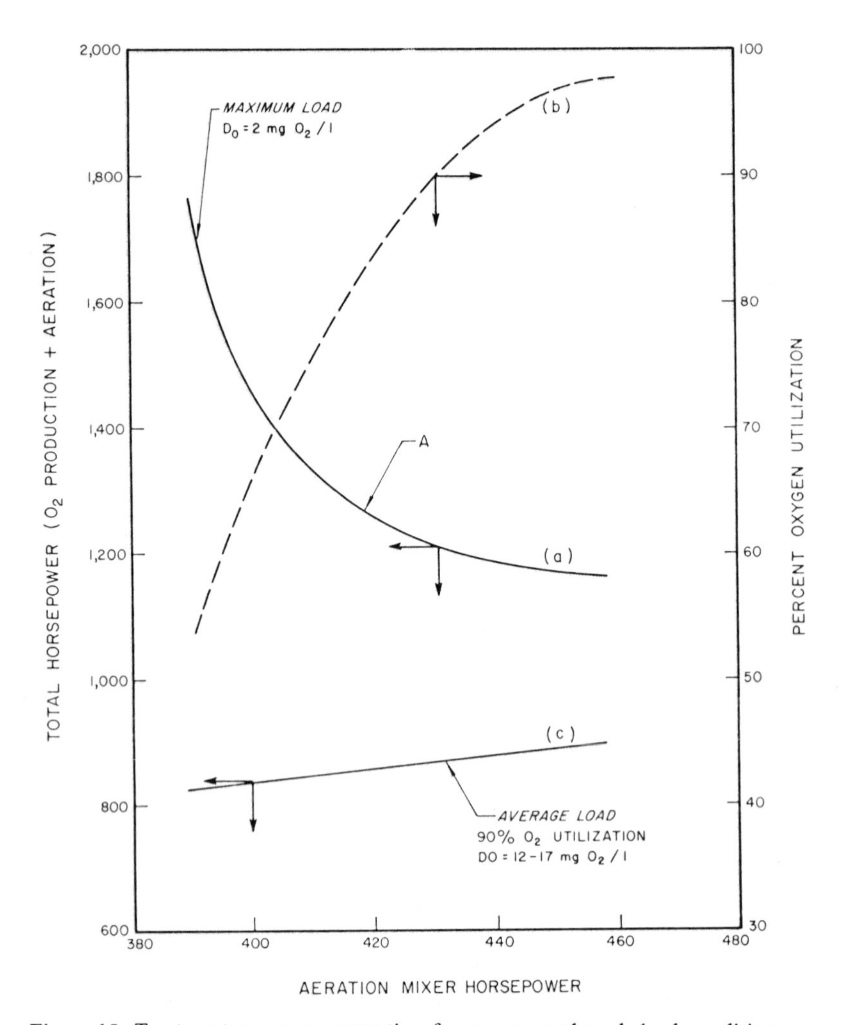

Figure 15. Treatment power consumption for average and peak-load conditions.

required to prevent the pH from decreasing below 6.0 to 6.4 (Figure 18). If effluent recycle is practiced in the filter, little change in pH results when organic nitrogen is present but the effluent quality has a somewhat higher COD due to the lower detention time in the unit. When no organic nitrogen is present in a wastewater with insufficient alkalinity, recycle does not aid system performance since no source of alkalinity is present in the reactors. Thus with respect to chemical effects, recycle should be excluded from consideration in design of anaerobic filter systems.

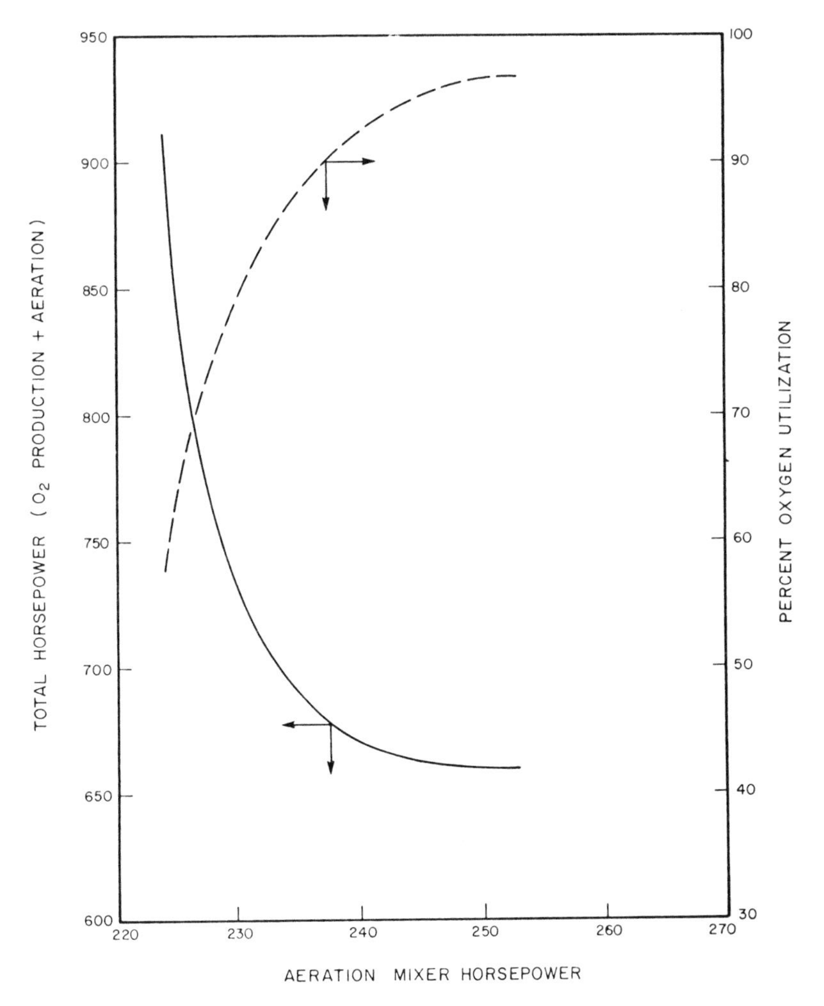

Figure 16. Variation of total horsepower with mixer horsepower to achieve DO = 4 at average load.

CONCLUSIONS

1. Proper evaluation of chemical changes occurring in biological systems provides additional insight to system operation.

2. In the case of the high-purity oxygen system, it allows rational design of the aeration equipment required for various loadings and provides the necessary input for optimization of design.

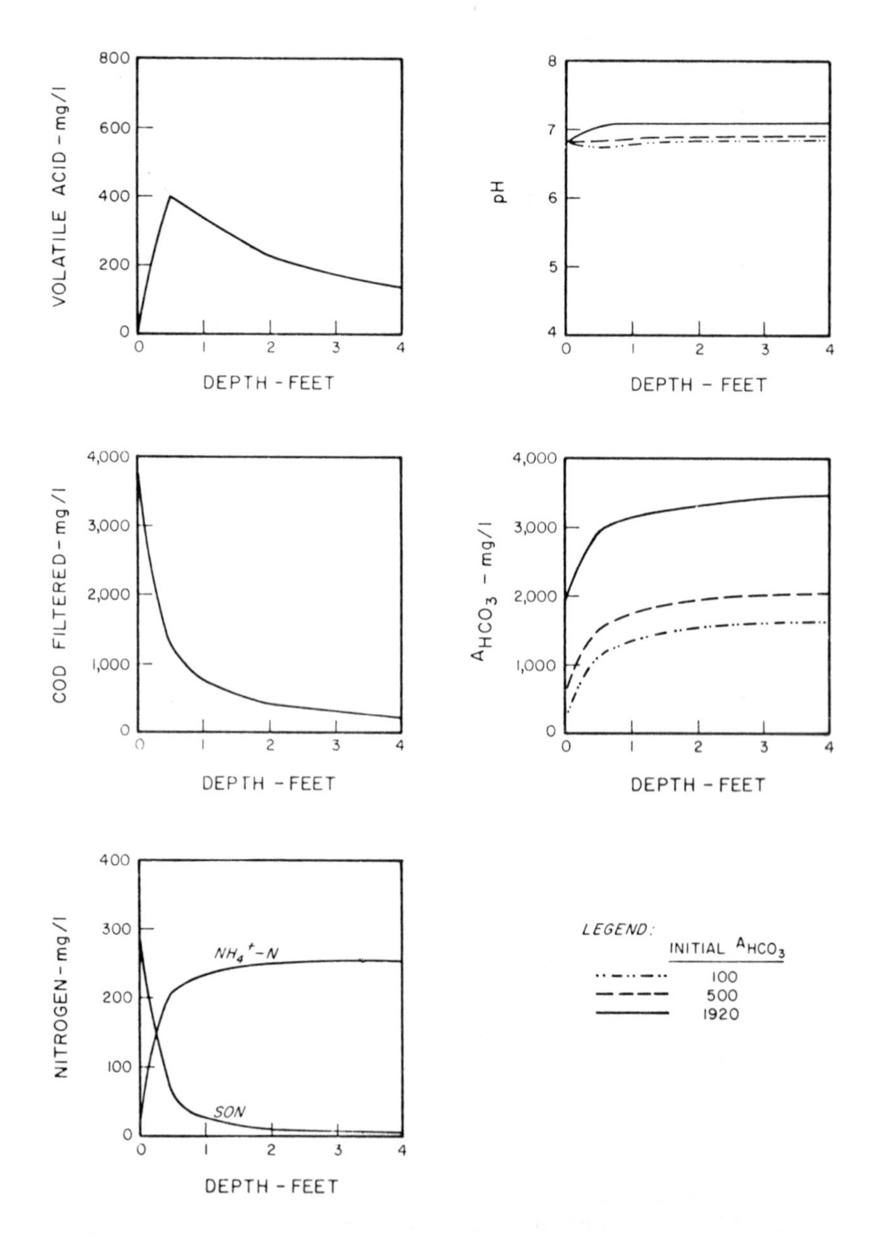

Figure 17. Effect of alkalinity on anaerobic filter-pH, SON = 300 mg/l.

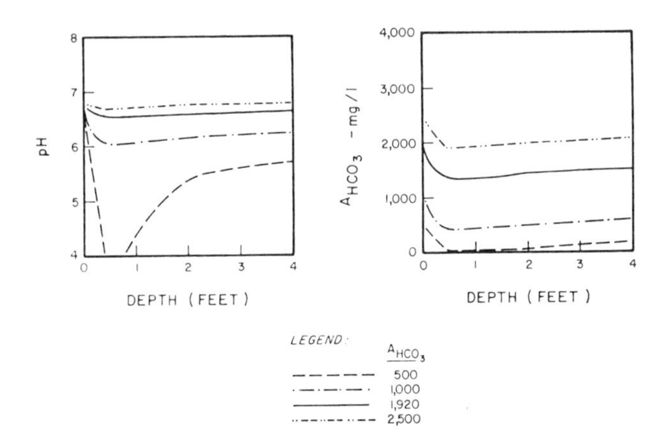

Figure 18. Effect of alkalinity on anaerobic filter pH, SON = 0.

3. With respect to the anaerobic filter system, it allows determination of required chemical additions to maintain desired system pH while accounting for the inherent alkalinity present in the wastewater as organic nitrogen.

4. After proper calibration, use of a mathematical model to evaluate the high-purity oxygen and anaerobic filter systems for various operational modes and waste characteristics provides an important tool for evaluation of design alternatives.

REFERENCES

1. Mueller, J. A., T. J. Mulligan and D. M. DiToro. *J. Environ. Eng. Div.*, ASCE 99:269 (1973).
2. Mueller, J. A., and J. L. Mancini. *Proc. Purdue Ind. Waste Conf.* 30:423 (1975).
3. Busby, J. B., and J. F. Andrews. *J. Water Poll. Control Fed.* 47: 1055 (1975).
4. Andrews, J. F., and S. P. Graef, in: *Advances in Chemistry Series* 105 (1971).
5. Andrews, J. F. *J. Sam. Eng. Div.*, ASCE 95:95 (1969).
6. Clark, R. H., and R. E. Speece. Paper Presented at 5th International Water Pollution Research Conference (1970).
7. Young, J. C., and P. L. McCarty. *J. Water Poll. Control Fed.* 41: R 160 (1969).
8. Young, J. C., and P. L. McCarty. Technical Report No. 87, Department of Civil Engineering, Stanford University, Palo Alto, CA (March 1968).
9. Haug, R. T. *Water Sew. Works* (February 1977), pp. 40-43.

INDEX